Handbook of
Semiconductor
Interconnection
Technology

edited by
GERALDINE COGIN SCHWARTZ
K. V. SRIKRISHNAN
ARTHUR BROSS
IBM Microelectronics
Hopewell Junction, New York

Handbook of Semiconductor Interconnection Technology

MARCEL DEKKER, INC.   NEW YORK · BASEL · HONG KONG

Library of Congress Cataloging-in-Publication Data

Handbook of semiconductor interconnection technology / edited by Geraldine Cogin Schwartz, K.V. Srikrishnan, Arthur Bross.
  p. cm.
  Includes bibliographical references and index.
  ISBN 0-8247-9966-6
  1. Semiconductors—Design and construction. 2. Electric contacts. 3. Semiconductors—Junctions. I. Schwartz, G. C. II. Srikrishnan, K.V. III. Bross, Arthur.
TK7871.85.H333   1997
621.3815'2—dc21   97-38710
                                                                                    CIP

The publisher offers discounts on this book when ordered in bulk quantities. For more information, write to Special Sales/Professional Marketing at the address below.

This book is printed on acid-free paper.

Copyright © 1998 by Marcel Dekker, Inc. All Rights Reserved.

Neither this book nor any part may be reproduced or transmitted in any form or by any means, electronic or mechanical, including photocopying, microfilming, and recording, or by any information storage and retrieval system, without permission in writing from the publisher.

Marcel Dekker, Inc.
270 Madison Avenue, New York, New York 10016
http://www.dekker.com

Current printing (last digit):
10 9 8 7 6 5 4 3 2 1

PRINTED IN THE UNITED STATES OF AMERICA

Preface

Interconnection technology is essential to both semiconductor logic and memory parts. The technical aspects of this technology include the deposition of thin layers of conductors and insulators, patterning them, and the establishment of electrically functional and reliable connections between wiring levels and device regions while isolating and insulating devices and wiring connections. In very large scale integration (VLSI) devices, in which millions of transistors are interconnected, the signal propagation delay from the interconnections is becoming equal to or greater than the devices themselves. The electrical resistivity and geometrical shape of the conductors as well as the dielectric constant of the insulators therefore become very important.

The technology roadmap for semiconductors by the Semiconductor Industry Association (SIA) states that the ability to connect devices will continue to lag considerably behind the ability to fabricate more devices in silicon for logic products and microprocessors. The SIA roadmap forecasts that from 1995 to 2010 the number of wiring levels will roughly double and the average interconnect length will increase 25 times, while the conductor cross-section will decrease 9 times. The SIA report asserts that "this presents a formidable challenge and will require large resources dedicated to both defect density reduction and reliability learning." The decrease in wiring pitch required for increased wirability of devices on a chip, while providing acceptable (i.e., low) resistance, requires that the shape of the metal conductor will change from being short and relatively wide as in today's chips to thinner and taller (rectangular) lines having an aspect ratio of greater than 1. According to the SIA roadmap, an average aspect ratio of 2.5 or higher for conductors will likely be in the 0.18 μm generation by the year 2000. The decrease in device size in dynamic random memories (DRAMs) similarly increases the aspect ratio of contacts drastically. As the critical image size decreases, the depth of field of exposure systems will be shallower so that planarization at every level will be required, thereby increasing the cost. The figure graphically shows this forecasted demand on interconnection technology. In order to reduce the signal propagation delay from the interconnection (RC delay), greater attention is being given to the

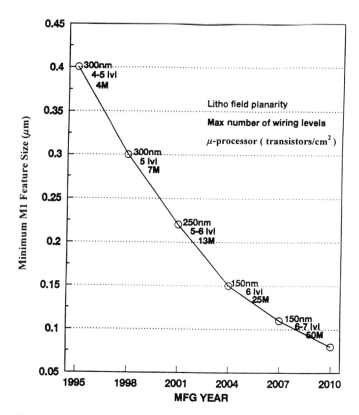

Interconnection roadmap. (Courtesy of the Semiconductor Industry Association.)

material and to the geometrical shape of the interconnection. It becomes obvious that future evolution of semiconductor technology will impose many requirements on the interconnection technology, thereby requiring an in-depth understanding of materials, processes, equipment, and finally manufacturability.

Semiconductor interconnection technology is structured in this book from a total perspective of cost, interconnection density, electrical performance, and reliability. With this view, the chapters were selected to cover technical issues that directly influence all of these needs. The first chapter is devoted to the description of equipment commonly used in manufacturing, their underlying physical principles, and advances in equipment. Equipment technology is fundamental to manufacturing and determines throughput and defects, thereby influencing yield and cost. Chapter II deals with common techniques used in the characterization of conductors and the dielectrics used in multilevel interconnection (MIC). It is hoped that this understanding of the equipment used in the processing of films, and techniques used for characterization, will lead to a greater appreciation of film structure and properties discussed in the subsequent chapters.

Chapter III is devoted to contacts, which is an important area associated with metal–silicon interfaces. This chapter reviews the physics of interfaces, different types of contacts, the materials used in the contacts, and finally common processes used in forming contacts. Contact studs—an extension of contacts that are increasingly used to achieve planarity, reliability, and wirability—are also reviewed. Chapter IV provides extensive coverage of both inorganic and organic insulators. Silicon dioxide, both undoped and doped, deposited by different processes, is thoroughly reviewed for its properties. Other inorganic insulators such as F-doped oxide, boron nitride, and silicon nitride are also discussed. Spin-on-glass and organic insulators are discussed because of their potential importance for gap-filling and providing a lower dielectric constant. Many new organic insulators have been synthesized to meet the demanding requirements for interconnect applications. The metallization chapter (V) provides an up-to-date review of sputtering and the evaporation of aluminum-based alloys, chemical vapor deposition, novel plasma deposition techniques, and plating. Aluminum, tungsten, and copper are extensively discussed. Chapter VI reviews the integration of discrete processes for fabricating a multilevel wiring structure and critically examines integration issues and alternative processes being used and developed. An extensive review of chemical mechanical polish technology is presented as it represents an important paradigm shift from exclusive clean-room-process-only thinking.

Chapter VII is devoted to reliability issues of thin conducting films and insulators and the challenge of increased interconnect complexity of future devices. The last chapter, VIII, reviews clean room technology and managing contamination from environment, equipment, and processes. Specific examples and discussions on interconnection processes are provided to illustrate contamination issues.

Geraldine Cogin Schwartz
K. V. Srikrishnan
Arthur Bross

Contents

Preface   *iii*

Contributors   *ix*

I. Methods and Principles of Deposition and Etching of Thin Films   1
   Geraldine Cogin Schwartz

II. Characterization   77
   Geraldine Cogin Schwartz

III. Semiconductor Contact Technology   161
   D. R. Campbell

IV. Interlevel Dielectrics   223
   Geraldine Cogin Schwartz and K. V. Srikrishnan

V. Metallization   301
   Geraldine Cogin Schwartz

VI. Chip Integration   363
   Geraldine Cogin Schwartz and K. V. Srikrishnan

VII. Reliability   473
   K. P. Rodbell

VIII. Contamination Control in Multilevel Interconnection
Manufacturing 507
A. Rapa and Arthur Bross

Appendix 567

Index *571*

Contributors

Arthur Bross*   **IBM Microelectronics, Hopewell Junction, New York**

D. R. Campbell   **Clarkson University, Potsdam, New York and CVC, Rochester, New York**

A. Rapa   **Jacobs Engineering, Phoenix, Arizona**

K. P. Rodbell   **IBM Research Division, Yorktown Heights, New York**

Geraldine Cogin Schwartz*   **IBM Microelectronics, Hopewell Junction, New York**

K. V. Srikrishnan   **IBM Microelectronics, Hopewell Junction, New York**

---

*Retired.

I
Methods and Principles of Deposition and Etching of Thin Films

Geraldine Cogin Schwartz
*IBM Microelectronics**
Hopewell Junction, New York

1.0 INTRODUCTION

This chapter covers many of the basic principles and methods used today in semiconductor manufacturing for depositing and etching both dielectric and conducting films. Some specialized techniques such as beam deposition and chemical mechanical polishing (CMP) will be covered in Chapter VI. A brief overview of deposition techniques can be found in Table I-1 and of etching in Table I-2.

2.0 EVAPORATION

Sputtering has almost completely displaced evaporation as a method of deposition because of its superior control of alloy composition, step coverage/hole fill by substrate biasing, ease of integration into "cluster tools," etc. However, since there are some applications, particularly for forming lift-off metal patterns, a brief review of the technique is included.

Evaporation is usually used for metal deposition but has been used to deposit some compounds (e.g., SiO, MgO). Early reviews of evaporation principles and equipment can be found in Holland (1961) and in Glang (1970); a later one is in Bunshah (1982). A review of some of the basics of high-vacuum technology can be found in Glang et al. (1970). Glang distinguished the steps into which the evaporation process may be broken: (1) transition from a condensed phase (solid or liquid) into a gaseous phase, (2) transport of the vapor from source to substrate at reduced gas pressure, and (3) condensation of the gas at the substrate.

The source that contains the evaporant must have a negligible vapor pressure at the operating temperature and must not react with the evaporant. There are many

---

*Retired.

TABLE I-1 Deposition Methods

Method	Materials deposited	General comments
Evaporation[1]	Pure metals Alloys Compounds	Need adequate vapor pressure Various support materials[2] Single/multiple sources for alloys High vacuum process Reactive evaporation
Sublimation	Metals Compounds	Used when very high evaporation temperature is needed
Sputtering[3]	Pure metals Alloys Compounds	Use of bias Can control/improve film properties Better control of stoichiometry of alloys Collimation; high source-to-substrate distance Magnetic enhancement; high density plasmas
CVD/PECVD[4]	Pure metals Alloys Dielectrics	Improved step coverage/fill Control of film properties by choice of reactants, operating parameters, and bias (PECVD) Cluster systems (e.g., dep/etch)
Plating[5]	Pure metals Alloys	Hole fill
Beams[6]	Metals Dielectrics	Directional deposition

Notes:
1. Source heating: resistance heaters, rf induction heaters, e-gun.
2. Crucibles, wires, foils.
3. Metals: dc or rf; dielectrics; rf; option of dc or rf reactive sputtering.
4. Includes high plasma density PECVD systems.
5. Electrolytic and electroless.
6. Beams discussed in Chapter VI.

types of sources and materials (Holland, 1961; Glang, 1970; Bunshah, 1982), e.g., crucibles of refractory oxides, nitrides, carbides, and metals, refractory wires and foils of various designs and shapes (Mathis Co.). Some materials, such as Cr, Mo, Pd, and Si, can be sublimed; this relaxes the temperature stability requirements for the source. Vaporization is accomplished by the use of resistance, induction, or electron bombardment heating; several configurations of electron guns (e-guns) are described by Bunshah; many types are available commercially. E-guns are now used most commonly, except where radiation damage may be a problem, e.g., causing flat-band shifts in FET devices. In properly controlled e-gun evaporation, a shell of solid material shields the molten mass from the crucible, preventing interaction between the evaporant and the hearth. Multiple-pocket crucibles are also available. They may be used for sequential evaporation of different films. Or, using several guns simultaneously, with appropriate control of the source temperatures, multiple component films of a desired composition may be deposited (Glang,

Deposition and Etching of Thin Films

TABLE I-2 Etching for Pattern Definition

Method	Materials etched	General comments
Wet chemical	All	Requires soluble product Almost always isotropic[1] Insoluble mask Mask adhesion important Mask profile not important Usually batch processing Inadequate for VLSI/ULSI
Sputtering ion-beam etch	All	Poor selectivity Vertical etching possible, faceting Angle-dependent yield Mask profile important Trenching Redeposition Substrate damage possible Relatively slow Alternate with deposition for gap-fill Ion beam: single wafer Sputter: batch; single wafer (cluster)
Reactive plasma (RIE, RSE, RIBE)	Almost all materials used for interconnections	Product: volatile or desorbed by ions High selectivity attainable Anisotropic/isotropic[2] Profile control Mask erosion Mask profile Mask adhesion is not as important Redeposition, trenching, substrate damage Aspect ratio dependent etch rates High etch rates attainable Batch; single wafer (cluster)

Notes:
1. Directionality possible in some cases: e.g., "slow" etches for Si which follow crystal planes; columnar structure resulting in vertical profile (molybdenum).
2. Anisotropy vs. isotropy: depends on many factors, i.e., reactants, etch parameters, ion energy, sidewall protection, etc.

1970). Alloy sources have also been used; the component ratio of the source is adjusted so that the deposited film has the required composition although the vapor pressures of the constituents are different. The source composition is usually determined empirically. Flash evaporation, in which small quantities of the constituents in the desired ratio are completely evaporated, is another way of depositing alloy films, and many kinds of dispensers have been used (Glang, 1970). However, whatever the evaporative technique, the control of the composition is rarely as reliable as that obtained by sputtering an alloy target.

Evaporation is carried out at very low pressures, e.g., 10^{-5} to 10^{-8} Torr. At these low pressures, the mean free path is very large compared to the source-to-substrate distance, so that the transport of the vapor stream is collisionless. The emission pattern of the evaporating species is directional; it is described by a cosine law: $dM/dA = M/\pi r^2 (\cos \varphi \cos \theta)$, which is illustrated in Fig. I-1. Thus the profile of the emitted flux can be visualized as shown in Fig. I-2.

The process of film growth on a substrate starts with the condensation of the vapor. Neugebauer (1970) outlined has four stages of film growth: nucleation and island growth, coalescence of islands, channel formation, and formation of a continuous film, as illustrated in Fig. I-3. A comprehensive discussion of the subject can be found in Neugebauer (1970).

Since thickness of the deposited film is greatest directly in line with the source and decreases in either direction, uniformity requires the use of planetary (rotating) substrate holders tailored to the particulate deposition requirements. Typical of the holders available commercially is the so-called normal-angle of incidence fixture (Fig. I-4a) and is used for high-uniformity and minimum step coverage, suitable for lift-off processes. Another fixture (Fig. I-4b), has additional planets that rotate at a higher speed and is designed for good step coverage as well as uniformity. Radiant substrate heating, using refractory wires or quartz lamps, is required because the properties of the deposited thin films are dependent on the deposition temperature. Temperature monitors and controllers are, therefore, also needed.

There are several kinds of thickness monitors (e.g., ionization-gauges and particle-impingement-rate detectors monitor the vapor stream). Crystal oscillators are used most frequently to measure the deposited mass; they utilize the piezoelectric properties of quartz. A thin crystal is part of an oscillator circuit so that the ac field induces thickness-shear oscillations whose frequency is inversely proportional to the crystal thickness; increasing the mass of deposit decreases the frequency. The crystal used has a specific orientation known as the AT cut, because this orientation exhibits the smallest temperature dependence. The thinner the crystal, the greater the sensitivity, if the mass deposited is small with respect to the

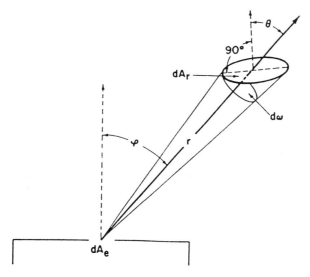

FIGURE I-1 Evaporation from a point source dA_s onto a receiving surface element dA_r.

Deposition and Etching of Thin Films 5

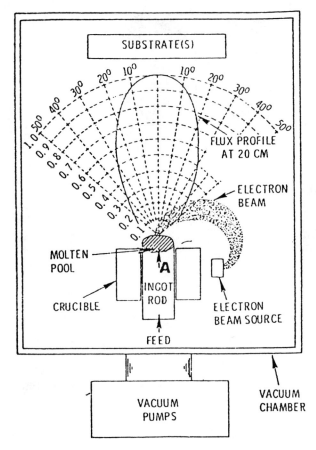

FIGURE I-2 Profile of emitted flux. (From Bunshak, 1989)

wafer thickness. For a quartz thickness of 0.28 mm, and an initial frequency of 6 MHz, the change of frequency/thickness is 81.5 Hz/μg cm^{-1} (Wagendristel and Wang, 1994). The availability and simplicity of use makes the crystal oscillator preferable to microbalances. Interferometry is used for transparent films. For metals, optical techniques such as light absorption, transmittance, and reflectance techniques, as well as resistance monitoring, have been used, but with less success. Thickness control is simply following the thickness monitor and stopping the process when the desired thickness is reached. Rate control is more complex; it requires adjustment of the source temperature which means that a measurement and feedback mechanism is required. The brochures supplied by equipment manufacturers are an excellent source of detailed information about the currently available evaporation systems and their operation.

In situ sputter cleaning, prior to evaporation of a metal film into a via hole, has the capability to remove a contaminant film responsible for high interfacial resistance (Bauer, 1994). When the lower surface is aluminum, the native oxide can be regrown quickly after sputter cleaning, due to the presence of residual water vapor. Sputtering of an Al electrode, before exposing the wafers to the plasma, is

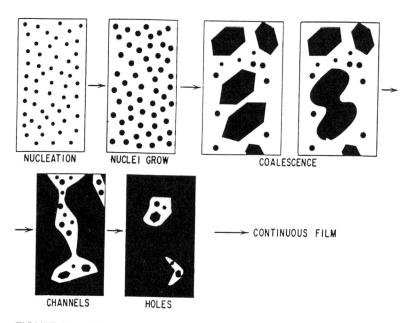

FIGURE I-3 Schematic of stages of film growth. (From Neugebauer, 1970)

an efficient way of reducing the partial pressure of water vapor, thereby eliminating the need for prolonged sputter cleaning. An implementation of this concept is shown in Fig. I-5.

3.0 CHEMICAL VAPOR DEPOSITION

3.1 INTRODUCTION

The term chemical vapor deposition (CVD), used without modifiers, will refer to the thermally activated reaction. Plasma and photon activation have also been used; these processes are called plasma-enhanced CVD (PECVD) and photon-enhanced CVD (sometimes referred to as LACVD, for laser-activated CVD) and are discussed later in this chapter. The term MOCVD refers to the use of a organometallic compound as a source gas in CVD. CVD processes have been used in the preparation of both metallic and insulating thin films as well as for depositing semiconductors. There are a number of reviews which may be consulted for more detailed information than can be covered here: Kern and Ban (1978), Sherman (1987), Mitchner and Mahawwili (1987), and Jensen (1989). Kodas and Hampden-Smith have compiled a book, *The Chemistry of Metal CVD Processes* (1994). In addition, there are individual papers collected in CVD symposia proceedings volumes of the Electrochemical Society. Discussion of specific CVD processes are postponed to the chapters covering the particular films.

Because of temperature restraints imposed by the interconnection metallization, high-temperature CVD processes cannot be used for interlevel dielectrics but have

Deposition and Etching of Thin Films

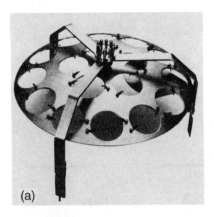

FIGURE I-4 Substrate holders in evaporators: (a) normal angle of incidence fixture, (b) planetary fixture.

been used for (usually) doped oxides to smooth the topography beneath the first interconnection level; they have been used to deposit W and various silicides.

3.2 PRINCIPLES

Film formation by chemical vapor deposition is a heterogeneous chemical reaction in which volatile reactants produce a solid film upon reaction at a hot surface. The sequential kinetic steps have been summarized by Jensen (1989) as follows: "(1) mass transport in the bulk gas flow region from the reactor inlet to the deposition zone, (2) gas-phase reactions leading to the formation of film precursors, (3) mass transport of film precursors to the growth surface, (4) adsorption of film precursors on the growth surface, (5) surface diffusion of film precursors to growth sites, (6) incorporation of film constituents into growing film, (7) desorption of volatile byproducts of the surface reaction, (8) mass transport of byproducts in the bulk gas flow region away from the deposition zone toward the reactor exit."

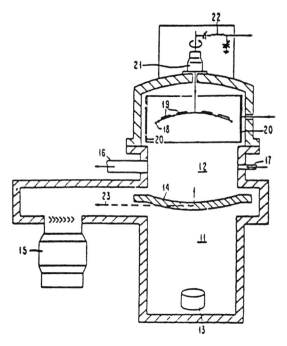

FIGURE I-5 Assembly for in situ sputter cleaning in an evaporator. (From Bauer, 1994)

It is clear that homogeneous gas phase reactions must be suppressed since they are responsible for the formation of "dust" particles which become incorporated into the growing film, making it hazy and defective.

In a thermally activated reaction, the dependence of rate on the temperature is given by the Arrhenius equation:

$$\ln(\text{rate}) = -\frac{E}{RT} + \text{constant}$$

where E is the energy of activation. However, if the deposition rate is controlled by the transport of the reactant, the rate will be approximately independent of temperature. In many CVD reactions, the two regimes are observed: (1) the surface-rate limited reaction (temperature controlled) and (2) the mass-transport limited reaction (temperature independent), as illustrated in Fig. I-6. In type 1 the surface reaction is fast relative to the transport of reactants. Temperature uniformity is critical for film uniformity for type 1 reactions. For type 2, flow across the wafer surface is critical.

3.3 REACTORS

3.3.1 Classification

One way of classifying CVD reactors is by the relative temperatures of the parts of the system: There is the hot-wall system in which the substrate and reactor walls are at the same temperature, and the cold-wall system in which the substrate is at

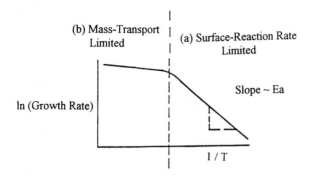

FIGURE I-6 Deposition rate vs. temperature for CVD process.

a higher temperature than the walls so that deposition occurs only on the substrate. There is the possibility of contamination by deposition on and subsequent flaking off the heated chamber walls. As pointed out by Kern and Ban (1978), the deposit is dense and adherent, so that if not permitted to become too thick the problem of flaking may not be severe, particularly since there is no thermal cycling. Also, since the wafers are stacked vertically, any flakes would not be likely to fall on them. In the cold-wall reactor, this source of contamination is negligible, but convection due to temperature differentials can arise (Carlsson, 1985).

Another classification scheme is the pressure at which the reactors were operated. The earlier classifications were atmospheric pressure, APCVD, and low pressure, LPCVD, which covers a pressure range of about 0.05 Torr to several Torr. More recently, particularly for the deposition of SiO_2 films, both subatmospheric (SACVD, ~600 Torr) and intermediate pressure (no acronym, ~60 Torr) have been used. At higher pressures, the rates of mass transfer of the volatile reactants and by-products and of reaction at the surface are about the same order of magnitude. Reducing the pressure increases the mass transfer rate so that reaction at the surface becomes the rate-limiting step. Reactor configuration greatly influences mass transport and thus is a critical factor for APCVD, but not for LPCVD. Uniform deposition is more easily achieved in LPCVD but the deposition rates are much lower than in APCVD.

Another way of classifying the reactor is by the deposition temperature: high temperature, HTCVD (~750–950°C), and low temperature, LTCVD (at temperatures below ~500°C).

It can be seen that there are many possible combinations for CVD reactor and process design.

3.3.2 Examples of Reactors

Winkle and Nelson (1981) described a cold-wall low-temperature (LT) APCVD reactor, made by Watkins-Johnson; it is used for depositing undoped and P-doped SiO_2 and is shown in Fig. I-7. At temperatures of ~350–450°C, deposition rates as high as ~1 μm/min were achieved using mixtures of O_2 and the appropriate hydrides. A feature of this reactor is the gas injector design (Fig. I-8) which improves surface reaction uniformities and coating efficiencies. The gas streams are

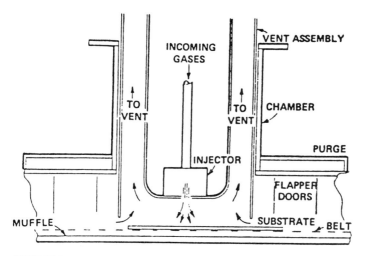

FIGURE I-7 Diagram of an APCVD reactor (Watkins-Johnson).

formed through laminar flow ports, thus preventing homogeneous gas phase reactions.

An example of a hot-wall LPCVD system is shown in Fig. I-9; this is typical of reactors used to deposit insulators and metals. Two versions of an experimental single wafer, LP cold-wall reactor designed for selective W deposition from WF6 and H_2 are shown in Fig. I-10a and b (Stoll and Wilson, 1986). In Fig. I-10a, the substrate is heated radiantly by means of quartz lamps and in Fig. I-10b, the wafer is heated on a hot plate. Another system (Heiber and Stolz, 1987) was an adaptation of a sputtering module equipped with a load lock, as shown in Fig. I-10c.

A commercially available, single-wafer, cluster-compatible, cold-wall LPCVD chamber is the Watkins-Johnson SELECT™ reactor shown in Fig. I-11, in which the wafer rests on a quartz cover and is heated by an encapsulated three-zone

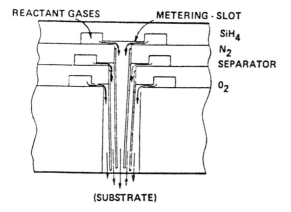

FIGURE I-8 Schematic of gas injector (Watkins-Jonson).

Deposition and Etching of Thin Films 11

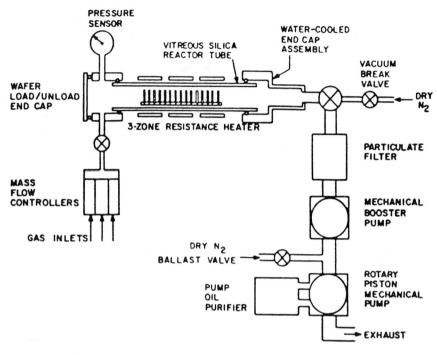

FIGURE I-9 Schematic of an LPCVD reactor. (From Kern and Schnable, 1979)

graphite heater. The system pressure is about 1 Torr, and was used to deposit BPSG from SiH_4 and O_2.

Another commercially available, load-locked, single-wafer, cold-wall reactor (Fig. I-12a) has been integrated into a "cluster tool" (Fig. I-12b), the Applied Materials Precision 5000 system. It has been used for blanket W deposition from WF_6 and H_2 at 10–80 Torr (Clark et al., 1991) and for SiO_2 (doped and undoped) from TEOS + O_3 (plus dopants) at 60 Torr or at 600 Torr (SACVD) (Lee et al., 1992); for these applications, the rf feed-through, shown in the diagram, is not used. This thermal CVD/PECVD reactor is covered by patents by Wang et al. (1989, 1991).

3.4 FILM PROPERTIES

The composition and purity of a film, its electrical and mechanical properties, the deposition rate and its uniformity are controlled by the many variables involved, and the interaction among them is complex and difficult to categorize. The reviews cited above should be consulted for more detailed information. Some discussion of specific films can be found in Chapters IV and V.

4.0 PHOTO-ENHANCED CVD

True photochemical processes depend on the fragmentation/activation of the reactant molecules, in the gas phase or on the surface, by photons. The advantage to

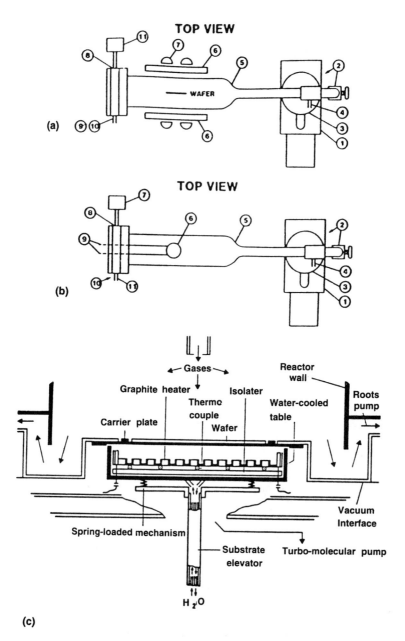

FIGURE I-10 Diagrams of experimental cold-wall CVD reactors: (a) tungsten filament lamp heating, (b) hot plate heating, (c) single-wafer cold-wall system with load-lock. [Parts (a) and (b) from Stoll and Wilson, 1986; part (c) from Heiber and Stolz, 1987]

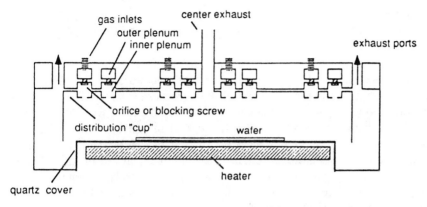

FIGURE I-11 Schematic of a single-wafer CVD system (Watkins-Johnson Select).

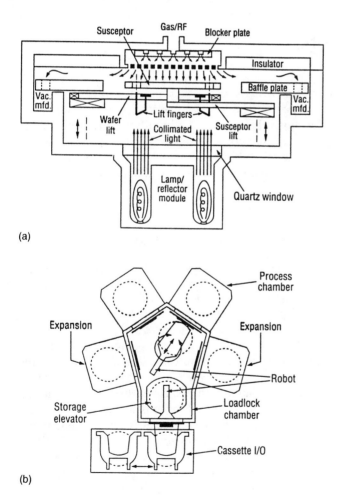

FIGURE I-12 Schematic of an Applied Materials Precision 5000 single-wafer CVD reactor: (a) side view of an individual chamber, (b) top view of the cluster system configuration.

photo-induced reactions is the absence of electromagnetic radiation and charged species which can induce damage in dielectric films. To quote Eden (1991): "Optical radiation can induce specific chemical reactions in the gas phase or at a surface. The selective production of atoms, radicals or other excited species in the vicinity of a surface independently of the substrate temperature, effectively decouples temperature from the number density of the species of interest. The introduction of photons allows one to drive the chemical environment far from equilibrium by selectively producing species not normally present in conventional CVD reactors." In some cases, radiation merely heats the surface and the process is, in reality, thermally activated LPCVD although the process may be localized to some degree if the light source is very narrow. If the light source simply heats the source gases, thermal fragmentation occurs as in conventional CVD processes. Ultraviolet (UV) and vacuum ultraviolet (VUV) lamps and lasers are used as energy sources.

Photo-CVD has not, at least up to now, been used in production because the deposition rates are low and therefore the process is expensive.

5.0 PLASMA PROCESSING
5.1 INTRODUCTION

Plasma processing has become essential for depositing and etching the materials used in building semiconductor devices. The low-plasma-density capacitively coupled discharge had been used exclusively in fabrication until recently, when high-plasma-density reactors became available commercially. Specific applications are discussed in Chapters IV to VI. The reliability issues associated with plasma processing, e.g., contamination, electrical damage, surface modification, etc., will be covered in the chapters on reliability and contamination.

5.2 CAPACITIVELY COUPLED rf GLOW DISCHARGE

When an increasing rf voltage is applied between electrodes in a low pressure (e.g., ~10–1000 mTorr) gas, ultimately the gas breaks down, i.e., it ionizes and current flows. A glow is observed. Adjacent to the electrodes are dark spaces, the sheaths, and a voltage drop occurs across the sheath regions. The glow region is virtually field-free; there are approximately equal numbers of positive and negative charges. Electron-impact dissociation produces not only the ions but photons, free radicals, and metastables (the neutral species) as well. *The potential in this plasma region is the most positive potential in the system.* Thus *all* electrodes (which may include the chamber walls) have a negative potential with respect to the plasma, and all are *bombarded by positive ions.* Sputtering of the surfaces is a source of contamination, making the choice of reactor materials an important issue (Vossen, 1979; Oehrlein, 1989). The relative potential (bias) developed at each electrode determines the ion bombardment energy; this is a function of their areas. If the electrodes have equal areas (*symmetrical* reactors), the voltage, i.e., the *dc bias*, is the *same on both* so that both are bombarded by ions of equal energy (Vossen, 1979) and the *plasma potential* is relatively *high*; the *point of attachment of the power is irrelevant.*

When the electrodes are *unequal* in area, the *dc voltage is higher on the smaller one* (Koenig and Maissel 1970); if the area of one electrode is very much smaller than the other (called *asymmetric* systems), the bias on the small electrode is approximately one half the peak-to-peak applied voltage. The high-bias electrode is usually called the cathode. The *plasma potential is low*; the potential of the plasma and that of the larger electrode are *approximately* equal; thus the ion bombardment energy at the very large electrode is low, but *not zero*. For convenience and safety, the smaller electrode is powered and the chamber (counter-electrode) is grounded. However, the *ion bombardment on a specific electrode does not depend on which electrode is powered* (Coburn and Kohler, 1987), although the plasma potential is higher when the larger electrode is powered.

In addition to the electrode potentials, the floating potential exists on all surfaces neither externally biased nor grounded; this is a function of the electron mass and temperature and on the ion mass and charge.

The plasma in these reactors is a *nonequilibrium plasma* in which the temperature of the electrons is much higher than the temperature of the gas. The plasma density is low ($\sim 10^9$–10^{11} cm^{-3}) and the fractional ion density (i.e., the ratio of ion to neutral species) is low ($\sim 10^{-6}$–10^{-3}).

5.2.1 Frequencies

A range of frequencies has been used, from 50 kHz to 2.54 GHz (microwave); 13.56 MHz (or multiples) is the most commonly used frequency (no interference with communications). However, Goto et al. (1992) preferred to treat the frequency as a process parameter and investigated the 10–215 MHz range; Martinu et al. (1989) used microwave (2.54 GHz) excitation. Colgan et al. (1994) suggested the use of very high-frequency capacitive discharges to obtain high plasma densities at low ion energy. An ultra-high-frequency (UHF), 500-MHz discharge has been used in conjunction with a new antenna (see below). In the low-frequency range, electrons and ions follow the electric field and the ions experience the full amplitude of the rf voltage, resulting in higher bombardment energy of the electrode. Above ~3MHz ions can no longer follow the field, as do the electrons. The ions interact only with the time-averaged field since it takes several rf cycles to cross the sheath. Therefore, the average energy is reduced and the electrode is bombarded with lower energy ions. However, at higher frequencies, energy coupling is more efficient so that, for a given power, the plasma densities are greater than at lower frequencies. At low frequencies, the peak energy of the ions is greater than at high frequencies but the energy distribution is broader (Bruce, 1981; Coburn and Kohler, 1987; Hey et al., 1990; Myers et al., 1994). At low frequencies, the angular distribution of the ions is more directional than at higher frequencies (Myers et al., 1994).

A two-frequency or dual excitation mode in which both frequencies are applied simultaneously is now used frequently for independent control of the substrate bias in both reactive plasma-enhanced etching and deposition. The excitation electrode may be powered using the higher frequency and the substrate electrode powered with the lower (using the proper filter networks) or both frequencies may be applied to the excitation electrode; the configurations are equivalent electrically. A variety of combinations have been used, e.g., 13.56 MHz/200 kHz (Tsukune et al., 1986),

13.56 MHz/450 kHz (van de Ven et al., 1988, 1990), 2.54 GHz/13.56 MHz (Martinu et al., 1989), 100 MHz/30 MHz (Goto et al., 1992). In the electrically equivalent mode of operation, both frequencies are fed to one electrode, e.g., 13.56 MHz/450 kHz (Hey et al., 1990), 13.56 MHz/350 kHz (Matsuda et al., 1995).

5.2.2 Reactor Requirements

Whatever the electrical configuration of the reactor, whether inert or reactive gases are used, and whether used for etching or for deposition, the reactors have many features in common. There are (1) the reaction chamber with the associated vacuum apparatus, e.g., pumps, pressure controllers/monitors, and the gas distribution systems with the appropriate control and monitoring equipment and (2) the glow discharge generation equipment consisting of the power source(s) and, where required, an impedance matching network for efficient power transfer. It should be noted that the power to the glow discharge may be significantly less than the input power due to (undetermined) losses in the matching network (Logan, 1990). There is optional equipment for following the processes: rate monitors, e.g., interferometers, grating patterns, etc., which can act as endpoint detectors in some cases. Particularly in reactive plasma etching processes, there are several other kinds of endpoint detectors, e.g., optical emission spectroscopy (OES), discharge impedance and pressure monitoring systems, etc., as well as plasma diagnostic equipment, e.g., OES, mass and laser-induced fluorescence spectrometry. Unfortunately much of this equipment is often not compatible with the configuration of reactors used for manufacturing.

Many of the newer reactors used in manufacturing are integrated into "cluster tools" or multichamber processing systems in which each chamber processes a single wafer (Singer, 1993, 1995). These provide not only load-locked entry into the first chamber and exit from the last but vacuum transfer between chambers as well. In situ plasma-cleaning capability is often a feature of such systems. The chambers in the system described in Fig. I-12b, into which a CVD reactor was incorporated, have been configured for plasma processing (PECVD and inert or reactive plasma etching) as well.

5.2.3 Capacitively Coupled Reactors

The most widely used reactors are what have been called planar or parallel plate diode reactors with an internal electrode capacitively coupled to an rf source. In *symmetrical* systems, used frequently in reactive-plasma-assisted processing, there are actually two flat electrodes inside a dielectric chamber; one is powered and the other, on which the wafers are placed, is grounded. These systems are based on the radial-flow reactor patented by Reinberg (1973) and are usually operated at relatively high pressures (several hundred mTorr). A schematic representation of the reactor is shown in Fig. I-13.

Asymmetrical reactors (Fig. I-14a) were developed earlier for sputter deposition and etching; there may be a counter electrode connected to the grounded chamber enclosure. In others, the grounded enclosure itself acts as the counter electrode, as shown in Fig. I-14b. Both are called planar diodes and are operated at relatively lower pressures (tens of mTorr). An axial configuration, the so-called hexode reactor, patented by Maydan (1981), is electrically equivalent to an asymmetric planar

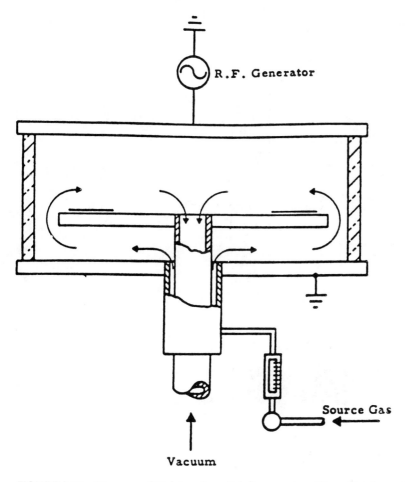

FIGURE I-13 Diagram of Reinberg's radial-flow reactor. (From Reinberg, 1973)

diode system. Diagrammatic representation of the hexode is shown in Fig. I-14c, and a drawing of the reactor, taken from the patent, is seen in Fig. I-14d.

The "flexible diode" (Ephrath, 1981), a planar reactor used for etching, has two independent matching networks and power supplies for biasing the two electrodes, as shown in Fig. I-15. The dual frequency reactors were described previously.

The triode system is an extension of the diode reactor; a third electrode has been added so that the substrate bias can be controlled essentially independently of the excitation energy, using the same or a different frequency. Triodes are used for bias sputtering to improve the properties of the deposited material. The chamber walls are grounded and the target and substrate are isolated. Two versions of a triode have been used for bias-sputtered oxide: the "tuned substrate" in which the bias in the substrate is controlled by a "tuning network" (Fig. I-16a) and the "driven" system in which both electrodes are powered by using rf two generators (Fig. I-16b) or a single power supply and a power splitting network (Fig. I-16c).

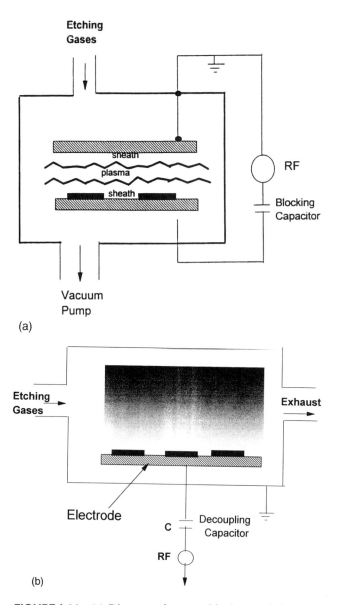

FIGURE I-14 (a) Diagram of a capacitively coupled asymmetric reactor. (b) Diagram of a capacitively coupled reactor used for RIE. (c) Diagram of a hexode reactor. (d) Drawing of an actual hexode reactor.

5.2.4 Magnetic Confinement

Magnetic confinement is used to obtain a *high plasma density*, *higher ion/neutral ratio* at relatively *low voltages* and at *lower pressure* than in an unconfined system. The magnetic field confines the electrons in the discharge. Bobbio (1989) reviewed the literature to that date.

Deposition and Etching of Thin Films

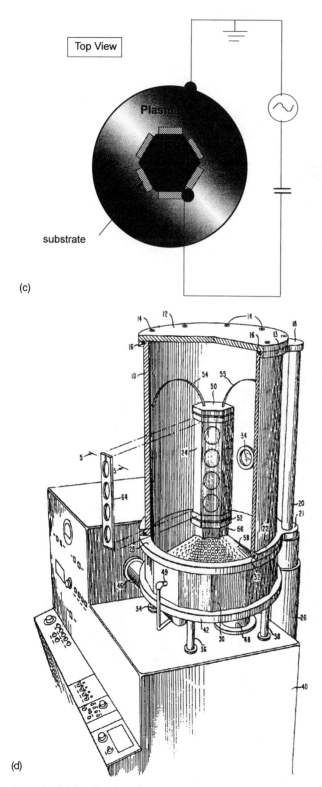

FIGURE I-14 Continued

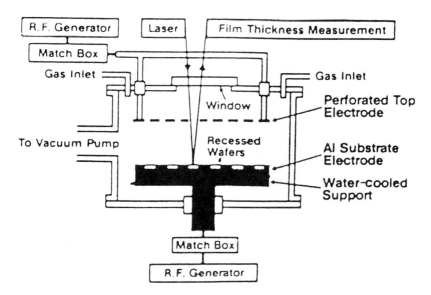

FIGURE I-15 Schematic of a flexible diode RIE reactor. (From Ephrath, 1981)

Magnetrons

Axial magnetic fields, used with a planar diode, increase the path length of the electrons and keep them away from the chamber walls. In magnetron sputtering systems, the object is to trap electrons near the target to increase their ionizing effect, thus increasing the deposition rate. The electric and magnetic fields are usually perpendicular (Chapman, 1980).

The many magnetron configurations, cylindrical, circular (sputter-gun and S-gun), and planar, used for sputtering have been described extensively in *Thin Film Processes* edited by Vossen and Kern (1978) and in *Glow Discharge Processes* (Chapman, 1980). A high-vacuum planar magnetron discharge, operating at pressures at or below 1 mTorr but with reasonable deposition rates, has been described by Asamaki et al. (1992, 1993).

Magnetron reactors, in which the magnetic field lines are parallel to the cathode surface, are used for RIE; these systems have been called MERIE (magnetically enhanced RIE) and MIE (magnetron ion etcher) systems. Various magnet configurations are shown in Fig. I-17: (a) planar, (b) band, (c) quadrupole, and (d) annular.

An example of a single-wafer magnetron RIE system (Schultheis, 1985) is shown in Fig. I-18.

Multipoles

Multipolar confinement or surface magnetic field confinement is one in which the chamber walls and sometimes an end wall are lined with strong permanent magnets arranged in an alternating N-S arrangement; they are used with several kinds of reactors.

The magnets produce a series of magnetic cusps around the wall in an effect forming a magnetic bottle (Mantei and Wicker, 1983; Mantei et al., 1985; Wicker and Mantei, 1985; Kuypers et al., 1988). The charged species are reflected by the

Deposition and Etching of Thin Films 21

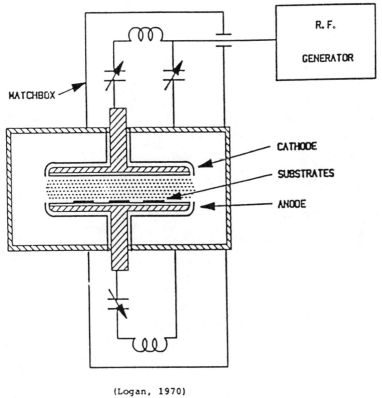

(Logan, 1970)

(a)

FIGURE I-16 (a) Schematic of a tuned anode sputtering system. (b) Schematic of a driven anode sputtering system with two generators. (c) Power-splitting rf drive for driven anode system.

magnetic mirror into the plasma, away from the walls. In one of the multipolar reactors using a hot filament discharge (Fig. I-19a), there is an increase in the plasma density of about a factor of 100 and a reduction of operating pressure to about 1 mTorr. Figure I-19b shows a multipolar microwave reactor.

Another use of multipoles is in the magnetically confined reactor (MCR). This is a triode etcher with 13.56 MHz applied to an annular electrode and 100 kHz applied to the wafer holder and the common top electrode is grounded. The multipoles are arranged around the chamber walls and embedded in the top electrode (Engelhardt et al., 1990; Engelhardt, 1991). A similar arrangement of a grounded cylindrical multipolar bucket, but using the same frequency at both electrodes in a triode reactor, was described by Singh et al. (1992).

Multipolar confinement has also been used with high-density discharges which are discussed below.

5.2.5 Hollow Cathode

A modification of a capacitively coupled reactor is the hollow cathode (HC) configuraton (Horwitz, 1989a; Gross and Horwitz, 1993), a diagram of which is shown

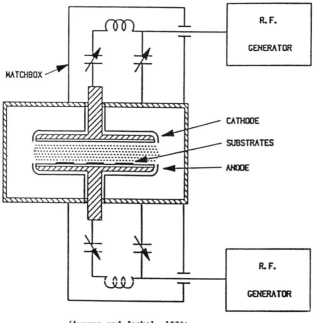

(Auvang and Jaekel, 1970)

(b)

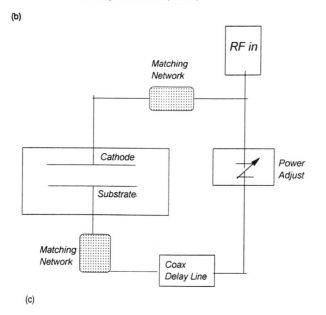

(c)

FIGURE I-16 Continued

Deposition and Etching of Thin Films

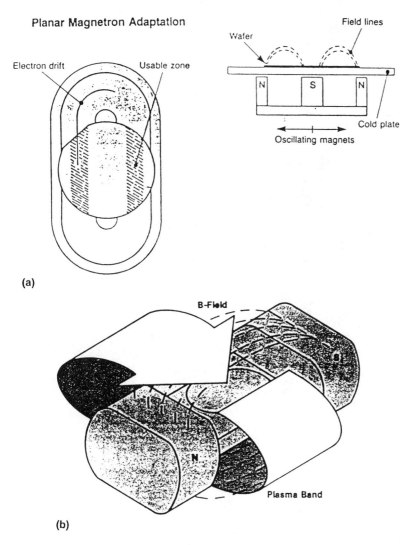

FIGURE I-17 (a) Diagram of a planar magnetron adaptation for magnetic RIE (Hinson et al., 1983). (b) Diagram of a band magnetron cathode (Hill and Hinson, 1985). (c) Diagram of a quadrupole magnetic configuration (Hill and Hinson, 1985). (d) Diagram of an annular permanent magnetic for a supermagnetron plasma etcher (Kimoshita et al., 1986).

Fig. I-20. In this configuration a high plasma density, low voltage discharge can be operated at low pressure. The confinement provided by the opposing rf-powered electrodes increases the utilization of the ions in the discharge and provides an "electron mirror" by which the secondary electrons are trapped. A more complete description of the hollow cathode can be found in a review by Horwitz (1989b). This kind of system has not been developed commercially.

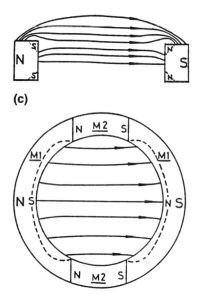

(c)

(d)

FIGURE I-17 Continued

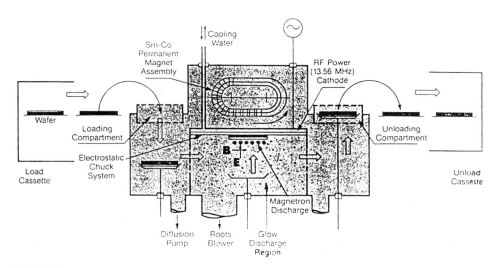

FIGURE I-18 Schematic diagram of a single wafer magnetron RIE system. (From Schultheis, 1985)

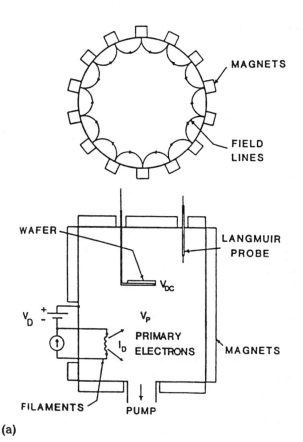

(a)

FIGURE I-19 (a) Diagram of a multipolar plasma reactor; (*t*) shows field lines, (*b*) A hot-filament reactor (Wicker and Mantei, 1985). (b) Diagram of a multipolar microwave plasma processing chamber (Asmussen, 1989).

5.3 TEMPERATURE EFFECTS

5.3.1 Heating

Bombardment by energetic ions heats a surface. In low-pressure environments, the heat transfer between the wafer and its holder is poor, unless a heat conducting medium (e.g., thermal grease, a moderate pressure of He) is interposed between. The temperature rise is proportional to the ion energy e.g., the input rf power in sputter etching (Schwartz and Schaible, 1981), or the sheath voltage in ion-driven reactive etch processes (Fortuno, 1986). If wafers are exposed at the same time to a plasma, and one is bombarded with the full ion energy and the other is not, the temperature of the first wafer is significantly higher.

5.3.2 Temperature Control

Since many of the recently developed processes require either cooling the wafer or keeping it at a constant, uniform, and reproducible temperature, wafer clamping techniques have been used in many reactors. At first the wafers were mechanically

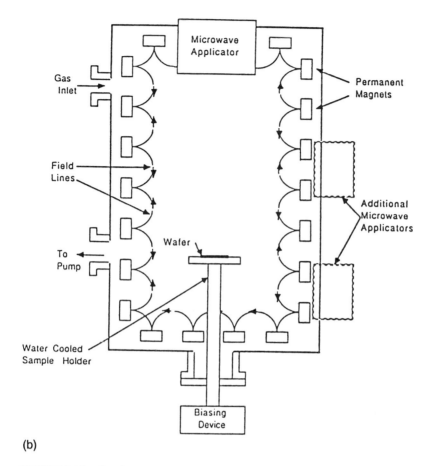

(b)

FIGURE I-19 Continued

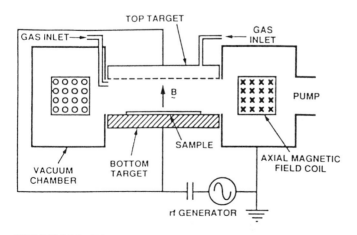

FIGURE I-20 Diagram of a hollow cathode system. (From Horwitz, 1989a).

clamped by a ring on the topside of the wafer to the temperature-controlled wafer holder (Hinson et al., 1983; Kanetomo et al., 1992), but the purely mechanical contact has often been found to be inadequate. An illustration of this kind of mechanically clamped wafer holder is shown in Fig. I-21.

The next advance was the use of several Torr of a heat-transfer gas, often helium forced across the backside of the mechanically clamped wafer (Wright et al. 1992). Heat transfer is dominated by the flow of gas. The transport of heat by a gas depends on the kind of gas (helium is best) and increases (up to some pressure) with increasing pressure; radiation is usually negligible. However, the topside ring covers a portion of the edge of the wafer, and, in addition, the ring may be responsible for increased contamination and nonuniformity.

Electrostatic wafer clamping appears to be a promising alternative. This technique, in which wafers can be clamped flat (i.e., without bowing), had been used earlier for e.g., holding a wafer during transport (Lewin, 1985; Kumagai, 1988; Nakasuji and Shimizu, 1992) and during lithographic processing (Clemens and Hong, 1991). It uses the attractive force between the charged plates of a capacitor to hold the wafer in place. Anodic oxides, deposited dielectrics, polymers, and ceramics, alone or in combination, have been used as the capacitor dielectric. The chuck can be installed in a plasma reactor, eliminating the need for a topside clamp, while maintaining the flow of gas across its back. This combination of electrostatic chuck (ESC) and back-side gas flow is the most commonly used, although ESCs which rely solely on mechanical contact have also been evaluated. Daviet et al. (1993) found that a soft polymeric dielectric was a better material than a hard inorganic one for this application. Nakasuji et al. (1994) introduced the concept of a figure of merit, defined as the product of the dielectric constant and breakdown strength, for dielectrics to be used in these chucks. A sputtered membrane of Ta_2O_5 had the maximum figure of merit compared with alumina or silica.

The attractive forces responsible for clamping may persist after the electrode voltage is removed, resulting in an increase in processing time and/or wafer breakage. According to Wright et al. (1995), the origins of the residual forces are permanent polarization of the dielectric in the chuck, slow mobile ions inside the dielectric, or charges trapped at the dielectric surface or back side of a wafer having a dielectric film on its surface. Avoiding or minimizing these effects is an important issue in chuck design.

It must therefore be emphasized that, whether used with or without back-side gas cooling, a well-designed ESC must clamp the wafer effectively during proc-

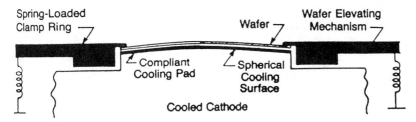

FIGURE I-21 Diagram of a mechanical clamp for wafer cooling. (From Hinson et al., 1983)

essing *and* must release ("declamp" or "dechuck") the wafer rapidly at the end of processing.

There are several configurations of ESCs in use and have been described by Field (1994). The configurations and equivalent circuits are shown in Fig. I-22: (a) unipolar, (b) bipolar, and (c) Johnsen-Rahbek (J-R). The J-R chuck makes use of the enhanced clamping effect obtained when using certain ceramics as the dielectric in the chuck. In these ceramics, some of the charge flows as a leakage current through the insulator or along its surface, so that charge accumulates at the wafer /chuck interface. Wright et al. (1995) noted that the conductive properties of the ceramic can be tailored to enhance both clamping and declamping.

For each type, Field has also discussed the advantages and disadvantages, such as declamping time and device damage. Wright et al. (1995) have also analyzed

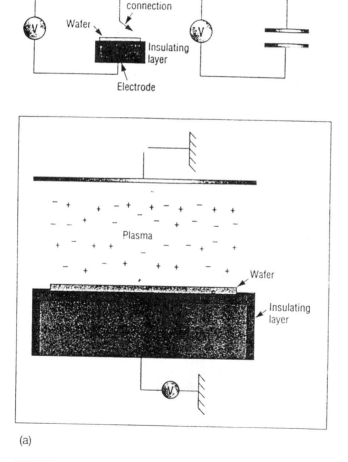

(a)

FIGURE I-22 Diagrams of electrostatic wafer clamps with equivalent circuits. (a) unipolar, (b) bipolar, (c) Johnsen-Rahbek configuration. (From Field, 1994)

Deposition and Etching of Thin Films

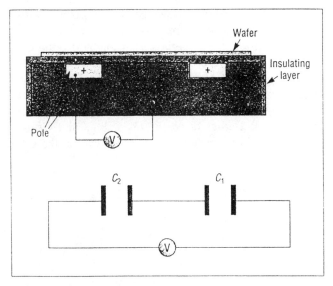

(b)

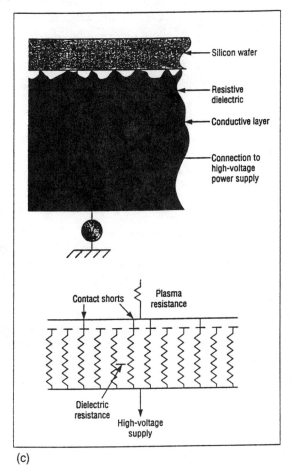

(c)

FIGURE I-22 Continued

the mono- and bi-polar chucks and, in addition, have addressed manufacturing issues of ESC: repeatability, reliability, material compatibility, and cost.

There are numerous patents describing the materials used for fabricating electrostatic chucks, their structures, as well as the methods of applying the voltage. Several do not state a particular application (Lewin and Plummer, 1985a; Lewin 1985; Ward and Lewin, 1987; Suzuki, 1987; Logan et al., 1991; Horwitz and Boronkay, 1992; Watanabe and Kitabayashi, 1992; Liporace et al., 1992; Hongoh and Kondo, 1993; Logan et al., 1993; Barnes et al., 1993; Collins and Gritters, 1994). Others are specifically for use in a plasma reactor (Nozawa, et al., 1993; Arami et al., 1994; Su et al., 1994).

5.4 SPUTTERING

5.4.1 Introduction

Sputtering is a physical process (Wehner and Anderson, 1970; Maissel, 1970; Vossen and Cuomo, 1978; Chapman, 1980; Logan, 1990; Wasa and Hayakawa, 1992) in which the positive ions in a glow discharge strike a surface and eject atoms from it by momentum transfer. About 1% of the incident energy goes into particle *ejection*, about 75% into *heating the bombarded surface*, and the rest dissipated by *secondary electrons* which *heat the substrate* (Vossen and Cuomo, 1978). Ejection occurs when the kinetic energy of the incoming ions exceeds the binding energy of the surface atoms of a solid. Sputtering is the result of a "collision cascade," a sequence of independent binary collisions; it is not a simple interaction between an incoming ion and a surface atom.

5.4.2 Sputter Deposition

Sputter deposition has almost always been carried out in a capacitively coupled reactor, often with magnetic enhancement, although recently high-density plasmas have been employed.

Sputtering Target

The solid from which the atoms are ejected is termed the "target." When used as a sputtering target for film deposition, a dense target is preferred, to eliminate the possibility of contamination, although for some materials only sintered, hot-pressed, or powder targets may be available. Since the target is heated by the bombarding ions, the backing electrode to which the target is bonded must be cooled, and the bonding material must be good heat-transfer medium that will not be a source of contamination. Shields, often called ground shields or dark space shields, surround the back of the target (placed at such a distance that no discharge will be initiated in that space) to suppress sputtering of the backing material; some shield configurations are illustrated in Chapman (1980).

Threshold Energy

The minimum ion energy required for sputtering is called the *threshold* energy; it depends on the heat of sublimation of the target material and is relatively insensitive to the nature of the bombarding ions. The sputtered material is usually monatomic although diatomic species (e.g., SiO from an SiO_2 target, Coburn et al., 1974) have been detected. In most sputtering processes, the ion source is solely an inert gas, most often argon, but in reactive sputtering (discussed more completely below) O_2 or N_2 is added, depending on the material to be deposited.

Yield

The sputtering yield is the number of atoms ejected for each incoming ion; it increases with ion energy, exponentially at lower energies and then linearly, reaching a plateau and finally decreasing at very high energy. In the low energy of exponential increase, the yields are very low, reaching ~0.1 in the energy range used in practical sputtering. Although it is often the case that the sputtering yield increases with increasing mass of the bombarding ion, as the oft-quoted results of Almen and Bruce (1961) for inert gas ion sputtering of copper indicate, this does not appear to be true for all substrates, as perusal of sputtering yield tables reveals (Vossen and Cuomo, 1978). Molecular ions dissociate into energetic atoms upon impact with the target surface and behave as though the individual atoms arrived separately. That is, an ion X_i^+ at an energy E has the same sputtering effect as i ions X^+ at energy E/i (Steinbruchel, 1984). The sputtering yield of neutral species is the same as the corresponding ion. The effect of the angle of incidence is discussed in a separate section.

Film Composition

The composition of the deposited film is usually the same as that of a homogeneous target. In the case of an alloy target, composed of atoms of different sputtering yields, an "altered layer" forms at the surface of the target. Initially, the component with the highest sputtering yield is preferentially removed, leaving the surface enriched with the lower sputtering yield component. At steady state, the composition of the material sputtered from the altered layer onto the substrate is the same as that of the bulk target. However, if there is significant preferential resputtering from the substrate surface and/or diffusion at the target surface, the composition of the deposited film will differ from that of the source. If one of the components of the target is volatile, ion-heating of the target may result in a difference in stoichiometry between the target and deposit; addition of the volatile component to the sputtering gas can compensate for this.

Effect of Operating Conditions

Raising the gas pressure increases the number of ions (ion current) for sputtering and although the energy of the ions decreases, the net result is an increase in deposition rate, because the yield decreases slowly with decreasing energy in the energy range used for sputtering. At some pressure, however, backscattering in the gas will result in a rate decrease.

The flow rate of the gas does not directly affect the deposition rate, but some contaminants (from the vacuum chamber or desorbed/sputtered from the target), which would be swept out in a high flow, do affect the rate, e.g., a small partial pressure of O_2 reduces the deposition rate of SiO_2 significantly (Jones et al., 1968). And by removing the contaminants it also reduces the probability of incorporating them into the growing film and degrading its properties.

Increasing the source-to-substrate distance lowers the accumulation rate but improves uniformity. The net accumulation rate decreases with increasing substrate temperature. The use of the term accumulation rate takes into account the fact that, in some instances, not all of the material sputtered from the target and arriving at the substrate remains on the surface; some of it may be resputtered or reemitted thermally.

Metals may be sputtered in a dc glow discharge but when insulators are exposed to a dc plasma, a positive charge accumulates on its surface preventing further positive ion bombardment. The use of rf sputtering, in which an rf potential is applied to a cooled metal electrode to which the insulating target is bonded, circumvents this problem. A grounded metal shield prevents sputtering from the edges of the metal electrode. In sputtering, the use of frequencies higher than 13.56 can be advantageous. At higher frequencies, the ion current increases but the ion energy decreases resulting in higher deposition rates at lower target voltages. Lowering the target voltage reduces the energy of the secondary electrons produced at the target, and substrate heating due to secondary electrons is also reduced.

Advantages of Sputter Deposition

There are a number of advantages to sputtering: (1) *controlled stoichiometry* of the deposit, (2) easy sputter cleaning of the substrates, (3) improved adhesion, (4) better control of film thickness, (5) use of *bias-sputtering* for improving the physical properties of the films and for step coverage/gap-fill. The improvement in film properties by the use of substrate bias can be related to the removal by the impinging ions of atoms trapped in nonoptimal surface sites and gap-fill/step coverage to the angle-dependence of the sputtering yield and perhaps by the elevated temperature resulting from ion bombardment heating.

Temperature Effects

Since many film properties are influenced by the deposition temperature, temperature control is desirable. Substrate holders may be cooled or heated by various techniques, but it must be emphasized that, since the substrates are heated by ion bombardment and secondary electrons, their temperatures may be different from that of the holder, unless a heat transfer medium is interposed between them. For a more extended discussion of the thermal history of substrates during sputter deposition and etching, the review by Lamont (1979) can be consulted. Accurate measure of the surface temperature is possible by the use of fluoroptic probes, but they are difficult to implement in a system used in manufacturing, so their use in feedback controls may not be possible. Monitoring and controlling the holder temperature is more feasible, but is only meaningful when there is excellent thermal contact between it and the wafer.

Reactive Sputtering

Reactive sputter deposition is one way of depositing an insulator in a dc sputtering system, although rf reactive sputtering is more common. In reactive sputtering, a metal target is sputtered in a mixture of an inert gas and the appropriate reactive constituent. One reason for preferring reactive sputtering is that metal targets are usually denser and more easily fabricated than compound targets. In addition, by changing the sputtering gas mixture, several compounds can be deposited using the same target. The reaction to form the required compound may occur on the target surface, in the gas phase (unlikely), or at the substrate. Since sputtering rates of metals are higher than those of oxides, it is best to adjust conditions so that reaction occurs at the substrate; this occurs at low reactive gas partial pressure and high target sputtering rates. This effect was utilized for high rate deposition of Al_2O_3, which has a particularly low sputter yield (Jones and Logan, 1989). The stoichiometry of the film is a function of the relative arrival rates at the substrate.

Conclusions

Although insulator films of excellent quality can be deposited by sputtering at "low" temperatures, defined as temperatures compatible with aluminum-alloy metallization, there are a number of disadvantages to sputter deposition as practiced traditionally. One is high cost. Batch systems were used, and although such systems can be very large, the throughput is quite low. Load-locks are difficult to implement in such systems so that flaking from the chamber walls during wafer load/unload become a major reliability problem when the flakes become incorporated into the film. The wafer temperature is a function of the input power. In order to increase the deposition rate to meet the demands of higher throughput, the power must be increased; this often results in an excessive rise in wafer temperature. He-backside temperature control of wafer temperature is impossible to implement in batch systems. Step coverage and gap-filling capabilities are limited (although they are superior to single-pass PECVD and early low-temperature CVD), and even marginal improvement requires much extended processing time. This last is probably responsible for the lack of interest in further development of advanced reactors and in the use of sputtering for insulator deposition.

Also, there have been major improvements in alternate insulator deposition methods with, apparently, costs lower than traditional sputtering. Better "low-temperature" CVD and PECVD reactors and processes have been developed. The dielectric and physical properties of the films have been improved significantly, although the properties of some of them are still inferior to sputtered films. Reasonable throughput multistep processes for gap-fill have been developed. In addition there has been great activity in the commercial development of high-density plasma reactors for gap-fill capability. Understanding of the interactions of reactor, process, and film properties is progressing. These, plus the integration of many of the processes into high throughput single-wafer, often integrated-chamber reactors have essentially eliminated sputtering for insulator deposition.

On the other hand, sputter deposition of metals has almost completely superseded the previously ubiquitous process of evaporation, despite the greater complexity of sputtering systems (the need for the networks in addition to the vacuum apparatus). For the most part, deposition is carried out in integrated systems, in which the individual chambers have been configured to meet the objectives of high film quality and gap-fill, e.g., bias, magnetron, high temperature, collimated, and long target-to-substrate sputtering with vacuum transport, among the various deposition chambers (if several metals are to be deposited sequentially), and to etching and annealing chambers where needed. In addition there are entrance and exit load-locks.

5.4.3 Sputter Etching: Ion Milling

Introduction

Etching by means of inert ion bombardment is usually performed in a capacitively coupled diode system which is essentially an rf sputtering system in which the wafers are placed on a holder that takes the place of the target in a deposition system. In some systems, the wafers are held against the upper electrode (cathode) facing the anode. A more convenient arrangement is one in which the wafers are placed on the lower electrode (cathode) and the chamber itself is the counter electrode.

Another method of bombarding a substrate with ions is in an ion-milling, or ion beam etching system which is shown in Fig. I-23 (Melliar-Smith, 1976). There are a number of other types of ion sources besides the Kaufman source shown in Fig. I-23. Among them are the Penning, the duoplasmatron, hollow anode, and glow discharge which have been described by Harper (1978).

Applications Other Than Pattern Transfer

Sputter etching in an inert plasma can be used for patterning, but is also used to clean surfaces before subsequent processing, e.g., *in situ sputter cleaning* which minimizes interface resistance by removing an insulating surface layer from a metal before deposition of a second metal. It has also been used to roughen a surface to enhance adhesion of a second layer.

Ion milling is used in depth profiling for Auger and XPS surface analysis and as part of the process in SIMS.

Another use of sputter etching is in a process called PECVD/sputter etch; still another is in bias sputtering and biased high-density-plasma deposition. These take advantage of the angle-dependent sputter yield in increasing the acceptance angle for incoming species. This will be discussed in detail in a later section.

Pattern Transfer

In this section, pattern transfer by sputter etching in a glow discharge or by ion milling will be treated. Patterning by sputter etching is carried out in an rf discharge by bombarding the masked substrates with positive ions formed by excitation of an inert gas (Davidse, 1971).

One of the theoretical advantages of sputter etching or ion milling is that, because of the directionality of the ions and the absence of chemical (isotropic)

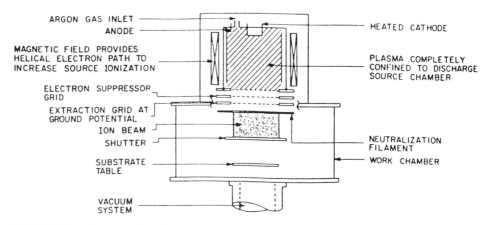

FIGURE I-23 Diagram of an ion beam etching system with Kaufman source. (From Melliar-Smith, 1976)

components, it is possible to etch without undercutting the mask and to form vertical edges. Undercutting is not observed, but often the end results are not vertical profiles. Masking by lithographic techniques is the first step. But since sputter etching is a physical process, there is not the same degree of selectivity which exists when using chemical reagents, such as solutions or reactive plasmas, since *all* materials *can be etched* by ion bombardment techniques. Thus *mask erosion* can be a significant problem particularly since some resists are among those materials with the highest sputtering yields. But thick resists degrade lithographic performance. Therefore it may be necessary to add to the complexity of the process by using a thinner resist layer to form a secondary mask in a material of somewhat lower sputter yield, such as an oxide.

In addition, the resist mask can flow and change shape because of the temperature rise due to ion bombardment.

Another problem is *redeposition* of sputtered material. Material deposited on the sidewalls of the masking pattern will alter the profile; grooves will be narrower and lines will be wider than the original mask (Lehmann, et al., 1977). The net result of faceting (to be discussed in the next section) and redeposition on the sidewalls can be seen in SEMs (Gloersen, 1975) and is shown diagrammatically in Fig. I-24.

Backscattered impurities can mask areas on the surface and due to the angular dependence of the sputtering yield, cones can develop on the surface during sputter

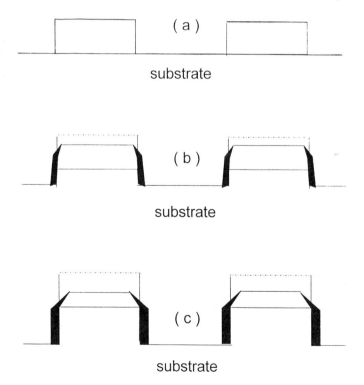

FIGURE I-24 Model of redeposition during ion bombardment of resist layer.

etching. The use of a "catcher plate," a series of concentric rings, with deep aspect ratios, bonded to the anode of a sputter etching system, reduces redeposition (Maissel et al., 1972). The catcher plate is shown in Fig. I-25a; as installed in an etcher, in Fig. I-25b.

Finally there is "trenching," enhanced etching at the sidewalls of an etched feature shown in Fig. I-26. This is a result of increased ion flux at the sidewall (forward reflection). Another factor in trenching is redeposition. The region close to the step will see a reduced solid angle (θ) for redeposition from above, while further out, a larger angle (θ') is apparent so that more redeposition (slower net etch rate) will occur (Melliar-Smith, 1976).

The problem of *faceting* the mask will be covered in the next section.

In ion milling there is an extra degree of freedom since the substrate (or the ion gun) can be rotated; this offers additional control of linewidths and profiles (Somekh, 1976).

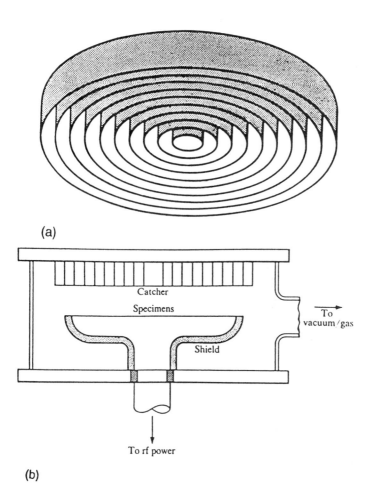

FIGURE I-25 (a) Sketch of a catcher plate, (b) installed in sputtering system. (From Maissel et al., 1972)

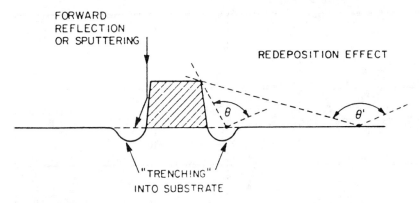

FIGURE I-26 Schematic diagram of "trenching" at sides of etched pattern. (From Melliar-Smith, 1976)

Etching for pattern transfer in inert plasmas has, for the most part, given way to reactive-plasma-assisted etching processes, although some of the problems described above are also problems in reactive-plasma etching.

5.5 ANGULAR DEPENDENCE OF THE SPUTTERING YIELD

Many processes for step smoothing, planarization, and gap-fill depend on the fact that the sputter etch rate (ion milling yield) usually depends on the angle of incidence of the ions. The change in rate with angle of incidence and the angle at which the maximum etch rate occurs differ for different materials as seen in Fig. I-27a (Oechsner, 1973; Somekh, 1976; Rangelow, 1983).

If the etch rate = $V(\theta)$ (μm/min), then the sputter yield, $S(\theta)$ (atoms/ion), is given by (Ducommun et al., 1975):

$$S(\theta) = \frac{n}{\phi} V(\theta) \cos \theta$$

where n = atomic density of target; ϕ = ion flux normal to surface ($\theta = 0$); and $\cos \theta$ accounts for the reduced current density at angles off normal. The relation between $S(\theta)$ and $V(\theta)$ is shown in Fig. I-27b.

The initial increase in rate from its value at normal incidence, V_0, as the angle of incidence is increased, is due to the fact that the probability of a collision resulting in an atom acquiring a component of momentum directed away from the surface increases with increasing angle of incidence; i.e., less of a directional change in momentum is required to eject an atom in the forward direction. Oblique incidence, particularly at higher energy, confines the action close to the target surface, enhancing sputtering.

At very high angles of incidence, the rate decreases because the incoming ion flux is spread over a larger surface area and the probability of purely elastic reflection of the incoming ions is increased at large angles. Stewart and Thompson (1969), Wehner and Anderson (1970), and Lee (1979) stress the latter points; they state that at the angle at which the rate is a maximum (θ_M), reflection of the ions

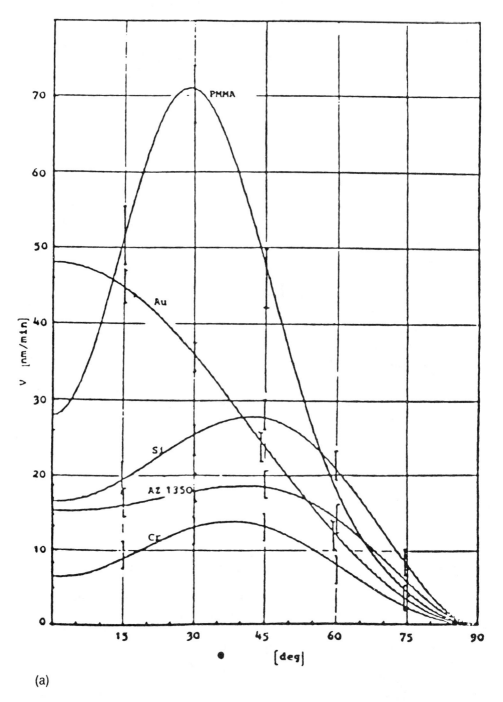

(a)

FIGURE I-27 (a) Etch rate vs. angle of ion incidence for several materials. (From Rangelow, 1983). (b) Relationship between rate V and yield S versus angle of ion incidence. (Du Commun et al., 1975)

Deposition and Etching of Thin Films 39

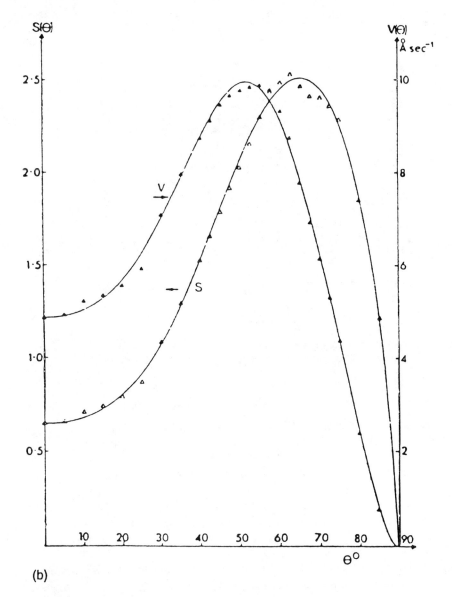

(b)

FIGURE I-27 Continued

from the potential barrier associated with the surface plane of atoms prevents penetration. θ_m, which has also been called the critical angle, is a function of ion energy, the atomic density in the target material, and the atomic numbers of both incoming ion and atom being sputtered.

Increasing the *atomic numbers* of either ion or atom *decreases* θ_m, as seen in Fig. I-28a and b, since both parameters increase the surface potential (Oeschsner, 1973).

Increasing the *ion energy increases* θ_m since a more energetic ion can more easily penetrate the surface potential barrier. At the glancing angle ($\theta = 90°$) the

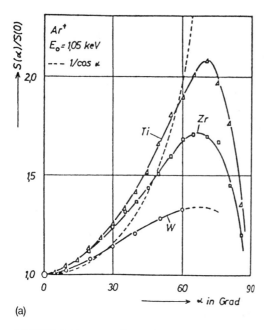

FIGURE I-28 (a) Effect of target mass on yield vs. angle of ion incidence. (b) Effect of ion atomic number on yield vs. angle of ion incidence. (From Oeshsner, 1973)

etch rate is essentially zero. S_m/S_0 was shown to decrease with increasing ion energy for Cu (Oeschsner, 1973) but increase for Si (Dimigen et al., 1976).

Thus it has been demonstrated that angled surfaces etch at a higher rate than horizontal ones when ions impinge normal to the horizontal surface; therefore sputter etching results in local planarization. The value of θ_m is important. When the angle of the edge of a structure being bombarded with ions is steeper than θ_m, a stable facet angle is formed, corresponding to θ_m; however, if the angle is less steep, there will be no change in angle. Erosion of a surface step is illustrated in Fig. I-29a (Stewart and Thompson, 1969).

The angular dependence of the sputter yield is responsible for the *increased gap-filling* capability in substrate-biased deposition processes, since the edges at the top of the gap become tapered which increases the acceptance angle of the incoming species.

However, the angular dependence can have an undesirable effect: It causes *faceting* of the edges of a resist mask, so that the edges pull back during etching. The final dimensions of the mask, therefore, will differ from those initially printed in the mask. Also, as the mask pulls back, the walls of the etched features will become tapered. The facet propagates and makes contact with the substrate surface; the two facets are propagated, as seen in Fig. I-29b. The angle of the upper facet depends on the etch rate of the substrate and the rate at which the substrate is exposed. The lower facet angle is determined by the maximum angle in the rate vs. angle curve of the substrate material (Smith, 1974).

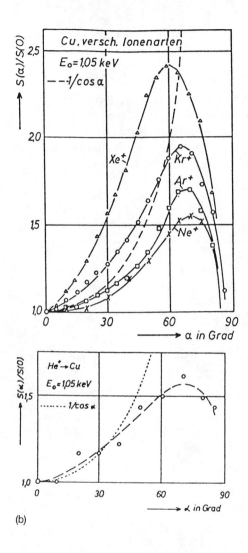
(b)

5.6 HIGH-DENSITY PLASMAS

5.6.1 Introduction

The limitations of the standard capacitively coupled plasma reactors for high rate anisotropic etching and deposition (gap-fill) have encouraged development of high-density, low-pressure systems which are now available commercially. There are the electron cyclotron resonance (ECR) and helicon plasma reactors, both of which are magnetically assisted; ECR is powered at a microwave frequency, the helicon uses an rf source. In addition is the nonmagnetic inductive discharge using an rf power source, which is known by several names: ICP (inductively coupled plasma), RFI (rf inductive) plasma, and TCP (transformer coupled plasma).

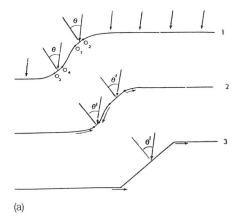

TREAT STEP as ASSEMBLY of PLANAR FACETS
ANGLE at which SPUTTERING YIELD IS MAX = θ'
IF θ = θ'
O_1 MOVES to LEFT, O_2 MOVES to RIGHT
O_3 MOVES to RIGHT, O_4 MOVES to LEFT
FINALLY
OBTAIN PROFILE in which SINGLE FACET
MAKING ANGLE of INCIDENCE θ'
MOVES ACROSS SURFACE

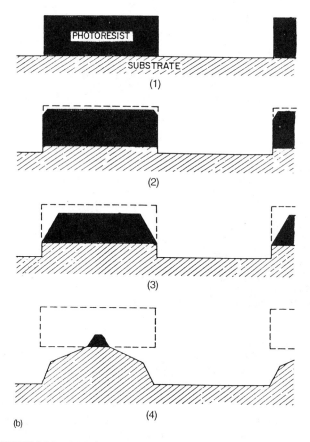

FIGURE I-29 (a) Diagram showing the erosion of a step by ion bombardment. (From Stewart and Thompson, 1969.) (b) Diagram showing facet formation during ion bombardment of a resist layer. (From Smith, 1974)

5.6.2 Electron Cyclotron Resonance (ECR)

This was one of the earliest of the high-density plasmas to be developed commercially for etching and deposition. It can produce a high-density plasma ($n_0 \geqq 10^{13}$ cm^{-3}) at low pressure. The use of an ECR reactor for RIE was first reported by Suzuki (1977) and its use for PECVD was introduced by Matsuo and Kiouchi (1982). An ECR reactor is also used as a high-rate sputtering system (Matsuoka and Ono, 1989).

Principles of ECR

An electron in motion in a uniform magnetic field undergoes circular motion transverse to the magnetic field direction; the frequency of motion is called the cyclotron frequency, $W_c = e_B/m_e$ (e = electron charge, B = magnetic field strength, m = electron mass). The magnetic energy is coupled to the natural resonant frequency of an electron gas in the presence of a static magnetic field. The resonance condition for energy transfer, i.e., for efficient transfer from the electromagnetic field to an electron exists when the electron undergoes precisely one circular orbit in one period of the applied field. For a microwave frequency of 2.54 GHz (a frequency in common use for many applications), electron cyclotron resonance occurs at a magnetic field of 875 Gauss. The very energetic electrons ionize the gas species creating a plasma. To obtain the highest plasma density, the microwave is launched into the resonance region from the direction in which the magnetic field is greater than the resonance field. The plasma densities obtainable in an ECR plasma are about 10 to 100 times that in an rf capacitively coupled plasma; the ion/neutral ratio is much higher, (~0.1), and the operating pressure is significantly lower (~0.1–1 mTorr). At higher pressures (~10 mTorr) the resonance cannot be established. Due to the low pressure of operation, there are fewer collisions in the plasma and in the sheath, resulting in greater directionality of the ions. Thus, impingement of active species is more normal to the surface than in the extended source systems, the capacitively coupled sputtering, PECVD, and RIE systems. Thus, there was an inference that the process was intrinsically directional (Machida and Oikawa, 1986). But it was realized quite quickly that substrate bias was needed for good gap-filling and directional etching. Therefore, for most applications an external bias (400 kHz–13.56 MHz) is applied to the substrate. In etching, the etch rate increases with increasing bias (Jin and Kao, 1991), and in deposition the film properties (Andosca et al., 1992), as well as good gap-fill, are a function of the applied bias. The ability to fill a gap depends on the proper balance between deposition and etching (e.g., Virmani et al., 1996). This subject is discussed more fully in Chapter VI.

In an ECR system, the ion energies are low, so it was assumed that there would be no substrate damage. And since the wafer holder was cooled, it was assumed that good films were being deposited at a very low temperature. It was neglect of the effect of substrate bias that led to these misconceptions. As mentioned previously, energetic ion bombardment of the wafer raises its temperature. A low-pressure gas is a poor heat conductor; therefore, unless additional heat transfer mechanisms are supplied, the wafer temperature must increase. In some of the early systems, in which there was no He-backside cooling, a temperature of about 500°C was reached during ECR PECVD deposition of SiO_2 (Schwartz, 1989). And ion bombardment damage is also a possibility. A diagrammatic representation of an ECR source is shown in Fig. I-30.

Divergent Field ECR

The configuration used most widely is the divergent field system; one commercially available version is shown in Fig. I-31. The microwave power is introduced into the evacuated plasma source chamber through a dielectric window. The horn and so-called target are temperature controlled surfaces used to limit extraneous deposition and thus control particle generation. The horn is part of the chamber wall and can be isolated from ground and powered by a 13.56 MHz generator for in-situ cleaning of the target and surrounding surfaces using NF_3 injected through the Ar/O_2 inlets. Usually solenoid coils surround the source chamber, although the use of permanent magnets has been described by Mantei and Dhole (1991) who used blocks of Nd-Fe-B (Fig. I-32a), by Shida et al. (1993) who placed concentric circles of Nd-Fe-B with successive opposite polarities on the ceiling of the reactor (Fig. I-32b), and by Getty and Geddes (1994) who used an array of permanent magnets arranged over the surface of the dielectric waveguide window.

The ECR position, the position at which resonance occurs, is most often within the source chamber, although advantages of locating it closer to the wafer surface have been reported (Fukuda et al., 1988). The plasma is extracted from the source chamber along *divergent* magnetic field lines; the magnetic flux density decays and approaches zero in the neighborhood of the wafer. As the electrons are extracted,

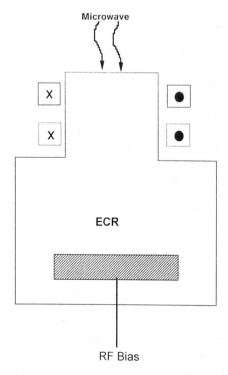

FIGURE I-30 Diagram of an ECR reactor with substrate bias.

Deposition and Etching of Thin Films

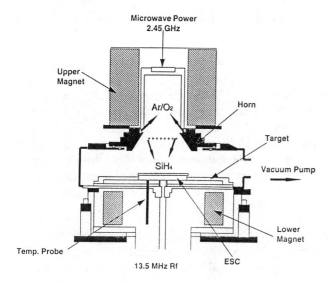

FIGURE I-31 Diagram of a commercial ECR reactor with substrate bias and solenoid coils. (From Dennison and Harshbarger, 1995)

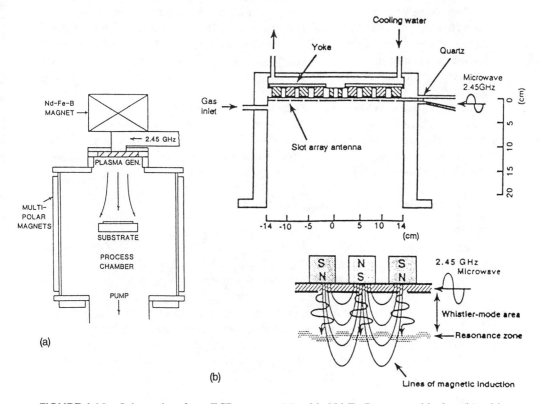

FIGURE I-32 Schematic of an ECR reactor (a) with Nd-Fe-B magnet blocks; (b) with concentric circles of permanent magnets and expanded diagram showing slot antenna for whistler launching. [Part (a) from Mantei and Dhole, 1991; part (b) from Shida et al., 1993]

an electrostatic potential is created which pulls the ions in the same direction toward the wafer.

Uniformity in a Divergent Field ECR Reactor

Multipolar magnets have been placed around the reaction chamber to confine the plasma (Mantei and Ryle, 1991; Nihei et al., 1992). Another approach was placing of a pair of solenoid coils beneath the substrate holder. The inner coil (current flowing in direction opposite to main coil current) generates a cusp magnetic field to make the plasma distribution at the wafer more uniform; the outer coil (current in same direction as main coil) generates a mirror magnetic field which confines the plasma, resulting in a narrow ion energy distribution (Matsuoka and Ono, 1987; Araki et al., 1990).

Other ECR Reactors

The other configurations are the microwave multipolar plasma (MMP) and the distributed (DECR) reactor (Burke and Pomot, 1988, 1989). MMP is the term used for surface wave excitation. In this system, a silica tube is inserted into an opening of a waveguide to create localized excitation and subsequent diffusion into a multipolar magnetic structure, as illustrated in Fig. I-33a.

In DECR, the ferrite multipole magnets create the resonant field for ECR excitation within the reactor chamber a few millimeters from each pole face. The microwave energy is applied by a set of tubular conductors placed around the chamber to distribute the plasma excitation around the chamber walls; one of these is shown in Fig. I-33b. Figure I-33c shows the principal plasma zones: the ECR cusps, the lobes, and the diffusion plasma. The lobes, developed by the alternating polarity of the multipoles, trap electrons and thus contribute additional ionizing regions.

Although DECR systems have been developed commercially, they do not appear to be used to any great extent in device fabrication. Reports about the use of microwave plasmas using multipolar confinement are etching: Arnal et al. (1984), Pichot (1985), Pomot et al. (1986), Cooke and Pelletier (1989); depositon: Cooke and Sharrock (1990), and Plais et al. (1990, 1992).

5.6.3 Radio Frequency Induction (RFI)

Introduction

Some of the earliest examples of glow discharges were those produced in tubes wound with coils (later used as "barrel" ashers and neutral-species-dominated reactive-plasma-assisted etchers). These were operated at high pressures. It was a matter of controversy whether these discharges were capacitive (plasma coupling to the ends of the coil) or inductive (induced electric field inside the coil). It was finally concluded that at low plasma densities (low power, high pressure) the discharge was capacitive (electrostatic origin, E discharge, faint glow); as the power is increased there is a transition to the inductive mode (electromagnetic origin, H discharge) (Amorin et al., 1991).

Principles

The low-pressure, high-density *inductive* discharges are those of interest since they meet the requirements for processing advanced devices. To quote Keller, "In an rf

Deposition and Etching of Thin Films

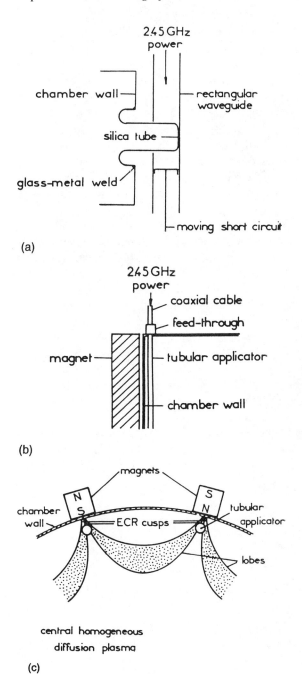

FIGURE I-33 Diagrams of microwave multipolar plasma configurations: (a) hybrid surfaguide excitation source, (b) tubular applicator of microwave energy, (c) tubular applicators surrounding a chamber showing cusps and diffusion plasma. (From Burke and Pomot, 1988)

induction system, power is coupled from the rf coil to the plasma which acts as a single turn secondary to a transformer." These plasmas are known by various names: RFI, radio frequency inductive (plasma); ICP, inductively coupled plasma; and TCP, transformer-coupled plasma. A review, "Inductive plasmas for plasma processing," by Keller (1996) gives details about the features of these systems.

Two coil configurations are used: cylindrical (wrapped around the source chamber) (e.g., Cook et al., 1990) and planar (electric-stove-coil) (Ogle, in a patent, 1990; Keller et al., 1993). These coils can also be fashioned to give hemispherical coil shapes, as disclosed in a patent by Benzing et al. (1995) and first described in the literature by Mountsier et al. (1994). The first two reactor configurations are shown schematically in Fig. I-34a and b and the last one is illustrated in Fig. I-34c.

Keller et al. (1993) used a 13.56 MHz power supply to drive the flat coil placed over a *thick* dielectric window; this reduces capacitive coupling. The length of the reaction chamber is small so that the plasma diffuses only a few mean-free paths from source to substrate; this contributes to a high efficiency of plasma production efficiency by reducing loss to the walls. It also minimizes the ion density loss characteristic of divergent systems. Multipolar confinement improved the uniformity. The *wafer* was held by He-backed electrostatic clamp and was *independently biased* using a 40.68-MHz *capacitively coupled* source. The plasma potentials are low; densities of $\sim 10^{12}$ cm^{-3} can be achieved at an operating pressure as low as 1 mTorr.

Applied Materials sells a system based on this concept which they call a decoupled plasma source (DPS); the chamber design and the frequencies used were not given (Ye et al., 1996).

A lower frequency, 0.46 MHz, was used in a hemispherical ICP source (Benzing et al., 1995; Tuszewski and Tobin, 1996) which was operated at 5 mTorr. Up to 2 kW of rf power was efficiently coupled to the plasma. The radial uniformity of the plasma was excellent.

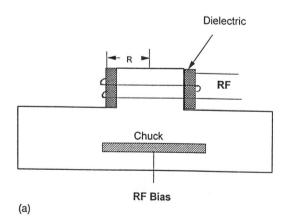

(a)

FIGURE I-34 Schematic diagrams of inductively driven sources: (a) cylindrical, (b) planar, (c) hemispherical. [Part (c) from Mountsier (1994)]

Deposition and Etching of Thin Films   49

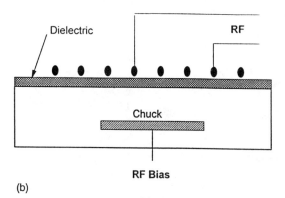

(b)

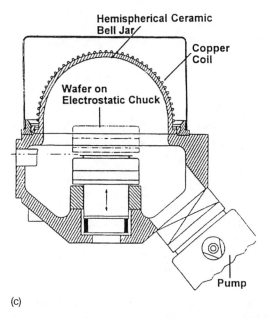

(c)

FIGURE I-34 Continued

5.6.4 Helicon Sources

These reactors, to be used for plasma processing, were first described by Perry and Boswell (1989) and Perry et al. (1991). A diagrammatic representation of a reactor using a helicon wave source is shown in Fig. I-35. A high-density plasma is generated in the source chamber by coupling 13.56 MHz rf to an antenna which surrounds the chamber. The source solenoid is required for coupling the rf source into a helicon mode in the plasma for efficient transfer of energy into the center of the plasma (Boswell, 1989). In this system, the plasma potential is low, and the substrate bias is controlled independently of the plasma excitation voltage. The plasma diffuses from the source to the reaction chamber where it is confined by the axial field produced by the chamber solenoid or by multipoles (Boswell et al.,

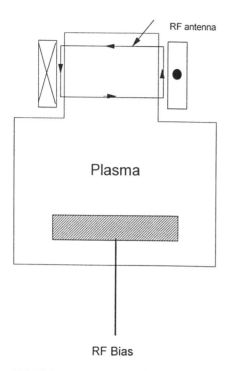

FIGURE I-35 Schematic diagram of a helicon reactor.

1989). Another option is to combine the source and process chambers or to place the wafer very near the source exit in order to increase the ion and neutral fluxes and reduce the spread in ion energy. At a pressure of 3 mTorr, average densities of $3-5 \times 10^{11}$ cm^{-3} were obtained in the reaction chamber for an input power of 500 W. Densities above 3×10^{13} cm^{-3} have been achieved in small helical discharge reactors using 2 kW rf power (Chen and Chevalier, 1992). Plasma-assisted etching in a helicon reactor has been called "resonant inductive plasma etching" or RIPE (Henry et al., 1992).

A more complete discussion of the physics and design of high-density plasma sources can be found in Lieberman and Lichtenberg (1994) and Lieberman and Gottsscho (1994).

5.6.5 Concluding Remarks about High-Density Reactors

ECR, inductively coupled, and helicon reactors are similar in their ability to achieve high plasma densities at low pressure, so that deposition and etching results are also similar. All can be configured with multipolar confinement. But ECR systems are more complex; it is said that the inductive and helicon systems are easier to maintain and less susceptible to failure. Such systems may well displace ECR reactors.

5.6.6 Ultra-High-Frequency (UHF) Source

In this reactor, a spokewise antenna is located on a quartz plate separating it from the vacuum chamber; this is illustrated in Fig. I-36. The ultra-high-frequency (500 MHz) power propagating between the spokes is efficiently coupled into the plasma both inductively and capacitively ("CM coupling"). The plasma is very uniform, has a high density (without the need for a magnetic field), a low ion and neutral temperature, and can be operated at higher pressures than can ECR plasmas, but collisional broadening is negligible (Samukawa et al., 1995; Samukawa and Nakano, 1996).

5.7 PLASMA-ENHANCED CVD (PECVD)

5.7.1 Introduction

PECVD is carried out in a nonequilibrium glow discharge. Impact dissociation by the high-energy electrons results in the formation of highly reactive species (largely free radicals) that normally are formed at high temperatures. It is possible to form materials with unique chemical, physical, and electrical properties in the highly reactive plasma (Hess, 1984). Among these are materials of importance to the semiconductor industry such as the oxides and nitrides of Si, Ti, Ta, etc., BN, SiBN, SiOBN, amorphous and polycrystalline Si, amorphous C, some metals, silicides, etc., and a variety of other materials used for applications not considered here (e.g., Nguyen, 1986).

Because of the complex mix of potential precursors (which makes possible the variety of PECVD films) and the large number of independent and interdependent operating parameters, control and reproducibility of films and processes are often difficult. Despite difficulties, the ability to form high-quality insulating films at substrate temperatures compatible with aluminum alloy-based metallization

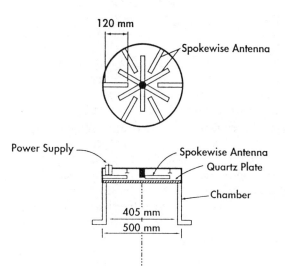

FIGURE I-36 Diagram of a UHF source with spoke-wise antenna. (Samukawa et al., 1995)

($\leq 450^\circ$C) has led to the extensive use of PECVD in semiconductor manufacturing. PECVD is used predominantly for depositing oxides for use as interlevel dielectric films and nitrides for passivation and moisture/Na^+ barriers and polish stops. Diffusion barriers, barrier layers for multilevel masks, and metals are some of the other less widely used applications.

5.7.2 Mechanisms

The kinetic reactions are the same as those for CVD, except that plasma-initiated radical formation replaces thermally activated precursor formation (Hess and Graves, 1989). The growing film is subjected to bombardment by energetic species in the plasma; this influences film properties and deposition characteristics (e.g., gap-fill, to be described in a later section). Ion bombardment has been cited as largely responsible for the excellence of the films deposited at a relatively low temperature at a relatively high rate (Hess, 1984; Claasen, 1987; Hey et al., 1990). However, there is the possibility of radiation damage to sensitive devices and methods for eliminating, or at least minimizing it are discussed elsewhere in this book.

The mechanisms of the reaction(s) occurring on the surface and in the gas phase prior to adsorption have been studied but will not be discussed in detail in this book. For more information a review by Hess and Graves (1989) may be consulted. Some of the models which may be of help in understanding the process or film properties will be discussed in the sections dealing with specific films. However, at times the experiments have been performed, for the sake of simplicity and ease of interpretation, using system configurations and/or operating conditions very different from those used in deposition for device applications, making universal application of the results and models somewhat questionable.

5.7.3 Reactors

PECVD reactors use capacitively coupled or high-density discharges; the basic properties and attributes of these plasmas have been described earlier in this chapter.

Capacitively Coupled Discharges

One of the earliest mentions of the deposition of silicon oxide films in an rf glow discharge was by Alt et al. (1963), but details of the system, beyond the use of a fused quartz reaction chamber, were omitted. Ing and Davern (1965) and Sterling and Swann (1965) described tubular ("barrel") reactors, powered by external rf coils or plates. These were found to be unsuitable for commercial application.

The use of PECVD films in semiconductor manufacturing can probably be dated from the introduction of the radial-flow capacitively coupled, parallel plate diode reactor with internal electrodes to which an rf source was connected, patented by Reinberg in 1974. This basic design, used for batch processing, had been used in many commercially available systems; some of the differences among them include gas-flow pattern, gas introduction method, electrode design, and rf frequency. One of these batch receptors is shown in Fig. I-37 in which the flow pattern of the gases, from center to edge, is in the direction opposite to that in the Reinberg reactor (Fig. I-13).

Deposition and Etching of Thin Films 53

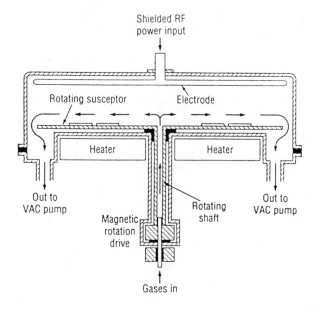

FIGURE I-37 Diagram of an Applied Materials batch PECVD reactor.

Reactors using magnetic enhancement of the discharge have been described (Kaganowicz et al., 1984).

A departure from the conventional design for batch processing is a hot-wall tubular system (resembling a hot-wall LPCVD reactor) with internal graphite electrodes, illustrated in Fig. I-38. It is also available commercially from ASM/PWS (Rosler and Engle, 1979).

Another variation was a stepped batch process reactor: a cassette-to-cassette linear, multiple-station reactor in which one fifth of the total film thickness was deposited at each station, using a vibrating track transfer mechanism. The mecha-

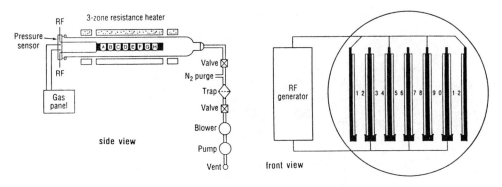

FIGURE I-38 Diagram of an ASM/PWS batch PECVD reactor. (From Rosler and Engle, 1979)

nism was not reliable and, in addition, was responsible for particulate contamination. It had a short commercial life (Rosler, 1991).

The early reactors were batch systems. The deposition rates were low, but the throughput was adequate because many wafers were processed simultaneously. However, defect levels were relatively high because deposits would form on all surfaces and could flake when they became too thick or were exposed to the atmosphere. Thus system cleaning was a major problem. The introduction of load-locks reduced the severity of the latter problem.

Among the next advances, using a parallel plate reactor, is the Applied Materials 5000, shown previously in Fig. I-12, used here with rf activation of the incoming gases. Processes in this system have been discussed by Law et al. (1987); Spindler and Neureither (1989); Bader et al. (1990). In this integrated system are the PECVD module, etch chambers with inert and reactive gases, and a low-temperature CVD reactor.

At about the same time there was a more successful reintroduction of the continuous or multiple-station system (van de Ven et al., 1987). Here, the configuration was circular, with cassette-to-cassette load/unload in a vacuum load-lock. In the first version, one seventh of the total thickness was deposited at each of seven stations, heated resistively; each station had its own showerhead for gas and rf distribution. An improved wafer transport mechanism reduced contamination. The system is shown in Fig. I-39. As the wafer size increased, fewer wafers could be accommodated and so a larger fraction of the total film was deposited at each station. This was among the first commercially available PECVD systems to use a two-frequency mode of operation (Martin, et al., 1988; van de Ven et al., 1990).

These systems used in situ cleaning as well as load-locks. Deposition rates are high, so that the throughput is comparable to batch systems.

5.7.4 High-Density Plasmas

These high-density systems, described in Section 5.6, have been available commercially for a relatively short time and have been used successfully for PECVD. Their use for depositing SiO_2 will be discussed in Chapter IV. They may replace

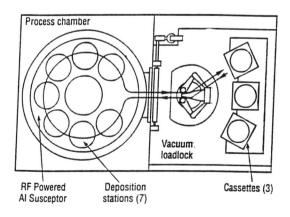

FIGURE I-39 Diagram of a Novellus multiple station chamber for PECVD.

Deposition and Etching of Thin Films   55

the older systems for the most advanced structures. Their widespread acceptance may depend on their reliability and the cost of replacement vs. performance.

5.7.5 Other Reactors

There are other systems which have not been used to any great extent in device fabrication, but are mentioned here for the sake of completeness. One is called remote PECVD (RPECVD), often used for basic studies of the mechanism of film formation. In these reactors, the N- or O-containing species are excited by an rf plasma, then transported out of the glow to react in the vicinity of the substrate with unexcited silane to form a nitride or an oxide (Helix et al., 1978; Meiners, 1982; Richard et al., 1985; Lucovsky et al., 1989, 1991). There are also the downstream or afterglow systems that operate on similar principles but use microwave excitation to produce the active species (Robinson et al., 1987; Spencer et al., 1987).

5.8 REACTIVE-PLASMA-ENHANCED ETCHING

5.8.1 Introduction

The use of substrate-biased reactive-plasma-assisted etching processes has almost completely replaced wet etching and sputter etching. Although sputter etching shares many of the substrate-biased reactive techniques, such as directionality of etching, compatibility with automation and vacuum processing, and relative cleanliness, the selectivity, enhanced rates and versatility of the reactive process has made it the process of choice for pattern transfer in semiconductor fabrication.

One of the major advantages of these processes is the ability to achieve anisotropic etching, i.e., etching without undercutting the mask, so that small features can be etched with fidelity, or as is sometime put, the "critical dimensions" defined by the mask can be maintained. However, anisotropy does not *always* mean a vertical profile. If the mask does not have vertical edges or if the mask edges shift during etching, due to temperature effects or lateral etching, then the resulting profile will be tapered.

The directionality of etching, which is a key advantage of this technique, does present a problem when the film to be etched has been deposited over a steep step, as shown in Fig. I-40. Since the thickness of the film to be etched is much greater at the sidewall, long overetches are required to clear the step, making great demands on the masking layer and requiring large film/substrate etch rate ratios.

These processes are carried out in nonequilibrium glow discharges of reactive gases, and in the same kinds of reactors as PECVD, e.g., capacitively coupled diodes, with and without magnetic enhancement, triodes, and high-density plasmas (ECR, RFI, and helicons). The basic steps are similar to those in PECVD: species generation, transport, adsorption, reaction, desorption of volatile species, and finally, pump-out. The *basic requirement* of plasma-assisted etching is the production of a reaction product that is *volatile* or can be *desorbed readily under the influence of ion bombardment*. The chemical reactivity of the radicals produced in the discharge results in higher etch rates and greater selectivity between the film to be etched and the substrate and mask than that obtained by physical sputter etching in inert (unreactive) plasmas. But chemical etching is isotropic. However, if an

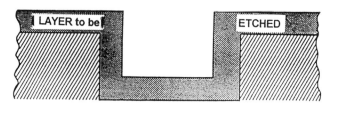

FIGURE I-40 Diagrammatic representation of anisotropic etching of a film deposited over a vertical step.

adequate bias is applied to the substrate, the ion bombardment responsible for the directionality of sputter etching contributes an anisotropic component which, in a sense, competes with or complements the isotropy of chemical etching in determining the shape of the etched feature. Both the adsorption of a reactive species on the surface and the desorption of a product from the surface, may be enhanced by ion bombardment (Oehrlein, 1989). Thus ion bombardment which is responsible for anisotropy also contributes to rate enhancement above that obtained by exposure to reactive gases only (Coburn and Winters, 1979).

5.8.2 Mechanisms

Several mechanisms have been proposed to account for the rate enhancement: (1) the reacted surface sputters with a higher yield than the unreacted surface (chemically enhanced physical sputtering) (Mauer et al., 1978; Dieleman, 1978; Haring et al., 1982; Dieleman et al., 1985); (2) ion bombardment damages the surface thereby increasing the reaction rate (Flamm et al., 1981; Greene et al., 1988; Oehrlein, 1993); and (3) ion bombardment supplies additional energy to the reaction layer, enhancing volatile compound formation and desorption (Tu et al., 1981). There are detailed reviews of mechanistic studies and models, which include spontaneous etching and sometimes contradictory effects of ion bombardment, e.g., Winters et al. (1983), Coburn and Winters (1985), and Coburn (1988). The applicability of each model appears to depend on the specific etchant/substrate system. In addition, some of the results obtained by reacting clean surfaces in a vacuum (as done in some mechanistic studies) may be different from those obtained in practical etching systems.

5.8.3 Etching Systems

The first commercially available "plasma etchers" were barrel reactors in which ion bombardment was minimal and etching was purely chemical (i.e., isotropic) (Irving, 1971; Abe et al., 1973). The advantage over etching in solution was the integrity of the mask substrate interface. In wet etching, the liquid, driven by capillary forces, often penetrated beneath the resist, lifting it and exposing regions of the substrate protected during the lithographic procedures. This does not occur in a plasma so that any undercut beyond that dictated by isotropy was a function only of the overetch time. The use of this configuration outlasted the pattern transfer application; it has been used extensively for resist ashing, i.e., oxidation in an O-radical plasma, although downstream etching may be replacing it (Boitnott, 1994).

5.8.4 Reactive Ion Etching (RIE) or Reactive Sputtering Etching (RSE)

RIE combines the chemical activity of the reactive species with the ion bombardment of physical sputtering. Among the first RIE reactors were the capacitively coupled sputter-etchers (asymmetric systems) retrofitted with reactive gases. These were low-pressure systems; the low pressure and high ion energy bombardment of the wafers (placed on the smaller electrode) favored anisotropy. Muto (1976) was one of the first to patent this process. He claimed vertical etching of masked surfaces placed on the cathode of a two-electrode planar reactor into which chemically reactive F- or Cl-containing gases were introduced for the purpose of rapid etching of thin film circuits or semiconductor chips. Matsuzaki and Hosakawa (1976) called it a method for sputter-etching silicon or its compounds, using a planar electrode and employing fluorohalogenhydrocarbon gas (including bromine and O_2, N_2, Ar, or air as diluents) for the manufacture of integrated circuits.

Symmetrical systems (based on the radial flow reactor) were also popular; they were labeled "plasma etchers" because the wafers were placed on the grounded electrode. However, since the system was symmetrical, the potential on each electrode was the same. The tendency toward isotropy was due, not to the grounding of the substrate electrode, but in part to the pressure regime in which they were operated, several hundred mTorr as compared with tens of mTorr in the RIE systems. In addition, the plasma potential was high due to the symmetry and plasma confinement between the closely spaced electrodes, so that the ion energy was lower than in RIE. The use of the so-called anisotropic etching gases, in which lateral attack (isotropic etching) was inhibited, extended the usefulness of these reactors (Ephrath, 1981). Their action was due to the formation in the plasma of species which (1) inactivated the etchant species (recombinants) or (2) formed a protective film on the sidewall (passivants). These systems were eventually abandoned in favor of the low-pressure/high-bias RIE systems, in which there was a greater latitude in the choice of etchants.

The distinction between RIE (cathode-coupled or diode etching) and what was known as "plasma etching" (anode-coupled) in what may appear to be a symmetrical reactor, because the internal electrodes are of equal area, was stated clearly by Mathad and Patnaik (1979): "In the diode mode, the power is applied to the wafer-carrying electrode while the rest of the system is grounded"; the ion energies at the wafer surface are high. "In the anode mode the power is applied to the upper electrode; the wafer-carrying electrode AND the reactor chamber are grounded and there is negligible ion bombardment of the wafer surface." These concepts were also illustrated by Toyoda et al. (1980).

The capacitively coupled RIE systems are being replaced, in turn, by the lower pressure, independently-biased, high-plasma-density RIE reactors which will be discussed later.

5.8.5 Choice of Etchants

The choice of etchants (and inert species) depends not only on the ability to form a volatizable product in its reaction with the substrate but on its potential for enhancing anisotropy and selectivity to mask and substrate (Coburn and Kay, 1979; Flamm and Donnelly, 1981). Mixtures of gases may be required to achieve the desired results. Inert additives are often used as diluents, but they may have other functions, such as plasma stabilization, cooling, alteration of the electron energy distribution, enhanced ion bombardment, and formation of metastable states for efficient energy transfer (Oehrlein, 1989); in metal etching, N_2 plays an active role (Ohno et al., 1989; Sato and Nakamura, 1982, Howard and Steinbruchel, 1991). H_2 (Heinecke, 1974) and O_2 (e.g., Flamm, 1981) are often used to interact with the etch-specific gases to modify the etch results. The nature and relative concentrations of the reactive species formed in the plasma from the source gases depend not only on the chemical composition of the gases but on the electron energy and energy distribution in the plasma, which in turn are a function of the system (e.g., low-density vs. high-density plasma) and the system parameters. A more detailed examination of some of these systems will be made in later chapters when some of the specific substrate/etchant interactions are discussed.

5.8.6 System Parameters

The physical parameters such as power, pressure, gas flow rate/pumping speed, and the reactor configuration in parallel plate reactors (e.g., chamber size and shape, relative electrode sizes and electrode spacing, the number and location of ports, etc.) affect the electrical properties of the glow discharge and thus the interaction with the source gases. Their influences on etching, individually and in concert, are important but beyond the scope of this book. Several recent reviews cover some of this material (Oehrlein, 1989; Manos and Flamm, 1989; Oehrlein and Rembetski, 1992).

Since, as pointed out by Oehrlein and Rembetski, there is no simple relationship between the process objectives and the physical variables, process optimization requires systematic investigation of the parameter space.

5.8.7 Profile Control

There are several aspects to profile control; one is suppression of undercut. Another is prevention of "bowing" and "trenching" or "dovetailing," illustrated by Arnold and Sawin (1991). Still another is deliberate profile shaping or edge tapering to form sloped sides to ensure better step coverage during subsequent deposition of overlying films, e.g., for etching via holes in the interlevel dielectric. The advances

Deposition and Etching of Thin Films

in step coverage and hole filling (discussed in Chapter VI) lessens the need for tapering so that the space-saving vertical profiles are now the process objective.

The extremes in profile, purely directional and purely isotropic, are illustrated in Fig. I-41. An isotropic profile results when the dominant etchants are neutral (chemical) species and there is no inhibition of this lateral component. Lateral etching may be minimized or even eliminated by what has been called sidewall passivation in which a chemically unetchable film is formed on all surfaces (Oehrlein and Rembetski, 1992). Only the horizontal surfaces are exposed to energetic ion bombardment which removes the film, exposing fresh surfaces to the etchant. There is essentially no ion bombardment on the vertical walls, so the film remains there. Therefore, vertical, but not horizontal etching proceeds. Passivating films have been observed in most material/etchant systems currently used. The inhibiting film may be formed from precursors created in the discharge by fragmentation of the source gas (Schaible and Schwartz, 1979), by redeposition from an eroded mask, or by both (Kawamoto et al., 1994). Another source of protection is the formation of a film, in the absence of ion bombardment, by a side reaction between an additive and a reaction product. Hirobe et al. (1987) found that the sidewall film was formed by materials sputtered from the cathode plate. A sidewall film, flaking off the edges of the etched feature, is seen clearly in Fig. I-42, which is an SEM of an Al film etched in CCl_4/Ar (Schaible and Schwartz, 1979).

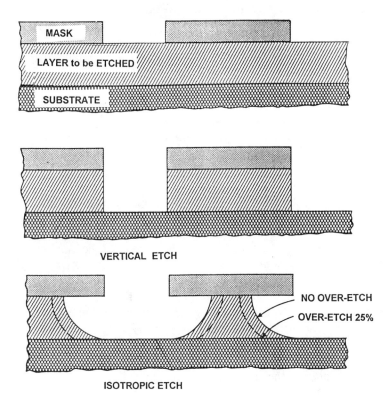

FIGURE I-41 Diagram showing the contrast between vertical and isotropic etch profiles.

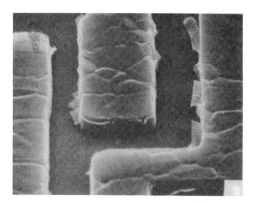

FIGURE I-42 SEM of a sidewall film formed during RIE of AlCu in CCl_4/Ar. (From Schwartz and Schaible, 1983)

Redeposition on the sidewalls of a marginally volatile etch product can also provide some protection. The competing processes of etching and deposition not only provide protection but result in a slight inward tapering of the edges, i.e., the sidewall film decreases in thickness with depth.

The shape of the etched feature can also be influenced by off-axis ions (Lii and Jorne, 1990; Nguyen et al., 1991) and by ions scattered from the edges of a mask (Ohki et al., 1987). Arnold and Sawin (1991) proposed that localized surface charging could cause surface potentials capable of skewing the directionality of the incoming ions resulting in ion fluxes to the walls of an etched feature which would contribute to bowing and trenching. Murakawa et al. (1992) invoked this same phenomenon to account for profile distortion, i.e., tilted edges.

Pulse-time modulation of the plasma in several high density reactors has been used for profile control of Si. For example, Boswell and Henry (1985) and Boswell and Poteous (1987) found that pulsing the plasma in a substrate-biased helicon plasma reactor, changed the profiles of Si etched in SF_6; the longer the duration of the pulse (e.g., 1000 ms at 20% duty cycle) the straighter the walls, (and the lower the etch rate and etch rate to SiO_2).

Modulation of chlorinated plasmas has been used in high-density reactors to achieve vertical and notch-free profiles, as well as highly selective and charge-free etching (e.g., Samukawa and Terada, 1994, Fujiwara, et al., 1996, Samukawa and Mieno, 1996). The improvements due to pulsing were attributed to the change of the flow of ions through the sheath region to the substrate surface (Mieno and Samukawa, 1995).

Tungsten was etched in pulsed SF_6 helicon plasmas (Petri et al., 1994). Two etching regimes were shown to exist. For short discharge-off periods, etching was limited by F adsorption on the surface (neutral-limited); for long discharge-off periods, etching was limited by the desorption rate (ion-limited).

A model was developed to interpret the results of a pulsed high-density plasma on metal etch. It dealt with pulsing the source power to control the ion and radical densities and pulsing the bias power to control the ion energy. It showed that higher selectivity to resist and minimized AR-dependent etching can be achieved at high etch rates. It also showed the same directionality and charge-up effects (Xie et al., 1996).

A time-modulated ECR plasma was used to control polymerization in etching SiO_2 to achieve high selectivity (Samukawa, 1993). In contact/via etching a time or power-modulated high-density plasma can be used to increase the selectivity of SiO_2 to Si by controlling the generation ratio between polymer precursor and etching species (Xie et al., 1996).

Other reports on pulsed plasmas can be found in papers by, e.g., Park and Economou (1990), Verdeyen et al. (1990), Ashida et al. (1995), Holland et al. (1996), Ahn et al. (1996), Sugai (1996).

A different approach was used by Tsujimoto et al. (1986) who proposed a "chopping method" to prevent lateral etching in ECR etching. The sample was "sequentially and independently exposed to a film forming gas and an etching gas, so that the side wall is covered as the bottom of the feature is etched." Shibata and Oda (1986) used a multistep process for etching Si by alternating RIE in a chlorinated plasma with surface oxidation. Another hole taper-etching procedure involved using an oxygen ion plasma in a biased ECR reactor; it was based on the incident-angle dependence of the etch rate and the etch selectivity of SiO_2 with respect to the metal in an O_2 plasma (Hashimoto et al., 1990).

There have been numerous approaches to tapered etching to form sloped edges of holes in an anisotropic etch process. These include tailoring the shape of the resist mask or the etch rate ratio of mask to substrate, facet formation, etc. which will be discussed in detail in the sections dealing with the specific substrates.

Profile evolution cannot be discussed in detail in this book. It has been the subject of many models and simulations and just a few of the references, in addition to those cited above, are given as examples: Smith et al. (1987), Arikado et al. (1988), Glowacki and Tkaczyk (1988), Cottler et al. (1988), Economou and Alkire (1988), Gross and Horwitz (1989), Jackson and Dalton (1989), Giffen et al. (1989), Shaqfeh and Jurgensen (1989), Pelka et al. (1989), Lii and Jorne (1990), Cottler and Elta (1990), Fujinaga et al. (1990), Fichelscher et al. (1990), McVittie and IslamRaja (1992), Singh et al. (1992), Shan et al. (1994), Hamaguchi and Dalvie (1994). Some models consider the two components of RIE, the chemical and the ion assisted, their interactions, the competing roles of etching and deposition, and changes in mask shape. Others concentrate on the effects of ion scattering, in the sheath and reflection off the sidewalls of the etched feature, or local surface charging which affects not only ion deflection but ion flux and energy distribution.

5.8.8 Masking

In wet chemical etching, the shape and thickness of the mask was not important; adhesion, particularly at the edges of the mask, was the principal requirement. In RIE, however, masks may etch at a significant rate due both to ion bombardment and chemical reaction. Therefore the choice of the mask material and its shape are very important; adhesion is a concern only if wet-chemical steps precede RIE. The interaction between mask shape and the etch rate ratio of mask to substrate is shown in Fig. I-43a for vertical and Fig. I-43b for tapered masks, assuming only vertical etching. Additional changes in the mask profile, due to lateral chemical etching, will be reflected in further (and possibly more complex) changes in the

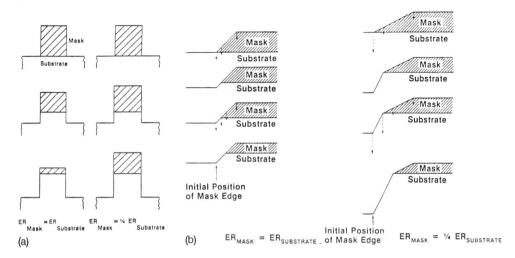

FIGURE I-43 Effect of mask shape and etch rate ratio of mask to substrate on the profile etched into the substrate, assuming only vertical etching. (a) Vertical mask; (b) tapered mask.

shape and dimensions of the etched feature. Another requirement is thermal stability since sample heating may occur due to ion bombardment and exothermic etching reactions. Heating can cause flow or reticulation of a resist mask, distorting it so that changes in its shape and dimensions result. Resist stabilization procedures have been developed to minimize these effects (Iida et al., 1977; Ma, 1980; Moran and Taylor, 1981; Hiraoka and Pacansky, 1982). Hard (unflowable) masks, such as SiO_2, Al_2O_3, MgO, have the additional advantage of having negligible etch rates in the processes for which they are used. The limitations of single-layer resist masks and the advantages and drawbacks of multilevel masks for RIE have been discussed by Tracy and Mattox (1982), Kinsbron et al. (1982), and Bushnell et al. (1986).

5.8.9 Loading Effect

A "loading effect" occurs in chemically reactive plasmas; it is observed as a change in etch rate in response to a change in the supply (consumption) of the reactant during an etch process (not as a result of changing the inlet supply). Loading effects are categorized as "global" or macroscopic loading or "local" or microscopic loading, but the assignment to a given classification is often made differently by different authors.

It was first observed as a *reduction* in the average etch rate as the total area of etchable material exposed to the plasma was increased; it is due to the average *consumption* of etchant species during etching (Schwartz et al., 1976; Mogab, 1977; Schaible et al., 1978). It is a macroscopic effect, known as "global" loading. Global loading affects production since the average etch rate decreases as the number of wafers etched simultaneously is increased; processing time will, therefore, depend on the number of wafers in the batch. If only a small fraction of the active species is consumed, the effect can be minimized; however, this would require a very large flow rate of the source gas and very efficient pumping of enormous volumes of gas. Another approach is to have the reactor fully loaded at all times, e.g., using an etch pallet made of the same material as that being etched or another material whose etch rate is the same. The use of single wafer etchers eliminates this problem and is a major advantage of these newer systems. However, the material of which a single wafer reactor is constructed can influence the etch rate. Watts and Varhue (1992) report a loading effect of a quartz microwave window or reactor liner on the etch rate of Si in SF_6.

The "local" loading effect is a microscopic effect; it is a universal problem and difficult to eliminate. One cause, which is perhaps subject to some degree of control, is due to nonuniformities in etch rates and film thickness. As the endpoint is approached, the area of the film being etched changes nonuniformly. As the underlying (nonreactive) substrate is exposed, nonuniformly, the local consumption of etchant decreases leading to increased availability of etchant for the neighboring film, *increasing* its etch rate. The result may be severe overetching, both into the substrate and, in some cases, undercut of a masked feature due to enhanced lateral etching.

Another aspect, unrelated to nonuniformities in thickness and average rate, is the "bull's-eye" etch pattern sometimes seen in etching Al and its alloys. The outermost regions of the wafer are adjacent to areas in which reactant is not consumed; thus the local reactive species concentration at the edges is high. As a result, the edges etch faster than do the central regions all of which are being etched and depleting the reactant supply.

A very important expression of local loading is the "pattern-factor effect," i.e., local *variations* in etch rates due to differences in the size, shape, and relative location of masked features, i.e., to differences in local consumption of etchant. Features of low-pattern density etch faster than those in high-density areas (Jones et al., 1990). The result may be undesirable differences in the size and shapes of etched features. Differences in chip lay-out may force changes in the details of the etch process. In addition, this effect may account for incomplete local etching (residues). In RIE of AlCu films in CCl_4/Ar, the narrower spaces cleared more rapidly than the wide ones, whereas in etching Cu in the same system/gases, they cleared more slowly (Schwartz and Schaible, 1983). In etching an Al alloy in Cl_2 in a biased ECR reactor, the microloading effect was opposite from that seen in the CCl_4 RIE system, i.e., narrow spaces etched more slowly. (Aoki et al., 1992). These differences have not been explained.

Giapis et al. (1990) discussed the issues of loading as uniformity control. They stated that uniformity is improved by reducing the gas pressure. Improvement of the macroscopic uniformity is attributed to the fact that the plasma is more homogeneous at low pressure and that the transport of reactants to the surface becomes diffusion limited and thus, independent of flow.

5.8.10 Feature Size Dependence of Etch Rate

Large trenches and holes etch more rapidly than smaller ones when features of different sizes are etched simultaneously. However, it was found that the rate depends on the *aspect ratio* (AR) and not on the absolute width of the opening. The normalized etch rate decreases linearly with increasing AR for AR $>$ ~2 (Chin et al., 1985). The phenomenon has been given several names: *RIE Lag* (Lee and Zhou,

1990, 1991), *aperture effect* (Abechev et al., 1991), *aspect-ratio dependent etching (ARDE)* (Gottscho et al., 1992). Fujiwara et al. (1989) and Sato et al. (1991) called it *microloading*. However, Gottscho et al. (1992), in an extensive review called "Microscopic uniformity in plasma etching," emphasize that microloading is a *misnomer* for the effect, since this effect results from microscopic transport within a single feature while microloading and macroloading arise from identical causes. Thus the AR effect can occur without microloading or vice versa.

Because the AR increases as etching proceeds, the etch rate decreases with time. Gottscho et al. point out that, to be sure the phenomenon is AR related, it is necessary to determine the etch rate as a function of time for several feature widths to distinguish it from other time-dependent etch rates which can arise due to polymer build-up or surface poisoning.

The AR effect has been observed in high-pressure plasma processes, in low-pressure batch RIE, and in low-pressure MERIE processes (Jones et al., 1990), as well as in ECR reactors (Fujiwara, 1989; Nojiri, 1989), in etching resists, oxides, and Si, and in a variety of etch gases.

Coburn and Winters (1989), using simple vacuum conductance arguments, concluded that in high AR features, conductance is adequate to allow the outflow of reaction products, but can be expected to limit the flow of reactive species to the bottom of the feature, thus explaining the etch rate dependence of high AR features. Jones et al. considered possible causes for the phenomenon: (1) transport limitations in narrow gaps, (2) polymerization, (3) geometric shadowing, (4) ion deflection by surface charges, (5) ion scattering in the plasma. They concluded that ion scattering in the sheath is the principle mechanism, as did Lii et al. (1990). Jones et al. stated that ion-driven processes are more sensitive to RIE lag and that processes in which the sheath width is small with respect to the mean free path in the plasma will exhibit less lag. Such processes would be those using electronegative gases (e.g., SF_6), very low pressures, or low bias, and that the scattering angle could be decreased by using light nonionizing diluents such as He.

Abachev et al. (1991) developed equations for the particle flux including particle reflection and adsorption (when the mean-free path was much larger than the width of the feature, i.e., width < 1 μm and pressure < 100 Pa) and concluded that limited ion and radical delivery were the main mechanisms. Both Sato et al. and Fujiwara et al. also concluded decreasing species concentration on the trench bottom was responsible.

Davis (1991) considered the image potential between an ion and the wall of an etched feature for small high AR features. He concluded that low-energy ions (<200 eV) are attracted to the walls resulting in loss of directionality and reduced vertical etch rates. This appears to counter the advice of Jones et al. (1990). Another model, based on polymer deposition while etching SiO_2 in selective (to Si) gases was proposed by Joubert et al. (1994a,b): (1) Ion scattering reduces the density and energy of ions at the bottom of a high AR trench or hole; (2) oxide etching moves from a regime in which the oxide is etched to one in which a thin fluorocarbon polymer can grow on the oxide surface slowing its etch rate; and (3) finally, as the AR increases (particularly in a hole) the reduction of the energy flux will result in deposition of polymer on the oxide, and etching may stop.

Gottscho et al. (1992) considered eight mechanisms which have been offered to explain ARDE: (1) Knudsen transport of neutrals, (2) ion shadowing, (3) neutral shadowing, (4) differential shadowing of insulating microstructures, (5) field curvature near conductive topography, (6) surface diffusion, (7) bulk diffusion, and (8) image force deflection and conclude that only the first four are consistent with AR scaling. "There may be many causes for ARDE; these may be reduced to neutral and ion transport phenomena which in turn are affected by gas phase collisions, surface scattering, and surface charging."

Sato et al. (1991) also observed a larger effect when etching Si with F-containing gases than with Cl-containing gases, which they attributed to enhanced ion-assisted etching with the heavier halogen, which runs counter to the argument of Jones et al. that ion-dominated etching processes are more susceptible. Fujiwara et al. (1989) and Nojiri et al. (1989) report that at very low pressures, in an ECR reactor, the effect is no longer observed, although other investigators have not found this to be true.

Since thick resists exaggerate the AR, finding mask materials or etch gases for which the etch rate of the mask is low, making thinner masks possible, helps to decrease the severity of the problem. Etching at the very low pressures possible in high-density plasmas may be a way of avoiding the problem. But Gottscho et al. (1992) state that other factors besides pressure are altered when the reactor geometry is changed, so that pressure reduction alone may not be adequate to account for the improvement in RIE lag.

Several explanations for RIE lag have been proposed but there has been no definitive resolution as to whether there is a single dominant mechanism or whether there are several working in concert.

A phenomenon called "reverse RIE lag," i.e., one in which small holes etch faster than large ones, has been observed in oxide etching (Dohmae et al., 1991; McVittie and Dohmae, 1992). This was attributed to ion reflections off sidewalls which depended on wall angle and ion energy, with reflection effects increasing at lower energies. When reflection is suppressed by using an overhang structure, "standard" lag results.

5.8.11 Temperature Effects

Initially, the need to keep the wafer temperature low during RIE was to prevent thermal damage to organic masks. Substrate heating was employed to assist in volatilization and prevention of redeposition of etch products during RIE of Cu.

More recently, very low temperature (as low as $\simeq 150°C$) RIE has been introduced in both capacitively coupled and ECR reactors, to improve selectivity and eliminate undercut (Tachi et al., 1988, 1991; Bestwick et al., 1990), reduce damage (Whang et al., 1992), microloading (Aoki et al., 1992; Sato et al., 1992) and post-RIE corrosion of Al (Aoki et al., 1992), while maintaining high etch rates. The low temperature limit is determined by the condensation temperature of the reactive gas. At the low temperatures, the chemical component of etch is reduced, thus suppressing lateral etching, without sidewall passivation (Whang et al., 1992), whereas the ion-assisted reaction is essentially independent of temperature. Giapis et al. (1990) demonstrated that both macroscopic and microscopic uniformity can be improved by etching at reduced temperature because etching occurs in an ion-activated, surface reaction limited regime. Thus etching is essentially independent of plasma geometry, gas flow and pressure, as well as reactor and mask materials.

6.0 ELECTROCHEMICAL DEPOSITION
6.1 ELECTROLESS PLATING

Electroless plating is an autocatalytic process; on a catalytic surface, metal ions are reduced to the metallic state:

$$\text{Metal}^{n+} + n \text{ electrons} \longrightarrow \text{Metal}^0$$

in the partial (cathodic) reaction in a solution of a reducing agent. The source of electrons for the reduction reaction is the partial (anodic) reaction in which the reducing agent is converted on the catalyst to the oxidation product:

$$\text{Reducing agent } (R) \longrightarrow \text{oxidized R (Ox)} + \text{electrons}$$

with an overall reaction:

$$\text{Metal}^{n+} + R \longrightarrow \text{metal}^0 + \text{Ox}$$

The electrons are transferred from the anodic to the cathodic sites on the conducting catalyst. Once the reaction is initiated and a film deposited, the reaction continues on the film surface. The kinetics and mechanism of electroless metal deposition have been discussed in some detail by Paunovic (1988), and Shacham-Diamond et al. (1995a) have reviewed electroless Cu plating as a metallization method for integrated circuit application.

Since metal deposits initially only on a catalytic surface, it can be grown selectively within a cavity by activating only the bottom surface. However, all surfaces may be activated for blanket deposition. Ion implantation (Kiang et al., 1992), L/O processing (Harada et al., 1986), evaporation (Dubin et al., 1993) sputtering (Schacham-Diamand et al., 1995), and focused laser patterning (Cole et al., 1988) are among the methods used to form an active substrate for electroless deposition. Whether the deposition is selective or blanket, the film grows from the activated surface; no voids or seams are formed when the metal is deposited in a hole or trench. Some examples of electroless deposition are: Ni on Si (Sullivan and Eigler, 1987), Ni on Al (Harada et al., 1986; Ting and Paunovic, 1989, Pd on silicides Ni (Sullivan and Eigler, 1987), Ni-Cu(P) on AlCuSi (Dubin, 1992), Ni on TiW (Dubin et al., 1993), and Cu on TiN-coated Cu (Dubin et al., 1995). The deposit called Ni is, in fact, a Ni alloy, either NiP (Feldstein, 1970) or NiB (Schmeckenbecker, 1971).

Si is an active surface for Ni, silicides are active for Pd, and AlCuSi is active for Ni-Cu(P), but activation is needed for deposition of Ni on silicides and on Al and for Cu on TiN-coated Cu. One procedure used for plating Ni on Al is sensitization in an acidic solution of $SnCl_2$, followed by activation in an acidic solution of $PdCl_2$ (Feldstein, 1973). Another pre-treatment before activation in $PdCl_2$ is immersion in a Cu displacement solution to form Cu particles on an Al surface (Schwartz, 1974). Immersion Cu has also been used to activate a TiN-coated Cu surface which then needs no further treatment before plating (Dubin et al., 1995).

Although deposition of the seed layer through a L/O has been used, it is not satisfactory, since the side walls become activated as well as the bottom resulting in overgrowth of the metal. This is illustrated in SEMs of via holes filled with electroless Ni. In Fig. I-44a, the activation was accomplished by immersion in a

Deposition and Etching of Thin Films

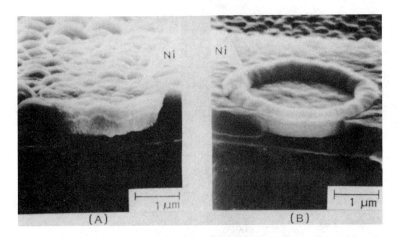

FIGURE I-44 SEMs of electroless Ni filling a via hole: (a) seeding with a PdCl$_2$ solution, (b) seeding using a lift-off process. (From Harada, et al., 1986)

PdCl$_2$ solution, and in Fig. I-44b, by deposition of Pd through a L/O mask (Harada et al., 1986). In the latter case, the rim surrounding the deposit is pronounced.

The conductivity of the Ni alloys is quite low so that its usefulness is limited. Cu is a better choice because of its lower resistivity but the pH of electroless Cu plating solutions is high enough to attack the underlying Al. With the advent of (clad) Cu interconnects, electroless Cu deposition becomes a feasible process. Electroless Cu plating will be discussed in greater detail in Chapter V.

Electroless deposition is attractive chiefly because of its low cost and simplicity of equipment and operation. Although there are potential problems, e.g., contamination from the solution and reaction by-products and inadequate or incomplete sensitization, some success has been reported by Dubin et al. (1995); the resistivity of the Cu film was close to the bulk value and there were no unfilled vias and submicron wires were robust and continuous.

6.2 ELECTROLYTIC PLATING (ELECTROPLATING)

The important advantage of electroplating is avoidance of contamination of the film by H$_2$, since the potential required for deposition of a noble metal (e.g., Cu) from a solution of its ions is less than that required for cathodic liberation of H$_2$. The disadvantage is that a conducting layer must be deposited along the entire surface to provide a path for the current.

Electroplating requires the use of an external anode(s); the wafer is the cathode. The electrolyte, in which the electrodes are immersed, contains the appropriate metal ion as well as any other constituents needed for improving the deposit. The "throwing power" of the bath, i.e., the ability to cover recesses, is one of its important properties. In the case of the small dimensions existing on a chip, the "microthrowing power" is important (Lowenheim, 1978). If the deposit is thicker over the peaks, the bath has poor microthrowing properties. If it is thicker in the recesses the bath is said to be *leveling*, which is desirable in this application. Leveling is achieved by organic additives.

The back side of the wafer must be protected. The process is usually carried out in a single-wafer cell. Pulsed-voltage plating has been employed to achieve good throwing power (Contolini et al., 1994). A "virtual anode," described in a patent by Poris (1993), was used to improve the current distribution at the cathode and to allow latitude in improving stress, film morphology, and step coverage, without degrading film uniformity.

REFERENCES

Abachev, M. K., Y. P. Baryshev, V. F. Lukichev, A. A. Orlikossky, K. A. Valiev, *Vacuum*, 42, 129 (1991).
Abe, H., Y. Sonobe, T. Enomoto, 1973, *Jpn. J. Appl. Phys.*, 12, 154 (1973).
Ahn, T. H., M. Ito, K. Nakamura, H. Sugai, *43rd Natl Symp. Amer. Vac. Soc.*, Abstr. PS-TuA7, 83, 1996.
Almen, O., G. Bruce, *Nucl. Instrum. Methods*, 11, 257 (1961).
Alt, L. L., S. W. Ing, K. W. Laendle, *J. Electrochem. Soc.*, 110, 465 (1963).
Amorin, J., H. S. Maciel, J. P. Sudano, *J. Vac. Sci. Technol.*, B9, 362 (1991).
Andosca, R. G., W. J. Varhue, E. Adams, *J. Appl. Phys.*, 72, 1126 (1992).
Aoki, H., T. Hashimoto, E. Ikawa, T. Kikkawa, *Jpn. J. Appl. Phys.*, 31, 4376 (1992).
Araki, H., T. Akahori, T. Tani, S. Nakayama, *The Sumitomo Search*, No. 44, 262 (1990).
Arami, J., Ito, T., US Pat. 5,275,683, 1994.
Arnal, Y., J. Pelletier, C. Pomot, B. Petit, A. Durandet, *Appl. Phys. Lett.*, 45, 132 (1984).
Arikado, T., K. Horioka, M. Sekine, H. Okano, Y. Horiike, *Jpn. J. Appl. Phys.*, 27, 95 (1988).
Arnold, J. C., H. H. Sawin, *J. Appl. Phys.*, 70, 5314 (1991).
Asamaki, T., T. Miura, G. Nakamura, K. Hotate, S. Yonaiyama, K. Ishibashi, N. Hosokawa, *J. Vac. Sci. Technol.*, A10, 3430 (1992).
Asamaki, T., T. Miura, K. Hotate, S. Yonaiyama, G. Nakamura, K. Ishibashi, N. Hosokawa, *Jpn. J. Appl. Phys.*, 32, 902 (1993).
Ashida, S., C. Lee, M. A. Lieberman, *J. Vac. Sci. Technol.*, A13, 2498 (1995).
Asmussen, J., chap. 11 in *Handbook of Plasma Processing Technology* (S. M. Rossnagel, J. J. Cuomo, W. D. Westwood, eds.), Noyes Publications, Westwood, N.J., 1989.
Bader, M. E., R. P. Hall, G. Strasser, *Solid State Technol.*, 5/90, 149 (1990).
Barnes, M. S., D. K. Coultas, J. C. Forster, J. H. Keller, US Pat. 5,207,437, 1993.
Bauer, H. J., *J. Vac. Sci. Technol.*, B12, 2405 (1994).
Benzing, J. C., E. K. Broadbent, J.K.H. Rough, US Pat. 5,405,480, 1995.
Bestwick, T. D., G. S. Oehrlein, D. Angell, *Appl. Phys. Lett.*, 57, 431 (1990).
Bobbio, S. M., *SPIE Vol. 1185*, 262 (1989).
Boitnott, C., *Solid State Technol.*, 10/94, 51 (1994).
Boswell, R. W., US Pat 4,810,935, 1989.
Boswell, R. W., A. J. Perry, M. Enami, *J. Vac. Sci. Technol.*, A7, 3345 (1989).
Boswell, R. W., D. Henry, *Appl. Phys. Lett.*, 47, 1095 (1985).
Boswell, R. W., R. K. Porteous, *J. Appl. Phys.*, 62, 3123 (1987).
Bruce, R. H., *J. Appl. Phys.*, 52, 7064 (1981).
Bunshah, R. F., in *Deposition Technologies for Films and Coatings* (R. F. Bunshah, ed.), Noyes Publications, N.J., 1982.
Burke, R. R., C. Pomot, *Solid State Technol.*, 2/88, 67 (1988).
Burke, R. R., C. Pomot, *Applied Surface Sci.*, 36, 267 (1989).
Bushnell, L. P. Mc., L. V. Gregor, C. F. Lyons, *Solid State Technol.*, 6/86, 133 (1986).
Carlsson, J. O., *Thin Solid Films*, 130, 261 (1985).

Chapman, B., chap. 6 in *Glow Discharge Processes*, John Wiley & Sons, New York, 1980.
Chen, F. F., G. Chevalier, *J. Vac. Sci. Technol.*, *A10*, 1389 (1992).
Chin, D., S. H. Dhong, G. J. Long, *J. Electrochem. Soc.*, *132*, 1705 (1985).
Claasen, W.A.P., *Plasma Chemistry Plasma Processing*, 7, 109 (1987).
Clark, T. E., M. Chang, C. Leung, *J. Vac. Sci. Technol.*, *B9*, 1478 (1991).
Clemens, J. T., Hong, S. Y., US Pat. 5,073,716, (1991).
Coburn, J., E. Taglauer, E. Kay, *Proc. 6th Intl. Vac. Cong.*, 501, 1974.
Coburn, J. W., E. Kay, *IBM J. Res. Develop.*, *23*, 33 (1979).
Coburn, J. W., *Physica Scripta*, *T23*, 258 (1988).
Coburn, J. W., H. F. Winters, *J. Appl. Phys.*, *50*, 3189 (1979).
Coburn, J. W., H. F. Winters, *Appl. Surf. Sci.*, *22/23*, 63 (1985).
Coburn, J. W., H. F. Winters, *Appl. Phys. Lett.*, *55*, 2730 (1989).
Coburn, J. W., K. Kohler, in "Proc. 6th Symp. Plasma Processing" (G. S. Mathad, G. C. Schwartz, R. A. Gottscho, eds.), *Electrochem. Soc. Proc. Vol. PV 87-6*, 13 (1987).
Cole, H. S., Y. S. Liu, J. W. Rose, R. Guida, *Appl. Phys. Lett.*, *53*, 2111 (1988).
Colgan, M. J., M. Meyyappan, T. R. Govindan, in "Proc. 10th Symp. Plasma Processing" (G. S. Mathad, D. W. Hess, eds.), *ECS PV 94-20*, 13 (1994).
Collins, K. S., E. A. Gritters, US Pat. 5,315,473, 1994.
Contolini, R. J., A. F. Bernhardt, S. T. Mayer, *J. Electrochem. Soc.*, *141*, 2503 (1994).
Cook, J. M., D. E. Ibbotson, D. L. Flamm, *J. Vac. Sci. Technol.*, *B8*, 1 (1990).
Cooke, M. J., J. Pelletier, *J. Electrochem. Soc.*, *136*, 1824 (1989).
Cooke, M. J., N. Sharrock, in "Proc. 8th Symp. Plasma Processing" (G. S. Mathad, D. W. Hess, eds.), *Electrochem. Soc. Proc. Vol. PV 90-14*, 538 (1990).
Cottler, T. J., M. S. Barnes, M. Elta, *J. Vac. Sci. Technol.*, *B6*, 542 (1988).
Cottler, T. J., M. E. Elta, *J. Vac. Sci. Technol.*, *B8*, 523 (1990).
Davidse, P. D., 1971, US Pat. 3,598,710, 1971.
Daviet, J.-F., L. Peccoud, F. Mondon, *J. Electrochem. Soc.*, *140*, 3254 (1993).
Davis, R. J., *Appl. Phys. Lett.*, *59*, 1717 (1991).
Denison, D., *Lam Research* (1995).
Dieleman, J., *J. Vac. Sci. Technol.*, *15*, 1734 (1978).
Dieleman, J., F.H.M. Sanders, A. W. Kolfschoten, P. C. Zalm, A. E. deVries, A. Haring, *J. Vac. Sci. Technol.*, *B3*, 1384 (1985).
Dimigen, H., H. Luthje, H. Hubisch, U. Convertini, *J. Vac. Sci. Technol.*, *13*, 976 (1976).
Dobkin, D. M., *J. Electrochem. Soc.*, *139*, 2573 (1992).
Dohmae, S.-i., J. P. McVittie, J. C. Rey, E.S.G. Shaqfeh, V. K. Singh, in "Patterning Science and Technology, II, Interconnection and Contact Metallization for ULSI" (W. Greene, G. J. Heffron, L. K. Whit, T. O. Herndon, A. L. Wu, eds.), *Electrochem. Soc. Prov. Vol. PV 92-6*, 163 (1992).
Dubin, V. M. *J. Electrochem. Soc.*, *139*, 633 (1992).
Dubin, V. M., S. D. Lopatin, V. G. Sokolov, *Thin Solid Films*, *226*, 87 (1993).
Dubin, V. M., Y. Shacham-Diamand, B. Zhao, P. K. Vasudev, C. H. Ting, *1995 VMIC*, 315 (1995).
Ducommun, J. P., M. Cantagrel, M. Moulin, *J. Mat. Sci.*, *10*, 52 (1975).
Economou, D. J., R. C. Alkire, *J. Electrochem. Soc.*, *135*, 941 (1988).
Economous, D. J., S.-K. Park, in "Proc. 8th Symp. Plasma Processing" (G. S. Mathad, D. W. Hess, eds.), *Electrochem. Soc. Proc. Vol. PV 90-14*, 185 (1990).
Eden, J. G., in *Thin Film Processes*, vol. II (J. L. Vossen and W. Kern, eds.), Academic Press, New York, 1991.
Englehardt, M., *Semiconductor International*, 7/91, 53 (1991).
Engelhardt, M., V. Grewal, S. Schwarzl, in "Proc. 8th Symp. Plasma Processing" (G. S. Mathad, D. W. Hess, eds.), *Electrochem. Soc. Proc. Vol. PV 90-14*, 470 (1990).

Ephrath, L. M., 1981, in "Semiconductor Silicon" (H. R. Huff, R. J. Kriegler, Y. Takeishi, eds.), *Electrochem. Soc. Proc. Vol. PV 81-5*, 627 (1981).
Ephrath, L. M., in "ULSI Sci. & Technol./1989" (C. M. Osburn, J. M. Andrews, eds.), *Electrochem. Soc. Proc. Vol. PV 89-9*, 394 (1989).
Ephrath, L. M., *IEEE Trans. Electron Devices*, ED-28, 1315 (1981).
Feldstein, N., *RCA Review*, 31, 317 (1970).
Feldstein, N., *Plating*, 60, 611 (1973).
Fichelscher, A., I. W. Rangelow, A. Stamm, *SPIE Vol. 1392*, 77 (1990).
Field, J., *Solid State Technol.*, 9/94, 21 (1994).
Flamm, D. L., V. M. Donnelly, *Plasma Chem. Plasma Process*, 1, 317 (1981).
Fortuno, G., *J. Vac. Sci. Technol.*, A4, 744 (1986).
Fujinaga, M., N. Kotani, T. Kunikiyo, H. Oda, M. Shirahata, Y. Asasaka, *IEEE Trans. Electron Devices*, 37, 2183 (1990).
Fujiwara, H., K. Fujimoto, H. Araki, Y. Tobinaga, *SPIE Vol. 1089*, 348 (1989).
Fujiwara, N., T. Maruyama, M. Yoneda, *Jpn. J. Appl. Phys.*, 35, 2450 (1996).
Fukuda, T., K. Suzuki, S. Takahashi, Y. Mochizuki, M. Onhu, N. Momma, T. Sonobe, *Jpn. J. Appl. Phys.*, 27, L1962, (1988).
Getty, W. D., J. B. Geddes, *J. Vac. Sci. Technol.*, B12, 408 (1994).
Giapis, K. P., G. R. Scheller, R. A. Gottscho, W. S. Hobson, Y. H. Lee, *Appl. Phys. Lett.*, 57, 983 (1990)
Giffen, L., J. Wu, R. Lachenbruch, G. Fior, *Solid State Technol.*, 4/89, 55 (1989).
Glang, R., chap. 1 in *Handbook of Thin Film Technology* (L. I. Maissel, R. Glang, eds.), McGraw-Hill, New York, 1970.
Glang, R., R. A. Holmwood, J. A. Kurtz, 1970, chap. 2 in *Handbook of Thin Film Technology* (L. I. Maissel, R. Glang, eds.), McGraw-Hill, New York, 1970.
Gloersen, P. G., *J. Vac. Sci. Technol.*, 12, 28 (1975).
Glowacki, P., Z. Tkaczyk, *Electron Technol.*, 21, 61 (1988).
Goto, H. H., H-D. Lowe, T. Ohmi, *J. Vac. Sci. Technol.*, A10, 3048 (1992).
Gottscho, R. A., C. W. Jurgensen, D. J. Vitkavage, *J. Vac. Sci. Technol.*, B10, 133 (1992).
Greene, W. M., D. W. Hess, W. G. Oldham, *J. Vac. Sci. Technol.*, B6, 1570 (1988).
Gross, M., C. M. Horwitz, *J. Vac. Sci. Technol.*, B7, 534 (1989).
Gross, M., C. M. Horowitz, *J. Vac. Sci. Technol.*, B11, 242 (1993).
Hamaguchi, S., M. Dalvue, *J. Vac. Sci. Technol.*, A12, 2745 (1994).
Harada, Y., Fushimi, S., Marokoro, H. Sawia, S. Ushio, *J. Electrochem. Soc.*, 133, 2428 (1986).
Haring, R. A., A. Haring, F. W. Saris, A. E. deVries, *Appl. Phys. Lett.*, 41, 174 (1982).
Harper, J.M.E., 1978, chap. II-5 in *Thin Film Processes* (J. L. Vossen, W. Kern, eds.), Academic Press, New York, 1978.
Hashimoto, C., K. Machida, H. Oikawa, *J. Vac. Sci. Technol.*, B8, 529 (1990).
Heinecke, R.A.H. US Pat. 3,940,506, 1976.
Helix, M. J., K. V. Vaidyanathan, B. G. Streetman, H. B. Dietrich, P. K. Chaterjee, *Thin Solid Films*, 55, 143 (1978).
Henry, D., J. M. Francou, A. Inard, *J. Vac. Sci. Technol.*, A10, 3426 (1992).
Hess, D. W., *J. Vac. Sci. Technol.*, A2, 244 (1984).
Hess, D. W., D. B. Graves, page 377 in *Advances in Chemistry* (D. W. Hess, K. Jensen, eds.), American Chemical Society Series #221, 1989.
Hey, H.P.W., B. G. Sluijk, D. G. Hemmes, *Solid State Technol.*, 4/90, 139 (1990).
Hieber, K., M. Stolz, *1987 VMIC*, 216 (1987).
Hill, M. L., D. C. Hinson, *Solid State Technol.*, 4/85, 243 (1985).
Hinson, D. C., I. Lin, W. Class, S. Hurwitt, *Semiconductor Intl.*, 10/83, 103 (1983).
Hiroaka, H., J. Pacansky, *J. Vac. Sci. Technol.*, 19, 1132 (1982).

Hirobe, K., K-i. Kawamura, K. Nojiri, *J. Vac. Sci. Technol.*, *B5*, 594 (1987).
Holland, L., *Vacuum Deposition of Thin Films*, Chapman and Hall, Ltd., London, 1966.
Holland, J. P., T. Q. Ni, M. S. Barnes, *43rd Natl. Symp. Amer. Vac. Soc.*, Abstr. PS-TuA8, 84, 1996.
Hongoh, T., M. Kondo, 1993, US Pat. 5,179,498, 1993.
Horwitz, C. M., *J. Vac. Sci. Technol.*, *B7*, 443 (1989a).
Horwitz, C. M., chap. 12 in *Handbook of Plasma Processing Technology* (S. M. Rossnagel, J. J. Cuomo, W. D. Westwood, eds.), Noyes Publications, Westwood, N.J., 1989b.
Horwitz, C. M., S. Boronkay, US Pat. 5,103,367, 1992.
Howard, B J., Ch. Steinbruchel, *Appl. Phys. Lett.*, *59*, 914 (1991).
Iida, Y., H. Okabayashim, K. Suzuki, *Jpn. J. Appl. Phys.*, *16*, 1313 (1977).
Ing, Jr., S. W., W. Davern, *J. Electrochem. Soc.*, *112*, 284 (1965).
Irving, S. M., K. E. Lemons, G. E. Bobos, 1971, US Pat. 3,615,956, 1971.
Jackson, S. C., T. J. Dalton, *SPIE Vol. 1185*, 225 (1989).
Jensen, K. F., p. 199 in "Microelectronics Processing, Chemical Engineering Aspects," *Advances in Chemistry*, vol. 221 (D. W. Hess, K. F. Jense, eds.), American Chemical Society, Washington, D.C., 1989.
Jin, M., K. C. Kao, *J. Vac. Sci. Technol.*, *B10*, 601 (1992).
Jones, F., J. S. Logan, *J. Vac. Sci. Technol.*, *A7*, 1240 (1989).
Jones, R. E., H. F. Winters, L. I. Maissel, *J. Vac. Sci. Technol.*, *5*, 84 (1968).
Jones, H. C., R. Bennett, J. Singh, in "Proc. 8th Symp. Plasma Processing" (G. S. Mathad, D. W. Hess, eds.), *Electrochem. Soc. Proc. Vol. PV 90-14*, 45 (1990).
Joubert, O., G. S. Oehrlein, Y. Zhang, *J. Vac. Sci. Technol.*, *A12*, 658 (1994a).
Joubert, O., G. S. Oehrlein, M. Surendra, *J. Vac. Sci. Technol.*, *A12*, 665 (1994b).
Kaganowicz, G., V. S. Ban, J. W. Robinson, *J. Vac. Sci. Technol.*, *A2*, 1233 (1984).
Katetomo, M., T. Kure, K. Tsujimoto, S. Kato, S. Tachi, in "Proc. 9th Symp. Plasma Processing" (G. S. Mathad, D. W. Hess, eds.), *Electrochem. Soc. Proc. Vol. PV 92-18*, 293 (1992).
Kawamoto, H., H. Miyamoto, E. Ikawa, in "Proc. 10th Symp. Plasma Processing" (G. S. Mathad, D. W. Hess, eds.), *Electrochem. Soc. Proc. Vol. PV 94-20*, 398 (1994).
Keller, J. H., J. C. Forster, M. S. Barnes, *J. Vac. Sci. Technol.*, *A11*, 2487 (1993).
Keller, J. H., *Plasma Sources Sci. Technol.*, *5*, 166 (1996).
Kern, W., V. S. Ban, p. 258 in *Thin Film Processes* (J. L. Vossen, W. Kern, eds.), Academic Press, New York, 1978.
Kern, W., G. L. Schnable, *IEEE Trans. Electron Devices*, *ED-26*, 647 (1979).
Kiang, M.-H., M. A. Lieberman, N. W. Cheung, *Appl. Phys. Lett.*, *60*, 2767 (1992).
Kinoshita, H., T. Isida, S. Ohno, *Proc. 8th Symp. Dry Process, IEEE Jpn.*, Tokyo, 11/86, 1986.
Kinsbron, E., W. E. Willenbrock, H. J. Levinstein, in "VLSI Sci. & Technol/1982" (C. J. Dell'Oca and M. W. Bullis, eds.), *Electrochem. Soc. Proc. Vol. PV 82-7*, 116 (1982).
Kodas, T., M. Hampden-Smith, *The Chemistry of Metal CVD Processes*, Weinheim, New York, 1994.
Koenig, H. R., L. I. Maissel, *IBM J. Res. Develop.*, *14*, 168 (1970).
Kumagai, H., US Pat. 4,733,632,1988.
Kuypers, A. D., E.H.A. Grannemanm, H. J. Hopman, *J. Appl. Phys.*, *63*, 1899 (1988).
Lamont, Jr., L. T., *Solid State Technol.*, 9/79, 107 (1979).
Law, K., J. Wong, D.N.K. Wang, *Tech. Proc. Semicon Japan*, 154 (1987).
Lee, Y. H., Z. H. Zhou, in "Proc 8th Symp. Plasma Processing" (G. S. Mathad, D. W. Hess, eds.), *Electrochem. Soc. Proc. Vol. PV 90-14*, 34 (1990).
Lee, Y. H., Z. H. Zhou, *J. Electrochem. Soc.*, *138*, 2439 (1991).
Lee, J. G., S. H. Choi, T. C. Ahn, P. Lee, L. Law, M. Galiano, P. Keswick, B. Shin, *Semiconductor Intl.*, 5/92, 116 (1992).

Lee, R. E., *J. Vac. Sci. Technol.*, *16*, 164 (1979).
Lehmann, H. W., I. Krausbauerm, R. Widmer, *J. Vac. Sci. Technol.*, *14*, 281 (1977).
Lewin, I. H., M. J. Plummer, US Pat. 4,502,094, 1985.
Lewin, I. H., US Pat. 4,554,611, 1985.
Lieberman, M. A., A. J. Lichtenberg, *Principles of Plasma Discharges and Materials Processing*, John Wiley & Sons, Inc., New York, 1994.
Lieberman, M. A., R. A. Gottscho, "Design of High Density Plasma Sources," in *Physics of Thin Films*, vol. 18 (M. H. Francombe and J. L. Vossen, eds.) Academic Press, San Diego, 1994.
Lii, Y.-J., J. Jorne, *J. Electrochem. Soc.*, *137*, 2837 (1990).
Liporace, J. W., Seirmarco, J. A., US Pat. 5,166,856, 1992.
Logan, J. S., J. M. Keller, R. G. Simmons, *J. Vac. Sci. Technol.*, *14*, 92 (1977).
Logan, J. S., *Thin Solid Films*, *188*, 307 (1990).
Logan, J. S., US Pat. 3,617,459, 1971.
Logan, J. S., R. R. Ruckel, R. E. Tompkins, Westerfield, Jr., R. P., US Pat. 5,055,964, 1991.
Logan, J. S., R. R. Ruckel, R. E. Tompkins, Westerfield, Jr., R. P., US Pat. 5,191,506 (1993).
Lowenheim, F. A., *Electroplating*, McGraw-Hill, New York, 1978.
Locuvsky, G., D. V. Tsu, S. S. Kim, R. J. Markunas, G. G. Fountain, *Appl. Surf. Sci.*, *39*, 33 (1989).
Lucovsky, G., D. V. Tsu, R. A. Rudder, R. J. Markunas, p. 565 in *Thin Film Processes*, vol. II (L. Vossen and W. Kern, eds.), Academic Press, Boston, 1991.
Ma, W. H.-L., IEDM Mtg., Washington, D.C., p. 574, 1980.
Machida, K., H. Oikawa, *J. Vac. Sci. Technol.*, *B4*, 818 (1986).
Maissel, L., 1970, chap. 4 in *Handbook of Thin Film Technology* (L. I. Maissel and R. Glang, eds.), McGraw-Hill, New York, 1970.
Maissel, L. I., C. L. Standley, I. V. Gregor, *IBM J. Res. Dev.*, *16*, 67 (1972).
Mano, D. W., D. L. Flamm, *Plasma Etching*, Academic Press, Boston, 1989.
Mantei, T. D., S. Dhole, *J. Vac. Sci. Technol.*, *B9*, 26 (1991).
Mantei, T. D., T. E. Ryle, *J. Vac. Sci. Technol.*, *B9*, 29 (1991).
Mantei, T. D., T. E. Wicker, in "Proc. 4th Symp. Plasma Processing" (G. S. Mathad, G. C. Schwartz, G. Smolinsky, eds.), *Electrochem. Soc. Proc. Vol. PV 83-10*, 125 (1983).
Mantei, T. D., T. E. Wicker, D. Kazmierzak, in "Plasma Synthesis and Etching of Electronic Materials" (R.P.H. Chang and B. Abeles, eds.), *Mat. Res. Soc. Symp. Proc. Vol.*, *38*, 409 (1985).
Martin, R. S., E. P. van de Ven, C. P. Lee, *1988 VMIC*, 286 (1988).
Martinu, L., J. E. Klemberg-Sapieha, M. R. Wertheimer, *Appl. Phys. Lett.*, *54*, 2645 (1989).
Mathad, G. S., B. Patnaik, *Electrochem. Soc. Ext. Abstr 603*, *PV 79-2*, 1510 (1979).
Mathis Co., D. Mathis Company bulletins.
Matsuda, T., M. J. Shapiro, S. V. Nguyen, *1996 DUMIC*, 22, 1995.
Matsuo, S., M. Kiuchi, *Jpn. J. Appl. Phys.*, *22*, L210 (1983).
Matsuoka, M., K.-i. Ono, *Appl. Phys. Lett.*, *50*, 1864 (1989).
Matsuoka, M., K.-i. Ono, *J. Appl. Phys.*, *65*, 4403 (1989).
Matsuoka, M., K.-i. Ono, *Appl. Phys. Lett.*, *50*, 1864 (1987).
Matsuoka, M., K.-i. Ono, *J. Appl. Phys.*, *65*, 4403 (1989).
Matsuzaki, R., N. Hosakawa, US Pat. 3,984,310, 1976.
Mauer, IV, J. L., J. S. Logan, L. B. Zielinski, G. C. Schwartz, *J. Vac. Sci. Technol.*, *15*, 1734 (1978).
Maydan, D., 1981, US Pat. 4,298,443, 1981.
McVittie, J. P., J. C. Rey, M. M. IslamRaja, in "Proc. 9th Symp. Plasma Processing" (G. S. Mathad and D. W. Hess, eds.), *Electrochem. Soc. Proc. Vol. PV 92-18*, 1 (1992).
McVittie, J. P., S-i. Dohmae, 1992, in "Proc. 9th Symp. Plasma Processing" (G. S. Mathad and D. W. Hess, eds.), *Electrochem. Soc. Proc. Vol. PV 92-18*, 11 (1992).

Meiners, L. G., *J. Vac. Sci. Technol.*, *21*, 655 (1982).
Melliar-Smith, C. M., *J. Vac. Sci. Technol.*, *13*, 1008 (1976).
Meyers, F. R., M. Ramaswami, T. S. Cale, *J. Electrochem. Soc.*, *142*, 1313 (1994).
Mieno, T., S. Samukawa, *Jpn. J. Appl. Phys.*, *34*, L1079 (1995).
Mitchener, J. C., I. Mahawili, *Solid State Technol.*, 8/87, 109 (1987).
Mogab, C. J., *J. Electrochem. Soc.*, *124*, 1262 (1977).
Moran, J. M., G. N. Taylor, *J. Vac. Sci. Technol.*, *19*, 27 (1981).
Mountsier, T. W., A. M. Schoepp, E. van de Ven, *Electrochem. Soc. Ext. Abstr 485*, PV-94-2, 770 (1994).
Muller, K. P., F. Heinrich, H. Mader, *Microelectronics Engr.*, *10*, 55 (1989).
Murakawa, S., S. Fang, J. P. McVittie, *IEDM 92*, 57 (1992).
Muto, S. Y., US Pat. 3,971, 684, 1976.
Nagy, A. G., *J. Electrochem. Soc.*, *132*, 689 (1985).
Nakasuji, M., H. Shimizu, *J. Vac. Sci. Technol.*, A10, 3573 (1992).
Nakasuji, M., H. Shimizu, T. Kato, *J. Vac. Sci. Technol.*, A12, 2834 (1994).
Neugebauer, C. A., chap. 8 in *Handbook of Thin Film Technology* (L. I Maissel and R. Glang, eds), McGraw-Hill, New York, 1970.
Nguyen, S. V., D. Dubuzinsky, S. R. Stiffler, G. Chrisman, *J. Electrochem. Soc.*, *138*, 1112 (1991).
Nguyen, S. V., *J. Vac. Sci. Technol.*, B4, 1159 (1986).
Nihei, H., J. Morkiawa, D. Nagahara, H. Enomoto, N. Inoue, *Rev. Sci. Instrum.*, *63*, 1932 (1992).
Nojiri, K., E. Iguchi, K. Kawamura, K. Kadota, *Ext. Abstr. 21st Conf. Solid State Dev & Mat.*, Tokyo, 153, 1989.
Nozawa, T., Arami, J., Okumura, K., US Pat. 5,255,153, 1993.
Oechsner, H., *Z. Physik*, *261*, 37 (1973).
Oehrlein, G. S., p. 221, in *Handbook of Plasma Processing Technology* (S. M. Rossnagel, J. Cuomo, W. D. Westwood, eds.) Noyes Publications, Westwood, N.J. 1989.
Oehrlein, G. S., *J. Vac. Sci. Technol.*, A11, 34 (1993).
Oehrlein, G. S., J. F. Rembetski, *IBM J. Res. Develop.*, *36*, 140 (1992).
Ogle, J. S., US Pat. 4,948,458, 1990.
Ohiwa, T., K. Horioka, T. Arikado, I. Hasegawa, H. Okano, *Jpn. J. Appl. Phys.*, *31*, 405 (1992).
Ohki, S., M. Oda, H. Akiya, T. Shibata, *J. Vac. Sci. Technol.*, B5, 1161 (1987).
Ohno, K., M. Sato, Y. Arita, *Jpn. J. Appl. Phys.*, *28*, L1070 (1989).
Paunovic, M., in "Proc. Symp. Electroless Deposition Metals & Alloys" (M. Paunovic, I. Ohno, eds.), *Electrochem. Soc. Proc. Vol. PV 88-12*, 3 (1988).
Pelka, J., M. Weiss, W. Hoppe, D. Mewew, *J. Vac. Sci. Technol.*, B7, 1483 (1989).
Perry, A. J., R. W. Boswell, *Appl. Phys. Lett.*, *55*, 148 (1989).
Perry, A. J., D. Venser, R. W. Boswell, *J. Vac. Sci. Technol.*, B9, 310 (1991).
Petri, R., B. Kennedy, D. Henry, N. Sadeghi, J.-B., Booth, *J. Vac. Sci. Technol.*, B12, 2970 (1994).
Pichot, M. G., *Microelectronic Engr.*, *3*, 411 (1985).
Plais, F., B. Agius, F. Abel, J. Siejka, M. Puech, G. Ravel, P. Alnot, N. Proust, *J. Electrochem. Soc.*, *139*, 1489 (1992).
Plais, F., B. Agius, F. Abel, J. Siejka, M. Puech, P. Alnot, in "Proc. 8th Symp. Plasma Processing" (G. S. Mathad, D. W. Hess, eds.), *Electrochem. Soc. Proc. Vol. PV 90-14*, 544 (1990).
Pomot, C., B. Mahl, B. Petit, Y. Arnal, J. Pelletier, *J. Vac. Sci. Technol.*, B4, 1 (1986).
Poris, J., US Pat. 5,256,274, 1993.
Rangelow, I. W., *J. Vac. Sci. Technol.*, A1, 410 (1983).
Reinberg, A. R., US Pat. 3,757,733, 1973.

Richard, P. D., R. J. Karkunas, G. Lucovsky, G. G. Fountain, A. N. Mansour, D. V. Tsu, *J. Vac. Sci. Technol.*, *A3*, 867 (1985).
Robinson, B., P. D. Hoh, P. Madakson, T. N. Nguyen, S. A. Shivashankar, *Mat. Res. Soc. Proc. Vol.*, *98*, 313 (1987).
Rosler, R. S., G. M. Engle, *Solid State Technol.*, *12/79*, 88 (1979).
Rosler, R. S., *Solid State Technol.*, *6/91*, 67 (1991).
Samukawa, S., *Jpn. J. Appl. Phys.*, *32*, 6080 (1993).
Samukawa, S., K. Terada, *J. Vac. Sci. Technol.*, *B12*, 3300 (1994).
Samukawa, S., Y. Nakagawa, T. Tsukada, H. Ueyama, *Jpn. J. Appl. Phys.*, *34*, Part 1, 6805 (1995).
Samukawa, S., T. Mieno, *Plasma Sources Sci. Technol.*, *5*, 132 (1996).
Samukawa, S., T. Nakano, *J. Vac. Sci. Technol.*, *A14*, 1002 (1996).
Sato, M., H. Nakamura, *J. Electrochem. Soc.*, *129*, 2522 (1982).
Sato, M., S.-c. Kato, Y. Arita, *Jpn. J. Appl. Phys.*, *30*, 1549 (1991).
Sato, M., D. Takehara, K. Uda, K. Sakiyama, T. Hara, *Jpn. J. Appl. Phys.*, *31*, 4370 (1992).
Schaible, P. M., W. C. Metzger, J. P. Anderson, *J. Vac. Sci. Technol.*, *15*, 334 (1978).
Schaible, P. M., G. C. Schwartz, *J. Vac. Sci. Technol.*, *16*, 377 (1979).
Schmeckenbecker, A. F., *Plating*, *58*, 905 (1971).
Schultheis, S., *Solid State Technol.*, *4/85*, 233 (1985).
Schwartz, G. C., unpublished, 1974.
Schwartz, G. C., L. B. Zielinski, T. Schopen, p. 122 in *Etching for Pattern Definition*, Electrochem. Soc., Princeton, N.J., 1976.
Schwartz, G. C., unpublished; oral presentation VMIC, State-of-the-Art-Symp., 1989.
Schwartz, G. C., P. M. Schaible, *J. Electrochem. Soc.*, *130*, 1777 (1983).
Schwartz, G. C., P. M. Schaible, in *Plasma Processing* (R. G. Frieser and C. J. Mogab, eds.) *Electrochem. Soc. Proc. Vol. PV 81-1*, 133 (1981).
Shacham-Diamand, Y., V. Dubin, M. Angal, *Thin Solid Films*, *262*, 93 (1995a).
Shacham-Diamand, Y, V. M. Dubin, C. H. Ting, P. K. Vasudev, B. Zhao *1995 VMIC*, 334 (1995b).
Shan, H., B. K. Srinivasan, D. W. Jillie, Jr., J. S. Multani, W. J. La, *J. Electrochem. Soc.*, *141*, 2904 (1994).
Shaqfeh, E. S. G., C. W. Jurgensen, *J. Appl. Phys.*, *66*, 4664 (1989).
Sherman, A., *Chemical Vapor Deposition for Microelectronics*, Noyes Publications, Westwood, N.J., 1987.
Shibata, T., M. Oda, *Ext. Abstr, 18th Conf. Solid State Devices & Mat.*, 725 (1986).
Shida, N., T. Inoue, H. Korai, Y. Sakamoto, W. Miyazaw, S. Den, Y. Hayashi, *Jpn. J. Appl. Phys.*, *32*, L1635 (1993).
Singer, P., *Semiconductor Intl.*, *8/93*, 46 (1993).
Singer, P. *Semiconductor Intl.*, *7/95*, 113 (1995).
Singh, V. K., E.S.G. Shaqfeh, J. P. McVittie, K. C. Saraswat, in "Patterning Science and Technol. II/Interconnection and Contact Metallization for ULSI" (W. Greenem, G. J. Hefferson, L. K. White, T. O. Herndon, A. L. Wu, eds.), *Electrochem. Soc. Proc. Vol. PV 92-6*, 163 (1992).
Singh, B., J. H. Thomas, III, V. Patel, *Appl. Phys. Lett.*, *60*, 2335 (1992).
Smith, H. I., *Proc. IEEE*, *62*, 1361 (1974).
Smith, R., S. J. Wilde, G. Carter, I. V. Katardjiev, M. J. Nobes, *J. Vac. Sci. Technol.*, *B5*, 579 (1987).
Somekh, S., *J. Vac. Sci. Technol.*, *13*, 1003 (1976).
Spencer, J. E., R. L. Jackson, J. L. McGuire, A. Hoff, *Solid State Technol.*, *4/87*, 107 (1987).
Spindler, O., B. Neureither, *Thin Solid Films*, *175*, 67 (1989).
Steinbruchel, Ch., *J. Vac. Sci. Technol.*, *B2*, 38 (1984).
Sterling, H. F., R.C.G. Swann, *Solid State Electronics*, *8*, 653 (1965).

Stewart, A.D.G., H. W. Thompson, *J. of Mat. Sci.*, *4*, 56 (1969).
Stoll, R. W., R. H. wilson, in "Proc. Symp. Multilevel Metallization, Interconnection, Contact Technologies" (L. B. Rothman and T. Herndon, eds.), *Electrochem. Soc. Proc. Vol. PV 87-4*, 232 (1987).
Su, J., G. W. Hills, P. Tsai, in "Proc. 10th Symp Plasma Processing," (G. S. Mathad and D. W. Hess, eds.), *Electrochem. Soc. Proc. Vol. PV 94-20*, 291 (1994).
Sugai, K. Nakamura, T. H. Ahn, *43rd Natl. Symp. Amer. Vac. Soc. Abstr.* –PS-TuA1, 83 (1996).
Sullivan, M. J., J. H. Eigler, *J. Electrochem. Soc.*, *104*, 226 (1987).
Suzuki, K., S. Okudaira, N. Sakudo, I. Kanomata, *Jpn. J. Appl. Phys.*, *16*, 1979 (1977).
Suzuki, Y., US Pat. 4,692,836, 1987.
Tachi, S., K. Tsujimoto, S. Okudaira, *Appl. Phys. Lett.*, *52*, 616 (1988).
Tachi, S., K. Tsujimoto, S. Arai, T. Kure, *J. Vac. Sci. Technol.*, *A9*, 796 (1991).
Temescal, Airco Coating Technology bulletins.
Ting, C. H., M. Paunovic, *J. Electrochem. Soc.*, *133* 2428 (1989).
Toyoda, H., H. Komiya, H. Itakura, *J. Electronic Materials*, *9*, 569 (1980).
Tracy, C. J., R. Mattox, *Solid State Technol.*, *6/82*, 83 (1982).
Tsujimoto, K., S. Tachi, K. Nimiya, K. Suzuki, *Ext. Abstr. 18th Intl. Conf. on Solid State Devices & Mat.*, Tokyo, 229 (1986).
Tsukune, A., Nishimura, K. Koyama, M. Maeda, K. Yanagida, *Electrochem. Soc. Ext. Abstr 185*, *PV 86-2*, 580 (1986).
Tu, Y.-Y., T. J. Chuang, H. F. Winters, *Phys. Rev. B. 23*, 823 (1981).
Tuszewski, M., J. A. tobin, *J. Vac. Sci. Technol.*, *A14*, 1096 (1996).
van de Ven, E. P., R. S. Martin, M. J. Berman, *1987 VMIC*, 434 (1987).
van de Ven, E. P., I.-W. Connick, A. S. Harrus, *1990 VMIC*, 194 (1990).
Verdeyen, J. T., J. Beberman, L. Overzet, *J. Vac. Sci. Technol.*, *A8*, 1851 (1990).
Virmani, M., V. Mahadev, T. S. Cole, *1996 DUMIC*, 139 (1996).
Vossen, J. L., W. Kern, eds., *Thin Film Processes*, Academic Press, New York, 1978.
Vossen, J. L., J. J. Cuomo, Chap. II-1 in *Thin Film Processes* (J. L. Vossen, W. Kern, eds.), Academic Press, New York, 1978.
Vossen, J. L., *J. Electrochem. Soc.*, *126*, 319 (1979).
Wagendristel, A. Y. Wang, p. 44 in *An Introduction to Physics and Technology of Thin Films*, World Scientific, Singapore, 1994.
Wang, D. N., J. M. White, K. S. Law, C. Leung, S. P. Umtoy, K. S. Collins, J. A. Adamik, I. Perlov, US Pat. 5,000,113, 1991.
Wang, D. N., J. M. White, K. S. Law, C. Leung, S. P. Umotoy, K. S. Collins, J. A. Adamik, I. Perlov, US Pat. 4,872,947, 1989.
Ward, R., Lewin, I.H., US Pat. 4,665,463, 1987.
Wasa, K., S. Hayakawa, *Handbook of Sputter Deposition Technology*, Noyes Publications, Westwood, N.J., 1992.
Watanabe, T., T. Kitabayashi, US Pat. 5,151,845, 1992.
Watts, A. J., W. J. Varhue, *Appl. Phys. Lett.*, *61*, 549 (1992).
Wehner, G. K., G. S. Anderson, chap. 3 in *Handbook of Thin Film Technology* (L. I. Maissel and R. Glang, eds.), McGraw-Hill, New York, 1970.
Whang, K. W., S. H. Lee, H. J. Lee, *J. Vac. Sci. Technol.*, *A10*, 1307 (1992).
Wicker, T. E., T. D. Mantei, *J. Appl. Phys.*, *57*, 1638 (1985).
Winkle, L. W., C. W. Nelson, *Solid State Technol.*, *10/81*, 123 (1981).
Winters, H. F., J. W. Coburn, T. J. Chuang, *J. Vac. Sci. Technol.*, *B1*, 469 (1983).
Wright, D. R., D. C. Hartman, U. C. Sridharan, M. Kent, T. Jasinski, S. Kang, *J. Vac. Sci. Technol.*, *A10*, 1065 (1992).
Wright, D. R., L. Cha, P. Federlin, K. Forbes, *J. Vac. Sci. Technol.*, *B13*, 1910 (1995).

Ye, Y., D. X. Ma, H. Hanawa, P. K. Loewenhardt, J. Shiau, A. Zhao, T. Nguyen, S. Arias, G. Z., Yin, *Electrochem. Soc. Ext. Abstr. 96-1*, 178, 241 (1996).

Xie, J. D. Kava, *43rd Natl. Symp. Amer. Vac. Soc., Abstr. PS-TuA6, 84 (1996).*

II
Characterization

Geraldine Cogin Schwartz
*IBM Microelectronics**
Hopewell Junction, New York

1.0 INTRODUCTION

In this chapter we enumerate and explain many of the most widely used techniques for characterizing the optical, mechanical, electrical, and chemical properties of thin films, examining structures fabricated from the films, and measuring some reactor properties. Finally some chemical analytical techniques are discussed.

Tables II-1a and II-1b summarize many of the characterization techniques and their applications.

2.0 OPTICAL CHARACTERIZATION OF DIELECTRIC FILMS

2.1 INTRODUCTION

The measurement of the refractive index, n, and the thickness, d, of thin dielectric films is used extensively in device processing. The value of n is an indication of the composition and stoichiometry of a film, but must be used in conjunction with other analytical techniques since, e.g., changes in density alter its value without a change in composition. The relationship between refractive index and density is given by the Gladstone–Dale equation:

$$\delta = K_1 (n - 1)$$

or the Lorentz–Lorenz formula:

$$\delta = \frac{K_2(n^2 - 1)}{n^2 + 2}$$

---

*Retired.

TABLE II-1a Characterization Techniques

Property	Technique	Comments
Thickness	Stylus	Any film; need sharp step
	Interferometry	Dielectric films
	Ellipsometry	Dielectric films
	XRFS	Any film; do not need step; layer thickness in composite film
	RBS	As in XRFS
	Resistance change	Measure metallic film growth
	SEM	Cross section
Composition	Wet chemical analysis	Elements; some groups[1]
	AES	Elements, some bonding[2]
	with sputter	Depth profile[3]
	XPS	Chemical bonding[4]
	with sputter	Depth profile[3]
	RBS	Elements; compound formation[5]
	XRFS	Elements
	Microprobe	Elements
	IR	Dielectrics; bonds
	SIMS	Elements[6]
	XRD	Phases, crystal structure

[1]Destructive. After sample dissolution, instrumental analysis, e.g., chromatography, colorimetry, spectrometry, etc.
[2]Beam damage (desorption, e.g., F); charging of dielectrics; small beam size, quantitative, detection limit can be excellent; sensitive to all elements but H_2, He; sample cannot decompose in high vacuum.
[3]Destructive; interface distortion.
[4]Nondestructive, large area for analysis, sensitive to all elements but H_2, good for organic films, some depth resolution with angle resolved XPS, can be quantitative; sample compatible with high vacuum.
[5]Do not need standards; depth profile without sputtering; better for high Z elements; large area for analysis.
[6]Destructive; very sensitive, detects all elements, isotopes, small beam, matrix effects, interface distortion.

where δ = density and K_1 and K_2 are constants. According to Pliskin (1977) the first is more applicable to SiO_2, since it is based on the assumption that the material is basically the same except for porosity. For SiO_2, assuming the pores are filled with air:

$$\delta = -4.784 + 4.785n$$

Used alone, n is a process monitor *only*. Thickness measurements are necessary for process calibration and control in etching and deposition. There are many techniques available; some of those for which commercially available equipment is available are described below. A more complete discussion of others can be found in Pliskin and Zanin (1970).

Characterization

TABLE II-1b Additional Characterization Techniques

Property	Technique	Comments
Resistivity	4-point probe	Sheet, via resistance
Breakdown field	MOS structure: *I-V*	Measure distribution
Dielectric constant	MOS structure: *C-V*	High-frequency Si surface in full accumulation
Dopants/impurities content	IR AES SIMS XRFS MICROPROBE	P, B, F, H, H_2O in dielectrics Limitations in Table II-1a
	Wet chemical Resonant nuclear reactions	Used to measure H in any film
Stress	Bending beam Wafer curvature	Automated laser probes; XRD
Adhesion	Scotch tape, peel scratch, blister	
Thermal stability	TGA, DTA, Anneal + IR Moisture evolution analyzer	Organic films Inorganic dielectric
Step coverage	SEM	Cross section best; on-line instruments
Surface topography	SEM Stylus Optical microscope	Tilted sample
Interface structure	TEM	Cross section

2.2 ELLIPSOMETRY

This technique is used to determine the refractive index and thickness of dielectric films and can also be used to determine the optical constants of a substrate. The method is based on measuring the change in the state of polarization of monochromatic light reflected from a substrate. The state of polarization is determined by the relative amplitude of the parallel, ρ_p, and perpendicular, ρ_s, components of radiation and the phase difference between them, $\Delta_p - \Delta_s$. Upon reflection, ρ_p/ρ_s (=tan Ψ) and $\Delta_p - \Delta_s$ change. The changes depend on the optical constants, n_2, k_2 of the substrate (remember that $n = n - ik$ where k = absorption coefficient) and the angle of incidence θ, the optical constants of the film n_1, k_1 and the film thickness d. The basic theory was developed by Drude and the relationships are given by Pliskin and Zanin.

The light is linearly polarized by the polarizer and elliptically polarized by the compensator. After reflection, a second polarizer acts as the analyzer. Measurements are made by rotating the polarizer and analyzer until the reflected beam from the sample is extinguished. The values of the polarizer and analyzer readings are then used to determine tan Ψ and Δ, and from them n_2 and d can be determined. In the early days, the values of n and d were determined by the use of graphs or tables or, eventually, a personal computer. Many of the newer commercially available ellipsometers are equipped with computers having the appropriate software for making the calculation. However, the absolute thickness is not measured, since the same values of tan Ψ and Δ recur regularly; this has been called thickness order periodicity and depends on the value of n_2. Thus a knowledge of the approximate film thickness is necessary for determining the order and thus, the absolute thickness. There are, however, thicknesses (i.e., values of tan Ψ and Δ) for which accurate values of n_2 are not obtainable (see Pliskin and Zanin, 1970, pp. 11–23, 24).

Figure II-1a shows the variation of Δ and Ψ as a function of n and d of transparent films on Si. A diagram of a simple ellipsometer is shown in Fig. 11-1b. There are many ellipsometers available commercially. Automatic ellipsometers reduced the time required for measurements significantly, although the earliest of them made no improvements in the measurements themselves. More recently several improvements on the basic system are being offered, together with automation. The automation includes not only automated rapid data acquisition and calculation but automatic wafer handling and positioning as well as the capability of mapping film thickness, as 2D contour maps or 3D representations.

Variable angle (of incidence) ellipsometers allow more accurate measurements of both single- and two-layer films as well as absolute thickness determination. Some ellipsometers also have the capability of using both linear and circularly polarized light. Spectroscopic ellipsometers use several wavelengths. They are said to measure very thin films more accurately than single-wavelength or multiple-angle ellipsometers. Absolute thickness can be determined and the thickness order periodicity is eliminated. Multiple films stacks can be measured. In addition, the dispersion of the refractive index can be determined. Rudolph, Gaertner, and Tencor are suppliers of ellipsometers.

2.3 INTERFEROMETRY

2.3.1 Principles of Optical Interference

The schematic diagram illustrating two-beam interference is shown in Fig. II-2. A light beam B_0 of wavelength λ is shown impinging, at an angle θ_1, on the surface of a transparent film of refractive index $n_{2\lambda}$, at wavelength λ, and thickness d. Part of the beam is reflected at the interface between medium 1 (usually air, $n_1 = 1$) and the film; this is beam B_{12}. Another part of the incident beam is refracted in the film at an angle θ_2 and then reflected at the interface between the film and the absorbing substrate ($n = n_3 - ik_3$); this is beam B_{23}. The path length between the two beams is $2n_{2\lambda}d \cos \theta_2$. If this difference is $N\lambda$, where N is an integer, the beams will be in phase and there will be constructive interference (i.e., maximum brightness); if N is a half integer, the beams are out of phase by 180° and interference

Characterization

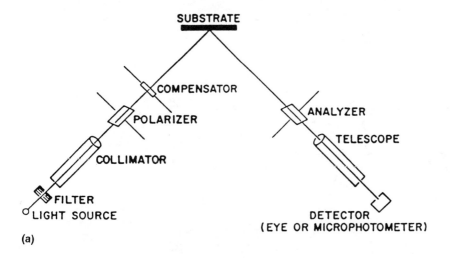

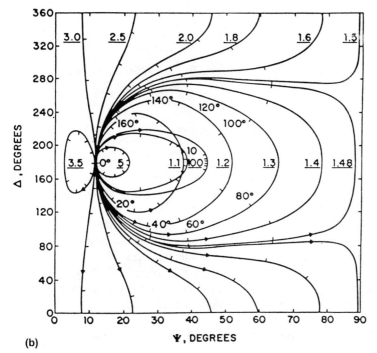

FIGURE II-1 (a) Schematic representation of an ellipsometer. (b) Variation of Δ and ψ as a function of refractive index and thickness. (Pliskin and Zanin, 1970)

is destructive (minimum brightness). Since $\sin \theta_1 = n_{2\lambda} \sin \theta_2$ (Snell's law), the conditions for maxima and minima are:

$$N\lambda = 2d\,(n_{2\lambda}^2 - \sin^2 \theta_1)^{1/2}$$

For normal incidence:

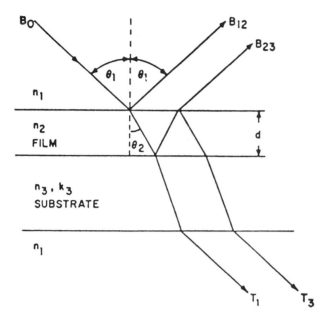

FIGURE II-2 Schematic diagram for two-beam reflection or transmission through a transparent substrate. (Pliskin and Zanin, 1970)

$$N\lambda = 2dn_{2\lambda} \quad \text{or} \quad d = \frac{N\lambda}{2n_{2\lambda}}$$

Minima are sharper than maxima and are used whenever possible.

When periodic variations in the reflected light occur, alternating bright and dark regions (fringes) are formed. Fringes can be formed by changing θ, d (non-uniformity or changes during deposition or etching), $n_{2\lambda}$, or λ. The fringe system obtained by thickness changes has been utilized for in situ measurement of the plasma-assisted etch or deposition rate of a dielectric film.

The change of n as λ changes is called the dispersion; Pliskin (1987) has compiled an extensive list of the refractive indices and dispersions of inorganic and organic dielectric films used in semiconductor fabrication.

2.3.2 Application to Measurement of Film Thickness

There are a number of instruments for thickness determination using interferometry. If the film thickness of known, then n_λ can be determined.

CARIS/VAMFO

In CARIS, constant-angle reflection interferometry, a spectrometer is used to vary λ at constant θ. In VAMFO, variable angle monochromatic fringe observation, θ is varied at constant λ. The equations for determining the d and n are given in Pliskin and Zanin (1970). If d is known, n can be calculated. The results have been obtained most readily using tables which give the film thickness as a function of the angle of incidence for a particular n, λ, and N. Where necessary, two wavelengths can be used for two successive measurements.

Multiple-Beam Interfermometry

Multiple-beam interferometry produces sharper fringes (Pliskin and Zanin, 1970). There are two techniques: FIZEAU fringes, generated by monochromatic light, which represent contours of equal thickness; this is useful for examining nonuniform films and has been called the TOLANSKY method. The other, fringes of equal chromatic order (FECO), is a more accurate technique. Collimated white light is used to illuminate a sample at normal incidence; the reflected light is dispersed by a spectrometer which varies λ to produce fringes. Pliskin and Zanin have discussed the principles of the method.

Commercially Available Instruments

Interferometric techniques are the basis for instruments available commercially. These systems include rapid automated data acquisition, computation on the basis of the value of n supplied by the user, wafer positioning/mapping, as well as capability of displaying thickness maps in 2D and 3D (e.g., Nonospec, Prometrix).

An adaptation of the VAMFO technique is the beam profile relectometer (BPR) (Willenborg et al., 1991); it is available commercially (Opti-Probe) with the usual automation capabilities. The claim for the system is that it has the speed of spectrometer measurements with the accuracy of ellipsometry, using a small spot size. A linearly polarized laser beam is focused through a high numerical aperture (NA) microscope objective. A bundle of light rays is thus incident on the surface of the sample, with the central ray normal to the surface. For a 0.9 NA lens, the angles of the bundle range from $0°$ to $\pm 64°$. Each ray undergoes interference in the film. A line of rays is P polarized relative to the plane of incidence and an orthogonal line is S polarized. The measurements are P- and S-interference profiles and the total reflected power enabling *both* n and d to be determined.

2.4 PRISM COUPLER

The principles of this method have been given by Tien et al. (1969), Tien (1971), and Wei and Westwood (1978), and the application of thin films on Si by Swalen (1976) and Adams et al. (1979).

A diagram of a prism coupler, used to measure d and n is shown in Fig. II-3a.

A highly refractive prism and a thin-film coated substrate are pushed into close contact. A laser source and a photodiode detector are mounted on a rotating platform. The angle of incidence of the laser light onto the prism face is varied. At the so-called coupling angles, the light is coupled into the film, decreasing the intensity of the reflected light, which is detected by the photodiode shown in the figure.

These angles are a function of n and d; by measuring two coupling angles, both may be computed. There is a minimum thickness, which depends on λ, for the coupling to occur.

Figure II-3b shows the reflected intensity as a function of the coupling angle. The technique has the particular advantage of allowing the determination of the value of n parallel to the surface of the film and perpendicular to it. This is done by using polarized light: the TM mode (electric vector parallel to the surface) yields the value of n in the plane; the TE mode (electric vector perpendicular to the surface), the value perpendicular to the surface.

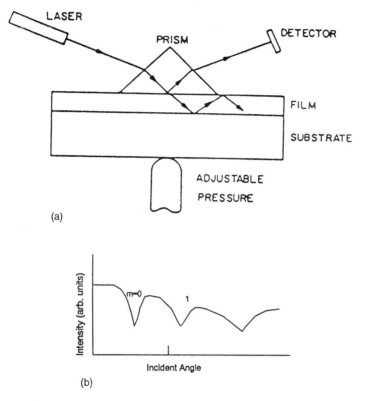

FIGURE II-3 (a) Schematic diagram of a prism coupler; (b) intensity of signal vs. incident angle. [Part (a) from Adams and Murarka, 1979]

2.5 OTHER

A new technique for measuring the thickness d and optical constants n_λ and k_λ of a thin film simultaneously is the analysis of reflectance or transmission measurements by the dispersion equations developed by Forouhi and Bloomer (1986, 1988) and patented by them in 1990. The "n&k Analyzer" consists of the data acquisition hardware, and the software for analysis and is supplied by a company called n&k Technology.

3.0 INFRARED (IR) SPECTROSCOPY

Infrared spectroscopy is a powerful technique and one used frequently for characterizing dielectric films.

3.1 ORIGIN OF IR BANDS

The vibrations of individual atoms within the molecule, e.g., stretching, bending, and rocking modes, absorb energy in the IR if there is a change in the dipole moment. IR spectroscopy consists of the detection and measurement of the position

Characterization

and intensities of the absorption of IR radiation. Transmission is preferred over reflection for reasons enumerated by Pliskin (1973) who also listed characteristics of the Si substrate required for the best results: high resistivity, doubly polished, and thick. Transmitted spectra are plotted as percent transmission on the linear scale of wave number, cm^{-1} (which is equal to $1/\lambda$ and λ is expressed in centimeters; the wave number is proportional to the frequency of the vibrating unit v, i.e., wave number = v/c). The position of the band identifies it; the exact position and the half-width can supply information about the quality of a film. Since the peak position can also vary with thickness (Pliskin and Lehman, 1965), equal thicknesses are required when making comparisons.

3.1.1 IR Bands

The position and identification of many of the IR bands which are often used are listed in Table II-2; they will be discussed in subsequent sections of this chapter as well as in Chapter IV.

3.2 IR SPECTROMETERS

3.2.1 Dispersion Spectrometers

Much of the early work was done using a dual-beam dispersion IR spectrometer, using a grating or a prism. A film-coated wafer and a bare wafer (the reference) are scanned continuously through the wavelength region of interest and the output

TABLE II-2 IR Frequencies and Their Identification

Band Position (cm^{-1})	Identification
1050–1100	Si—O STRETCH
~810	Si—O BENDING
~450	Si—O ROCKING
3650	H-BONDED Si—OH
~880	Si—H in O-DEFICIENT SiO_2
~2260	Si—H in O-DEFICIENT SiO_2
3400–3300	ABSORBED H_2O
~2160	Si—H (in NITRIDE)
850	Si—N
3350	N—H
930	Si—F
1370	B—O
~670	B—O
~920	B—O—Si
720	B—O—B
~1050–950	P—O
~1350–1300	P=O
2976	Si—CH_3

is displayed on graph paper which is driven in synchronism with the dispersing system of the monochrometer. Alignment is critical. Sensitivity is a function of the scan speed; greater sensitivity requires longer data collection time. This increases the hazards of component drift and, in some cases, interaction of the film with the ambient. This last can be minimized by flushing the sample compartment with dry N2. Analysis of the output graph is usually manual, but the data can be fed into a computer. A diagram of a typical spectrometer is shown in Fig. II-4.

3.2.2 Fourier-Transform IR (FTIR) Spectrometers

A typical Fourier-transform IR spectrometer is diagrammed in Fig. II-5; it is available from several manufacturers and is now the instrument of choice. In this method, the entire frequency (wavelength) range of interest is passed through the spectrometer simultaneously which produces an output signal containing all the frequencies. This interferogram is fed into a computer (which is an integral part of the apparatus) which performs a mathematical operation, called a Fourier transform. The output is a spectrum of intensity vs. frequency. Data collection is rapid, reducing the potential problems of system component drift and wafer instability, but it requires separate scanning of sample and reference.

3.3 APPLICATION TO SiO$_2$

An IR spectrum of a film of interest in semiconductor fabrication, one of a thermally grown SiO$_2$ film (often used as a basis for comparison with other deposited oxides), is shown in Fig. II-6.

One important application of IR spectroscopy has been the study of deposited SiO$_2$ films. For example, the position of the Si—O stretching band at ~1050–1100 cm^{-1} is shifted to lower frequencies (lower wavenumber, higher wavelength) and

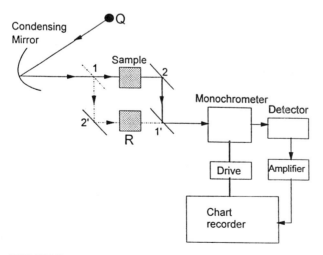

FIGURE II-4 Diagram of a recording IR spectrometer.

Characterization

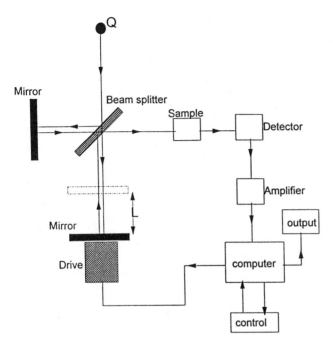

FIGURE II-5 Diagram of an FTIR system.

the band is broadened by porosity, strain, and oxygen deficiency. An example of the shift due to densification is shown in Fig. II-7.

However, Pliskin (1973) points out that one cannot assume that any band changes are due to only one property change. This caveat was illustrated by the example of an oxide which has strained bonds and is initially porous, so that the IR band due to the Si—O stretch is broad and peaked at a lower frequency. Such oxides react very readily with water to form Si—OH groups, relieving the strain. The band will become sharper and will be shifted to higher frequencies, although no densification has occurred.

This was verified by Machonkin and Jansen (1987). The effect of oxygen deficiency was demonstrated clearly by Pai et al. (1986), as seen in Fig. II-8.

3.3.1 Detection/Measurement of Impurities

The identification and measurement of impurities such as —OH, H_2O, by IR spectroscopy (Pliskin, 1967) has been a very important technique in evaluating deposited SiO_2 films and is relied upon frequently. The effect of moisture absorption on the Si—O band and the spectrum of Si—OH and H_2O are illustrated in Fig. II-9. The other Si—O bands, at ~810 cm^{-1} (bending) and at ~450 cm^{-1} (rocking), are not usually examined since they do not provide additional information.

The identification of Si—OH as H-bonded (3650 cm^{-1}) is based on the fact the band due to free SiOH occurs at higher frequencies and is quite sharp. The

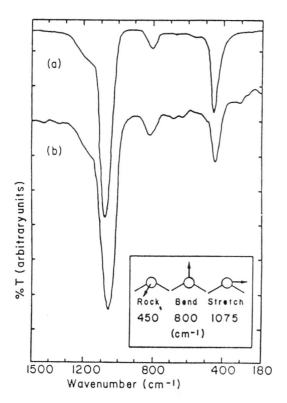

FIGURE II-6 IR absorption spectra of (a) thermally grown SiO_2 film, (b) oxide deposited at 500°C. Inset: O atom motion for rocking, bending, and stretching vibrations. (From Lucovsky and Tsu, 1987)

band for absorbed water occurs at 3400–3300 cm^{-1}. The change in the spectrum as water and "loosely" bound OH (i.e., silanol groups having a common silicon atom or are near neighbors and form water readily) are removed by heating, is shown in Fig. II-10 (Pliskin, 1973, 1977).

In incompletely oxidized silicon oxide films, absorption bands have been observed at ~2260 cm^{-1}, and at ~880 cm^{-1}; these have been assigned to Si—H (Pliskin, 1973). Lucovsky and Tsu (1987) characterize these bands, also seen by them in O-deficient oxide films as the bond-stretching (2250 cm^{-1}) and bond-bending (870 cm^{-1}) vibrations of the Si—H bond. Lucovsky and his colleagues have relied heavily on many of these findings in their IR studies of remote plasma deposition of silicon oxides (e.g., Lucovsky et al., 1986, 1989; Lucovsky and Tsu, 1987; Pai et al., 1986; Theil et al., 1990).

Other impurities, such as CO and CO_2, have been detected by IR spectroscopy (Pliskin et al., 1969).

3.3.2 Detection/Measurement of Dopants

The presence of P in PSG or BPSG films is detected at ~1350–1300 cm^{-1} (P=O) and at ~1050–950 cm^{-1} (P—O) (Hurley, 1987). Incorporation of P causes the

Characterization

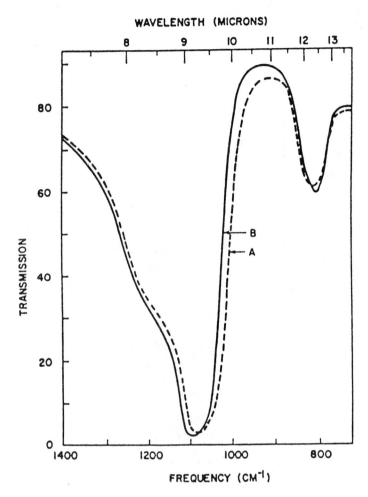

FIGURE II-7 IR spectra showing effect of steam densification of a pyrolytic SiO_2 film. Spectrum A: as-deposited oxide; spectrum B: oxide after densification. (From Pliskin and Lehman, 1965)

Si—O band to shift to a higher frequency (Pliskin and Lehman, 1965). Adams and Murarka (1979) reported that the P content of PSG films, determined by IR, agreed (to within ±0.3 wt. % P) with the results of chemical analysis, neutron activation, and microprobe techniques.

Other additives have been detected and estimated by IR, for example, B in BSG or BPSG. The B bands in BSG or BPSG are at 1370 cm^{-1} and ~670 cm^{-1} (B—O), and at ~920 cm^{-1} (B—O—Si) (Pliskin, 1967; Rojas et al., 1992). Wong (1976) mentions a B—O—B band at 720 cm^{-1}. The B content in silica glass has been determined by the ratio of the absorbance maximum of the B—O band at ~1370 cm^{-1} to that of the Si—O band at ~1070 cm^{-1} (Rojas et al., 1992). The Si—F band in F-doped SiO2 appears at ~930 cm^{-1}.

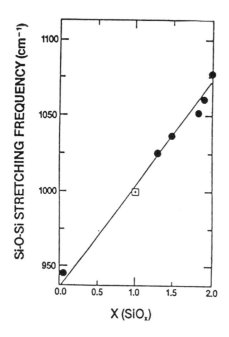

FIGURE II-8 Si—O—Si stretching frequency vs. oxygen composition x in SiO$_x$ films. (From Pai et al., 1986)

The correlation of band intensities with concentration for other many bands has been made using the same techniques used for P concentration.

3.4 APPLICATION: SILICON NITRIDES

IR spectroscopy has played an important role in the characterization of silicon nitride films: Si—N, N—H, and Si—H bands. Lanford and Rand (1978) correlated the spectral intensities of the Si—H and N—H bands with the H content as determined by nuclear reaction (see below). An IR spectrum of a "typical PECVD nitride film is given in Fig. II-11.

The use of multiple internal reflection (MIR) of IR radiation has also been used to study the H content and the effects of annealing of very thin films of LPCVD silicon oxy-nitride (Stein, 1976) and CVD nitride (Stein and Wegener, 1977). The many reflections amplify the signal from very thin films; this is illustrated in Fig. II-12.

3.5 QUANTITATIVE MEASUREMENTS OF SPECTRA

The absorption band intensities in transmission spectra are approximately proportional to the thickness of a film, i.e., they obey Beer's Law,

$$I/I_0 = e^{-\alpha x} \quad \text{or} \quad 2.303 \log I_0/I = \alpha x$$

where I is the intensity of light after passing through the film, I_0 the intensity of

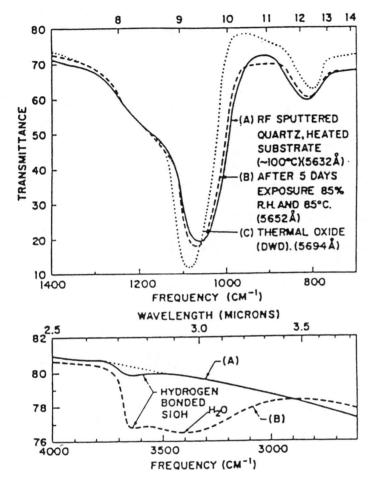

FIGURE II-9 IR spectra showing effect of moisture on a sputtered SiO_2 film. (From Pliskin, 1967)

light with no sample present, x the film thickness, and α the absorption coefficient. Although the integrated intensity is the best measure of the absorption intensity, a reasonable approximation is given by the intensity at the peak maximum, the optical density (OD).

Quantitative analysis of the silanol and water content of SiO_2 films are related to the band intensities:

$$W = -14A_{3650} + 89A_{3330} \, (2.2/\rho)$$

$$S = 179A_{3650} - 41A_{3330} \, (2.2/\rho)$$

where W = wt% water (including H_2O from "easily" removed silanol; S = wt% OH as silanol; ρ = film density; and A_v = O.D./μm film at frequency v (Pliskin, 1973, 1977).

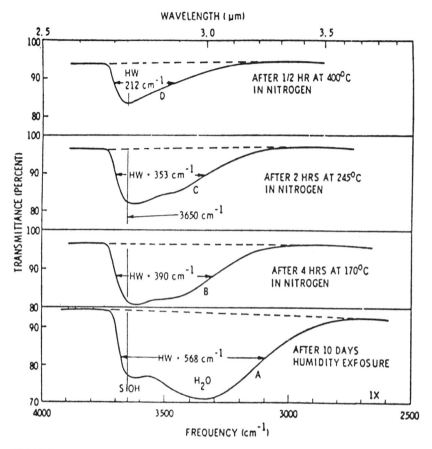

FIGURE II-10 IR spectra showing effect of heating in a dry ambient to remove water and "loosely" bound hydroxyl from a CVD oxide film. (From Pliskin, 1977)

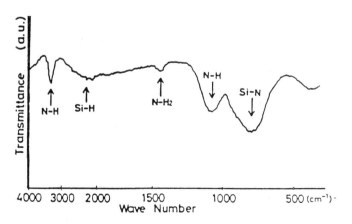

FIGURE II-11 A typical IR spectrum of a PECVD SiN film. (From Mura Kami et al., 1988)

Characterization

INTERNAL REFLECTION SPECTROSCOPY

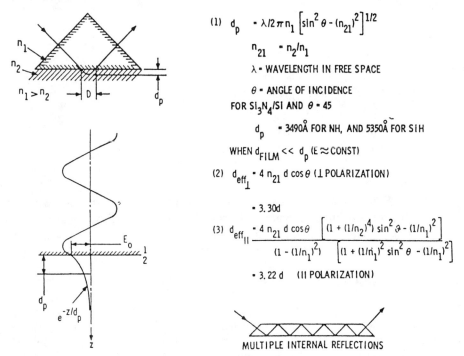

FIGURE II-12 Illustration of the internal reflection technique for obtaining IR absorption data on thin films. (From Stein and Wegener, 1977)

4.0 RESISTIVITY OF METAL FILMS

4.1 INTRODUCTION

The electrical resistivity of a metallic conductor is one of its most important properties. Lowering the resistance of the interconnections decreases signal propagation delay. Thus the trend toward Cu, away from Al and its alloys, although the resistivity of the barriers in which Cu must be encapsulated, increases the total resistance and must be taken into account when considering the net advantages of a shift in metallization.

4.2 SHEET RESISTANCE

4.2.1 Definition

The resistance of a rectangularly shaped section of a film, shown in Fig. II-13, measured parallel to the surface of the film is:

$$R = \left(\frac{\rho}{t}\right)\left(\frac{l}{w}\right)$$

where ρ is the resistivity of the film, t the thickness, l the length, and w the width of the conductor sample.

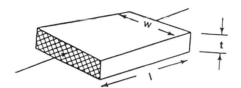

Resistance = (resistivity/thickness)(length/width)
R = (ρ / t)(l / w)

FIGURE II-13 Definition of sheet resistance.

If $l = w$, $R = \rho/t$ and this quantity is called the sheet resistance R_s; R_s is independent of the size of the square and depends only on the film thickness and is expressed as ohms/square, $\Omega/\square$. Thus if the thickness is known, the resistivity can be obtained from the measured sheet resistance.

4.2.2 Measurement

Probes

A four-terminal method is required since it eliminates the effect of contact resistance between the film and the probe. Current is fed to the ends of the sample and the voltage drop across several *squares* is measured, as seen in Fig. II-14a, and the resistance determined: $R = V/I$. To eliminate the need for fabricating special samples, a linear array of equally spaced probes can be placed on a long thin metal film deposited on an insulator, as shown in Fig. II-14b; current is fed to the outer probes and the voltage drop across the inner probes is measured to yield R_s (Maissel, 1970):

$$R_s = \frac{4.532V}{I}$$

The smaller the probe spacing, the better the resolution. A square array, shown in Fig. II-15, is used for higher resolution. In this case (Maissel, 1970):

$$R_s = \frac{V}{I}\left(\frac{2\pi}{\ln 2}\right) = \frac{9.06V}{I}$$

Eddy Currents

Eddy currents can be induced in a conductor when it moves through a nonuniform magnetic field or in a region where there is a change in magnetic flux. Measurement of such currents is the basis for noncontact resistivity measurement, a desirable technique since it avoids the possibility of probe damage.

There are commercially available systems which can measure the sheet resistance of a film at many points on the surface, rapidly, and have the software for plotting a resistance contour map of the surface. The Tencor "OmniMapNC110" uses a very small coil and thus can measure the resistance on a small blanket area directly on a product wafer.

Characterization

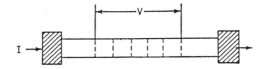

(a)

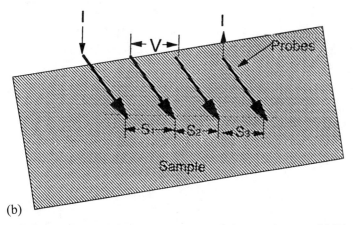

(b)

FIGURE II-14 (a) Direct measurement of sheet resistance. (b) Diagram of an in-line four-point probe for measuring sheet resistance.

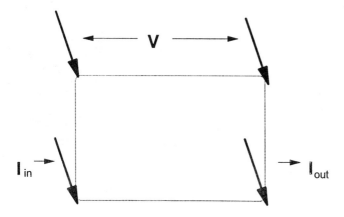

FIGURE II-15 Square probe array for measuring sheet resistance.

If the sample is other than a straight strip, the resistance calculation is more complex, as demonstrated by Hall (1967).

At very low-film thicknesses, the resistivity decreases rapidly with increasing thickness, reaching a constant value (in some cases, the bulk resistivity). However, the resistivity of thin films is often greater than the bulk resistivity, due to impurity inclusion and structural defects. If any change is seen when annealing single-metal films, it is a reduction of resistivity. Alloy films may behave differently, due to morphological changes, etc. The addition of other constituents (alloying) increases the resistivity.

4.2.3 Temperature Dependence

The resistivity of metal films decreases with decreasing temperature; the temperature coefficient of resistance (TCR), α_T, is given by:

$$\alpha_T = \frac{R_1 - R_2}{R_T(T_1 - T_2)}$$

where $T_1 > T > T_2$.

However, if the films are very thin and not continuous, TCR may have a negative value.

4.3 VIA RESISTANCE

4.3.1 Introduction

Low interlevel (via) resistance (metal to metal contact) is another essential ingredient for a fast, reliable MLM device. The subject of contact resistance (metal to semiconductor) will be covered in Chapter III.

4.2.2 Measurement

The accurate method for determining the via resistance, and the effect of various processing conditions on it, is to measure individual vias (preferably of different sizes) using a "double L" (Kelvin) structure, shown in Fig. II-16. This is a four-

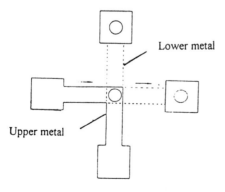

FIGURE II-16 "Double L" test structure used for measuring via resistance.

point measurement which eliminates the influence of probe resistance. An assessment of the via yield for a given process is obtained from via chains; the larger the number in the chain, the better the assessment for a dense chip.

4.3.3 Contributors to Via Resistance

Even a clean metal to metal contact will exhibit some resistance beyond that due to the number of squares in the area because of current crowding as the current flows from the interconnect into a constricted via hole. Steep via slopes (resulting in thinned metal over the step) and misaligned vias will also add to the resistance; the use of via plugs eliminates the first problem and improved lithographic alignment procedures can minimize the second. The problem remaining, to be discussed in greater detail, is that of an interfacial film.

Some of the common films are organic or carbonaceous and are due to redeposition of organic materials during via etch of a polymer dielectric (Day and Senturia, 1982; Smith et al., 1983) or some SOGs (Shacham-Diamond and Nachumovsky, 1990). Another is specific to Al-based conductors. It is the existence of a film of aluminum oxide at the interface. For example, the native Al_2O_3 on a freshly formed Al surface exposed only to the atmosphere is approximately 3 nm thick. Wildman and Schwartz (1982) reported that the via resistance due to an Al_2O_3 layer increased by about one order of magnitude for every 0.2 nm of oxide. Water desorbed from SOG interlevel films during via etch oxidizes the Al surface. Via "cleaning," e.g., using a buffered HF (BHF) or BHF/NH_3 dip resulted in thick porous oxide layer (Wildman and Schwartz, 1981); after sintering, however, the via resistance was not abnormally high, probably due to the porous nature of the film which allowed Al interdiffusion during heating. There are, however, many reports of the advantages of an HF dip for removing the oxide if this step is followed by sintering (e.g., Smith et al., 1983). Precautions must be taken to ensure that the BHF is rinsed very rapidly in huge volumes of water to prevent dissolution of the metal which occurs in partially diluted BHF (MacIntyre, 1968).

4.3.4 Reduction of Via Resistance

In situ Ar^+ sputter cleaning (Bauer, 1980) or ion milling (Petvai, 1978), if done properly, can produce a clean interface requiring no sintering. Bauer found that the presence of Al on the fixturing or on Al blanks in unused positions on the dome (Bauer, 1980), or even better, sputtering an Al getter electrode before sputter cleaning (Bauer, 1994), suppressed the reoxidation of the cleaned surface by absorbing the moisture in the system. If the cleaned surface can be reoxidized before the next level of metal is deposited, the procedure will not be successful even though the sputter clean time was sufficient to remove the initial thickness of oxide. However, Tomioka et al. (1989) reported that during sputter cleaning, redeposition of the insulator which surrounds the via left a thin insulating film at the interface.

Alternatives to Ar^+ sputter etching have been proposed. Exposing an oxide covered Al surface to a CF_4 plasma, reduces the O coverage, replacing it with F (Chu and Schwartz), but AlF_3 is also an insulator and may cause problems. Takeyasu et al. (1994) avoided the potential problem of redeposition of the Al/Al_2O_3 on the side walls of the via by in situ RIE of the oxide in BCl_3/Ar before deposition of an Al plug by CVD.

Another alternative is deposition of a thin layer of Ti onto the lower oxidized Al surface (followed immediately by the thick Al-based next metal layer) will consume some of the oxide by dissolving it in the lattice, not by chemical reduction. Since there is a limit to the amount of oxide the Ti layer can consume, the surface oxide thickness (which may have been thickened by previous processing) can be reduced to that of the native oxide by a dip in phosphochromic etch (H_3PO_4/CrO_3/H_2O). However, the concentration of CrO_3 is critical: Too little and H_3PO_4 will attack the Al; too much and certain regions of the chip will be dissolved (Shankoff et al., 1978). Another alternative was proposed by Horie et al. (1984): Immediately after deposition, a film of $MoSi_2$ was deposited onto the lower Al surface to prevent oxidation of the Al surface.

5.0 THICKNESS

5.1 INTRODUCTION

Although some of these techniques are applicable to dielectric films as well as to metals, they are almost always employed only on blanket films of metals; the optical measurements are preferred for the dielectrics. The method used most frequently in the recent past is stylus profilometry but x-ray fluorescence spectroscopy (XRFS) is now coming into use on the manufacturing line; these will be described below.

5.2 EARLY METHODS

Some of the older methods preferred, perhaps, when stylus profilometry was in its infancy and held to be unreliable and apt to scratch and deform soft films are:

1. Gravimetric: density, area must be known.
2. Beta backscattering: measure amount of backscatter from films of beta particles emitted from radioactive source; film and substrate must have large difference in atomic numbers; standards required.
3. X-ray absorption and emission: require expensive equipment and standards.
4. Electrical resistance using four-point probe technique: requires accurate knowledge of resistivity which must be constant in the required thickness range. Since this often depends on deposition conditions, it is useful for process control, if not for absolute measurement.

5.3 STYLUS PROFILOMETERS

The newest stylus profilometers are viewed as reliable if used properly. There are several suppliers as well as several models from each. The systems have data collection and processing hardware and software including the capability for automated processing.

Initially they were used simply to measure the thickness of a deposited film. For best results, a sharp-edged narrow groove was etched into the film, rather than making a step using a shadow mask during deposition. This had two benefits: better resolution from the steep edge and the ability to level properly using the two steps.

Characterization

The stylus method is also used to measure step heights of both metals and insulators on patterned wafers, e.g., after RIE. The sizes of the features and the spaces between them have decreased, requiring greater resolution both vertically and horizontally. The new systems provide improved viewing capability for locating measurement sites.

Stili are available in many sizes down to submicron radii. Although it is more cost-effective to use a stylus with a large radius because it is less expensive, more rugged, and less likely to cause damage, if the stylus tip is larger than the grove being measured, it will not reach the bottom and register the wrong (too small) depth. The spherical nature of the stylus tip rounds the profiles and broadens them; the effect is more pronounced the larger the tip radius. Decreasing the stylus size increases the downward force of the stylus; this may cause deformation of some surfaces leading to erroneous measurements and lack of repeatability. The equipment provides for the adjustment of the force. If, in trying to avoid damage the force is set too low, the stylus may hop over features.

The horizontal resolution is increased by increasing the number of measurements made per scan as well as decreasing the scan speed; different models provide different capabilities. The scan length is also important; the smaller the scan length, the larger the number of measurements. Profilers are supplied by Veeco (Dektak) and Tencor.

The instruments are also used to measure surface roughness.

5.4 X-RAY FLUORESCENCE SPECTROMETRY (XRFS)

Routine thickness measurement is a more recent application of this technique; the method is discussed in Section 15.

6.0 DIELECTRIC CONSTANT OF DIELECTRICS

6.1 INTRODUCTION

The term dielectric constant as used here (ε) is more accurately called the relative permittivity of an insulator and is one of its most important characteristics. In applications in which minimizing signal delay and cross talk is of paramount importance, (i.e., in the interconnections) ε should be as low as possible. There are other applications (not considered as a topic in interconnection technology) for which charge storage is important, so that ε should be as high as possible.

6.2 MEASUREMENT

The value of ε, at different frequencies and temperatures, is determined from capacitance-voltage (*C-V*) measurements using a parallel plate capacitor:

$$C = \frac{\varepsilon \varepsilon_0 A}{d}$$

where C = capacitance, ε_0 = permittivity of free space, A = area, d = dielectric thickness. Nicollian and Brews (1982) discuss the subject of *C-V* techniques in detail; Mego (1990) has published brief guidelines for interpreting *C-V* data.

The capacitor may be a dielectric sandwiched between two metal electrodes (MIM) or between a metal electrode and a silicon substrate (MIS), which is called an MOS structure when the dielectric is SiO_2. To use the MIM structure, the insulator must be etched to make a contact to the lower electrode. The MIS or MOS capacitor is often preferred; this avoids any problems of surface roughness in the lower metal film. In this case, a metal film is deposited on the back of the wafer for contact unless the silicon is very heavily doped (i.e., has very low resistivity). The upper electrode is usually deposited through a shadow mask. Unless the electrode area is very large, the sample is sometimes subjected to a light etch to remove any "halo," formed by scattering beneath the mask, to eliminate any error in the electrode area. Mercury probes are used at times, to avoid evaporating the electrodes, but control of the area of the electrode is rather poor. But their use can be a convenience in another way. When it becomes necessary to heat the dielectric, e.g., to study moisture evolution and absorption, the evaporated metal dots must be etched off before heat treatment and redeposited before making the next measurement.

The dielectric constant is almost always measured at high frequency, i.e., ≥ 10 kHz, to avoid the problem of surface inversion in MIS or MOS capacitors, in which the surface layer of the substrate Si becomes conductive.

6.3 PRECAUTIONS IN MEASUREMENTS

When using an MIS or MOS capacitor for measurement of ε, it is essential that the Si surface be in full accumulation. In accumulation, the majority carriers are accumulated at the Si surface; as a result, the capacitance of the Si layer is negligible so that only the capacitance of the dielectric is measured. For p-type Si, a negative voltage is applied to the metal electrode; for n-type the polarity is reversed. Taking a full C-V curve will ensure that this condition is met. Another precaution is avoiding formation of a *pn* junction at the surface by, e.g., doping an n-type wafer with B. This appeared to have occurred during deposition of *BN* (Nguyen et al., 1994) and may also occur during deposition of BSG and BPSG films.

6.4 DISSIPATION FACTOR

Since insulators have a finite parallel resistance R, the total impedance Z is found from the diagram of ωC versus $1/R$ as shown in Fig. II-17, where $\omega = (2\pi)$ frequency. The value of $\sin \delta$ is a measure of the energy absorbed in the insulator. For most insulators, R is very large (the loss is very small) so that $\sin \delta \sim \tan \delta = 1/\omega RC$ which is readily measured on an appropriate bridge. The quantity $\tan \delta$ is called the dissipation loss factor or loss tangent; it is very small ($\leq 10^{-3}$) in "good" inorganic dielectrics such as SiO_2. Organic films are often electrically "leaky"; thus $\tan \delta$ may have an appreciable value.

The change in dissipation factor has been used to follow the cure cycle of several thin polyimide films; in this case it reached a minimum value (~ 0.01) when the films were fully cured but increased upon further heating when degradation set in (Rothman, 1980). The values of the dissipation factors were even lower (<0.004) for other properly cured polyimide films (Samuelson and Lytle, 1984).

Characterization

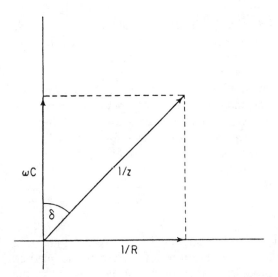

FIGURE II-17 Dissipation factor: Vector diagram of a practical capacitor showing δ (tan δ = loss tangent).

7.0 BREAKDOWN STRENGTH
7.1 MEASUREMENT

A parallel plate capacitor is also used for the current-voltage (I-V) measurements in determining the breakdown strength of an insulator. A dc voltage is applied to a large number of individual capacitors sequentially and the current is measured. The voltage is either ramped (increased continuously) or stepped until a current spike (indicating a self-healing breakdown) is observed or until the capacitor fails catastrophically (becomes shorted). The voltage when breakdown occurs is called the breakdown voltage, and the field (V/cm) the breakdown strength. When the catastrophic breakdown criterion is used, the distribution of breakdown voltages of the individual capacitors is usually narrower and the field higher. The distribution depends on the electrode area (larger area, lower strength) rate of change of the voltage (slow increase, lower strength), and may be sensitive to the nature of the lower electrode, its surface, and its method of preparation (Patrick et al., 1991). They also reported that for PECVD films deposited and measured under identical conditions, there was a wide range in the breakdown voltage distribution and its maximum value, making characterization of a dielectric film by this technique questionable.

7.2 APPLICATION

The low-field breakdowns (not the maximum strength) are the important ones for assessing the reliability of an insulator; this subject will be discussed in greater detail in the chapter on reliability. In device operation, the voltages are low and the thickness of the interlevel insulators as well as the width of the dielectric between adjacent conductors is relatively large, so that the operating fields are quite low. (This is in contrast to the fields experienced by the insulator in an FET device.)

A narrow distribution of breakdown strengths is taken as an indication of the homogeneity of the film. The very low-field breakdowns are usually attributed to defects; a preponderance of them is an indication of a very poor dielectric. The poor quality may be due to, e.g., improper formulation, application, or curing of a spin-on insulator, or a poor choice of deposition parameters for PVD, CVD, or PECVD films, but the influence of surface preparation cannot be ignored.

8.0 ADHESION

8.1 INTRODUCTION

A loss of adhesion or delamination results when the bonding forces at the interface between two films is weakened. This may be the result of surface contamination, surface modification by a previous processing step, or it may be inherent to the specific layers, e.g., Cu on SiO_2, some photoresists on hydrophilic surfaces, etc. Local concentration of stress at edges and bends (stress risers) may lead to adhesion loss. Poor adhesion of a film which has a high compressive stress may result in blistering; in a film in which the stress is highly tensile, delamination and cracking result, particularly where there is a stress riser.

The adhesion energy is the difference between the free energy of an interface formed between two surfaces when they are brought together and the original surfaces.

8.2 MEASUREMENT

There are several techniques for measuring adhesion. Comparisons are valid when different systems are evaluated by a single technique, but results from one kind of test are not readily comparable to those obtained by a different method. However, relative adhesion values are useful. Several systems are of interest: adhesion of an insulator (1) to a metal, (2) to itself, (3) to another insulator, and (4) to the semiconductor substrate as well as adhesion of a metal to various insulators.

8.2.1 "Scotch Tape Test"

Perhaps the most common and certainly the simplest test of adhesion is the *scotch tape test*, in which a piece of adhesive tape is attached to the surface of a film and pulled. It is qualitative and subjective but does screen for some minimum level of adhesion. If the top layer peels, it is usually concluded that the adhesion is too poor to warrant any further testing, and that further work is necessary either to change the materials or the process of applying the film, or to modify the underlying surface, e.g., by roughening, plasma cleaning, using an adhesion promotor, etc.

The causes of poor adhesion are sometimes determined by examination of each of the surfaces of the failed bond by analytical techniques such as XPS or AES. These same methods are used to assess the effectiveness of surface treatments designed to improve adhesion (Bacchetta et al., 1994).

Characterization

8.2.2 Peel Test

A more quantitative test is the *90° peel test*, which can be used when one of the films is thick and ductile. It has been used widely and successfully to measure the adhesion between a polymeric film and metals or inorganic layers but not between two polymeric layers. The concept of the test is shown in Fig. II-18. This technique is not practical for measuring the adhesion between thin metallic films and inorganic insulators because of the difficulty in attaching the film and initiating peel.

The force required to peel a unit length of film, i.e., to separate the film from the substrate and thereby produce two surfaces from the interface, is used to determine the adhesion stress. The puller can be advanced at a programmable speed and has a force sensor to measure the peel force. At steady state, the peeling force is expressed in g/mm (or equivalent units).

It may be necessary to use films for the test that are significantly thicker than those used in a real structure. Since the film is subjected to a steady stretching or deforming force during peel, it is important that the deformation energy be small compared to the total peel energy to avoid the use of corrections which can introduce errors.

8.2.3 Scratch Test

In the *scratch test*, a smoothly rounded tip is drawn across the surface; the vertical load applied to the point is gradually increased until a clear channel is produced (Ahn et al., 1976). The action of the tip involves plastic deformation of the substrate which produces a shearing force at the film substrate interface. Hamersky (1969) pointed out that the accuracy depends on the point radius, and larger radii were more reliable. Chapman (1974) noted that optical transparency was not a suitable measure of loss of adhesion since the film could become thin and translucent without detachment. A more sophisticated scratch test used a diamond *microin-*

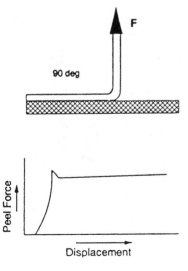

FIGURE II-18 Peel test arrangement for measuring stress.

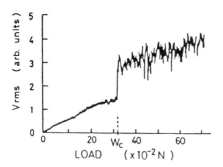

FIGURE II-19 RMS signal vs. load curve for MgO coated surface showing fracture during loading. (From Baba et al., 1986)

denter (*microtribometer*) (Baba et al., 1986; Otterman et al., 1991). In this technique as the *stylus* traversed the sample, the signal generated in a coil by a moving magnet was recorded by a true rms voltmeter. The technique has been used to measure the adhesion of several metal and oxide thin films. A typical result (for MgO) is shown in Fig. II-19 (Baba et al., 1986) in which the critical load for the fracture was $0.32N$ (adhesion force 1.7×10^9 Pa). This method avoids the error cited by Chapman since it does not depend on visual examination. In his review of adhesion test methods for thin hard coatings, Valli (1986) has discussed the scratch test in detail.

8.2.4 Blister Technique

Another method is the *blister* or *bulge technique*, illustrated in Fig. II-20. A window is etched in the substrate to expose the interface between it and a thin film. A

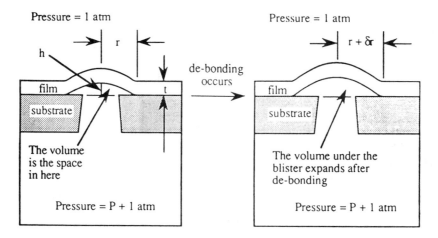

FIGURE II-20 Diagram of the blister test for adhesion. *Left*, before; *right*, after debonding. (From Sizemore et al., 1993)

Characterization

pressure difference is applied to the film which is deformed; at a critical pressure the radius of the blister expands and the film peels, i.e., debonds. Gent and Lewandowski (1978) and Sizemore et al. (1993) are among those who have analyzed and modeled the test. The crack extension force can be obtained from the measurement of the critical pressure and volume. The test has been applied to films such as polymers, silver, CVD diamond, and silicon nitride. Small et al. (1993) discussed possible causes for inconsistencies in results and suggest improvements, e.g., the use of initially flat reflective samples of well-defined size and geometry whose fabrication they describe.

8.2.5 Stretch Deformation

A *stretch deformation* method for direct measurement of the adhesion energy was developed by Ho and Faupel (1988). A test sample, consisting of a polymer substrate clamped at both ends and a metal overlayer, is shown as the inset in Fig. II-21. The adhesion energy is measured from the difference in the load vs. elongation curves between film/substrate (Cu/PI) and substrate (PI) structures, as shown in Fig. II-21. Microscopic examination, to determine the onset and end point of delamination, is required in order to extract the adhesion energy; these are at the points ε_1 and ε_2 in the curve. The adhesion energy $\gamma_{ad} = 1/b_o$ (integral from ε_1 to ε_2 of $[F(\varepsilon) - F_s(\varepsilon)]d\varepsilon$) where b_o is the original film width, F the applied load, and s denotes the substrate. Valli (1986) has mentioned a number of additional tests: ultracentrifugal and ultrasonic tests, acoustic imaging, and laser spallation.

9.0 FILM STRESS

9.1 INTRODUCTION

Tensile films may fail by delamination or by cracking and the crack may be propagated into the substrate. Very highly *compressive* films may buckle and eventually delaminate. Thouless (1991) has reviewed these failure mechanisms and the conditions that determine which one will operate. Thus measurement of film stress is an important part of process development, guiding the choice of materials and processes.

When a surface is coated with a tensile film (by convention the stress is +) the coated side is concave; it is convex when the stress is compressive (−). Film stress is expressed as dynes/cm^2 or as MPa (100 MPa = 10^9 dynes/cm^2).

9.2 STRESS MEASUREMENT

Stress in dielectric films, such as deposited oxides and nitrides of Si, organic films, thermally grown SiO$_2$, and stress in metal films have been measured by the methods described below. Murarka (1994) has discussed the origin of stress in metallic films and analyzed the stresses in multilayer metal films.

9.2.1 Cantilevered Bending Beam

In this method, illustrated in Fig. II-22, the deflection of one end of a coated beam is measured while the other end is clamped. The stress, σ, is given by:

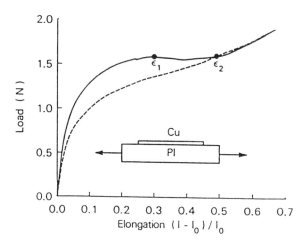

FIGURE II-21 Stretch deformation test: load vs. elongation curves for Cu/PI (solid line) and PI (dashed line). Insert shows sample geometry for adhesion test. (From Ho and Faupel, 1988)

$$\sigma = \frac{E_s t_s}{3L^2 \, t_f (1 - v) \, \delta}$$

where E_s is Young's modulus of the substrate, v Poisson's ratio of the substrate, L the length of the beam (substrate), t_s and t_f the thicknesses of the substrate and film, and δ the deflection. The term $E/(1 - v)$ is called the elastic biaxial modulus. Single-crystal Si is the substrate used most commonly, but others, e.g., Ge, GaAs, and quartz, have also been used, particularly in determining the elastic constants of deposited films, as discussed below. Table II-3a lists the values of $E/(1 - v)$ for these substrates. The length of the beam is much greater than its breadth. The sensitivity of the method depends on the detection systems used to observe the movement; Campbell (1970) has described several of them: microscopic observation, use of a contactometer, electromechanical, capacitive, or inductive device, a hot wire, or a Michelson interferometer. Abermann (1992) measured the stress during the growth of thin films on a quartz bending beam substrate under UHV conditions.

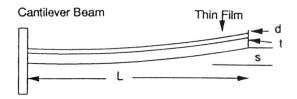

FIGURE II-22 Cantilever beam used to measure film stress.

TABLE II-3a Substrates: Value of $E/(1 - v)$

Material	$E/(1 - v)$ in units of 10^{12} dyne/cm^2
Si (100)	1.805[a]
Si (111)	2.290[a]
Ge (100)	1.420[a]
Ge (111)	1.837[a]
GaAs (100)	1.239[a]
GaAs (111)	1.741[a]
Fused quartz	0.854[b]
Bulk silica	0.88[c]
Thermal oxide	0.90[c]

[a] Brantley (1973).
[b] Handbook of Tables for Applied Engineering Science, p. 138.
[c] Quoted in Carlotti et al. (1996).

9.2.2 Change of Curvature

This method is most widely used for determining the stress in a thin film. The change in curvature of a circular substrate (usually a silicon wafer), resulting from the deposition of a thin film, is measured. This has the advantage of simplicity of sample preparation. The stress is given by:

$$\sigma = \frac{E_s}{6(1 - v)} \frac{t_s^2}{t_f} \left(\frac{1}{R_f} - \frac{1}{R_s} \right)$$

where R_s is the radius of curvature before film deposition, and R_f the radius after deposition.

The curvature has been measured by many methods, e.g., a light-section microscope (Glang et al., 1965), Newton's rings (Irene, 1976) an optically levered laser technique (Sinha et al., 1978), a visible light reflection method (Kobeda and Irene, 1986). Meng et al. (1993) measured the intrinsic stress of AlN during sputter deposition onto silicon substrates in an UHV chamber using a scanning laser beam reflection method.

Many investigators now use the commercially available automated laser probe systems (e.g., Flexus). 3D stress maps, outputs of multiple scans in a stress measurement system, are valuable for visualizing stress nonuniformities not apparent in individual scans (Blech and Robles, 1994). Automated stress measurements are an option with surface profilers made by, e.g., Veeco/Dektak.

The radius of curvature of a single-crystal substrate (usually a silicon wafer) on which a thin film has been deposited can also be measured by x-ray techniques. Both one- and two-crystal measurements have been used (Hearn, 1977). The one-crystal technique is the faster and simpler one but is less sensitive by an order of magnitude. A schematic of the two-crystal arrangement is shown in Fig. II-23 (Goldsmith et al., 1983).

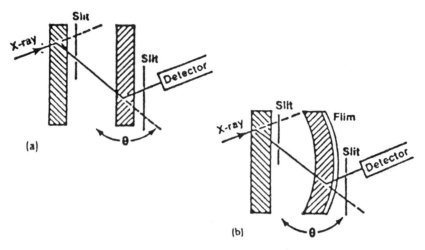

FIGURE II-23 Schematic of double crystal camera arrangement for measuring stress (a) with unstrained crystal, (b) with a strained crystal. (From Goldsmith et al., 1983)

In Fig. II-23a, both crystals are unstrained and each aligned to obtain, in transmission, Bragg diffraction from a set of planes. The crystals can be translated with no change in the x-ray intensity since the Bragg angle is constant across the crystals. In Fig. II-23b, the crystal in the second position is curved due to the stress generated by the thin film. The crystals are again aligned to obtain Bragg diffraction. But now translation results in the decrease of the x-ray intensity because of the curvature. The Bragg angle θ must be adjusted during translation to maintain maximum intensity. This adjustment is used to measure the radius of curvature of the crystal, R:

$$R = \frac{L}{\Delta \theta}$$

where L is the scan length.

Flinn (1989) has reviewed the principles and application of wafer curvature techniques for stress measurements in thin films.

9.2.3 Deflection Technique

Stress can also be determined by measuring the deflection, δ, of a circular disk at a distance r from the center of the disk, before and after deposition using, e.g., a noncontact fiber optic probe (Schaible and Glang, 1969), a Mikrokater thickness gauge (Choi and Hearn, 1984) and a wafer deflection (stress) gauge (Chen and Fatemi, 1986). In this last case the stress is given by:

$$\sigma \frac{\delta}{r^2} \left[\frac{E_s}{3(1-v)} \right] \left(\frac{t_s^2}{t_f} \right)$$

The assumptions for all these measurements are $t_s \gg t_f$, the substrate is

Characterization

linearly elastic, homogeneous, uniformly thick, thin compared to its radius, and the deflection small compared to the wafer thickness. It is also assumed that there is excellent adhesion between film and substrate.

9.2.4 Other Methods

X-ray diffraction techniques have been used to determine the stress in a thin film by measuring the strain in a single-crystal substrate induced by the deposition of a film upon it; the strain produces a change in the lattice parameters. For measurements made perpendicular to the film plane, the stress is computed from (Hoffman, 1966):

$$\sigma = \frac{E}{2v}\left(a_0 - \frac{a}{a_0}\right)$$

where a_0 and a are the lattice constants of the bulk and the strained crystal; in the film plane.

$$\sigma = \frac{(E/1 - v)(a - a_0)}{a_0}$$

Polycrystalline x-ray techniques have also been used for stress analysis of metal films, deposited on thermally oxidized silicon, by measuring the d-spacing of a single reflection for several orientations of a sample. The slope of d versus $\sin^2 \psi$ (ψ = tilt angle) determines the stress (Shute et al., 1989). Further discussion of these and other x-ray techniques can be found in, e.g., Norton (1968), Flinn and Waychunas (1988), and Vreeland et al. (1989).

A number of specialized stress measuring techniques have been devised. For example, Ku et al. (1991) monitored the stress as W was sputtered onto x-ray mask membranes, by measuring the resonant frequency of the membrane. The membrane was driven to vibrate using a sine wave applied to a concentraic ring underneath the membrane. The oscillations were detected by a fiber-optic sensor. Wu et al. (1994) described a microstress measuring apparatus based on polarized phase shifting and image processing which gives the stress distribution and surface deformation on an entire region, in a small region, and in any direction on a wafer. They report the minimum stress measured to be of the order of 10^6 dynes/cm^2.

9.3 TEMPERATURE DEPENDENCE

Stress is temperature dependent. In the absence of any chemical or structural changes in a film, the stress at any temperature δ_t is given by (Sunami et al., 1970, 1985):

$$\delta_t = \sigma_d + (T_t - T_d)\left[\frac{E_f}{(1 - v_f)}\right](\alpha_s - \alpha_f)$$

where σ_d is the film stress at the deposition temperature (intrinsic stress), α_s and α_f are the *average* values of the coefficient of thermal expansion (or TCE) of the substrate and film between temperatures T_d and T_t (since they are temperature dependent), and E_f and v_f are Young's modulus and Poisson's ratio for the film. This stress, due to the thermal mismatch between film and substrate, can result in cracking or loss of adhesion of the film when it is subjected to heat treatment.

The temperature dependence of stress is given by:

$$\frac{d\sigma}{dT} = \frac{E_f}{(1 - v_f)} (\alpha_s - \alpha_f)$$

The relative magnitudes of α_s and α_f can be determined from the sign of $d\sigma/dT$ (Smolinsky and Wendling, 1985), but the absolute value of α_f, however, can be obtained only when $E_f/(1 - v_f)$, is known. $E_f/(1 - v_f)$ is called the "composite elastic constant" (Brantley, 1973), "plane-strain modulus" (Fitch et al., 1989), or "elastic biaxial modulus" (Carlotti et al., 1996). Since stress = $E/(1 - v) \times$ strain, the modulus can be determined from the slope of the stress vs. strain curve.

Both this modulus and the thermal expansion coefficient have been determined in a variety of ways for many of the films used in device fabrication, such as silicon oxides (see below), nitrides (e.g., Tokuyam et al., 1967; Retajczyk and Sinha, 1980; Ambree et al., 1993), and silicides (Retajczyk and Sinha, 1980).

Jacodine and Schlegal (1966) used a beam for stress measurements and an "unsupported SiO_2 window as a balloon and measured the strain as a function of the air pressure inflating the balloon."

An easier technique is the determination of stress vs. temperature, i.e., $d\sigma/dT$, for films deposited on two different substrates; both α_f and $E/(1 - v)$ can be obtained by solving the two equations simultaneously since the value of α_s for each of the substrates is known. Sunami et al. (1970) and Bouchard et al. (1993) used Si and quartz; Blaauw (1983) and Ambree et al. (1993) Si and GaAs; and Retajczyk and Sinha (1980) Si and sapphire.

Determination of stress-strain curves for SiO_2 using IR spectra is based on the assumption that the Si—Si distance provides an *atomic* parameter for the strain and that stress relaxation occurs through changes of the Si—O—Si bond angle. The center frequency of the Si—O bond stretching vibration provides a measure of that angle (Flitch, 1989a). The strain obtained from IR spectra is *relative to a relaxed oxide*, and is calculated from the shifts in the IR bond-stretching frequency due to stress in the film (Nakamura et al., 1986; Fitch et al., 1989a,b,c; Ambree et al., 1993). The strain parameter (ε) is given by $(f - f_r)/f_r$, where f is the measured IR frequency of the Si—O bond in the stressed oxide and f_r is the frequency for the relaxed oxide, 1078.5 cm^{-1}, (Flitch et al., 1989a,b,c). Carlotti et al. (1996) pointed out that "this method is a probe of the local SiO_2 structure and is well adapted for thermally grown oxide but not for CVD oxides which contain various chemical bonds." Nevertheless, for thermally grown oxide, the value measured by IR (Fitch et al., 1989a) was significantly lower than that for silica, which it should resemble quite closely. Correcting for the thermal mismatch between Si and SiO_2 resulted in a somewhat better agreement (Flitch et al., 1989b); they stated that the discrepancy could be explained by the fact that modulus values reported for thin films are typically low by ~50% compared with bulk values (Flitch, 1989a).

Recently, the Brillouin light scattering technique, by which the elastic constants of the film can be measured, was used to determine separately the values of E and v from which $E/(1 - v)$ was calculated (Doucet and Carlotti, 1995; Carlotti et al.,

1996); their value for a thermally grown oxide film was in excellent agreement with that of bulk silica.

Table II-3b lists the value of $E/(1 - v)$ for several oxide films. The CVD and PECVD films studied have not been described completely by the various investigators; they may differ in their preparation methods and their properties; e.g., dopant concentration, silanol content, etc. may vary. Thus, differences among the results may be attributed to differences among the films as well as to the method of measurement.

The thermal mismatch stress is determined by the value of α_f relative to that of the substrate. There is an appreciable mismatch between Si and SiO_2; $\alpha(SiO_2) < \alpha Si$ so that the stress becomes more tensile as the films are heated. Silicon nitride (CVD and PECVD) is a better thermal match to Si than is the oxide.

TABLE II-3b Oxides: Value of $E/(1 - v)$

Oxide	$E/(1 - v) \times 10^{11}/cm^2$	Source/technique
CVD high rate	10	Sunami et al. (1970)/ 2 substrates
CVD low rate	7.5	
PSG; $P_2O_5/SiO_2 = 0.04$	9.8	
BSG; $B_2O_3/SiO_2 = 0.29$	16	
Thermally grown	4.7	Flitch et al. (1989c)/IR
PECVD (N_2O/SiH_4); annealed	10 ± 1	Ambree et al. (1993)/ 2 substrates
PECVD (N_2O/SiH_4); annealed	3 ± 1	Ambree et al. (1993)/IR
LPCVD	8.7 ± 0.5	Bouchard et al. (1993)/ 2 substrates
PECVD	8 ± 3	
LPCVD PSG (4 wt% P)	7 ± 1	
PECVD PSG (4 wt% P)	9 ± 2	
LPCVD BPSG (4 wt% P, 3 wt% B)	4 ± 1	
PECVD BPSG (4 wt% P, 3 wt% B)	6 ± 2	
PECVD	8.0	Doucet/Carlotti (1995)/ Brillouin scattering
APCVD	7.1	
PECVD PSG	7.4	
APCVD BPSG	7.1	
Siloxane SOG; annealed	1.9	
PECVD (SiH4)	8.0	Carlotti et al. (1996)/ Brillouin scattering
PECVD (TEOS)	8.5	
ECR	8.0	
Si-rich	7.7	
PSG	7.4	
Thermally grown	9.0	

By following the changes in stress while increasing and then decreasing the temperature, any hysteresis can be detected. When hysteresis occurs it indicates changes in the composition or structure of the film (e.g., Ramkumar et al., 1993).

Systems, e.g., Flexus, are available commercially in which stress can be measured, in a controlled ambient, as the temperature is cycled. Some of the other work reporting on the stress variations during thermal cycling can be found in, e.g., Shimbo and Matsuo (1983), Bushnan et al. (1990), and Wu and Rosler (1992).

10.0 THERMAL PROPERTIES
10.1 INTRODUCTION

The thermal coefficient of expansion (TCE), thermal conductivity, and thermal stability are the properties of interest.

10.2 COEFFICIENT OF THERMAL EXPANSION, TCE OR α_f

As mentioned in Section 9.3, differences in the TCE of a film and its substrate result in thermal stresses during processing and/or use and this may affect the yield of the process and the reliability of the structure. The value of TCE for a thin film may differ from that of the bulk material. The TCE of films has almost always been measured on coated substrates, as discussed in Section 9.3. Burkhardt and Marvel (1969), however, measured the TCE of a freely suspended sputtered silicon nitride film over a wide temperature range using a cathetometer to measure the distance between two incised marks. This method eliminated the need for a reference TCE.

The TCE of bulk samples has been determined by measuring the lattice expansion during heating by x-ray diffraction. The values of TCE for many materials used in semiconductor devices are known and are available in various handbooks; a comparison of some materials is given in Fig. II-24 (Brown, 1970).

10.2.1 Thermal Mismatch Stress

The stress due to thermal mismatch between a film and the substrate has already been covered (Section 9.3).

10.3 THERMAL CONDUCTIVITY

The thermal conductivity of metals is high, that of insulators is low. Figure II-25 (Brown, 1970) illustrates these differences.

The poor thermal conductivity of insulators is, therefore, a concern since the heat generated by the devices and by the I^2R losses of the metal interconnections during device operation must be dissipated through the insulator.

10.4 THERMAL STABILITY
10.4.1 Organic Films

The thermal stability of organic materials, the temperature above which the film degrades or decomposes, is one of the most important factors in determining the

Characterization

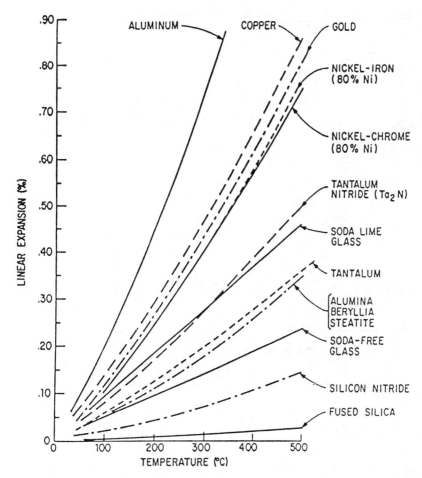

FIGURE II-24 Linear expansion as a function of temperature for various materials. (From Brown, 1970)

usefulness of such films as interlevel insulators. Thermal stability is measured using *thermogravimetric analysis* (*TGA*) in which the weight loss is monitored as the sample is heated at a constant rate in an inert ambient. Figure II-26a is an illustration of TGA.

A significant weight loss occurs above 450°C with catastrophic loss above 500°C. However, as shown in Fig. II-26b, if the sample is held at 450°C for extended times, there is insignificant weight loss.

Chemical and physical transformations in organic films are often detected by the following techniques.

In *differential thermal analysis* (*DTA*), the difference in temperature ΔT between a sample and a reference with the same characteristics is observed while both are being heated or cooled simultaneously. This technique should be able to detect any phenomenon that is accompanied by a change in enthalpy (heat content), i.e., ΔH. Polymerization, degree of cure, oxidation, cross-linking, polymer–polymer

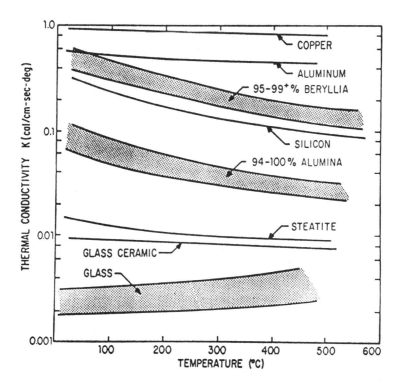

FIGURE II-25 Thermal conductivities of various materials as a function of temperature. (From Brown, 1970)

reactions, and thermal degradation are some of the chemical reactions studied by DTA.

Differential scanning calorimetry (DSC) is used in order to determine calorimetric data directly. In DSC, the temperatures of a sample and a reference are maintained at a fixed temperature as the temperature is changed; the variation in power required to maintain this equality in temperature during a transition is measured (Carroll, 1972). Carroll discusses these techniques in detail and demonstrates how to extract the information required from the experimental data.

Among the other techniques mentioned by Carroll are thermodilatometric analysis (TDA) in which the sample length is monitored during heating or cooling and is useful for determining the glass transition temperature and thermal volatilization analysis (TVA) used to study degradation of a polymer to volatile products by pressure measurements.

10.4.2 Inorganic Films

Instability has been a problem with some inorganic insulators as well. Some examples are the evolution of H_2 from PECVD SiH_xN_y (SiN) and H_2O from SOG, (doped and undoped) CVD and PECVD films during heating, usually during processing. The H in SiN_xH_y, as the formula indicates, is incorporated during depo-

Characterization **115**

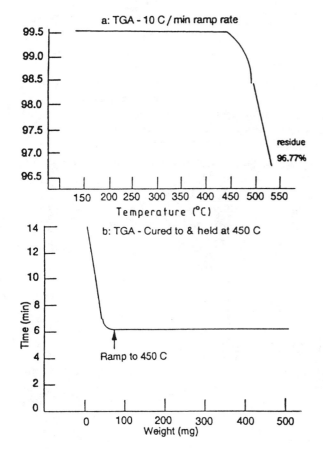

FIGURE II-26 Thermal gravimetric analyses (TGAs) of PI. (a) Percent weight loss vs. temperature (10°C/min ramp rate); (b) isothermal weight loss at elevated temperature (450°C).

sition; the H_2O in the various oxides appears to be due to absorption after deposition.

Annealing SiNxHy above the deposition temperature results in stress relaxation (Budhani et al., 1988; Kelm and Jungnickel, 1991), i.e., the compressive stress in the film is reduced and may in some cases become tensile, i.e., prone to crack (Claasen, 1987). Stein et al. (1979) have attributed stress relaxation to the loss of chemically bound H, i.e., breaking of Si—H and N—H bonds. On the other hand, Budhani et al. attribute it to the loss of unbonded, molecular H_2. Another manifestation of H evolution is blisters in the nitride films and void formation in the Al metallization beneath it (Peek and Wolters, 1986; Kikkawa et al., 1987). Diffusion of hydrogen from the SiNxHx into a gate oxide beneath it results in voltage shifts (Xie et al., 1989).

In the case of SOG, the manifestation of water evolution has been high via resistance (e.g., Wolters and Heesters, 1990) and metal blistering (Hirashita et al., 1990). Cox (1990) stated that moisture evolution from oxides poses significant reliability problems. Kato et al. (1992) reported that if oxides that absorbed water were not outgassed thoroughly, the grain growth of Al sputtered over such films was suppressed and hillock growth promoted. Other manifestations of instability are cracking, delamination, or reactions with underlying/overlying layers.

IR analysis is one way of tracking moisture content and evolution. Several other kinds of moisture analysis systems have been described. Tompkins and Tracy (1989) used an RGA (residual gas analyzer) to measure the partial pressure of water as the SOG film was heated by a lamp. Cox et al. (1990) describe a system built around a commercially available unit, the DuPont 902H Moisture Evolution Analyzer. The water desorbed from various oxides is swept by a carrier gas into an electrolytic cell of the detector where it is absorbed by anhydrous phosphoric acid and electrolyzed; the current is monitored and integrated to calculate the total amount of water electrolyzed. Kato et al. (1992) used a gas-conductance method to study desorption from SOG and several oxides: the pressures, P_1 and P_2, in two chambers separated by an orifice were measured. The effective outgasing rate Q (Torr l/s) is given by

$$Q = C_o (P_1 - P_2)$$

where C_o is the conductance of a calibrated orifice.

11.0 AUGER ELECTRON SPECTROSCOPY (AES)

11.1 INTRODUCTION

This is a *surface* analytical technique which can detect all elements heavier than He and is suitable for the *analysis of very small areas*.

11.2 PRINCIPLES

When an energetic beam of electrons or photons irradiates a surface, it dislodges an inner shell (core) electron. After the vacancy is created, an electron from an outer shell can replace the ejected electron (deexcitation) and a second characteristic electron, the "Auger" electron, is ejected:

$$e^- + M \longrightarrow M^{2+} + e^-_{Auger}$$

The *kinetic energy* of this electron is *characteristic* of the element. Although an electron beam ($\sim$1–10 keV) is usually used for excitation, the energy of the Auger electron is independent of the way the initial vacancy was created. The process of Auger emission is shown schematically in Fig. II-27.

11.3 THE AUGER SPECTRUM

The Auger spectrum is a plot of the number of electrons emitted, N (proportional to atomic concentration) vs. kinetic energy, E, but it is usually plotted as a derivative spectrum, dN versus E, to facilitate the identification of the Auger electron peaks and suppress the background of the inelastically scattered electrons from the primary beam. The derivative spectrum is produced by modulating the energy selected by the analyzer and using a lock-in amplifier to detect the signal (Bindell, 1988);

Characterization

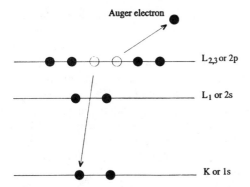

FIGURE II-27 Schematic diagram of Auger (AES) deexcitation processes. (From Mullenberg, 1979)

the difference between the two is shown in Fig. II-28. More recently, due to improvements in instrumentation, N versus E spectra have become satisfactory.

The Auger peaks due to the various *elemental constituents* of a film can be used for *compositional analysis*. There is also some contribution from the chemical state of the element since changes in the valence electrons influence the binding energy of the core electrons. Therefore the exact energy peak position and/or its shape may provide some chemical information; in general the interpretation is difficult. XPS (discussed below) is preferable for determining chemical shifts. Nev-

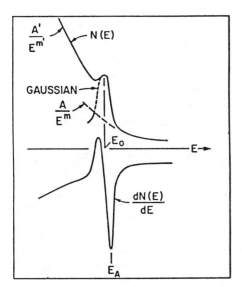

FIGURE II-28 A hypothetical AES spectrum, $N(E)$, containing a continuous background, A/E^m, a Gaussian peak and a low-energy step of form A'/E'^m. The energy of the most negative excursion of the derivative corresponds to steepest slope of $N(E)$.

ertheless, there has been some success in using AES for this purpose. For example, Madden (1981) analyzed the shape of the Si derivative signal from a thin film of PECVD silicon nitride and concluded that Si—Si, Si—H, and Si—N bonds were present. Wildman and Schwartz (1982), in a study of interfacial resistivity, determined the thickness of the surface oxide on Al from the ratio of the chemically shifted Auger peaks of Al^{+3} (Al_2O_3) to Al^0 (metal), a method first described by Chang and Boulin (1977). They used the information to calibrate the area under the oxygen curve in the depth profile in terms of thickness. In this way, the thickness of thinner-than-air-grown oxides coated in vacuo by an Al film to prevent further oxidation, could be determined by Auger profiling. High-resolution spectra showing this chemical shift are seen in Fig. II-29. In the top spectrum both peaks are evident, and the estimate of the Al_2O_3 thickness is given. In the bottom spectrum, the Al^0 peak has disappeared indicating that the surface oxide is very thick; the shift in the peak position due to charging is also apparent.

Although the electron beam can penetrate into the film, only the Auger electron produced in the top 10–30 A of the film contributes to the signal. This depth is called the escape depth and is the region from which electrons are ejected without loss of energy.

11.4 APPLICATIONS

One of the applications of AES is shown in Fig. II-30. The top spectra are of an Al surface and the bottom of a $TiAl_3$ surface; (a) shows the spectra of the surfaces as received (contaminated), and (b) the spectra after a brief sputter cleaning. The distortion due to charging is evident at the lower-energy part of the spectra of Fig. II-30a. Also seen are the plasmon loss peaks in the spectrum of the cleaned $AlTi_3$ surface; they are more prominent in the metallic state and their absence indicates the presence of an oxide.

An ion gun is usually an integral part of the apparatus so that compositional variations within a film can be detected and measured by alternating sputtering and probing the surface to produce a depth profile; this is usually plotted as the atomic % of an element vs. sputter etch time. However, during ion milling, an originally sharp interface becomes broadened due to surface roughness, ion-induced topography, angle of incidence effects, cascade mixing ("knock-on"), and ion-induced diffusion, as illustrated in Fig. II-31. Preferential sputtering and surface segregation are effects that must also be considered.

Despite the problems mentioned above, depth profiles are widely used and provide valuable information. Figure II-32 shows depth profiles for the Al and $AlTi_3$ samples of Fig. II-29a.

If the sputter etch rate were known accurately, the sputter time could be converted into depth. It should be noted that the depth scale in Fig. II-32 is given as the sputter equivalent of Å of SiO_2; since Al_2O_3 sputters more slowly than SiO_2, the thickness of the oxide layer is overestimated. However, the thickness comparison between the two samples is valid, i.e., there is significantly more oxide on the surface of the $AlTi_3$ sample than on the Al sample.

In addition, since the sputter etch rate depends on the composition of the film, so that, in general, it is not a trivial conversion from sputter time to thickness. It

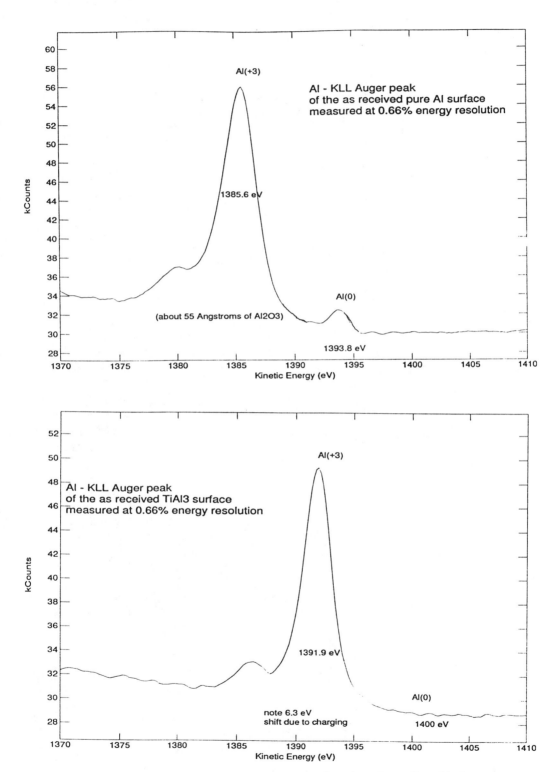

FIGURE II-29 High-resolution AES spectra of contaminated surfaces. *Top*, Al; *bottom*, AlTi$_3$. (H. S. Wildman, IBM Analytical Services)

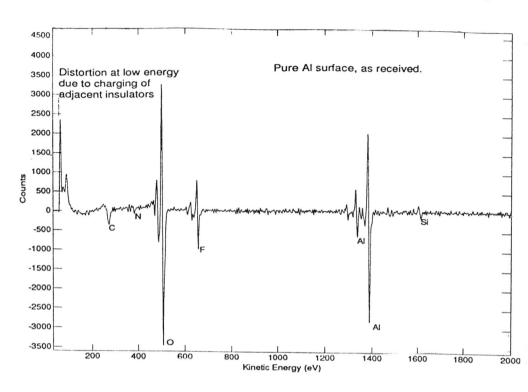
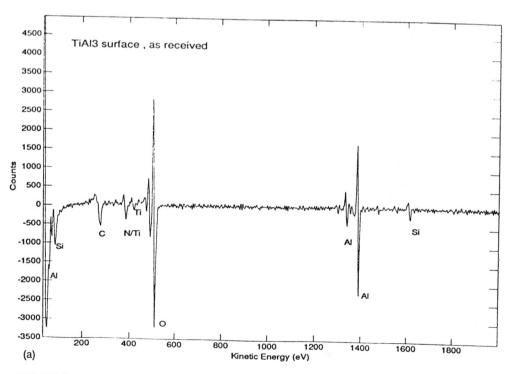

FIGURE II-30 AES spectra of Al (upper) and TiAl₃ (lower) surfaces, (a) as received, i.e., contaminated, and (b) after sputter cleaning. (H. S. Wildman, IBM Analytical Services)

Characterization

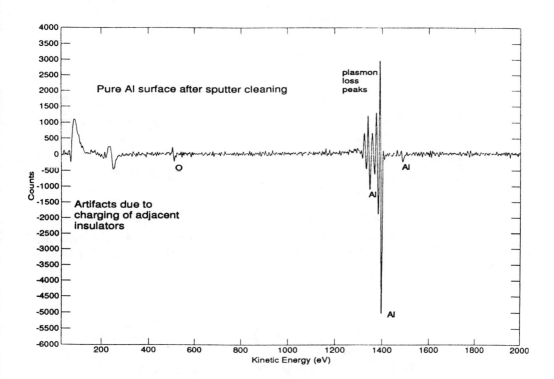

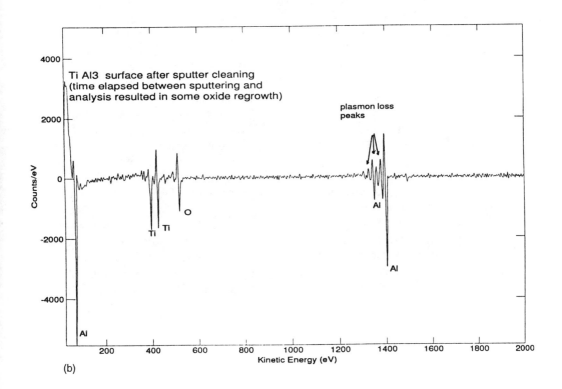

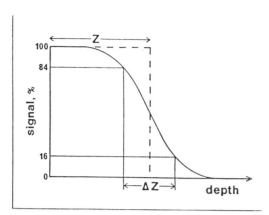

FIGURE II-31 Depth resolution in sputter-depth profiling; true profile indicated by dashed line. ΔZ corresponds to depth separation between points on profile corresponding to 84% and 16% of layer intensity.

is possible to ion mill a known standard to obtain an approximate time vs. depth scale for a given compound.

The sensitivity of AES depends on the relative Auger electron efficiencies of the various elements, on the primary beam energy and current, the equipment, and the inelastic electron background. Thompson et al. (1985) give the approximate sensitivity of AES as 10^{-10} g/cm² for the surface layer and the atomic fraction in the bulk as 10^{-3}.

The conversion of the number of electron emitted is complicated by the influence of the matrix on the backscattered electrons and escape depth. External standards are usually used; if the composition of the standard is close to that of the test sample, the elemental composition can be determined directly from the ratio of the Auger yields (Feldman and Mayer, 1986). The measured surface concentrations may not be identical to those in the bulk, however.

Additional advantages of AES are the ability to monitor several elements at once, high sensitivity particularly to low-mass impurities such as oxygen or carbon (common contaminants of surfaces and interfaces), rapid data collection, easily focused to small spots with the possibility of scanning AES (SAM) to produce "maps" of the surface composition, the ability to vary beam energy to operate under optimum conditions, and commercially available equipment (e.g., Perkin-Elmer, JEOL). Disadvantages include surface degradation, stimulated desorption, and charging of insulator surfaces by the electron beam (which was illustrated above) and the difficulty of quantitative analysis and interpretation of the Auger chemical shift. In addition, the equipment is expensive (Thompson et al., 1985).

Characterization

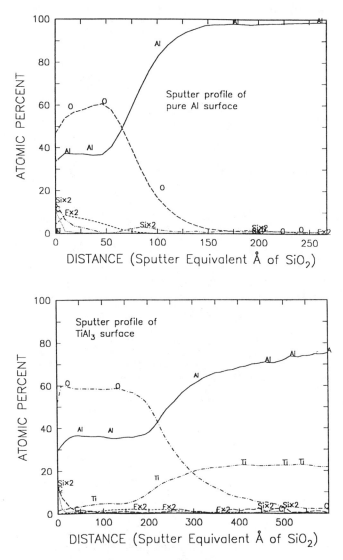

FIGURE II-32 AES depth profiles of the samples of Fig. II-29a; Al (*upper*), AlTi$_3$ (*lower*). (H. S. Wildman, IBM Analytical Services)

12.0 X-RAY PHOTOELECTRON SPECTROSCOPY (XPS)

12.1 INTRODUCTION

This is another *surface* analytical technique, analogous to AES, in that a flux of energy results in the ejection of electrons, of characteristic energy, from within a small escape depth.

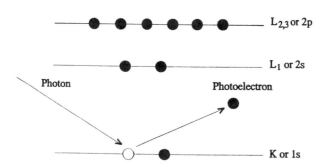

FIGURE II-33 Electronic transition involved on XPS. (From Mullenberg, 1979)

12.2 PRINCIPLES

In XPS, also called electron spectroscopy for chemical analysis (ESCA), a beam of low-energy x-rays (often the K-α line of Al, 1.487 keV) is used to probe the sample surface. All the photon energy is absorbed and interacts with the inner shell electrons, causing photoemission of an electron. The electronic transition involved in XPS is shown in Fig. II-33.

From the measured kinetic energy of the emitted electron (E_{kin}), its binding energy (E_B) can be calculated from the known photon energy ($h\upsilon$) and the difference in work function between the sample and the spectrometer ($\Delta\phi$):

$$E_K = h\upsilon - E_B - \Delta\phi$$

12.3 XPS SPECTRUM

An XPS spectrum is the plot of the number of electrons detected (proportional to the relative abundance of the species) vs. binding energy. Superimposed on the XPS spectrum are Auger transitions, as illustrated in Fig. II-34a, which shows the XPS spectrum of Cu.

The exact binding energy for an electron in a given element depends on the chemical environment of that element because the configuration of the valence electrons (due to chemical interaction) will influence the lower lying (core) electrons, changing their binding energies (Feldman and Mayer, 1986). Thus both the *element* and the *type of bonding* in the *surface* constituents can be determined. If several valence states of the element are present, they can be distinguished from each other. For example, as shown in Fig. II-34b, the silicon peak from silicon bonded to oxygen is shifted in energy from that in elemental silicon and the relative proportions of each can be determined.

12.4 APPLICATIONS

XPS is used extensively to detect and measure surface compounds or adsorbates and their alteration due to chemical and physical processes, e.g., oxidation of a metallic surface, removal of surface contaminants, etc. It has also been used to study the interfacial reactions during the deposition of metals on polyimide (Ho et

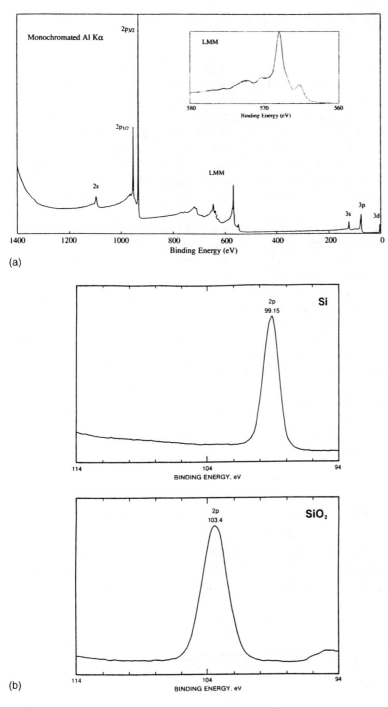

FIGURE II-34 (a) The XPS spectrum of Cu, showing superimposed Auger transitins. (b) Chemical shift in binding energy of the Si 2p line; *top*, elemental Si, *bottom*, Si in SiO_2. (From Shastain, 1992)

al., 1985). Depth profiling, by alternating ion milling and surface probing, is an option, as it is for AES (with the same concerns about interface broadening, etc.).

Angle-resolved XPS, i.e., reducing the angle of the detector relative to the surface of the sample, increases the surface sensitivity. This is illustrated in Fig. II-35, which is the XPS spectrum of a Si sample with a coating of approximately one monolayer of SiO_2.

Changing the electron emission angle varies the effective electron escape depth and thus makes possible depth profiling which is nondestructive and yields an absolute depth scale in terms of the escape depth. The technique is restricted to very thin layers, about three times the escape depth (about 50 Å). Changing the kinetic energy of the emitted electrons by varying the energy of the photon source also varies the escape depth and thus is an alternate method of depth profiling (Hoffman, 1990).

XPS has been used to analyze patterned samples as well as flat surfaces. The topography of a periodic structure together with angle-resolved XPS was used to cause geometrical shadowing for selective area analysis after RIE using the parameters of selective etching of SiO_2 over Si (Oehrlein et al., 1988). Matsuura et al. (1991) used XPS analysis of the sidewalls of polysilicon features etched in Cl_2 and Cl_2/N_2 plasmas to clarify the nature of sidewall protection. The photoelectrons were collected both parallel and perpendicular to the etched periodic structures.

The principle advantage of XPS is its ability to provide *chemical* bonding information. As in AES, neither hydrogen nor helium can be detected. There is less surface damage from the x-rays than from an electron beam and there are no charging effects; however, the spatial resolution is inferior to that of AES and scanning is not feasible. It is not a very sensitive technique; detection limits are about 0.1 to 1.0 at%.

A good reference for a more complete coverage of AES and XPS is *Practical Surface Analysis*, edited by Briggs and Seah (1990).

12.5 ULTRAVIOLET PHOTOELECTRON SPECTROSCOPY (UPS)

This is a technique related to XPS, but of no use for elemental analysis. In UPS the electrons are excited from the valence band by photons of much lower energy, e.g., 20–40 eV. However, Ho et al. (1985) used UPS to observe the valence states which are directly involved in bond formation at the interface between metals and polyimide.

13.0 SECONDARY ION MASS SPECTROSCOPY (SIMS)

13.1 INTRODUCTION

This is the *most sensitive* of the analytical techniques and is used frequently to determine the *very low concentrations* and *depth profiles* of dopants and or contaminants found in semiconductor devices, as well as of more abundant species. It can be used to detect and measure *all* elements. It is capable of high lateral resolution, and has excellent mass discrimination.

Characterization

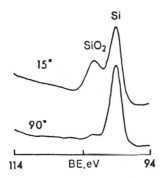

FIGURE II-35 Increase in sensitivity obtained by changing the angle of the XPS detector with respect to the sample. (From Mullenberg, 1979)

13.2 DESCRIPTION

In SIMS, a focused beam of ions is used to bombard the surface and sputter it; excited neutral species and singly or multiply charged ions (+ or −) are emitted. The secondary ions are detected and counted using an energy filter and a mass spectrometer; an electron multiplier increases the sensitivity. O_2^+ is the primary ion used most commonly for electronegative ions, Cs^+ for the negative ones. The relative abundance of the sputtered species indicates the composition of the layer being removed. If there is preferential sputtering, the surface concentration is rearranged so that the total yield gives the bulk composition. However, analysis of the surface layer at this point would not represent the bulk composition. The detection limit has been quoted as 10^{14} to $10^{17}/cm^3$ depending on the element; the sensitivity is greater for the lighter elements (Bindell, 1988). The use of electron-neutralizing beams has made possible the analysis of insulators. The influence of sputtering effects on the depth profile is the same as in AES and XPS, i.e., broadening of the interface, ion mixing, etc. For this reason, primary ions of the lowest possible energy are used; altering the incident ion angles can also improve the depth resolution.

The instrument is designed to detect only those ions emitted from the central portion of the crater. A schematic of the process is shown in Fig. II-36.

The secondary ion yields vary widely from element to element. They are also sensitive to matrix effects which influence the yield of secondary ions. For example, the secondary ion yield from Si could vary over 3 orders of magnitude depending on the oxygen concentration (Feldman and Mayer, 1986). Thus the variation in ion yield may not allow accurate determination of the relative concentration of the species in the film. Ion implanted samples are used as standards for analysis.

When the conditions are adjusted so that the sputtering rate is very slow, the process has been called static SIMS (SSIMS).

In contrast to AES and XPS, SIMS is capable of analyzing for and profiling hydrogen (Lundquist et al., 1982). There are, however, problems with detection of trace amounts. Magee and Botnick (1981) enumerated them and discussed ways of overcoming them so that H could be detected in Si < 50 ppm atomic.

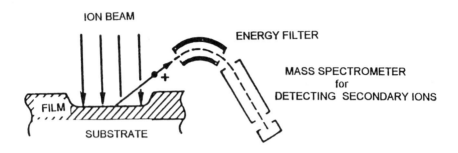
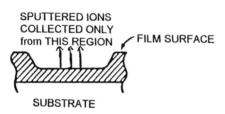

FIGURE II-36 Schematic of SIMS apparatus showing crater formation by incoming ions and detection of the secondary sputtered ions by a mass spectrometer. The lower figure illustrates the fact that the signal is collected only from the center of the crater.

13.3 RELATED METHODS

13.3.1 Secondary Neutral Mass Spectrometry (SNMS)

In this technique sputtered neutrals are used for compositional analysis. This avoids some matrix effects. The sensitivity of SNMS and SIMS for detecting low concentrations of impurities is about the same (detection limit ~1 ppm).

13.3.2 Laser Ionization Mass Spectrometry (LIMS) / Laser Microprobe Mass Analyzer (LMMA)

Another related technique is laser ionization mass spectrometry (LIMS) or laser microprobe mass analyzer (LMMA) in which a high-energy, finely focused laser pulse volatilizes and ionizes the region of interest. A time-of-flight mass spectrometer is used to identify the ions based on the mass-to-charge ratio. Molecular species can also be analyzed by reducing the laser power so that the species are desorbed without decomposition (Singer, 1986).

14.0 ELECTRON MICROPROBE

This technique is used for elemental analysis.

14.1 BASIS OF METHOD

Analysis by the electron microprobe is based on the detection and measurement of the characteristic x-rays produced when a material is excited by energetic electrons.

Characterization

Since the electron beam can be finely focused, the technique is capable of small-area analysis (~1 μm) and mapping; the sampling depth is of the same order so that x-rays generated in the substrate may interfere if films ≲1 μm thick are probed.

14.2 MODES OF OPERATION

There are two modes of operation depending on the type of detector used: energy dispersive (EDS) or wavelength dispersive (WDS). For EDS a Si(Li) detector, protected by a thin window, and kept at liquid nitrogen temperature, is used. Only $Z > 5$ elements can be detected. However, the radiation from all the elements in the sample is detected simultaneously. WDS involves x-ray diffraction from an analyzer crystal. The resolution and range of element detection ($Z \geq 4$) are better than in EDS, but the elements are detected sequentially, increasing the data collection time. Figure II-37 compares the results from the two modes of operation. According to Smith and Hinson (1986), the detectability limit is 50–1000 ppm, and the accuracy is 2–3%, with suitable standards (and with reasonable care). Figure II-38 shows a microprobe line scan (i.e., a plot of the concentration of each element in the sample vs. its position on the sample). This capability is a feature of some probes.

The analysis may be carried out either in a standalone system or in a scanning electron microscope (SEM) equipped with the proper detectors.

15.0 X-RAY FLUORESCENCE SPECTROMETRY (XRFS)

15.1 INTRODUCTION

XRFS is a method of *elemental analysis* in which identification is made by measuring the wavelength or energy of characteristic x-rays emitted from the atoms in a sample. The lowest atomic number element that can be measured by this technique is carbon (fair sensitivity) and even boron (> several %). The x-rays can be excited by electron irradiation, but this is an inefficient process with most of the energy converted into heat. There is the potential for vaporizing or melting the

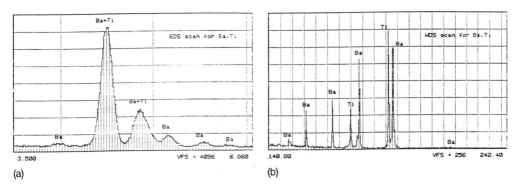

FIGURE II-37 Microprobe analyses of benitoite (a mineral used as a standard). (a) Energy dispersive x-ray spectrum (EDS). (b) Wavelength dispersive spectrum (WDS). (From Falcon, *IBM Analytical Services*)

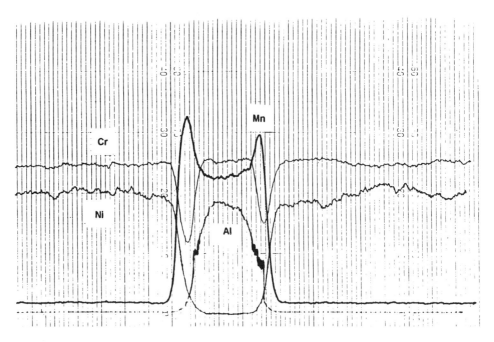

FIGURE II-38 A microprobe line scan (about 15 μm across) of a precipitate in stainless steel. (From Falcon, *IBM Analytical Services*)

samples. It is preferable, therefore, to use primary x-rays to excite the secondary characteristic x-ray spectrum.

15.2 SPECTROMETERS

Wavelength-dispersive spectrometers, most commonly employed, use the diffracting property of a single crystal to separate the polychromatic beam emitted by the specimen.

The energy-dispersive spectrometer uses a Si(Li) detector to give a spectrum of voltage pulses directly proportional to the spectrum of x-ray photon energies emitted; a multichannel analyzer collects and records the pulses according to their energies. XFR instruments of both kinds are available commercially.

There are several specialized instruments: total reflection XRF (TRXFR) for measuring trace components, synchrotron source XRF (SSXRF) for high resolution, and the proton excited XRF (PIXE) whose great sensitivity shortens analysis time and is used for trace elements (Jenkins, 1988).

15.3 MEASUREMENT OF CONCENTRATION

XRFS is very valuable for its ability to detect/identify and measure low concentrations of elements. Standardization is required and the specific procedures used depend on whether a single or multiple element sample is being analyzed. For the measurement of Ar in sputtered SiO_2, one standardization method was measuring the K_α and Cl_α of a KCl film of known thickness, since K and Cl bracket Ar, to infer the argon mass/argon K_α net counts (Lloyd, 1969). Background and matrix effects must be taken into account for any quantitative analysis. Solid samples can be analyzed directly, powders are fused, and solutions are usually concentrated before analysis. There has been an attempt to use an intensity concentration algorithm instead of standardization procedures but it does not appear to be used widely.

15.4 APPLICATIONS

Examples of the use of XRF are the measurement of the P concentration in PSG and BPSG films, (Grilletto, 1977; Levy et al., 1985; Madden et al., 1989) the argon content of sputtered SiO_2 films (Lloyd, 1969; Hoffmeister and Zuegel, 1969; Schwartz and Jones, 1970), the composition of alloys such as permalloy and AlCu, and the detection of trace contaminants. It was also used to analyze the deposit at points across a substrate to determine the relationship between the distribution of material on the substrate and the region of the multicomponent target from which it was sputtered (Schwartz et al., 1969). XRF is now used to measure film thickness more accurately and faster than stylus techniques (Dax, 1996). It is also possible to identify and measure the thickness of individual components of a multilayer film. Another more recent application has been the rapid detection of low concentrations (1–5 ppb) of transition metals in semiconductor processing chemicals (e.g., HF). The analysis, called "dried residue XRF," can be performed by relatively unskilled personnel right on the manufacturing line (Gupta et al., 1996).

16.0 HYDROGEN ANALYSIS

16.1 INTRODUCTION

The identification of H incorporated into SiO_2 and plasma nitride (SiNxHy) films by IR spectroscopy as Si—H, Si—OH, N—H has been mentioned. This section deals with the quantitative analysis of H and its depth distribution in the films; this information has been used to calibrate the IR bands.

16.2 RESONANT NUCLEAR REACTIONS

16.2.1 $^1H + ^{15}N$

The narrow isolated resonance in the nuclear reaction (Lanford et al., 1976)

$$^1H + {}^{15}N \longrightarrow {}^{12}C + {}^4He + \gamma \text{ ray (4.43 MeV)}$$

and was employed to determine the H content of PECVD SiNxHy films (Lanford and Rand, 1978). To carry out the measurement, the sample is bombarded with a beam of 6.385 MeV $^{15}N^{2+}$, since there is an appreciable probability for a reaction only at that energy. The number of γ rays emitted is proportional to the H-concentration at the surface. As the energy is increased, there is no further reaction at the surface but as the ions are slowed passing through the film, resonance will occur at some depth, and the yield of γ rays is proportional to the H concentration at that depth. By measuring the γ-ray yield versus ^{15}N energy, a profile of H concentration vs. depth is obtained. The depth resolution is 5–10 nm and the sensitivity is better than 1 part in 1000 or $\sim 2 \times 10^{19}/cm^3$. The H content of the plasma nitride films studied was in the range of 20–25 at%. They were able to calibrate the IR absorption bands corresponding to Si—H (2160 cm^{-1}) and N—H (3350 cm^{-1}) and determined that the Si—H band had about 1.4 times the absorptivity of the N—H band. Another study of plasma nitrides covered films containing a wider range of H content, 4–39 at% (Chow et al., 1982). Xie et al. (1989) extended the measurement to the comparison of LPCVD and PECVD silicon nitrides. Xie et al. (1988) used the technique to determine the H concentration profiles in as-deposited and annealed CVD PSG.

16.2.2 $^1H(^{19}F, \alpha\gamma)^{16}O$

Another resonant nuclear reaction for measuring H-concentration profiles is

$$^1H(^{19}F, \alpha\gamma)^{16}O$$

which shows a strong resonance at 0.83 MeV (Leich and Tombrello, 1972). According to Lanford et al. (1976), it has a better yield but a poorer depth resolution than the $^1H + {}^{15}N$ reaction ($\sim$200 A) and has a limited range ($\sim$0.4 μm).

16.2.3 $^1H(^{15}N, \alpha\gamma)^{12}C$

Leich and Tombrello (1972) also mention another resonance which may be useful for measuring proton profiles:

$$^1H(^{15}N, \alpha\gamma)^{12}C$$

which has a strong resonance 0.40 MeV; the estimated resolution is about 50 A, and the maximum depth $\sim$3 μm.

16.3 PROTON–PROTON SCATTERING

Still another method for quantitative determination of H is proton–proton (p-p) scattering (Cohen et al., 1972). It was used to measure H profiles in PECVD nitride (Paduschek and Eichinger, 1980) who preferred it to the $^1H(^{19}F, \alpha\gamma)^{16}O$ reaction because a tandem accelerator was necessary to generate a sufficient yield of 19F ions with the required high energy (16–18 MeV). This p-p scattering method has sensitivity in the parts/million range, but at the expense of some loss in depth resolution. In addition, it requires thin samples since it is a transmission technique. This method was used to measure H concentration in PECVD nitride films, using a standard mylar film for calibration.

16.4 FORWARD-SCATTERING ELASTIC RECOIL DETECTION (ERD)

This method (Doyle and Peercy, 1979), which is related to Rutherford backscattering spectroscopy (which will be discussed below), has also been employed. Gujrathi and Bultena (1992) and Godet et al. (1992) coupled ERD with the time-of-flight (TOF) detection system for better depth resolution and increased depth range.

Characterization 133

A beam of 2.5 MeV He$^+$ impinges on the sample; the recoiled H atoms and forward scattered He$^+$ appear in the same direction; the He$^+$ are absorbed in a Mylar film and the H atoms detected and measured. EDR has been used to measure the H content of PECVD silicon nitride (Takahashi et al., 1987) and of LPCVD silicon oxynitride films (Kuiper et al., 1988).

16.5 OTHER TECHNIQUES

Secondary ion mass spectrometry (SIMS), which was discussed above, has been used for H analysis and profiling. Mass spectrometry has also been used for the quantitative analysis of the H content of plasma silicon nitride (Yoshimi et al., 1980).

Recently, nuclear magnetic resonance (NMR), with a detection limit of 10^{19} H/cm^3, has been used to measure H incorporated during the growth of an SiO$_2$ film and that resulting from contamination of the surface during processing (Levy and Gleason, 1993).

17.0 RUTHERFORD BACKSCATTERING SPECTROMETRY (RBS)

17.1 INTRODUCTION

RBS yields quantitative information about *elemental composition* and *quantitative depth profiles* and is used in the analysis of both metals and dielectrics.

17.2 PRINCIPLES/DESCRIPTION

The basis for the measurement is the scattering of impinging ^{4}He ions by atoms (of mass > mass of ^{4}He) in a solid. It is inherently a quantitative technique since the scattering cross sections and He stopping cross sections in all elements required for the analysis are known. Another advantage is the speed of data collection. The disadvantage is the limitation in the detection of light elements; this can be overcome by using a low mass substrate such as C. Another weakness is the lack of specificity; two elements of similar mass cannot be distinguished if they appear together in a sample. In addition, the sample must be uniform laterally and in depth (Chu et al., 1978).

When a collimated beam of monoenergetic high-energy (MeV range) doubly ionized ^{4}He strikes a sample, a small fraction of the ions undergoes elastic collisions with the atoms in the sample. The large coulomb repulsion forces between them result in elastic scattering of the ^{4}He ion, as shown in Fig. II-39. The energy of the scattered ^{4}He ion identifies the atom on the *surface* of the sample since $E_1 = KE_0$, where E_1 is the energy after scattering by surface atoms, E_0 is the energy of the bombarding particle, and K is called the *kinematic factor*; K has been tabulated for ^{4}He ions scattered at various angles for all elements. An example is given in Fig. II-40.

The probability that a collision will result in a scattered particle is given by the *scattering cross section*, which is proportional Z^2, where Z is the atomic number. Thus, high Z atoms are detected with greater sensitivity than low Z atoms. However, high Z atoms are more difficult to distinguish from each other. These

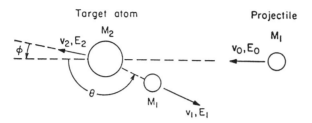

FIGURE II-39 Schematic representation of an elastic collision between projectile of mass M_1 and velocity v_0 and energy E_0 and target mass M_2 initially at rest. After collision, projectile and target mass have velocities v_1, E_1 and v_2, E_2. (From Chu et al., 1978)

points are illustrated in Fig. II-41 (Magee and Hewitt, 1986), which shows expected energy and yields for various elements.

The use of high-energy resonance back scattering (Li et al., 1995), i.e., the use of incident beams for which elastic scattering is resonant, increases the sensitivity for light atoms such as O, C, and N by enhancing the elastic-scattering cross section.

A particle loses energy as it travels *through* a solid, losing energy before scattering occurs from the back surface and thence to the detector, as illustrated in Fig. II-42. The loss of energy gives depth information. The values of energy loss through matter are given in tables of *stopping cross sections*. Therefore, the differences in energy can be quantitatively related to the thickness of the film.

Since a multichannel analyzer is used to detect the scattered particles, several elements can be detected simultaneously. This is shown in the diagrammatic example of an RBS of a film which contains equal numbers of a heavy atom M and

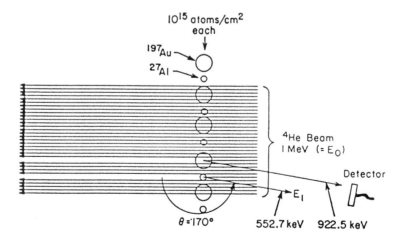

FIGURE II-40 The kinetic factor K_M gives ratio of energy after (E_1) and before (E_0) an elastic collision of the projectile (^{4}He) with atom of Au (197 amu) and Al (27 amu). (From Chu et al. 1978)

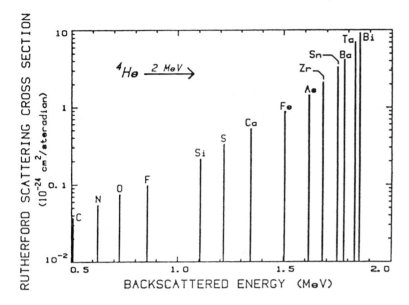

FIGURE II-41 Graphical representation of how both Rutherford scattering cross section (σ) and surface backscattering energy (E_1) vary for elements across the periodic table. (From Magee and Hewitt, 1986)

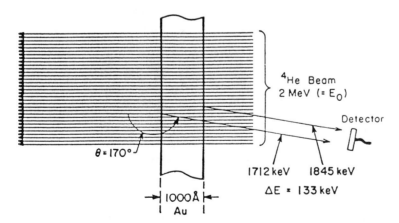

FIGURE II-42 Energy loss as particle passes through dense medium; particle scattered at rear surface of film has less energy when detected that particle scattered at front surface. (From Chu et al., 1978)

a light atom n (Fig. II-43). Although the number of atoms of each kind is the same, the yield of the heavier atom is greater than that of the lighter one. The signal due to the light mass is at a lower energy than that of the heavier one.

17.3 APPLICATIONS

Among the applications of RBS are (1) the determination of the composition and thickness of metallic, (2) dielectric films, (3) expitaxial layers, and (4) surface

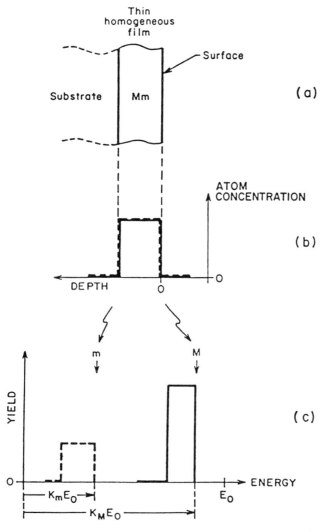

FIGURE II-43 Translation of concentration profiles to signals in backscattering spectrum from thin homogeneous film of binary compound Mm with elements of a heavy M and a light m atomic mass. In lower diagram profiles appear as two separate signals. Light mass → low energy/yield, heavy mass → high energy/yield. (From Chu et al., 1978)

Characterization

impurities, (5) the study of thin-film reactions of all kinds (e.g., silicide formation, oxidation, interdiffusion), and (6) implant profiles. Figure II-43 is a diagrammatic illustration of (1). Experimental spectra are shown in Figs. II-44 and II-45. Figure II-44 is an illustration of (4), the spectrum of an Al film contaminated with F (Chu and Schwartz, 1976). Figure II-45 illustrates (5), Fig. II-45a shows the formation of Ni2Si (Tu et al., 1975), and Fig. II-45b the oxidation of tantalum silicide (Gomez-SanRoman et al., 1995).

For a more complete discussion of RBS, the reader is referred to Chu et al. (1978).

17.4 FORWARD RECOIL SCATTERING

Forward recoil scattering, FRES or ERD, described in the section on H analysis, is a relative of RBS and can be performed in an RBS chamber. The method can be used for any atoms lighter than the projectile.

18.0 SCANNING ELECTRON MICROSCOPE (SEM)

18.1 INTRODUCTION

The SEM may be the most widely used of the modern analytical instruments since there are microscopes designed to be installed in the fabrication facility and used by the personnel there, for inspection and measurement. There are, in addition, instruments designed for highly skilled microscopists. The SEM is used to study the surfaces of thin films as well as the structures fabricated by the various pro-

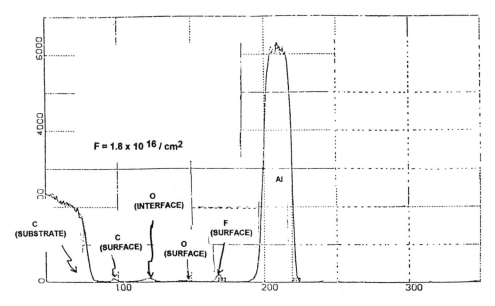

FIGURE II-44 Backscattering spectrum for 2.0-MeV ^{4}He$^+$ incident on Al (deposited on a C substrate instead of oxidized Si to eliminate interferences) after exposure to a CF_4 plasma during overetch of an overlying SiO_2 film.

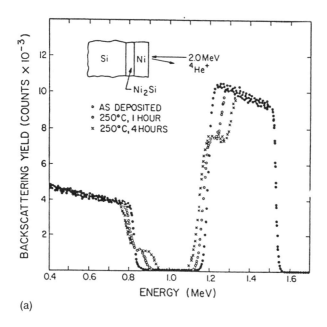

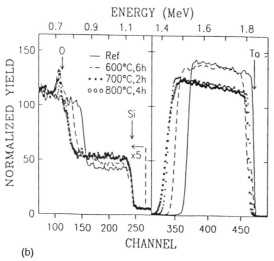

FIGURE II-45 (a) Backscattering spectra for 2.0-MeV $^4He^+$ incident on a multilayer sample of Ni on Si, before and after heat treatment; it shows the formation and growth of a Ni_2Si layer at the Ni—Si interface, forming an intermediate layer between the Ni and Si, as indicated by the shoulders developed in the Ni and Si portions of the spectra. (From Tu et al., 1975) (b) Backscattering spectra for 2.5-MeV $^4He^+$ incident on $TaSi_{1.4}$/Si samples showing the extent of plasma oxidation and migration of Si resulting from heating at different temperatures for different times. The reference spectra indicate the theoretical surface energy positions. (From Gomez-SanRoman et al., 1995)

cesses discussed in this book. Some of the newer models allow for extensive examination of entire wafers, but in many cases only small pieces can be accommodated.

18.2 APPLICATION

For many, but not all, applications it has completely superseded the optical microscope because the small dimensions of the semiconductor devices require much higher magnification. The SEM offers not only greater resolution and depth of field, but also may have the ability to perform some analytical functions, such as EDS and WDS (described above). With some modifications of the SEM, electron-induced current (EBIC) and voltage (EBIV) may be observed to study semiconductor devices. Many of the smaller SEMs do not perform well at the low accelerating voltage required to eliminate image distortion due to charging of insulated regions; therefore samples are coated with a thin conducting layer. This may hide buried features, e.g., voids in a metal coated with an insulator which are seen easily in the optical microscope.

18.3 OPERATION

The electron-specimen interactions are shown in Fig. II-46 (Wells et al., 1974). Figure II-47 is a diagram of an SEM. In an SEM an electron gun generates a narrow electron beam with a high brightness. W or LaB_6 (for thermonic emission) or a field emission gun (for high brightness and long operating life) are used. As the beam is rastered across the sample surface, secondary electrons are emitted from the surface (from the escape depth) and create the image displayed on the

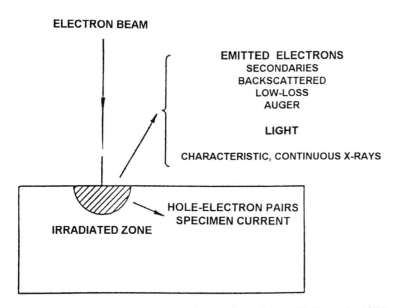

FIGURE II-46 Electron-specimen interactions. (From Wells et al., 1974)

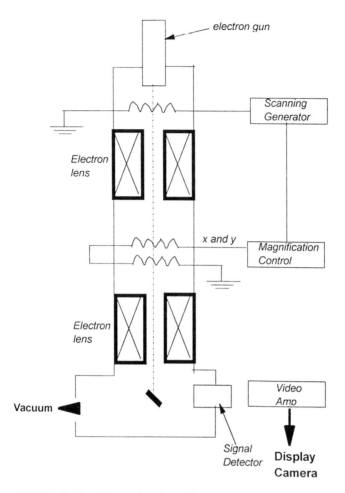

FIGURE II-47 Schematic of an SEM.

CRT; this image is the one used most widely. Tilting the sample increases the secondary electron signal; this is illustrated in Fig. II-48.

More complete topographical information can be gleaned from tilted samples, but this requires a large sample chamber if entire wafers are to be examined.

Backscattered primary electrons originate from a much larger volume within the depth of the sample. In this mode, the higher-density regions appear brighter, so it contributes additional information, e.g., observation of Cu segregation to the grain boundaries in the AlCu alloy film.

18.4 EXAMINATION OF CROSS SECTIONS

Examination of cross sections is one of the most valuable SEM operations. Both cleaved and polished sections are used, but artifacts can occur more readily using

Characterization

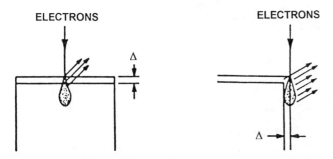

FIGURE II-48 Enhancement effect of specimen tilt on secondary electron emission in an SEM (Δ = escape depth).

the more easily prepared cleaved sections. Often the samples are decorated with an appropriate etchant to reveal features such as growth seams and grain size, but the fact that defects are often exaggerated must be kept in mind. It is often difficult to isolate specific structures when they are not periodic; thus time-consuming polishing may be necessary. A more complete description and discussion of scanning electron microscopy can be found in references such as Wells et al. (1974).

19.0 TRANSMISSION ELECTRON MICROSCOPE (TEM)

The TEM is not used as widely, or as easily, as SEM; it is largely a research instrument but at times it is the only technique capable of solving structure and processing problems encountered in device fabrication.

19.1 OPERATION

A diagrammatic representation of a TEM, equipped for x-ray and energy-loss spectroscopy as well, is shown in Fig. II-49 (Reimer, 1984). TEM uses a high-energy electron beam which impinges upon and through a very thin sample; the entire imaged specimen area is illuminated simultaneously. Electron diffraction patterns are produced, so that structural and chemical information can be obtained. Both flat (plan-view) and vertical (cross-section) samples can be examined.

However, unless the very large (and rare) very high voltage TEM is used, only very thin samples can be examined; this requires tedious and lengthy sample preparation and mounting. Mechanical or electrochemical polishing, chemical etching, and ion milling are methods used to attain the desired thickness.

19.2 SCANNING TEM (STEM)

In the STEM, the specimen is scanned in a raster, point by point, with a small electron probe.

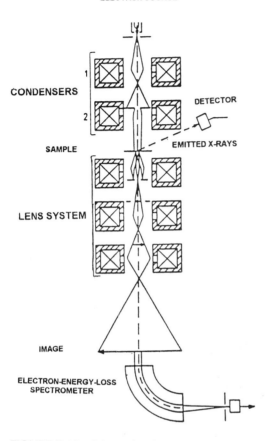

FIGURE II-49 Schematic of a TEM. (From Reimer, 1984)

20.0 FOCUSED ION BEAM (FIB)
20.1 INTRODUCTION

Ion beams, focused to submicron dimensions (0.04–0.1 μm), have been used in many ways in the development of IC fabrication processes. Commercial FIB equipment is available (Seiko Instruments, Inc.; JEOL).

20.2 DESCRIPTION

An FIB system consists of an ion source (a liquid metal) and an ion optical column and sample stage. It resembles an e-beam exposure system, with ions replacing electrons. The FIB system can deliver the desired dose with 0.1-μm accuracy, aligned to existing features on a wafer. Ion milling with inert ions, reactive-ion-induced deposition and etching for mask and circuit repair are done with elemental sources (Melngailis, 1987, 1988). FIB deposition has been modeled by Overwijk and van den Heuvel (1993). To test their model, they used the deposition of W (from tungsten hexacarbonyl directed at the surface through a nozzle close to it) and a raster-scanned Ga^+ FIB source. As might be expected, redeposition and loss of sputtered material is important.

Alloy sources are used, with mass separation (since several species of ions are sometimes emitted by the source) for implantation without masking. Thus FIB can be used for process optimization by implanting neighboring devices, which may have different geometries, with different dopant doses and for dose gradation within a device. Another possible use is in for lithography (replacing e-beam exposure) since there is little or no proximity effect and writing speed is higher (Melngailis, 1988).

20.3 APPLICATIONS

Nikawa (1991) has reviewed the application of the FIB techniques mentioned above, to failure analysis. FIB is used to prepare maskless, "clean," accurate microscopic cross sections as well as multiple cross sections in a very small area. Secondary electrons are emitted as a result of ion impingement, making it possible to view the cross section in situ (by tilting the sample stage) using scanning ion microscopy (SIM). Other applications are preparation for further failure analysis by changing a circuit (cutting and connecting metal lines), making holes in a dielectric layer for electron beam probing, depositing probe pads, and marking for electron spectroscopy observation. A recent application is micromachining for TEM specimen preparation (Szot et al., 1992). Nikawa et al. (1989) have listed the advantages of FIB vs. SEM and TEM for cross-section examination.

An important application of the SIM function of the FIB is the observation of aluminum microstructure and thus grain size distributions (Nikawa et al., 1989; Nikawa, 1991; Pramanik and Glanville, 1990). In the SIM image, the strong contrast reflects the crystallographic orientation of each grain; this has been confirmed by comparison with TEM images. Thus the true microstructure can be obtained without the extensive sample preparation required for cross-sectional SEM or for TEM examination which are the alternatives, since what appear to be grain boundaries upon optical and SEM observation of an aluminum surface are often merely surface shapes since the surface is covered by the native oxide.

Some examples of the potential of FIB for fabrication of very small features for device fabrication have been reported. In one, in which 100-nm-wide refractory metal lines were formed on Si, Koshida et al. (1990) exposed MoO_3 (a high-contrast negative resist for exposure to Ga^+ FIB), developed after exposure in alkaline solution, and reduced to Mo in H_2 at 800°C. The limiting resolution was determined by the FIB diameter. Yasuoka et al. (1990) fabricated microcontact holes in the SiO_2 insulating the Ge substrate.

21.0 THERMAL WAVE MODULATED OPTICAL REFLECTANCE IMAGING (TW)

21.1 INTRODUCTION

The technique is often called simply thermal wave imaging. Its principle use is the detection of voids within a metallic conductor which is beneath an overlying dielectric layer.

21.2 DESCRIPTION

A small (0.8 μm) region of a sample is irradiated by a modulated Ar ion laser beam; the light is partially absorbed by the substrate. Periodic heat waves flow into the sample. If there are voids in the conductor, the heat flow is impeded so that the temperature of that region is higher than the regions which are void-free. Optical reflectance is a function of the surface temperature and it is the modulated reflectance of the irradiated area that is measured using a HeNe laser probe. The laser beams remain stationary, and the sample rastered beneath them (Smith, et al., 1990a).

21.3 APPLICATION

The technique is very often used to detect stress voids and can also be used to detect electromigration voids, microcracks, and precipitates in Al alloy metallization (Smith et al., 1990b) as well. The technique has been combined with surface imaging by laser deflection, which is sensitive to surface topography, to detect and measure roughness, hillocks, and scratches (Smith, 1991).

The technique is nondestructive (does not require removal of the dielectric film), rapid, and can detect defects smaller than the probe size. A commercially available apparatus is available (Therma-Wave, Inc.).

Although high energy, backscattered electron imaging, using an STEM at 120 kV (or above), has been also used to detect small (~0.1 μm) voids in Al metallization beneath overlayers, as does TW. However, only small samples can be used (Follstaedt et al., 1991). In addition the instrument is much more complex than the TW apparatus and probably requires much more expertise in operation.

22.0 X-RAY DIFFRACTION (XRD)

22.1 INTRODUCTION

XRD is a method for determining the *arrangement of atoms* in a substance. Since the wavelength of x-rays is comparable to atomic spacings, diffraction can occur. X-ray diffraction is usually used during *process development*.

22.2 APPLICATIONS

Among the applications are identifying the components of a thin film, e.g., an intermetallic phase in deposited or annealed alloy films ($CuAl_2$ in AlCu) or as

Characterization

the product of reaction between adjacent films during heat treatment (Ti/Al ⟶ TiAl$_3$) or identifying which silicide is formed when a metal reacts with the substrate during heat treatment. Other uses are distinguishing between a single crystal or polycrystalline structure, between polycrystalline and amorphous structures, and determining texture (preferred orientation) in polycrystalline films. X-ray methods are also used in stress measurement.

22.3 DIFFRACTION PATTERN

An x-ray diffraction pattern consists of a series of intensity peaks vs. angle; each peak corresponds to specific atomic spacings (d). Bragg's law shows the relationship between the angle measured for each peak and the corresponding spacing:

$$n\lambda = 2d \sin \theta$$

where n is an integer (constructive interference), λ the wavelength of the radiation, d is the spacing between adjacent planes in a crystal, and θ is the angle which the primary and diffractive beam make with the plane; θ is called the Bragg angle. The diffraction pattern is a characteristic of a material and therefore can be used as a basis for identification.

The powder method is used most frequently; it depends on the diffraction of a collimated monochromatic beam from many randomly oriented crystallites. The earlier method recorded the spectrum on film. Today, the powder pattern is recorded using a counter diffractometer shown in Fig. II-50. "There are various methods for detecting the radiation: Geiger counters, gas proportional counter, or crystal scintillation counters. During the measurement, the counter turns at an angular velocity ω and the sample turns at a velocity of $\omega/2$ to maintain focusing conditions and a simple absorption geometry. The 2θ values for each reflection are read directly from such a recording" (Warren, 1990). Each line, in its position and intensity, represents a set of crystal planes, unique to the material. Identification is made by matching the powder pattern with one in the Powder Diffraction File. Every component in a mixture can be identified since each has a characteristic pattern, not altered by the presence of others.

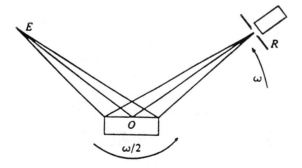

FIGURE II-50 Schematic representation of the counter diffractometer as used for recording powder patterns. (From Warren, 1990)

A diffractometer recording of the powder pattern of plated Cu is shown in Fig. II-51.

Detailed information about the fundamentals of x-ray crystallography, equipment, etc. can be found in various texts, e.g., Brown (1966), Wormald (1973), and Warren (1990).

23.0 WET CHEMICAL METHODS

These may be divided into characterization techniques and analytical techniques.

23.1 CHARACTERIZATION

The evaluation and characterization of SiO_2 and plasma silicon nitride thin films is commonly carried out in HF-based solutions. One of these is the so-called preferential or P etch, 15 parts of 49% HF, 10 parts of 70% nitric acid, and 300 parts water (Pliskin and Gnall, 1964), although dilute HF and buffered HF (NH_4F/HF mixtures of varying ratios) solutions are also used. Often the ratio of the etch rate of the oxide film in question is compared to that of thermally grown SiO_2. The etch rate (or etch rate ratio) is an indication of composition, bonding, and density, but *must* be used in conjunction with other tests to avoid erroneous conclusions. For example, increased density and excess silicon decrease the etch rate, whereas bond strain increases it (Pliskin, 1970); without other knowledge, one could not

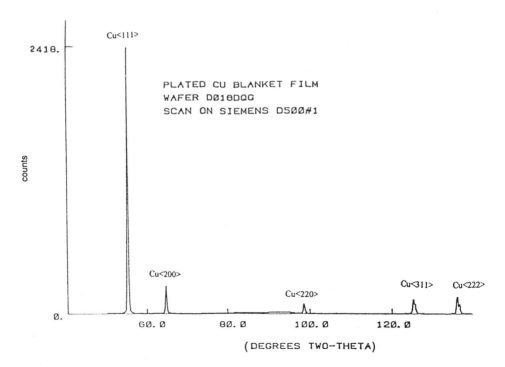

FIGURE II-51 Powder pattern of a plated Cu film. (Courtesy of IBM)

conclude which properties were responsible for the measured rate. The P etch rate of PSGs tracks the phosphorous content. The etch rate of plasma nitrides in HF-based etchants correlates well with their H content which in turn influences the stress, density, and electrical properties of the films.

The etch rate in an alkaline solution has been used to characterize the completeness of the imidization reaction in curing polyimides (Ginsburg and Susko, 1984).

23.2 ANALYSIS

The wet portion of the analysis, these days, consists of the dissolution of a sample and when necessary, preparation for further instrumental analysis. The instrumental methods are discussed in another section of the chapter.

One use of this technique has been in the analysis of the P content of PSG films. In two of the methods, all the phosphorous is converted to orthophosphate and in one case, reacted to form a colored solution which is analyzed by spectrophotometry (Hughes and Wonsidler, 1987) and in another, coprecipitated as beryllium phosphate and counted by x-ray fluorescence (Adams and Murarka, 1979). The solutions have also been analyzed, without any further treatment, using ion chromatography of both phosphate and phosphite ions (Houskova et al., 1985) or of orthophosphate ion (Hughes and Wonsidler, 1987), or inductively coupled plasma-atomic emission spectroscopy (Levy and Kometani, 1987). Houskova et al. (1985) reported that the ion chromatography method did not account for all the P in the sample (compared to colorimetry).

These same methods have been used to determine the B content of BPSG and BSG.

24.0 CHROMATOGRAPHY

This is a technique for the separation and identification of components in a mixture using the interaction between the species in the mixture (the mobile phase) and a stationary material. The mobile phase may be a gas or a liquid. The liquid may be nonpolar or polar (ionic). Most often, the stationary phase is a resin, packed in a column, but there are other chromatographic methods (e.g., Miller, 1975). Separation of ionic species by this method is called ion chromatography (IC). It has been used for analyzing processing chemicals, processing residues, and materials incorporated into multilevel metal structures. They must be converted into a form suitable for IC; this may require dissolution, extraction from a surface, or concentration. The "unknown" is introduced at the top of the column followed by the "eluent" (an ionic solution) and both migrate through the column.

Both suppressor and nonsuppressor ion chromatographs are available commercially; a description of their operation can be found in Smith and Chang (1983). The affinity of the column material for the species in question determines the time it takes for each to reach the end of the column. The resulting chromatogram is a series of peaks, one for each component, separated by time intervals. The emerging species are identified and their concentrations measured by mass spectrometry, photometry conductivity, absorption of radiation, etc.

25.0 OTHER ANALYTICAL TECHNIQUES

25.1 NEUTRON ACTIVATION

An example of this technique is the determination of the P content of PSG. The films were irradiated using a thermal neutron flux and the beta activity from P^{32} counted, using P-doped Si as a reference.

25.2 INDUCTIVELY COUPLED PLASMA ATOMIC EMISSION SPECTROSCOPY (ICP-AES)

In this technique, the material to be analyzed is vaporized or an aerosol solution of it is injected into a plasma and a spectrometer used to measure the peak intensities of the atomic emission lines in the plasma. The intensities are calibrated using standard solutions of the elements being analyzed. The method was improved by the use of internal standards (Cargo and Hughes, 1989).

26.0 THERMOMETRY

26.1 INTRODUCTION

The temperature of the wafer during processing is a critical parameter. In reactive-plasma-assisted etching it determines the etch rate and profile of both the substrate and the mask; in addition, too high a temperature causes mask distortion in any plasma-assisted etching process. In deposition, by CVD, PECVD, sputtering, temperature influences the accumulation rate and at times the reaction mechanism and stop coverage. Film properties such as structure, chemical reactivity (e.g., etch rate), stress, optical and electrical properties, and in many cases contaminant incorporation are a function of the temperature during deposition.

Many processes are carried out at low pressures where thermal conduction is poor. Thus, unless a suitable heat transfer medium is interposed between the wafer and its holder, there can be a substantial difference in temperature between them so that measuring the temperature of the holder is meaningless, making wafer temperature measurements necessary.

26.2 MEASUREMENT

Many devices for measuring wafer temperature have been suggested.

26.2.1 Thermocouples

Thermocouples, securely welded to the wafer, can give accurate results, e.g., in an evaporator (with a stationary wafer dome), or in rapid thermal processing or in a CVD reactor, but they are perturbed by an rf field. This effect can be eliminated by appropriate circuitry, but substrate heating through the thermocouple wires raises the temperature of the monitor wafer which is, therefore, not representative of any other wafer in the system.

26.2.2 Interferometers

A fiber-optic Mach-Zehnder interferometer was described by Hocker (1979). Another technique is the use of optical fiber Fabry-Perot interferometers (Yeh et al., 1990). Still another is infrared laser interferometry which measures the temperature-dependent optical pathlength from the front to the back of wafer (Donnelly et al., 1992). The usefulness in a plasma was demonstrated by measuring the wafer temperature during etching of Si in Cl_2 in a helical-resonator-plasma reactor. A thermometer based on changes in reflectivity, at optical wavelengths, of metals and semiconductors has also been described (Guidotti and Wilman, 1992). Its use in a plasma was shown by measuring the temperature of product wafers during rf sputter deposition of SiO_2; it requires that a hole be drilled in the substrate holder so that the probe can see the back side of a polished wafer.

26.2.3 Pyrometry

Infrared pyrometry overcomes the rf problems, but must be used with caution. The detector requires calibration using an emittance representative of the radiating surface. This infers that the emittance must remain constant during processing. This can be accomplished for etching, using a special monitor wafer (Schwartz and Schaible, 1981), but is not useful for deposition. A two-color fiber-optic IR detector (Krotchenko and Matthews, 1986) is independent of variations in emissitivity but presented problems for measuring temperatures below 400°C.

26.4 FLUOROPTIC THERMOMETRY

Probably the most widely used technique is fluoroptic™ thermometry not only because of its ability to measure wafer temperature *during* plasma processing but particularly because of the commercial availability of the apparatus (Luxtron). One measurement is based on the temperature dependence of the uv-light-induced optical emission of a rare earth phosphor. The first phosphors were europium-activated lanthanum or gadolinium oxysulfide. A high vacuum port is required for the fiber-optic probe which transmits the uv light to illuminate the phosphor which emits sharp lines whose relative intensities change with temperature; this temperature-related fluorescence is fed out through the probe to a detector. "By isolating two appropriate emission lines and determining the ratio of their intensities, a signal-independent measure of phosphor temperature can be obtained" (Wickersheim and Sun, 1985). It is used in one of two ways: (1) the phosphor is embedded into the tip of the probe which is either bonded to the wafer or held in close proximity to it (Egerton et al., 1982; Hussla et al., 1987; Castricher et al., 1987; Hess, 1993) or (2) the phosphor is coated on the back of the wafer and the tip of the optical probe placed against it (Nakamura et al., 1988). The basic configurations are shown in Fig. II-52a and b).

Another system uses a "phosphor, magnesium fluorogermanate activated with tetravalent manganese; it is illuminated briefly with a pulse of blue light; the rate of decay of the fluorescence is measured and correlated unambiguously with temperature" (Wickersheim and Sun).

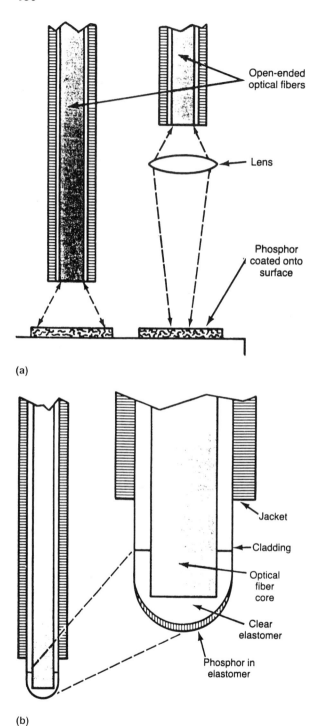

FIGURE II-52 (a) Remote fluoroptic thermal probe; phosphor coated on surface of sample. (b) Contact fluoroptic thermal probe.

Characterization

However, the temperature measurement techniques require extra ports (often not available in production reactors) for probes, or preparation of monitor wafers, so that they do not appear to be suitable for use during production processing, but are useful in the preliminary design experiments. A more complete discussion of these techniques can be found in *Optical Diagnostics for Thin Film Processing* (Herman, 1996).

27.0 PLASMA DIAGNOSTICS

27.1 OPTICAL DIAGNOSTICS FOR PLASMA PROCESSING

27.1.1 Introduction

These techniques are used to *monitor* a plasma *within the reaction chamber* (1) during *process development*, to study species formed from the reactants and etch products, the changes produced by variations in the processing parameters (e.g., power, pressure, mode of operation, etc.) which, together with physical examination of the processed wafers, lead to an understanding of the mechanism of the process and ultimately to its optimization, and (2) *during processing*, for detection of the endpoint, impurities, and particulates as well as for monitoring etch uniformity.

27.1.2 Optical Emission Spectroscopy (OES)

This is the most commonly used optical technique. A description of monochromators and light detectors, some typical spectra and practical examples, as well as some problems, such as light absorption by windows in the reactor, spectral resolution and second-order diffraction, can be found in a review by Selwyn (1993).

Many of the excited species formed in a plasma emit light; OES is used to detect and identify *excited* (not ground-state), *light emitting* species, which constitute a small fraction (~5%) of the species in the plasma. Not all activated species emit light; they may be deactivated by collisions or are metastables with no optical transitions.

27.1.3 Actinometry

Actinometry is the term applied to the process by which the concentration of the ground state species (the chemical reactants) is derived from the optically detected species. It has been used in studying reaction mechanisms and in process evaluation. The emission intensity of a trace concentration of a second, nonreactive species is used to compensate for the emission intensity of the reactant species due, not to changes in the ground state population, but to changes in the electron density or energy distribution. A widely used application is the determination of the concentration of ground state F atoms in a discharge containing a source of F (Coburn and Chen, 1980, 1981; d'Agostino et al., 1981a,b; Gottscho and Donnelly, 1984):

$$(F^*) = k_f(F)$$

where (F^*) = number of excited species
(F) = number of ground state F atoms
k_f = rate of electron-induced excitation

A similar relation holds for Ar emission, the actinometer

$$(Ar^*) = k(Ar)$$

If the two species have similar electron cross sections, then

$$\frac{(F^*)}{(Ar^*)} \sim \frac{(F)}{(Ar)}$$

Since (Ar) is constant and known, by measuring the intensities of both F^* and Ar^* at the appropriate wavelengths, (F) is obtained and can be monitored as operating conditions are changed.

In addition to the condition stated above about cross sections, two other criteria must be met for the valid application of actinometry: (1) Both species must be produced by electron impact excitation of the ground state species; (2) deexcitation must be chiefly by light emission (Gottscho and Miller, 1984). Plasma reactions of interest for which actinometry is not valid are, e.g., excited species production by electron-induced dissociation, $M_2 + e \rightarrow M^* + M$, and collisional quenching, $M^* + X \rightarrow M + X$.

27.1.4 OES for Endpoint Detection

In the simplest cases, this may be based on the (1) consumption of reactant species (when the reaction is complete, their concentrations rise reaching a limiting value) or (2) the production of etch products (when the reaction is complete, their concentrations diminish and ultimately become zero). A uniform etch rate results in a sharp change in signal. Selwyn (1993) discusses in detail some of the factors which may influence the choice of an OES endpoint strategy.

27.1.5 Laser-Induced Fluorescence (LIF)

This technique is used chiefly in research. In LIF, a laser beam, impinging on the plasma, is absorbed by the plasma; this induces an electronic transition to an upper state. Subsequent relaxation and emission follow rapidly. LIF directly probes ground state species; it is very sensitive and selective and is capable of high spatial resolution (Gottscho et al., 1983; Miller, 1986). When this technique is available, actinometry is not needed. However, commercial reactors do not have the number of ports (3) required to use LIF.

High-resolution LIF measurement of the line profiles has been used to measure the ion and neutral temperatures in an ECR reactor (Nakano et al., 1991).

Laser light scattering is used to detect particles generated in the plasma.

27.2 PLASMA PROBE TECHNIQUES

27.2.1 Introduction

Langmuir probes and microwave interferometers are used in the development of a plasma reactor or process to acquire basic understanding, but not as process monitors. They complement optical techniques but are, for the most part, *research* "tools."

27.2.2 Langmuir Probes

These probes are considered "intrusive, a source of contaminants and causing perturbations in the plasma. The theory used to interpret the *I-V* characteristics is relatively complicated. The probe consists of a small electrode in contact with the plasma; the simplest is a bare wire. The probe current is measured as its potential is increased as both positive and negative bias are applied to the probe. The ion density, electron temperature, floating potential, and plasma potential can be extracted from the measurements. Some of the many applications of Langmuir probe measurements have been characterization of etching processes (e.g., Steinbruchel, 1983; Kopalidis and Jorne, 1992), of divergent ECR reactors configurations using inert and reactive gases (Forster and Holber, 1989; Shatas et al., 1992), and of a multipolar ECR reactor (Hopwood et al., 1990, Forster et al., 1992). Ashtiani et al. (1992) have described other probe configurations: the emissive (heated) probes which is said to have advantages for measuring the plasma potential in a magnetic field and a dual probe assembly of both a conventional and an emissive probe. Herskowitz (1995) has a complete description of these probes.

27.2.3 Microwave Interferometer

The microwave interferometer is a nonintrusive method. It measures the electron density along the path of the probing microwaves. The microwaves, 10–100 GHz, are split into two beams, one of which is sent through the plasma. This wave combines with the reference beam in a mixer; the two beams interfere and the result is a measurement of the phase shift caused by the plasma. The shift depends on the electron plasma density and the microwave frequency. Unless the microwave frequency ($\omega/2\pi$) is greater than the electron plasma frequency, the wave is reflected. This means that the method can be used when the electron density is in the range 10^9–10^{13}, and the device >10 cm. One application of the method has been the characterization of an ECR reactor (Rossnagel et al., 1991). A detailed description can be found in Breun (1995).

REFERENCES

Abermann, R., in *Thin Films: Stresses and Mechanical Properties III* (W. D. Nix, J. C. Bravman, E. Arzt, L. B. Freund, eds.), *Mat. Res. Soc. Proc. Vol. 239*, 25 (1992).
Adams, A. C., S. P. Murarka, *J. Electrochem. Soc.*, 126, 334 (1979).
Adams, A. C., D. P. Schinke, C. D. Capio, *J. Electrochem. Soc.*, 126, 1539 (1979).
Ahn, J., K. L. Mittal, R. H. MacQueen, p. 134 in *Adhesion Measurement of Thin Films, Thick Films and Bulk Coatings*, ASTM Special Technical Publication 640 (K. L. Mittal, ed.) Philadelphia, Penn., 1976.
Ambree, P., F. Kreller, R. Wolf, K. Wandel, *J. Vac. Sci. Technol.*, B11, 614 (1993).
Ashtiani, K. A., J. L. Shohet, F. S. B. Anderson, D. T. Anderson, J. B. Friedmann, *Plasma Chemistry & Plasma Processing*, 12, 161 (1992).
Baba, S., A. Kikuchi, A. Kinbara, *J. Vac. Sci. Technol.*, A4, 3015 (1986).
Bacchetta, M., L. Bacci, M. S. Marangon, G. Queirolo, P. Sonego, L. Zanotti, *1994 VMIC*, 259 (1994).
Bauer, H. J., *Proc. 8th Intl. Vac. Cong.*, Cannes, France, 226 (1980).
Bauer, H. J., *J. Vac. Sci. Technol.*, B12, 2405 (1994).

Bhushan, B., S. P. Murarka, J. Gerlach, *J. Vac. Sci. Technol.*, *B8*, 1068 (1990).
Bindell, J. B., p. 534 in *VLSI Technology*, 2nd ed. (S. M. Sze, ed.), McGraw-Hill, New York, 1988.
Blaauw, C., *J. Appl. Phys.*, *54*, 5064 (1983).
Blech, I., S. Robles, *Solid State Technol.*, *9/94*, 75 (1994).
Bouchard, H., A. Azelmad, J. F. Currie, M. Munier, S. Blain, T. Darwell, in "Thin Films: Stresses and Mechanical Properties IV" (P. H. Townsend, T. P. Weihs, J. E. Sanchez, Jr., P. Borgensen, eds.), *Mat. Res. Soc. Proc. Vol.*, *308*, 63 (1993).
Brantley, W. A., *J. Appl. Phys.*, *44*, 534 (1973).
Breun, R. A., p. D3.1:1 in *Handbook of Thin Process Technology* (D. A. Glocker and S. Shah, eds.), Institute of Physics Publishing, Bristol, United Kingdom, 1995.
Briggs, D., M. P. Seah, eds., *Practical Surface Analysis*, Vol. 1, 2nd ed., John Wiley & Sons, New York, 1990.
Brown, J. G., *X-rays and Their Application*, Plenum Press, New York, 1966.
Brown, R., chap. 6, in *Handbook of Thin Film Technology*, (L. I. Maissel, R. Glang, eds.), McGraw-Hill, New York, 1970.
Budhani, R. C., R. F. Bunashah, P. A. Flinn, *Appl. Phys. Lett*, *52*, 284 (1988).
Burkhardt, P. F., R. F. Marvel, *J. Electrochem. Soc.*, *116*, 864 (1969).
Campbell, D. S. pp. 12–25 in *Handbook of Thin Film Technology* (L. I. Maissel, R. Glang, eds.), McGraw-Hill, New York, 1970.
Cargo, J. T., M. C. Hughes, *J. Electrochem. Soc.*, *136*, 1239 (1989).
Carlotti, G., L. Doucet, M. Dupeux, *J. Vac. Sci. Technol.*, *B14*, 3460 (1996).
Carroll, B., p. 243 in *Physical Methods in Macromolecular Chemistry*, vol. 2, Marcel Dekker, Inc. New York, 1972.
Castricher, G., P. M. Banks, T.-M. Pang, P. Bauman, H. Grunwald, I. Hussla, G. Lorenz, H. Stoll, H. Ramisch, *Microelectronic Engr.*, *6*, 559 (1987).
Chang, C. D., D. M. Boulin, *Surface Science*, *69*, 395 (1977).
Chapman, B., *J. Vac. Sci. Technol.*, *11*, 106 (1974).
Chen, Y. S., H. Fatemi, *J. Vac. Sci. Technol.*, *A4*, 645 (1986).
Choi, M. S., E. W. Hearn, *J. Electrochem. Soc.*, *131*, 2442 (1984).
Chow, R., W. A. Lanford, W. Ke-Ming, R. S. Rosler, *J. Appl. Phys.*, *53*, 5630 (1982).
Chu, W.-K., J. M. Mayer, M. A. Nicolet, *Backscattering Spectrometry*, Academic Press, New York, 1978.
Chu, W.-K., G. C. Schwartz, unpublished, 1976.
Claasen, W. A. P., *Plasma Chemistry & Plasma Processing 7*, 109 (1987).
Coburn, J. W., M. Chen, *J. Appl. Phys.*, *51*, 3134 (1980).
Coburn, J. W., M. Chen, *J. Vac. Sci. Technol.*, *18*, 353 (1981).
Cohen, B. L., C. L. Fink, J. H. Degnan, *J. Appl. Phys.*, *43*, 19 (1972).
Cox, J. N., G. Shergill, M. Rose, J. K. Chu, *1990 VMIC*, 419 (1990).
d'Agostino, R., F. Cramarossa, S. De Benedictis, G. Ferraro, *J. Appl. Phys.*, *52*, 1259 (1981a).
d'Agostino, R., V. Colaprico, F. Cramarossa, *Plasma Chemistry & Plasma Processing*, *1*, 365 (1981b).
Day, D. R., S. D. Senturia, *J. Electronic Mat.*, *11*, 441 (1982).
Dax, M., *Semiconductor International*, *8/96*, 91 (1996).
Donnelly, V. M., D. E. Ibbottson, C.-P. Chang, *J. Vac. Sci. Technol.*, *A10*, 1060 (1992).
Doucet, L., G. Carlotti, in "Thin Films: Stresses and Mechanical Properties V" (S. P. Baker, C. A. Ross, P. H. Townsend, C. A. Volkert, P. Borgensen, eds.), *Mat. Res. Soc. Proc. Vol.*, *356*, 215 (1995).
Doyle, B. L., P. S. Peercy, *Appl. Phys. Lett.*, *34*, 611 (1979).
Egerton, E. J., A. Nef, W. Millikin, W. Cook, D. Baril, *Solid State Technol.*, *8/82*, 84 (1982).
Feldman, L. C., J. W. Mayer, *Fundamentals of Surface and Thin Film Analysis*, North-Holland, New York, 1986.

Flitch, J. T., G. Lucovsky, E. Kobeda, E. A. Irene, *J. Vac. Sci. Technol.*, *B7*, 153 (1989a).
Flitch, J. T., C. H. Bjorkman, G. Lucovsky, F. H. Pollak, X. Yin, *J. Vac. Sci. Technol.*, *B7*, 775 (1989b).
Flitch, J. T., C. H. Bjorkman, J. J. Sumakeris, G. Lucovsky, in "Thin Films: Stresses and Mechanical Properties" (J. C. Bravman, W. D. Nix, D. M. Barnett, D. A. Smith, eds.), *Mat. Res. Soc. Proc. Vol.*, *130*, 289 (1989c).
Flexus, Santa Clara, California.
Flinn, P. A., G. A. Waychunas, *J. Vac. Sci. Technol.*, *B6*, 1749 (1988).
Flinn, P. A., in "Thin Films: Stresses and Mechanical Properties," (J. Bravman, W. D. Nix, D. M. Barnett, D. A. Smith, eds.), *Mat. Res. Soc. Proc.*, *Vol. 130*, 41 (1989).
Follstaedt, D. M., J. A. van den Avyle, A. D. Romig, Jr., J. A. Knapp, in "Materials Reliability Issues in Microelectronics" (J. R. Lloyd, F. G. Yost, P. Ho, eds.), *Mat. Res. Soc. Proc. Vol. 225*, 225 (1991).
Forouhi, A. R., I. Bloomer, *Phys. Rev. B*, *34*, 7018 (1986).
Forouhi, A. R., I. Bloomer, *Phys. Rev. B*, *38*, 1865 (1988).
Forouhi, A. R., I. Bloomer, US Pat. 4,905,170, 1990.
Forster J., W. Holber, *J. Vac. Sci. Technol.*, *A7*, 899 (1989).
Forster J., C. C. Klepper, L. A. Berry, S. M. Gorbatkin, *J. Vac. Sci. Technol.*, *A10*, 3114 (1992).
Gent, A. N., L. H. Lewandowski, *J. Appl. Polymer Sci.*, *33*, 1567 (1987).
Ginsburg, R., J. R. Susko, in *Polyimides* (K. L. Mittal, ed.), Plenum Press, New York, 1984.
Glang, R., R. A. Holmwood, R. L. Rosenfeld, *Rev. Sci. Inst.*, *36*, 7 (1965).
Godet, C., P. R. i Cabarrocas, S. C. Gujrathi, P. A. Burret, *J. Vac. Sci. Technol.*, *A10*, 3517 (1992).
Goldsmith, C., P. Geldermanns, F. Bedetti, G. A. Walker, *J. Vac. Sci. Technol.*, *A1*, 407 (1983).
Gomez-San Roman, R., R. Perez-Casero, J. Pierrem, J. P. Enard, J. M. Martinez-Duart, *J. Vac. Sci. Technol.*, *A13*, 54 (1995).
Gottscho, R. A., G. P. Davis, R. H. Burton, *J. Vac. Sci. Technol.*, *A1*, 622 (1983).
Gottscho, R. A., T. A. Miller, *Pure Appl. Chem.*, *56*, 189 (1984).
Gottscho, R. A., V. M. Donnelly, *J. Appl. Phys.*, *56*, 245 (1984).
Grilletto, C., *Solid State Technol.*, 2/77, 27 (1977).
Guidotti, D., J. G. Wilman, *J. Vac. Sci. Technol.*, *A10*, 3184 (1992).
Gujrathi, S. Bultena, *Nucl. Instr. & Methods in Phys. Res.*, *B64*, 789 (1992).
Gupta, P., S. H. Tan, Z. Pourmotamed, C. Flores, R. McDonald, *43rd Natl. Symp. Amer. Vac. Soc.*, Abstr. #MS+As-ThM, 163 (1996).
Hall, P. M., *Thin Solid Films*, *1*, 277 (1967).
Hamersky, J., *Thin Solid Films*, *3*, 2673 (1969).
Hearn, E. W., p. 273 in *Advances in X-ray Analysis* (H. F. McMurdie, C. S. Barrett, J. B. Newkirk, C. O. Rund, eds.), Plenum Press, New York, 1977.
Herman, I. P., *Optical Diagnostics for Thin Film Processing*, Academic Press, New York, 1996.
Hershkowitz, N., p. D3.0:1 in *Handbook of Thin Process Technology* (D. A. Glocker and S. Shah, eds.), Institute of Physics Publishing, Bristol, United Kingdom, 1995.
Hess, D. W., in "Proc. of Symp. Highly Selective Dry Etching & Damage Control" (G. S. Mathad, H. Horiike, eds.), *Electrochem. Soc. Proc. Vol. PV 93-21*, 1 (1993).
Hirashita, N., I. Aikawa, T. Ajioka, M. Kobayakawa, F. Yokoyama, Y. Sakaya, *28th IEEE/IRPS*, 216 (1990).
Ho, P. S., P. O. Hahn, J. W. Bartha, G. W. Rubloff, F. K. LeGoues, B. D. Silverman, *J. Vac. Sci. Technol.*, *A3*, 739 (1985).
Ho, P. S., F. Faupel, *Appl. Phys. Lett.*, *53*, 1602 (1988).
Hocker, G. B., *Applied Optics*, *18*, 1445 (1979).

Hoffman, R. W., p. 211 in *Physics of Thin Films*, vol. 3 (G. H. Hass and R. E. Thun, eds.), Academic Press, New York, 1966.
Hoffmeister, W., M. A. Zuegel, *Thin Solid Films*, *3*, 35 (1969).
Hopwood, J., D. K. Reinhard, J. Asmussen, *J. Vac. Sci. Technol.*, *A8*, 3103 (1990).
Horie, H., T. Fukano, T. Ito, *Fujitsu Sci. Tech. J.*, *20*, 39 (1984).
Houskova, J., K.-K. N. Ho, M. K. Balazs, *Semiconductor Intl.*, *5/85*, 236 (1985).
Hughes, M. C., D. R. Wonsidler, *J. Electrochem. Soc.*, *134*, 1488 (1987).
Hurley, K. H., 1987, *Solid State Technol.*, *3/87*, 103 (1987).
Hussla, I., K. Enke, H. Grunwald, G. Lorenz, H. Stoll, *J. Phys. D: Appl. Phys.*, *20*, 889 (1987).
Irene, E. A., *J. Electron Mater.*, *5*, 287 (1976).
Irving, I. P., *Optical Diagnostics for Thin Film Processing*, Academic Press, New York, 1996.
Jenkins, R., *X-ray Fluorescence Spectrometry*, John Wiley & Sons, New York, 1988.
Kato, T., H. Ashida, S. Hosada, Y. Ishimaru, K. Watanabe, *1992 VMIC*, 79 (1992).
Kelm, G., G. Jungnickel, *Materials Science & Engineering*, *A139*, 401 (1991).
Kikkawa, T., H. Watanabe, T. Murata, *Appl. Phys. Lett.*, *50*, 1527 (1987).
Kobeda, E., E. A. Irene, *J. Vac. Sci. Technol.*, *B4*, 720 (1986).
Kopalidis, P. M., J. Jorne, *J. Electrochem. Soc.*, *139*, 839 (1992).
Koroychenko, V., A. Matthews, *Vacuum*, *36*, 61 (1986).
Koshida, N., H. Wachi, K. Yoshida, M. Komuro, N. Atoda, *Jpn. J. Appl. Phys.*, *29*, 2299 (1990).
Ku, Y.-C., L.-P. Ng, R. Carpenter, K. Lu, H. I. Smith, *J. Vac. Sci. Technol.*, *B9*, 3297 (1991).
Kuiper, A. E. T., M. F. C. Willemsen, L. J. Van Ijzendoorn, *Appl. Phys. Lett.*, *53*, 2149 (1988).
Lanford, W. A., H. P. Trautvetter, J. F. Zielger, J. Keller, *Appl. Phys. Lett.*, *28*, 566 (1976).
Lanford, W. A., M. J. Rand, *J. Appl. Phys.*, *49*, 2473 (1978).
Leich, D. A., T. A. Tombrello, *Nuclear Instruments & Methods*, *108*, 67 (1973).
Levy, R. A., S. M. Vincent, T. E. McGahan, *J. Electrochem. Soc.*, *132*, 1472 (1985).
Levy, R. A., T. Y. Kometani, *J. Electrochem. Soc.*, *134*, 1565 (1987).
Levy, D. H., K. K. Gleason, *J. Electrochem. Soc.*, *140*, 797 (1993).
Li, J., F. Moghadam, L. J. Matienzo, T. L. Alford, J. W. Mayer, *Solid State Technol.*, *5/95*, 61 (1995).
Lloyd, J. C., p. 601 in *Advances in X-ray Analysis*, vol. 12, Plenum Press, New York, 1969.
Lucovsky, G., D. V. Tsu, *J. Vac. Sci. Technol.*, *A5*, 2231 (1987).
Lucovsky, G., P. D. Richard, D. V. Tsu, S. Y. Lin, R. J. Markunas, *J. Vac. Sci. Technol.*, *A4*, 681 (1986).
Lucovsky, G., J. T. Fitch, D. V. Tsu, S. S. Kim. *J. Vac. Sci. Technol.*, *A7*, 1136 (1989).
Lundquist, T. R., R. P. Burgner, P. R. Swann, I. S. T. Tsong, *Appl. of Surf. Sci.*, *7*, 2 (1982).
Machonkin, M. A., F. Jansen, *Thin Solid Films*, *150*, L97 (1987).
MacIntyre, M. W., unpublished, 1968.
Madden, M., J. N. Cox, B. Fruechting, J. Matteau, *Solid State Technol.*, *8/89*, 53 (1989).
Madden, H. H., *J. Electrochem. Soc.*, *128*, 625 (1981).
Magee, C. W., E. M. Botnick, *J. Vac. Sci. Technol.*, *19*, 47 (1981).
Magee, C. W., L. R. Hewitt, *RCA Review*, *47*, 162 (1986).
Maissel, L. I., chap. 13 in *Handbook of Thin Film Technology* (L. I. Maissel and R. Glang, eds.), McGraw-Hill, New York, 1970.
Matsuura, T., H. Uetake, T. Ohmi, J. Murota, S. Ono, in "ULSI Science and Technology/1991" (J. M. Andrews and G. K. Celler, eds.), *ECS PV 91-11*, 236 (1991).
Mego, T. J., *Solid State Technol.*, *5/90*, 159 (1990).
Melngailis, *J. Vac. Sci. Technol.*, *B5*, 469 (1987).
Melngailis, *SPIE Vol. 923*: Electron-Beam, X-Ray and Ion-Beam, Technology: Submicrometer Lithographies VII, 72 (1988).

Meng, W. J., J. A. Sell, G. L. Eesley, T. A. Perry, in "Thin Films: Stresses and Mechanical Properties IV" (P. H. Townsend, T. P. Wells, J. E. Sanchez, Jr., P. Borgesen, eds.), *Mat. Res. Soc. Symp. Proc. Vol. 308*, 21 (1993).

Miller, J. M., *Separation Methods in Chemical Analysis*, John Wiley & Sons, New York, 1975.

Miller, T. A., *J. Vac. Sci. Technol.*, *A4*, 1768 (1986).

Mullenberg, G. E., ed. *Handbook of X-ray Photoelectron Spectroscopy*, Perkin-Elmer Corporation, Physical Electronics Division, Eden Prairie, Minn., 1979.

Murakami, K., T. Takeuchi, K. Ishikawa, T. Yamamoto, *Appl. Surf. Sci.*, *33/34*, 742 (1988).

Murarka, S. P., p. 66 in *Metallization, Theory and Practice for VLSI and ULSI*, Butterworth-Heinemann, Boston, 1994.

Nakamura, M, R. Kanzawa, K. Sakai, *J. Electrochem. Soc.*, *133*, 1167 (1986).

Nakamura, M., T. Kurimoto, H. Yano, K. Yanigida, in "Symp. Dry Process" (J.-I. Nishizawa, Y. Horiike, M. Hirose, K. Suto, eds.), *Electrochem. Soc. Proc. Vol.*, *PV 88-7*, 78 (1988).

Nakano, T., N. Sadeghi, R. A. Gottscho, *Appl. Phys. Lett.*, *58*, 458 (1991).

Nguyen, S. V., T. Nguyen, H. Treichel, O. Spindler, *J. Electrochem. Soc.*, *141*, 1633 (1994).

Niccolian, E. H., J. R. Brews, *MOS Physics and Technology*, John Wiley & Sons, New York, 1982.

Nikawa, K., K. Nasu, M. Murase, T. Kaito, T. Adachi, S. Inoue, *27th IEEE/IRPS*, 43 (1989).

Nikawa, K., *J. Vac. Sci. Technol.*, *B9*, 2566 (1991).

Norton, J. T., p. 401 in *Advances in X-Ray Analysis*, vol. 11 (J. B. Newkirk, G. R. Mallett, H. G. Pfeiffer, eds.), Plenum Press, New York, 1968.

Oehrlein, G. S., K. K. Chan, M. A. Jaso, *J. Appl. Phys.*, *64*, 2399 (1988).

Otterman, C., N. Tadakoro, Y. Tomita, K. Bange, in "Thin Films: Stresses and Mechanical Properties IV" (P. H. Townsend, T. P. Wells, J. E. Sanchez, Jr., P. Borgesen, eds.), *Mat. Res. Soc. Proc. Vol. 308*, 627 (1993).

Overwijk, M. H. F., F. C. van den Heuval, *J. Appl. Phys.*, *74*, 1762, 237 (1993).

Pai, P. G., S. S. Chao, Y. Takagi, G. Lucovsky, *J. Vac. Sci. Technol.*, *A4*, 689 (1986).

Paduschek, P., P. Eichinger, *Appl. Phys. Lett.*, *36*, 62 (1980).

Patrick, W. J., G. C. Schwartz, J. D. Chapple-Sokol, K. Olsen, R. Carruthers, *J. Electrochem. Soc.*, *139*, 2604 (1992).

Peek, H. L., R. A. M. Wolters, *1986 VMIC*, 165 (1986).

Petvai, S. I., R. H. Schnitzel, R. Frank, *Thin Solid Films*, *53*, 111 (1978).

Pliskin, W. A., R. P. Gnall, *J. Electrochem. Soc.*, *111*, 872 (1964).

Pliskin, W. A., S. J. Zanin, chap. 11 in *Handbook of Thin Film Technology* (L. I. Maissel and R. Glang, eds.), McGraw-Hill, New York, 1970.

Pliskin, W. A., *J. Vac. Sci. Technol.*, *14*, 1064 (1977).

Pliskin, W. A., *J. Electrochem. Soc.*, *134*, 2819 (1987).

Pliskin, W. A., p. 280 in *Measurement Techniques for Thin Films* (B. Schwartz and N. Schwartz, eds.) Electrochem. Soc., New York, 1967.

Pliskin, W. A., R. G. Simmons, R. P. Esch, p. 524 in *Thin Film Dielectrics* (F. Vratny, ed.), Electrochem. Soc., New York, 1969.

Pliskin, W. A., p. 509 in *Semiconductor Silicon* (H. R. Huff and R. R. Burgess, eds.), Electrochem. Soc., New York, 1973.

Pliskin, W. A., *J. Vac. Sci. Technol.*, *14*, 1064 (1977).

Pliskin, W. A., H. S. Lehman, *J. Electrochem. Soc.*, *112*, 1013 (1965).

Pramanik, D., J. Glanville, *Solid State Technol.*, *5/90*, 77 (1990).

Ramkumar, K., S. K. Goshh, A. N. Saxena, *J. Electrochem. Soc.*, *140*, 2669 (1993).

Reimer, L. *Transmission Electron Microscopy*, Springer-Verlag, Berlin, 1984.

Retajczyk, T. F., A. K. Sinha, *Thin Solid Films*, *70*, 241 (1980).

Rojas, S., R. Gomarasca, L. Zenotti, A. Borghesi, A. Sassella, G. Ottaviani, L. Moro, P. Lazzeri, *J. Vac. Sci. Technol.*, *B10*, 633 (1992).

Rossnagel, S. M., K. Schatz, S. J. Whitehair, R. C. Guarnieri, D. N. Ruzic, J. J. Cuomo, *J. Vac. Sci. Technol.*, *A9*, 702 (1991).
Rothman, L. B., *J. Electrochem. Soc.*, *127*, 2216 (1980).
Samuelson, G., S. Lytle, p. 751 in *Polyimides* (K. L. Mittal, ed.), Plenum Press, New York, 1984.
Schaible, P. M., R. Glang, p. 577 in *Thin Film Dielectrics* (F. Vratney, ed.), Electrochem. Soc., New York, 1969.
Schwartz, G. C., R. E. Jones, L. I. Maissel, *J. Vac. Sci. Technol.*, *6*, 351 (1969).
Schwartz, G. C., R. E. Jones, *IBM J. Res. Develop.*, *14*, 52 (1970).
Schwartz, G. C., P. A. Schaible, in "Plasma Processing" (R. G. Frieser, C. J. Mogab, eds.), *Electrochem. Soc. Proc. Vol. PV 81-1*, 133 (1981).
Selwyn, G. S., *Optical Diagnostic Techniques for Plasma Processing*, AVS Monograph Series, American Vacuum Society, New York, 1993.
Shacham-Diamand, Y., Y. Nachumovsky, *J. Electrochem. Soc.*, *137*, 190 (1990).
Shankoff, T. A., C. C. Chang, S. E. Haszko, *J. Electrochem. Soc.*, *125*, 467 (1978).
Shastain, J., ed., *Handbook of X-ray Photoelectron Spectroscopy*, Perkin-Elmer Corporation, Eden Prairie, Minn., 1992.
Shatas, A. A., Y. Z. Hu, E. A. Irene, *J. Vac. Sci. Technol.*, *A10*, 3119 (1992).
Shimbo, M., T. Matsuo, *J. Electrochem. Soc.*, *130*, 135 (1983).
Shute, C. J., J. B. Cohen, D. A. Jeanotte, in "Thin Films: Stresses and Mechanical Properties" (J. C. Bravman, W. D. Nix, D. M. Barnett, D. A. Smith, eds.), *Mat. Res. Soc. Symp. Proc. Vol. 130*, 29 (1989).
Singer, P. H., *Semiconductor International*, *7/86*, 46 (1986).
Sinha, A. K., H. J. Levenstein, T. E. Smith, *J. Appl. Phys.*, *49*, 2423 (1978).
Sizemore, J., D. A. Stevenson, J. Stringer, in "Thin Films: Stresses and Mechanical Properties IV" (P. H. Townsend, T. P. Wells, J. E. Sanchez, Jr., P. Borgesen, eds.), *Mat. Res. Soc. Symp. Proc. Vol. 308*, 165 (1993).
Small, M. K., J. J. Vlassak, S. F. Powell, B. J. Daniels, W. D. Nix, in "Thin Films: Stresses and Mechanical Properties IV" (P. H. Townsend, T. P. Wells, J. E. Sanchez, Jr., P. Borgesen, eds.), *Mat. Res. Soc. Symp. Proc. Vol. 308*, 159 (1993).
Smith, W. J., C. Welles, A. Bivas, F. G. Yost, J. E. Campbell, *28th IEEE/IRPS*, 200 (1990a).
Smith, W. J., C. G. Welles, A. Bivas, *Semiconductor Intl.*, *1/90*, 92 (1990b).
Smith, W. J., in "Materials Reliability Issues in Microelectronics" (J. R. Lloyd, F. G. Yost, P. Ho, eds.), *Mat. Res. Soc. Proc. Vol. 225*, 291 (1991).
Smith, P. K., T. O. Herndon, R. L. Burke, D. R. Day, S. D. Senturia, *J. Electrochem. Soc.*, *130*, 225 (1983).
Smith, J. F., D. C. Hinson, *Solid State Technol.*, *11/86*, 135 (1986).
Smith, F. C., Jr., R. C. Chang, *The Practice of Ion Chromatography*, John Wiley & Sons, New York, 1983.
Smolinsky, G., T. H. F. Wendling, *J. Electrochem. Soc.*, *132*, 950 (1985).
Stein, H. J., V. A. Wells, R. E. Hampy, *J. Electrochem. Soc.*, *126*, 1750 (1979).
Stein, H. J., *J. Electronic Mat.*, *5*, 161 (1976).
Stein, H. J., H. A. R., Wegener, *J. Electrochem. Soc.*, *124*, 908 (1977).
Steinbruchel, Ch., *J. Electrochem. Soc.*, *130*, 648 (1983).
Sunami, H., Y. Itoh, K. Sato, *J. Appl. Phys.*, *41*, 5115 (1970).
Swalen, J. D., M. Tacke, R. Santo, J. Fisher, *Optical Communications*, *18*, 387 (1976).
Szot, J., R. Hornsey, T. Ohnishi, S. Minagawa, *J. Vac. Sci. Technol.*, *B10*, 575 (1992).
Takahashi, M., M. Maeda, Y. Sakakibara, *Jpn. J. Appl. Phys.*, *26*, 1606 (1987).
Theil, J. A., D. V. Tsu, M. W. Watkins, S. S. Kim, G. Lucovsky, *J. Vac. Sci. Technol.*, *A8*, 1374 (1990).
Therma-Wave, Inc., Fremont, Calif.

Thompson, M., M. D. Baker, A. Christie, J. F. Tyson, *Auger Electron Spectroscopy*, John Wiley & Sons, New York.
Thouless, M. D., *J. Vac. Sci. Technol.*, A9, 2510 (1991).
Tokuyama, T., Y. Fijii, Y. Sugita, S. Kisnino, *Jpn. J. Appl. Phys.*, 6, 1252 (1967).
Tien, P. K., *Appl. Opt.*, 10–11, 2395 (1971).
Tien, P. K., R. Verich, R. J. Martin, *Appl. Phys. Lett.*, 14, 291 (1969).
Tomioka, H., S.-I. Tanabe, K. Mizukami, *27th IEEE/IRPS*, 53 (1989).
Tompkins, H. G., C. Tracy, *J. Electrochem. Soc.*, 136, 2331 (1989).
Tu, K. N., W. K. Chu, J. W. Mayer, *Thin Solid Films*, 25, 403 (1975).
Valli, J., *J. Vac. Sci. Technol.*, A4, 3007 (1986).
Veeco, Santa Barbara, Calif.
Vreeland, J., T., A. Dommann, C.-J., Tsai, M.-A. Nicolet, in "Thin Films: Stresses and Mechanical Properties" (J. C. Bravman, W. D. Nix, D. M. Barnett, D. A. Smith, eds.), *Mat. Res. Soc. Proc. Vol. 130*, 3 (1989).
Warren, B. E., *X-Ray Diffraction*, Dover Publ., Inc., New York, 1990.
Wei, J. S., W. D. Westwood, *Appl. Phys. Lett.*, 32, 819 (1978).
Wells, O. C., A. Boyde, E. Lifshin, A. Rezanowich, *Scanning Electron Microscopy*, McGraw-Hill, New York, 1974.
Wickersheim, K., M. Sun, *Res. & Dev.*, 11/85, 114 (1985).
Wildman, H. S., G. C. Schwartz, unpublished, 1981.
Wildman, H. S., G. C. Schwartz, *J. Vac. Sci. Technol.*, 20, 396 (1982).
Willenborg, D. L., S. M. Kelso, J. L. Opsai, J. T. Fanton, H. Rosencwaig, *SPIE Vol. 1594*, 322 (1992).
Wolters, R. A. M., W. C. J. Heesters, *1990 VMIC*, 447 (1990).
Wong, J., *J. Electronic Materials*, 5, 113 (1976).
Wormald, J., *Diffraction Methods*, Clarendon Press, Oxford, England, 1973.
Wu, G., L. Xu, W. Chen, G. Zhang, Z. Li, Y. Hao, J. Fang, *1994 VMIC*, 221 (1994).
Wu, T. H. T., R. S. Rosler, *Solid State Technol.*, 5/92, 65 (1992).
Xie, J. Z., S. P. Murarka, X. S. Guo, W. A. Lanford, *J. Vac. Sci. Technol.*, B6, 1756 (1988).
Xie, J. Z., S. P. Murarka, X. S. Guo, W. A. Lanford, *J. Vac. Sci. Technol.*, B7, 150 (1989).
Yasuoka, Y., K. Harakawa, K. Gamo, S. Namba, *Jpn. J. Appl. Phys.*, 29, L1221 (1990).
Yeh, Y., C. E. Lee, R. A. Atkins, W. N. Gibler, H. F. Taylor, *J. Vac. Sci. Technol.*, A8, 3247 (1990).
Yoshimi, T., H. Sakai, K. Tanaka, *J. Electrochem. Soc.*, 127, 1853 (1980).

III
Semiconductor Contact Technology

D. R. Campbell
*Clarkson University, Potsdam, New York
and CVC, Rochester, New York*

1.0 INTRODUCTION

The quality of the electrical contacts within an integrated circuit is critical to the performance of the chip. In view of both its technological importance and the many ways possible to form contacts, no other single area of circuit technology appears so extensively investigated, with the possible exception of gate oxides. As shown in Fig. III-1, there was the problem of the interaction between Al and Si in the earliest practical silicon contacts in which Al, and later AlSi, was the interconnection metallization. An extensive annealing or sintering process was required to reduce the Si suboxide by formation of Al oxide, thereby providing a measure of direct contact between Al and Si. In the course of this annealing, Si would also diffuse into the overlying Al at weak points in the porous oxide that partially separated Al from Si. As the metallurgical limitations of this simple structure emerged, improvements were made by introducing platinum silicide (PtSi) as the first Si-contacting layer. Platinum was sputtered into a "contact hole," formed in the overlying SiO_2 layer by patterning and etching, and then sintered. Platinum reacted readily with Si, more easily reducing the native oxide on Si and forming either ohmic or rectifying (Schottky barrier) contacts of consistently high quality. Reaction between the overlying Al metallization and the silicide again led to "junction penetration," so that diffusion barriers were required to prevent interdiffusion and subsequent reaction. A typical contact structure including both silicide and a diffusion barrier is shown in Fig. III-2.

The purpose of the following review of contact technology is to chronicle the significant steps in developing state-of-the-art Si contacts. This review follows contact development, from the early beginnings of VLSI circuit technology (in the late 1960s) to the present. It draws from both patent and technical literature sources.

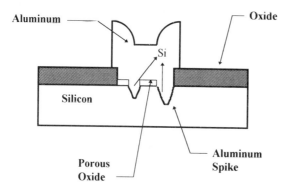

FIGURE III-1 Typical early contact structure for aluminum metallization on silicon that underwent Al spiking as a result of sintering. Note Al metal directly contacts Si surface.

2.0 IMPORTANCE OF CONTACT TECHNOLOGY

Contact technology accounts for a substantial amount of the time and resources of chip interconnection technology, that is, the back end of the line (BEOL) processing, and the majority of the reliability problems. Problems with ohmic contacts range from early, fatal electrical malfunctions such as incompletely formed or electrically "open" contacts and penetrated junctions, etc., to slower emerging, but ultimately equally serious, metallurgical problems. (An electrically "open" contact is typically one in which the contact hole in the insulating layer is not etched deep enough to reach the silicon surface.) The latter may involve progressive kinetic phenomena such as interdiffusion, phase formation, and voiding, all of which can contribute to electromigration divergences and cause very resistive, and even opened, interconnects. Since Schottky diodes (rectifying contacts) are resistive structures as compared to ohmic contacts, they are less sensitive to some of the factors that contribute to excessive resistance in ohmic contacts. The presence of metallic oxides or other poorly conducting interlayer contaminants at the metal-to-metal interfaces that are included in the finished contact structure are examples of such. However, they are still highly sensitive to the condition of the metal-silicon

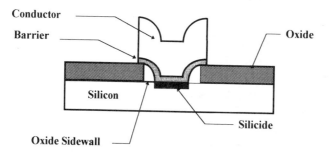

FIGURE III-2 Contact structure complete with conductor, diffusion barrier, and silicide contact to Si. Sidewalls formed on vertical surfaces of contact opening provide for a smooth transition from the level of the interconnection to the surface of Si.

interface and to any change in this interface during processing or operation. These conditions influence the barrier height of the diode (to be discussed later) and therefore its operating characteristics. Successful circuit applications of these diodes require stable I-V characteristics for the transistors with which they combine in various circuits. Figure III-3 illustrates a typical use for a Schottky barrier diode (SBD) as a clamp on the emitter current port of a biopolar transistor. This prevents the transistor from going into saturation and thereby increasing the time required for switching between logic states.

3.0 ELECTRICAL ASPECTS OF SILICON CONTACTS

Two properties, barrier height, ϕ_B, and contact resistance, R_c, are essential parts of any discussion of the electrical characteristics of a metal-silicon contact. The physics of contacts appear in several texts (Grove, 1967; Muller, 1977; Sze, 1969a) and a brief review is included to aid discussions of material choices for contacts. Barrier height, ϕ_B, as the name suggests, is the energy or potential barrier for the passage of electrons between the metal contact and the single crystal silicon device. Together with the width of the barrier, W, these two material parameters determine whether the contact is rectifying or ohmic in nature. Since the contacts are part of an electrical circuit, contact resistance directly influences circuit operation. Low R_c values provide small resistance-capacitance (RC) time delays. For high R_c values, the reverse applies.

3.1 BARRIER HEIGHTS

Figure III-4 shows the energy band diagram for n-type Si. Contact with a metal layer through an intervening, insulating layer represents the most general case. However, in Fig. III-4, an interfacial layer is omitted in the interests of simplicity. This removes any potential influence of such an interfacial layer to the subsequent discussion, except to assume that the electrons can flow through it and that the

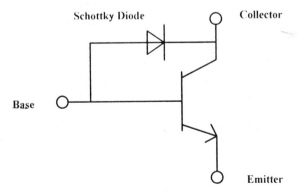

FIGURE III-3 Circuit with Schottky barrier diode used as an emitter current clamp which prevents the transistor from reaching saturation and avoids subsequent decrease in switching speed. The diode is connected between the base and collector terminals. (After Tada et al., 1967)

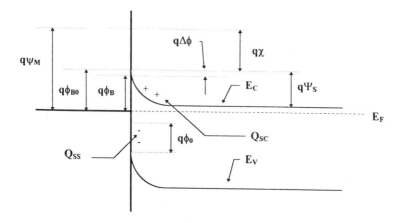

ψ_M -- work function of metal
φ_B -- barrier height of metal semiconductor barrier
φ_B0 -- asymptotic value of φ_B at zero electric field
φ_0 -- energy level at surface
Δφ -- image force barrier lowering
χ -- electron affinity of semiconductor
ψ_S -- built-in potential
Q_ss -- surface state density on semiconductor
Q_SC -- space charge density in semiconductor

FIGURE III-4 Energy band diagram for n-type Si in contact with a metal layer. (After Crowley et al., 1965)

density of interface states, N_{ss}, is a property of the silicon surface and not of the metal or interfacial layer. The figure legend shows the symbols representing the relevant interface energies (Crowley et al., 1965). Nicollian et al. (1978) have shown that the barrier height, ϕ_B, can be given by the following two expressions:

A. *True Schottky barrier* ($N_{ss} = 0$, that is, no semiconductor interface states)

$$\phi_B = \phi_m - \chi - \Delta\phi \tag{III-1}$$

For this case, the barrier height depends on the work function of the metal, ϕ_m, the electron affinity of the semiconductor, χ, and $\Delta\phi$ is the image force barrier lowering. Since χ depends on the Fermi level, E_F, which in turn depends on the doping level, N_D, then barrier height in this regime should also depend on the doping level.

B. *Bardeen barrier* ($N_{ss} = \infty$ or infinity, that is, semiconductor interface states dominate)

$$\phi_B = \frac{E_g}{q} - \phi_0 - \Delta\phi \tag{III-2}$$

This corresponds to most cases encountered in silicon contact technology in which the Fermi level is pinned by interface states at a level of $q\phi_0$ above the valence band edge. It should be noted that the work function of the metal does not appear here but the barrier height is also weakly dependent on doping level because ϕ_0 depends on χ, etc., as described for case A above. The remaining undefined parameter in the above expression, E_g, stands for the semiconductor (e.g., Si) band gap. For circuit applications, values of the diode barrier heights are either high such as PtSi at ~0.9 eV, or "low" (actually mid gap) such as TiW at ~0.5 eV. In the first instance, the application is typically a clamp to prevent a bipolar transistor from going into saturation, and in the second, it is a diode element in a logic circuit where its lower barrier height permits the diode to "turn on" at lower forward bias and, for the same reason, conduct higher currents through smaller contacts, i.e., use less Si "real estate."

3.2 ANALYZING DATA

Typically, experimental data are interpreted according to the expression:

$$J_f = A^{**}T^2 \exp\left(\frac{-q\phi_B}{kT}\right) \exp\left(\frac{qV_f}{nkT}\right) \quad \text{(III-3)}$$

Where J_f is the forward current density, A^{**} the effective Richardson constant, T the temperature, k the Boltzman constant, q the electronic charge, V_f the forward applied voltage, and n the ideality (see below). The effective Richardson constant depends on the *effective* mass of the majority charge carrier and will therefore vary with dopant type, crystallographic orientation, type of semiconductor, etc. A typical value for A^{**} for electrons in metal-silicon systems is 110 A/cm^2 K^{-2} (Sze, 1969b). Measurements of I_f versus V_f, plotted as log I_f versus V_f, typically yield two parameters, ϕ_B and n, according to the expressions:

$$\phi_B = \left(\frac{kT}{q}\right) \ln\left(\frac{A^{**}T^2}{J_0}\right) \quad \text{(III-4)}$$

and

$$n = \left(\frac{q}{kT}\right) \frac{\partial V_f}{\partial (\log I_f)} \quad \text{(III-5)}$$

The barrier height is determined by finding the saturation current density, J_0, by extrapolating the linear portion of the log I_f versus V_f plot to its intercept at zero forward voltage. The ideality, n, is introduced into Eq. (III-3) to allow for a deviation from purely thermionic emission such as tunneling, bias dependence on ϕ_B, or recombination currents. The presence of any of these factors can increase n beyond its "ideal" value of unity. The ideality is determined from the inverse of the slope of a plot of log I_f versus V_f as indicated by Eq. (III-5) above.

A plot of barrier height versus a weighted average of the work functions of Si and the metal overlayer, as initially calculated and reported by Freeouf (1981), is shown in Fig. III-5. Freeouf used the quantity $\phi_S \approx (\Psi_M \Psi_{Si}^4)^{1/5}$ to represent the barrier height of a hypothetical interfacial silicide layer presumed to exist between the indicated silicide phase and the single crystal silicon. Here ϕ_S is the barrier height of the silicide layer in contact with Si and Ψ_M and Ψ_{Si} are the metal and silicon work functions, respectively. This layer is intended to represent the interfacial layer structure and provides a reasonable fit to the data with the possible

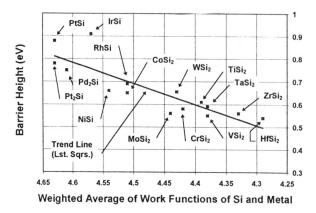

FIGURE III-5 Plot of barrier height $\phi_s \sim (\psi_M \psi_{Si}^4)^{1/5}$, a weighted average of work functions of Si and the metal overlayer (indicated by subscript M). (After Freeouf et al., 1981)

exception of the IrSi phase, which is somewhat more above the theoretical prediction, although it appears to lie within a few standard deviations of the trend line. Other authors, notably Andrews et al. (1975) and Ottaviani et al. (1980), have made other correlations with structure-bonding characteristics such as the heat of formation of the silicide and the melting points of compositionally similar eutectic phases, respectively.

3.3 CONTACT RESISTANCE

It is appropriate to start our discussion of contact resistance by examining the doping level (N_D) dependence of the conduction of the forward biased metal and n-type semiconductor interfaces. As shown in Fig. III-6, there are three distinct regimes corresponding to three mechanisms. In descending order of doping level, they are (1) field emission or tunneling, where the excitation is due to the electric field alone; (2) thermionic field emission, where conduction takes place by a combination of thermionic emission and tunneling; and (3) thermionic emission, where thermal excitation of electrons over the barrier is the predominant process. The silicon doping level controls which regime dominates, by determining the barrier width. Figure III-7 shows the case of the field emission region, (1), corresponding to high doping levels as well as the two other regimes indicated above (Yu, 1970).

The expression for specific contact resistance, ρ_c, is

$$\rho_c = \left(\frac{\partial V}{\partial J}\right)_{V \to 0} \tag{III-6}$$

where V is the applied bias and J is the forced current density. The units of ρ_c are $\Omega \text{ cm}^2$.

The corresponding expressions for the three regimes discussed above are:

Semiconductor Contact Technology

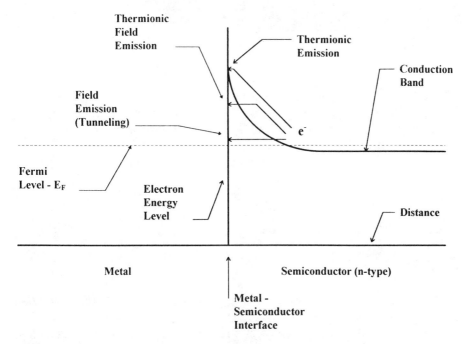

FIGURE III-6 Three regimes of specific contact resistance. (F, field emission; T-F, thermionic field emission; T, thermionic emission.) (After Nicollian et al., 1978)

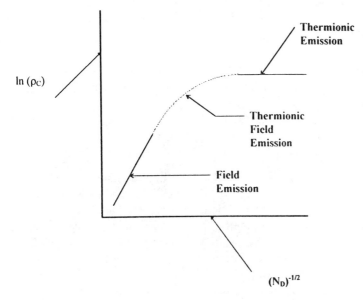

FIGURE III-7 Specific contact resistance, ρ_c, as a function of doping level, $N_D^{-1/2}$. (FE, field emission; T-FE, thermionic field emission; TE, thermionic emission.) Note that in the field emission regime, ρ_c has a linear dependence of ρ_c on $N_D^{-1/2}$, and in the thermionic emission regimen, ρ_c is independent of $N_D^{-1/2}$. (After Yu, A.Y.C., et al., 1970)

1. Field emission (tunneling; highest N_D)

$$\rho_c \approx \exp\left(\frac{\phi_B}{\sqrt{N_D}}\right) \tag{III-7}$$

where N_D is the doping concentration at the contact. In this case, conduction is entirely through the narrow barrier by quantum mechanical tunneling. This corresponds to the linear region of the curve at low values of $(N_D)^{-1/2}$ as plotted in Fig. III-7.

2. Thermionic field emission (intermediate N_D)

$$\rho_c \approx \exp\left\{\frac{\phi_B}{\sqrt{N_D}[\coth(E_\infty/kT)^{-1}]}\right\} \tag{III-8}$$

where E_∞ is defined by the expression

$$E_\infty = \left(\frac{q\hbar}{2}\right)\left(\frac{N_D}{m^*\varepsilon}\right)^{1/2} \tag{III-9}$$

and q is the electron charge, $\hbar$ is Planck's constant, m^* is the effective mass of the charge carriers, and ε is the dielectric constant. In this regime, carriers flow past the barrier by a two-step process consisting of (a) thermal excitation from the band level up to a higher energy level where the barrier width is narrow enough to support tunnel currents; and (b) tunneling through this narrower portion of the barrier. This corresponds to the dotted portion of the curve in Fig. III-7.

3. Thermionic emission (lowest N_D)

$$\rho_c \approx \exp\left(\frac{\phi_B}{kT}\right) \tag{III-10}$$

Thermionic emission occurs when carriers fully surmount the energy barrier by thermal excitation. The progression of ρ_c behavior indicated by cases 1 to 3, corresponds to decreasing concentration of dopant in Si, i.e., for case 1, the doping is heavy ($N_D \sim 10^{20}$ cm^{-3}), for case 3 it is lighter ($N_D \approx 10^{17}$–10^{15} cm^{-3}). In this instance, ρ_c is independent of N_D and corresponds to a constant value of ρ_c as indicated in Fig. III-7.

The values of ρ_c are typically determined at the highest obtainable dopant concentrations, that is, at the dopant solubility limit in Si, as the lowest possible ρ_c values are sought for optimal circuit operation. This corresponds to case 1 and therefore ρ_c will depend strongly on the doping level. Consequently, concentration levels of dopants are usually reported along with the measured values of ρ_c. Another notable characteristic of the field emission regime is the absence of any significant temperature dependence. A plot of $\ln \rho_c$ versus $\phi_B/\sqrt{N_D}$ is shown in Fig. III-8, which includes both calculated and experimental values of contact resistivity versus doping concentration for PtSi/Si interfaces (Nicollian et al., 1978). Due to the large differences in the barrier heights of PtSi to n-Si (0.85 eV) and p-Si (0.25 eV), one anticipates a large difference in ρ_c, as indicated by the tunneling calculations. However, the actual data indicate the ρ_c values for n-Si and p-Si are nearly the same,

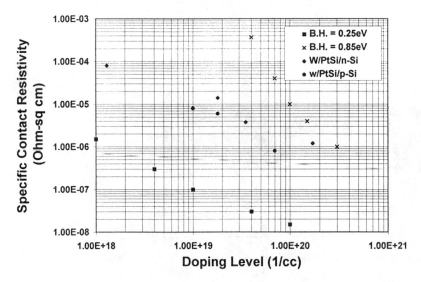

FIGURE III-8 Plot of log (ρ_c) versus log (N_D). Data points plotted as solid squares represent calculations from a tunneling model assuming a PtSi/p-Si contact, and crosses represent the calculations for PtSi/n-Si contacts. The experimental data is plotted as solid diamonds for W/PtSi/n-Si and as solid circles for W/PtSi/p-Si. Experimental measurements are much less sensitive to barrier height (0.25 eV on p-Si and 0.85 eV on n-Si) than are model calculations because of the redistribution of dopants during silicide formation. Concentrations are indicated in scientific notation where, e.g., 1E-08 is equivalent to 1×10^{-8}. (After Chang et al., 1971)

that is, Eq. (III-7) is not strictly obeyed. It was eventually discovered (for example, see Witmer et al., 1982, 1983) that this occurs due to dopant redistribution or the so-called "snow-plow" effect that increases the effective dopant level in the immediate vicinity of the contact, thereby lowering ρ_c below the anticipated value. This will be discussed again in a later section.

4.0 MATERIAL ASPECTS

Proper selection and control of material properties are essential for a successful contact technology, that is, for forming an electrical contact with acceptable electrical characteristics. Relative to the more refractory semiconductors (Si, polysilicon) and inorganic dielectrics (SiO_2 and Si_3N_4) used to build Si devices, the metals used for contacts and interconnections are quite reactive at BEOL processing temperatures of 400–450°C. Strategies that enhance the stability of contacts include alloying, layering of materials in a particular sequence, utilizing diffusion barriers, etc. These strategies are employed to retard rates of reactions to suitably low values so that Si contacts and the devices they access remain functional over the projected lifetime of the chip. Control of the kinetics of metallurgical reactions is therefore key to building high yielding and reliable circuits. A typical FET contact structure that might be used in present-day integrated circuits is shown in Fig. III-9. This

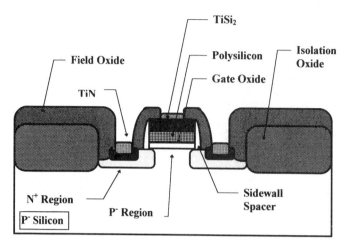

FIGURE III-9 Advanced contact structure featuring self-aligned source and drain, self-aligned silicide ($TiSi_2$) for ohmic contacts, and self-aligned TiN diffusion barriers to prevent Al spiking. The TiN barriers are formed by ion implantation of N into $TiSi_2$. (After Chin et al., 1995)

structure includes such advanced features as self-aligned silicide (so-called "salicide") device contacts and self-aligned TiN barriers formed by nitrogen implantation into $TiSi_2$.

4.1 METAL/SILICON REACTIONS

Earlier reviews (Rosenberg et al., 1978; Lau et al., 1978) discussed the related phenomena of Al spiking or penetration and solid state epitaxial regrowth of Si in contact regions. Because of the appreciable solid state solubility of Si into Al as evident from the Al-Si phase diagram (Fig. III-10, Murray et al., 1984), sintering of pure Al in contact with Si will cause dissolution of Si into Al until the solid solubility requirement is met. Si reacts with Al locally, at weak points in a residual oxide film on the Si surface. As Si dissolves and migrates into Al, Al stays in contact with the recessing Si surface to maintain minimal interfacial energy. This results in penetration of Al, taking the form of a "spike" or inverted pyramid. The detailed morphology will depend on the crystallographic orientation of the Si wafer. Both Totta et al. 1967 and Lane 1970 have given early accounts of Al penetration. A schematic of these effects on an early device structure appeared in Fig. III-1.

One example of the serious consequences this causes for bipolar devices is shorting of the emitter to the collector, that is, penetration through the base region of a bipolar transistor as shown in Fig. III-11. Also, upon cooling of a sintered contact, Si can regrow on the contact surface as hillock-like structures, because the amount of Si dissolved in Al at sintering temperatures exceeds the solubility limit upon cooling. With reference to Fig. III-12, it is instructive to consider the Al contact as divided into two regions: (1) the Al volume that lies at or within one diffusion length from the Si interface, that is, within $\sqrt{D_{Si}t}$, where D_{Si} is the dif-

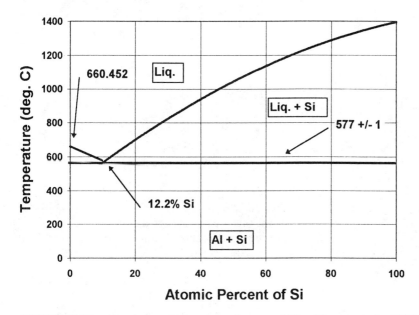

FIGURE III-10 Aluminum-silicon phase diagram. (After Murray et al., 1984)

fusivity of Si in Al at the sintering temperature and t is the duration of the sintering, and (2) the remainder of the Al land that lies beyond $\sqrt{D_{Si}t}$. In region (1), excess Si was observed to precipitate into epitaxial, Al-doped, rectifying structures of p-type Si on n-type Si surfaces, but in region (2), Si precipitates within the Al as small crystallites, and preferably near a free surface (McCaldin et al., 1972). The

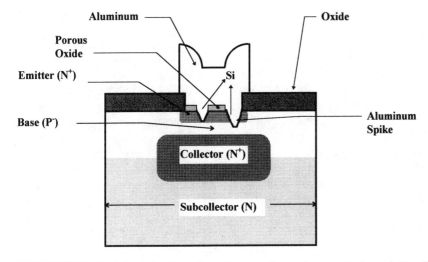

FIGURE III-11 Incipient shorting of emitter to collector by penetration of Al spike through base.

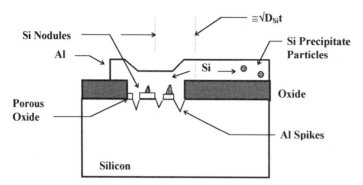

FIGURE III-12 Formation of Si nodules and precipitates in Al metal lines. For distances of less than one diffusion length (of Si in Al), the preferred nucleation morphology is as nodules of Si at the Al/Si interface. For distances greater than one diffusion length, the preferred nucleation morphology is as Si precipitate particles within the Al metal.

presence of this Al-doped, p-type Si on the contact surface in region (1) alters the barrier height and therefore the I-V properties of Schottky diodes. Early reports of these electrical effects are Chino (1973), Basterfield et al. (1975), and Card et al. (1975), Card (1976), and Reith et al. (1976). Rosenberg et al. (1978), reviewed the effects of the duration of sintering and of aging of sintered diodes. Sintering produced changes in barrier height and ideality, as shown in Figs. III-13 and III-14. The changes were interpreted as due to the formation of penetration pits followed by the formation of a p-type layer on the contact. The p-type layer causes the increase in the barrier height. Aging of contacts at temperatures lower than the sintering temperature lowered the barrier height because aging causes migration of the Al dopant in the regrown Si to internal sinks such as dislocations where they were no longer electrically active.

Inoue et al. (1976) observed the migration of Si into overlying Al lands directly by using the methods of electron microprobe analysis combined with Auger spectroscopy to investigate both lateral and through-the-thickness distributions of Si in the Al. They found a Si rich layer within the Al immediately above the contact area and a relatively constant concentration of Si within the Al lying over SiO_2 by using through-the-thickness profiling. Lateral profiling showed a high Si concentration at the edges of the contact holes. Increased sintering time increased penetration, as expected. Blair et al. (1977) conducted subsequent studies of the influence of vacuum ambients on Si contacts. They concluded that common vacuum chamber contaminants such as H_2O, O_2, and CO, at pressures up to 5×10^{-7} torr (6.7×10^{-5} Pa), did not degrade the electrical characteristics of evaporated Al/Si contacts. Another comparison of several methods of reducing polysilicon dissolution and contact resistance at Al/polysilicon interfaces led Naguib et al. (1978) to advocate use of Al-2wt%/Si metallization to saturate the Al "sink," as it were, with Si "solute." They suggested that alloying was the most practical technique to minimize the undesired effects.

Faith et al. (1984) undertook to eliminate hillocks on the metal surfaces by using fast heat pulses. They sintered Al-1%Si metallizations using 5-s-duration

Semiconductor Contact Technology

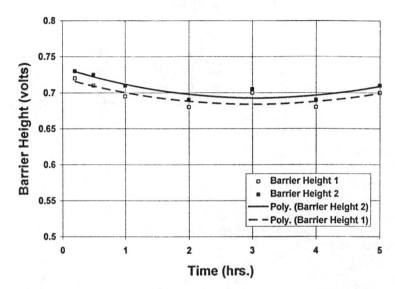

FIGURE III-13 Effect of sintering time on barrier height of Schottky barrier diode device. There are two sets of data displayed, indicated as 1 and 2 in the accompanying legend. The "poly" labels in the legend indicate that the drawn curves are least-squares fit to second-order polynomials. (After Reith, 1976)

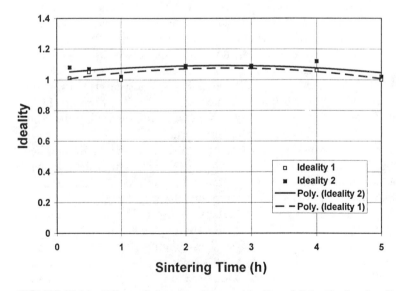

FIGURE III-14 Effect of sintering time on ideality of Schottky barrier diode device. There are two sets of data displayed, indicated as 1 and 2 in the accompanying legend. The "poly" labels in the legend indicate that the drawn curves are least-squares fit to a second-order polynomials. (After Reith, 1976)

pulses at 490–540°C and compared the results with those obtained by annealing in a furnace at 425°C for 30 min. Depending on the heat pulse temperature, the contact resistances and diode leakages were comparable to those obtained using standard furnace annealing, but did not result in undesired hillocks. Wu (1985) also claimed to achieve reduced penetration and provide diodes with excellent characteristics by using rapid pulse heating. Sintering resulted in two orientations for the epitaxial growth of Al spikes in the metallurgical system consisting of Al-1%Si/TiW/Si (Chang, 1988). These were (1) Al[011]//Si[$\bar{1}$23] and Al(200)//Si(1$\bar{1}$1); and (2) Al[001]//Si($\bar{1}$12] and Al(200)//Si(1$\bar{1}$1) (italics indicate the negative indices). Using finite element methods, Ohdomari et al. (1990) have calculated the spatial distribution of compressive stress occurring in the Al in contact regions. Upon simulated heating, they found the stress to be highest at corners, i.e., at the contact openings and edges, due largely to thermal expansion mismatch between Al, SiO_2, and Si. The authors then argued that these compressed areas will be sites for greater Si precipitation upon cycling back to room temperature. They based this prediction on a model of Si particle precipitation that was derived from classical nucleation theory and incorporated the Gibbs-Thompson effect. The results of the model were that the Si solubility was higher in the compressive stress field existing at sintering temperature, but as the sample cooled and the stress reduced, the Si accumulated in the compressive field could migrate elsewhere and cause a higher growth rate of precipitates.

Hirashita et al. (1988) observed the solid-phase epitaxial growth of Si at contact holes through the Al-Si alloy directly by transmission electron microscopy. They obtained clear analytical evidence for the mechanisms previously inferred from electrical and post-sintering structure analysis. Since attempts to remove native oxides from the bottom of contact holes are never totally successful, Nogami et al. (1994) suggested an alternative procedure based on their observation that a very thin silicon oxide film can actually prevent Si solid phase epitaxial growth onto the Si contact. They could obtain low contact resistance for both p- and n-type Si by sputtering AlSi onto the thin oxide surface. The thinness of the oxide film apparently supported sufficient tunneling current to provide for an acceptable contact resistance. A difficulty of this process is the ability to consistently form native oxides that are sufficiently thin to support the necessary tunneling currents, but at the same time, thick enough to suppress Si(Al) regrowth. Additionally, if the oxide layer is *too* thick it will form a current-blocking interface.

A novel method of reducing the formation of Si nodules at the Al/Si interface, patented by Wong (1992), involved modification of grain size within the thickness of the contacting metallization by changing sputtering conditions. In the simplest embodiment, a bilayer of Al was formed that had relatively large grains for the portion directly on Si, and smaller grains in the next Al layer. This structure created a divergence in the diffusion flux of Si in Al by abruptly reducing the density of high diffusivity paths in the layer next to Si. The resulting flux divergence was responsible for Si precipitation on this micro-structural interface, instead of at the Si surface. Also, a thin layer of interfacial oxide between the two Al layers created by the interrupted deposition would have a similar effect.

4.2 SILICIDE FORMATION

Silicides derive some of their importance to the formation of Si contacts due to their ability to (1) reduce residual oxides on Si, thereby providing the contact area with spatially uniform electrical properties, and (2) form discrete, intermetallic phases that bind metallic components in thermally stable compounds, thereby preventing junction poisoning from metal diffusion. Summary accounts of the kinetics of silicide formation appeared in earlier reviews (Tu et al., 1978; Murarka, 1983). The reader may consult these sources for excellent fundamental treatments of the silicide phase formation process. Included here is a brief discussion of silicide formation, but mainly in the context of contact formation.

4.2.1 Platinum Silicide

Amouroux et al. (1971) provided one of the early accounts of the use of PtSi for device contacts by observing the formation of these layers on Si semiconducting devices. The study was part of the development of a PtSi/Mo/Au contact metallization for a microwave transistor. They presputtered the Si surface to remove contamination, deposited a 100–500 Å layer of platinum by sputtering, and formed the PtSi layer by annealing at 450–550°C, which was below the Si-Pt eutectic point. Then they etched away the unreacted Pt using aqua regia. Rand et al. (1974) pointed out that the resistance of PtSi to the etch was provided by a thin, approximately 100 Å, layer of SiO_2 that forms on the surface of the silicide during annealing (Fig. III-15). Removal of this oxide would expose the PtSi to an etchant that could readily dissolve it, so its existence is essential to the process.

4.2.2 Palladium Silicide

Ho et al. (1982) investigated the matallurgical stability of $Al/Pd_2Si/Si$ contacts by mapping the phase diagram of Al-Pd-Si at 400°C. Reaction kinetics relating to rates of Pd_2Si decomposition and Al reaction and penetration were examined for both polycrystalline and epitaxial Pd_2Si. Changes in the electrical characteristics were explained by Si regrowth from silicide dissociation. The contact resistance of $Mo/Pd_2Si/Si$ structures to both pn^+ and np^+ junctions using a Kelvin, 4-terminal method was reported by Singh et al. (1986). They reported values of ρ_c of

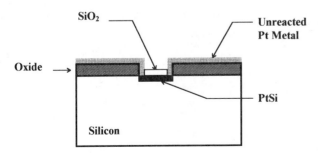

FIGURE III-15 Formation of PtSi by blanket deposition and sintering of Pt on contact structure. A layer of SiO_2 protects the silicide from removal during strip of unreacted Pt in aqua regia etch.

0.28–0.31 × 10^{-6} Ω-cm² for p^+ contacts that were 1.3 μm in diameter and the corresponding n^+ contacts had ρ_c values ranging from 0.15–0.19 × 10^{-6} Ω-cm². The authors further studied the thermal stability of these values and suggest the use of these contacts for VLSI provided temperatures of 475°C are not exceeded.

4.2.3 Titanium Nitride/Titanium Silicide

$TiSi_2$ was once undisputed as the contacting metal for MOSFET structures because of its high temperature stability, its compatibility with self-aligned processing, its relatively high conductivity when compared to other silicides, and its compatibility with TiN barriers. Presently, $CoSi_2$ is preferred in some applications according to Murarka et al. (1993a). This stems from less sensitivity to processing conditions below 900°C and greater thermodynamic stability relative to $TiSi_2$.

The lowest resistivity form of titanium silicide is the C54 phase of $TiSi_2$ which forms at 800°C and has a resistivity of 12–15 $\mu\Omega$-cm. In addition, it is possible to convert a portion of the silicide layer to nitride, providing a barrier against reaction with Al. A relatively common method for doing this is annealing of a $TiSi_2$ layer in the presence of NH_4 vapor as shown schematically in Fig. III-16. Norstroem et al. (1983) showed that bilayers of TiN/Ti deposited by magnetron sputtering will react to form $TiSi_2$. In the same step, the TiN layer is transformed from its deposited composition to the true stoichiometric compound, TiN. The nitride layer is an effective diffusion barrier against Al up to 550°C. Ye et al. (1989) described a novel technique to form shallow junctions with low resistance silicide contacts in MOSFETs for VLSI. It involved first siliciding the gate and source-drain region in a self-aligned manner and next, implantation of Ti through the metal layer to transform the Si into $TiSi_2$ by ion mixing. Figure III-17 is a schematic illustration of this self-aligned process.

$TiSi_2$ forms two phases, a low-temperature phase designated as C49 and a high temperature phase designated as C54. While both are $TiSi_2$, the C49 phase forms at temperatures of 625–675°C with a resistivity of 60–65 $\mu\Omega$-cm. The C54 phase is typically formed in a second anneal, following the C49 phase formation, at temperatures at or exceeding 800°C, and has a much lower resistivity of only 10–15 $\mu\Omega$-cm. These steps are illustrated in Fig. III-18.

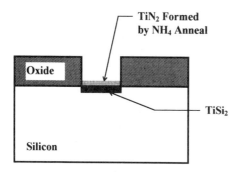

FIGURE III-16 Nitriding of $TiSi_2$ contact by sintering in NH_4 atmosphere.

Semiconductor Contact Technology

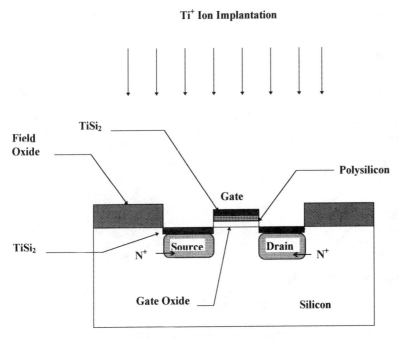

FIGURE III-17 Self-aligned, TiSi$_2$ contact structure formed by ion mixing. (After Ye et al., 1989)

Yu et al. (1989) reported methods of optimizing the formation of TiS$_2$/polysilicon structures to improve the contact resistance; these included annealing at 800°C to form the low-resistivity C54 phase and ion implantation of As ions to mix the silicide/polysilicon interface before a second anneal. Arsenic implantation doses of less than 10^{16} cm^{-2} into the polysilicon did not inhibit formation of the TiSi$_2$ layer. Perea et al. (1990) have characterized TiSi$_2$ contacts to n^+ and p^+ Si for contact sizes of 0.3 μm × 0.2 μm and found well-behaved electrical and structural characteristics.

However, problems did arise because of the interaction of TiSi$_2$ with ion-implanted dopants. Mitwalsky et al. (1990) found that unacceptably high contact resistances occurred if B and As dopants were implanted into Si before TiSi$_2$ formation took place. They measured both the sizes of particles and depth distributions of precipitates of metal-dopant compounds of TiAs and TiB$_2$. A schematic of the precipitate particles that develop within the silicide layer is shown in Fig. III-19. The authors concluded that TiSi$_2$ is neither a good candidate to be in contact with implanted silicon that undergoes a high-temp heat treatment, nor as a preformed silicide layer that receives ion implanted dopants for subsequent out-diffusion into silicon. The reasons for this conclusion can be seen in Fig. III-20a,b. The out-diffusions of As and B ions from the implanted silicide sources are clearly suppressed with respect to the concentrations and diffusion depths achievable with polysilicon as a source. In particular, the resulting concentrations of dopants in Si are orders of magnitude less. Obviously, the formation of intermetallics within the

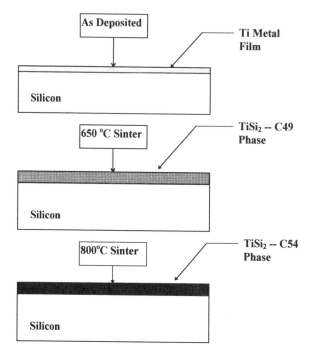

FIGURE III-18 Schematic indicating annealing sequence needed to form C49 and C54 phases of TiSi$_2$.

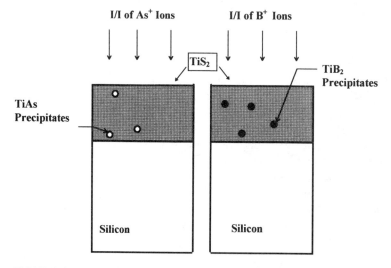

FIGURE III-19 Schematic of formation of intermetallic precipitates in ion implanted TiSi$_2$. After implantation of dopant ions, a sintering step produces the intermetallic precipitates of TiAs and TiB$_2$, for As$^+$ and B$^+$ implants, respectively.

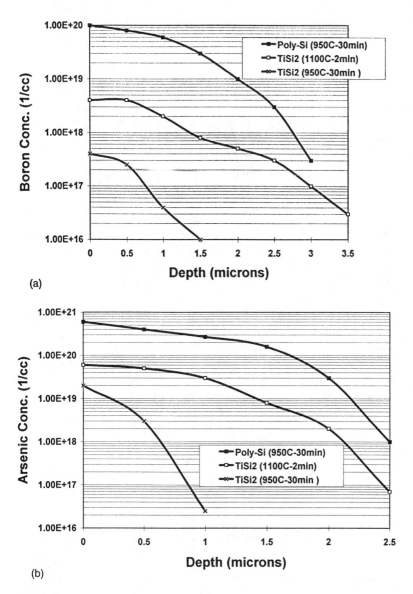

FIGURE III-20 Out diffusion profiles of dopants from polysilicon and $TiSi_2$, (a) Boron profiles, (b) arsenic profiles. For both (a) and (b), the anneals of polysilicon take place at 950°C for 30 min and for $TiSi_2$, at either 950°C for 30 min or 1100°C for 2 min.

silicide layer is thermodynamically favored over dissolution of dopants into single crystal Si.

Concerns for the potential problems created by intermetallic compounds have stimulated several recent investigations. Probst et al. (1991) investigated compound formation including TiAs, TiB_2, TaAs, and TaB_2 for both $TiSi_2$ and $TaSi_2$ films and assessed its impact on dopant diffusion and contact resistance. They noted several "detrimental consequences" which were (1) reduction in the solubility of As and

B; (2) retardation of the outdiffusion of dopant into Si or polysilicon; and (3) higher contact resistances since precipitation reduces the carrier concentration from the usual 10^{20} cm^{-3} range to only the 10^{18}–10^{17} cm^{-3} range. Despite limitations of using TiSi$_2$ films in post silicide, ion implant applications, this contacting material has been extensively investigated from both structural and electrical perspectives. Saito et al. (1993) used selective TiSi$_2$ to form ohmic contacts to very shallow junctions. Using XPS analysis, Lee et al. (1993) investigated these structures as potential ohmic contacts and barriers for submicron devices. In situ annealing in an atmosphere of NH$_3$ formed the TiN/TiSi$_2$ bilayer. Both TiN and TiSi$_2$ layers had stable, crystallographic structures when annealed at 800°C. Boron atoms redistributed both within the TiN layer and at the silicide/Si interface. Liuah et al. (1993) investigated various physical and electrical characteristics of TiSi$_2$/(100)Si including microstructures, sheet resistance, Schottky barrier height, contact resistance, and junction current leakage. By focusing their investigation mainly to temperatures at or below 450°C, they were able to measure SBD characteristics on silicide interfaces with a thin, amorphous, "interlayer" of Si that was either As or B doped. The amorphous interlayer (a-Si) is formed by low-temperature diffusion between the silicide and monocrystalline Si (c-Si); it affects ρ_c measurably. Barrier heights were 0.52–0.54 eV for amorphous n-Si and 0.59–0.57 eV for amorphous p-Si. Values of ρ_c were lowest when an a-Si layer was present, and they reported values of 1.4×10^{-7} Ω-cm^2 for n-type a-Si and 3×10^{-7} Ω-cm^2 for p-type a-Si. The lowering of ρ_c from c-Si values might be caused by an atomic redistribution of dopant from the thin, a-Si layer to the c-Si interface as the a-Si layer formed. Chou et al. (1993) also investigated the influence of boron dopant on TiSi$_2$. They found that the C49 phase formed at 700°C and that almost all the film was transformed into the C54 phase at 800°C. As the concentration of impurities increased, the resistance of the silicides increased and the thickness decreased. Revva et al. (1994) deposited alternating layers of Ti and Si by electron beam evaporation and used them to form TiSi$_2$/Si contacts. They measured the barrier heights for these multilayered structures at progressively increasing sintering temperatures and compared them to conventional, single-layer structures. Although at low sintering temperatures, the multilayered film structure had higher barrier heights, at higher temperatures, the ϕ_B values for single and multilayered structures are nearly identical.

4.2.4 Nickel and Cobalt Silicides

Although there is little or no use of nickel silicide in manufacturing integrated circuit products, significant learning about phase formation, epitaxy, etc., arose from initial studies of Ni phases and these apply to the structurally similar Co silicide phases. Hoekelek et al. (1978) studied an Al/NiSi/Si structure on n-type (111) Si. The diodes consisted mainly of the NiSi phase that formed a Schottky diode with a barrier height of 0.62 eV. Upon heating in contact with a thin Al overlayer, the NiSi layer was transformed into NiAl$_3$, with a consequent increase in the barrier height to 0.76 eV. They observed stable electrical characteristics and an absence of penetration for this layer up to 500°C. As mentioned earlier, CoSi$_2$ has been studied extensively as a replacement for TiSi$_2$, because of its comparably low resistivity (14–18 μΩ-cm) and its suitability for forming self-aligned silicides, for which the term "salicides" had been coined (Murarka, 1986; Murarka et al., 1987;

Murarka, 1993b). A thin film of Co reacts readily with Si or poly-Si but negligibly with SiO_2. Based in part on this property, Murarka et al. (1987) developed a two-step annealing process. This involved heating to form metal rich silicide at a temperature below 450°C, followed by removing the unreacted metal in a wet etchant. Subsequently the silicide is heated to over 600°C to form the low-resistance disilicide ($CoSi_2$) layer. Alternatively, Broadbent et al. (1989) used rapid thermal heating to form $CoSi_2$. J. H. Werner et al. (1993) investigated the interface structure and the Schottky barrier heights of buried $CoSi_2/Si(001)$ produced by high dose ion implantation of Co^+ ions followed by rapid thermal annealing. The layers so formed showed excellent epitaxy and smooth {001} interfaces. Both 6-fold and 8-fold coordinated Co atoms existed at the interfaces. Barrier height measurements indicated higher values for the lower or more deeply implanted layers of $CoSi_2/n$-$Si(001)$ than for the shallower layers. Stress induced the precipitation of dopants diffused into Si from both $CoSi_2$ and $TiSi_2$ layers according to La Via et al. (1993). Both rapid thermal annealing and conventional annealing of As, P, and B ions, implanted through the silicide layer, resulted in precipitation of the dopants at the silicide/Si interface. The authors speculated that the high tensile stress, induced by the silicide layer on the surface Si region, and its subsequent influence on the solid solubility of the dopant caused the precipitation.

Chiou et al. (1994) studied the thermal stability of Cu-$CoSi_2$ contacts to p^+n shallow junctions and concluded that the structure had poor thermal stability and a TiW diffusion barrier was highly desirable.

4.2.5 Snow Plow Effect

Specific contact resistance values for various silicides do not correlate well with Eq. (III-7) because of low temperature diffusion of dopants during silicide formation. Witmer et al. (1983) electrically analyzed the profile of implanted As and then the redistribution of As which was caused by the formation of Pd_2Si. He and his coworkers correlated these measurements with previous ion channeling observations made by Witmer et al. (1982) of this same effect. "Snow-plow effect" is the name typically applied to this phenomenon. The increase in As concentration beneath the silicide/Si interface that is shown schematically in Fig. III-21 reduced the contact resistance, and was suggested as a way to adjust barrier heights. One obvious benefit this has is compensation for the anticipated increase in ρ_c if a contacting silicide or metal layer with a high ϕ_B needs to be used, because of a reduction in the ratio ($\phi_B/\sqrt{N_D}$), as predicted by Eq. (III-7). Rejection of dopants, in this case As, from the region transformed to silicide will increase N_D from initial implanted levels thereby lowering this ratio and ρ_c values.

4.3 DIFFUSION BARRIERS

4.3.1 TiW Films

Diffusion barriers are an indispensable part of a reliable contact structure; they prevent junction penetration by Al or poisoning by Cu. Fortunately, effective barriers were found nearly in concert with the recognized needs, and several of these barrier concepts and related material properties are described in this section.

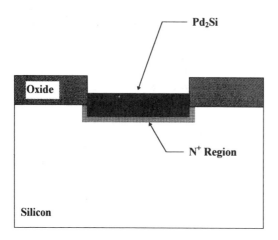

FIGURE III-21 Snow plow effect indicating accumulation of N^+ dopant occurring during formation of Pd_2Si phase.

Ghate et al. (1978) reported the earliest study pointing out the merits of TiW as a diffusion barrier for Al metallization. They used a pseudo-alloy of TiW (10:90 wt%) in structures of Al/TiW/Si and looked for sheet resistance changes upon annealing as evidence for metallurgical reactions. After annealing, the resistivity of the Al layer had increased less than 10%. This relatively small change is most likely due to limited diffusion of Ti and/or W into the Al. Auger profiling of Si/TiW/Al structures indicated that the barrier remained intact, i.e., there was no penetration of Al to the TiW/Si interface after annealing at temperatures of 450, 500 and 550°C, which suggested application to Al/TiW/PtSi contact structures in integrated circuits.

The reaction kinetics of Al/$Ti_{22}W_{78}$ and A-2%Cu/$Ti_{22}W_{78}$ thin film couples were measured by Olowolafe et al. (1985) using Rutherford back scattering. The TiW samples were sputter deposited and the Al and Al-2%Cu layers were evaporated in a separate system. The surface of the $Ti_{22}W_{78}$ was either sputter cleaned or intentionally contaminated with air to produce an interfacial oxide. The accumulation of Ti on the surface of the overlying Al or Al-2%Cu film provided a measure of the extent of the reaction with the sputter cleaned, $Ti_{22}W_{78}$ barriers. The quantity of accumulated Ti had a linear dependence on the square root of the reaction time for the four temperatures investigated (530, 500, 475 and 450°C), indicating a diffusion controlled process. The barriers that were exposed to air prior to deposition of Al or Al-2%Cu resisted any reaction when heated for 500 or 550°C for 10 h but did eventually react at 600°C. The activation energy derived from this data was 2.4 eV and was the same for both Al and or Al-2%Cu. Although it is not possible to definitively determine the rate limiting mechanism, one can argue that this energy is much too high to be the diffusion of Ti through Al (or Al-2%Cu), which should be in the range of 1.0–1.5 eV. The relatively high energy observed suggests grain boundary diffusion within the refractory barrier. As it turns out, this is more than an idle speculation because in a companion study, Palmstrom et al. (1985) used TEM to show that the $Ti_{22}W_{78}$ film was penetrated by Al at grain boundaries. Therefore the activation energy observed by Olowolafe et al. (1985) is very likely the diffusion energy for Al along grain boundaries in the $Ti_{22}W_{78}$ film. Once Al penetrated the barrier film, Ti could easily migrate to the surface with its much lower energy. This lower energy should not be rate limiting and therefore not detectable by the measurements used. Palmstrom et al. (1985) also detected the intermetallics $TiAl_3$ and WAl_{12} in the reacted films.

Babcock et al. (1986) investigated Al penetration into TiW contacts on Si. They determined a barrier height of 0.51 eV to n-Si and showed that structures with 20 at% Ti were stable up to 500°C and 30 min. Increasing the Ti content to 30 at% made them stable up to 600°C. The barrier properties of TiW were improved by sputtering it in an $Ar-N_2$ atmosphere; a reduced reaction of W with overlying Al layers was also noted. The investigation by Wolters et al. (1986) covered both Al/TiW:N/Si contact structures and Si/Al-Si/TiW:N/Al multilevel structures. Although the incorporation of N in TiW increased the resistivity of this barrier film, the diffusion barrier properties were improved and the resulting structures still had acceptable dry etching and contact resistance characteristics.

Grove et al. (1989) annealed a W/Ti bilayer in a N_2 atmosphere to create structures such as $Al/W/TiN_y/TiSi_z$ on both n^+ Si and p^+ Si, thus forming a reliable contact diffusion barrier that was characterized by several profiling techniques. Auger profiles showed the stability of these structures after they were annealed at 650°C for 20 min. Thalapaneni (1993) patented the combination of Ti and TiW films for use as an improved barrier and electrical contact to Si. The Ti layer reduced the contact resistance and helped to block the diffusion of Al along TiW grain boundaries. Additionally, a plasma etch, which took about 250 Å of Si from the contact surface, removed the damage created by the more energetic, anisotropic etch. The anisotropic etch was used initially to open the contacts and to access higher dopant concentration that occurs a few hundred Angstroms beneath the Si surface. The barrier properties of TiW, applied to both Al- and Au-based conductors have been generally reported to be satisfactory. A recent study by Evans et al. (1994) has uncovered a potential problem for Au metallizations. Barrier properties of 3%Ti-W films sputtered in N_2 exhibited a catastrophic failure mode that correlates with the occurrence of an A15 structure or β structure in the TiW film. These structures are known to occur in thin films of several refractory, bcc metals (W, Ta) and are undesirable because of their metastability and high resistivity relative to the bcc phases. They transform to the bcc phase upon annealing creating high stress and loss of adhesion. However, addition of N to the TiW film to form TiW:N layers appeared to reduce the problem to negligible proportions.

4.3.2 TiN and TiN_xO_y Films

Kohlhase et al. (1987) investigated the electrical and barrier stability properties of Al-Si/TiN/Ti structures and reported promising attributes for application to 4 Meg DRAM memory circuits. Kumar et al. (1987) reported an extensive study of the barrier properties of TiN, in which a Ti target was sputtered in $Ar-N_2$, varying the applied bias during deposition. Without applied bias, films appeared dark brown; with bias they were gold. Using structures such as $Al/TiN/TiSi_2$, they found that the gold-colored films were superior as both diffusion barriers and had lower R_c values as compared to the brown films. Films were also sputtered in $Ar-N_2-O_2$

without bias; addition of O_2 to the Au colored films increased R_c but decreased the failure rate due to penetration during sintering. The phenomena of the in situ growth of an Al_xO_y layer by reaction of Al with oxides present in the TiN:O film is likely responsible for the augmentation of the barrier properties by incorporation of oxygen into the TiN film. For this reason, barriers for Al particularly benefit from incorporation of oxygen or intentional contamination of the barrier surface by air exposure. Figure III-22 contains an illustration of the contact structure following Al_xO_y formation.

Inoue (1990) patented the use of the augmented barrier properties of TiN:O films to prevent the reaction between Al and Si. The barrier layers were 50–200 nm thick, fabricated by sputtering in gas mixtures that contained were 1–5 vol % O_2, maintaining the Si substrate temperature at 350–550°C. The films had a resistivity of 100 $\mu\Omega$-cm which is too high for wiring applications but quite suitable for barrier applications.

TiN barriers sputtered without collimation do not effectively protect contacts with high aspect ratios (i.e., AR > 1.0), according to Pintchovski et al. (1992). The authors concluded that CVD TiN processes were necessary to address the higher ARs associated with 0.35-μm technologies. Yokoyama et al. (1991) investigated the barrier layer properties of TiN films formed by LPCVD, using $TiCl_4$ as the precursor, for application to high AR contact holes. These investigators found good step coverage in contact holes with AR = 1.8 and TiN films of nearly stoichiometric proportions. They also found that they could reduce the Cl content of the films from 5.7 at% at 500°C to <1 at% at 700°C. Annealing in H_2 at 1000°C also reduced the Cl concentration of films deposited at 500°C by a factor of 2. Films formed at 700°C had resistivities of 80 $\mu\Omega$-cm, which is less than films formed at lower temperatures, leading to speculation that the resistivity may be sensitive to the Cl content.

Frahani (1989) reported on the rapid thermal processing (RTP) of Ti films in the presence of NH_3 to form TiN barriers. They found that only relatively thin

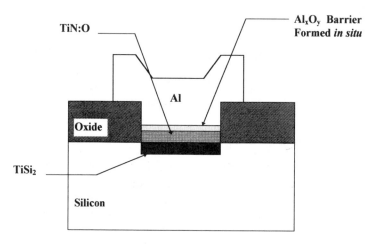

FIGURE III-22 Formation of Al_xO_y layer by reaction of Al with TiN:O barrier. Barrier enhancement by incorporation of oxygen occurs due to the formation of Al_xO_y layer upon sintering.

layers of TiN (200–240 Å) formed in a single step nitridization process and that this occurred over the relatively narrow temperature range of 600–610°C. Effective diffusion barriers required nitridization and the presence of $TiSi_2$ under the TiN as well.

4.3.3 Other Barrier Films

Neppl et al. (1984) reported on the properties of $TaSi_2$ for Al interconnect metallization. They studied the properties of both cosputtered amorphous and polycrystalline films using Schottky and ohmic contacts. When the silicide films contained an excess of Ta, they served effectively as barriers up to 475°C. Investigators Singh et al. (1986) studied the reaction and interdiffusion between Al-Si and Mo films. Thermal annealing of the coupled films resulted in the formation of the $MoAl_{12}$ phase as a nonplanar front. The kinetics of the phase formation was diffusion-controlled, as shown by a parabolic time dependence; the activation energy, E_a, was 5.9 eV. The presence of Si precipitates at the Mo/Al-Si interface explained the high value of E_a and the occurrence of incubation periods before parabolic behavior. Kolawa et al. (1991) investigated the diffusion barrier properties of sputtered Ta-based films. These investigators showed that Ta films were relatively less effective as barriers for Cu since they failed at temperatures of 500°C. In comparison, amorphous binary ($Ta_{74}Si_{26}$) and ternary ($Ta_{36}Si_{14}N_{50}$) films prevented Cu migration up to temperatures of 650 and 900°C, respectively. Reverse diode characteristics were sensitive indicators of barrier effectiveness in this study.

Farahani et al. (1994) evaluated the utility of Nb as a diffusion barrier to prevent Al-0.5%Cu migration into Si. They concluded that Nb was an effective diffusion barrier and that it also significantly enhanced the electromigration resistance of the conducting metallurgy. Structural analysis showed the presence of the $NbAl_{13}$ phase.

Eizenberg et al. (1984) investigated titanium carbide films, with various carbon contents, for their suitability as diffusion barriers between Al and Si; the optimal film had the composition $Ti_{3.1}C$. The film was preannealed at 750°C to form a Ti_xC outer layer, and an inner layer (adjacent to Si), that was a mixture of several phases of titanium silicide (Ti_5Si_3, TiSi, $TiSi_2$). These structures were stable in the presence of Al films for heat treatments at 550°C for 30 min. At temperatures of 600°C, the barriers failed with the formation of Al_4C_3, suggesting that the TiC layer had effectively decomposed.

Eizenberg et al. (1985) also studied the thermal stability of $Al/Pd_xW_{100-x}/Si$ contact systems using AES, RBS, XRD, and the I-V properties of Schottky diodes. They found that W rich alloys, such as $Pd_{20}W_{80}$, provided contact stability even after annealing at 550°C for 30 min. The Pd from the alloy migrated to the Si surface to form a thin Pd_2Si layer and to the interface with Al to form Al-Pd intermetallic compounds. The authors cited that the advantages of this structure was providing both a diffusion barrier and a shallow silicide contact. Takeyama et al. (1993) cited an even higher thermal stability, up to 600°C, in an investigation of $Al/Al_{12}W/W_2N/Si$ structures. Part of the rationale for this structure was the commonality of at least one element across an interface for the two bordering film materials. The intention was to create a lower free energy state and consequently reduce the driving force for metallurgical reactions. Wang et al. (1994) have favored the use of a bilayer structure of $TaSi_xN_y/TaSi_x$ in Si contacts. They showed that the upper layer provided an effective diffusion barrier against Al penetration, while the bottom layer provided low sheet resistance, Ω_s, and a low ρ_c.

5.0 OHMIC CONTACTS

5.1 METAL/SILICON

Ohmic contacts are usually characterized by a value of ρ_c, in, as stated earlier, units of Ω-cm². Contact resistance, R_c, is expressed in units of ohms and is given, therefore, by the expression:

$$R_c = \frac{\rho_c}{A} \tag{3.11}$$

where A is the area of a contact device port, such as the emitter, base, or collector of a bipolar transistor or perhaps the source or drain of a field effect transistor. For those contacts that carry the highest current in a circuit, designers seek the lowest possible resistance values. Examples are the emitter and collector in a bipolar logic circuit or the source and drain in a CMOS memory chip. It is common practice to implant these resistance-sensitive contacts with a dose that at least equals the solubility limit for the particular dopant in Si. This corresponds to As or P concentrations exceeding about $1–2 \times 10^{20}$/cm³, and similarly, for p-type Si with B as the dopant. As Table III-1 indicates, the lowest values of ρ_c attainable are approximately in the range of $1–3 \times 10^{-7}$ Ω-cm². For a 1-μm² contact, the corresponding R_c values are 10–30 Ω, and for 0.5 μm², the values are 40–120 Ω, etc. Due to its much lower current carrying requirements and the higher resistance of the base structure of the transistor, the base contact can function with a lower doping level and, consequently, higher value of R_c. This circumstance is probably responsible for the general lack of experimentally determined values of ρ_c for these contacts. As shown in Fig. III-23, poly-Si is usually used as a circumfrential base contact in bipolar transistors and as part of the defining structure of the emitter contact. Due to grain boundary segregation, boron doped poly-Si can have a carrier concentration of only 1/10 or so of the actual atomic doping level, N_D. This would tend to raise ρ_c, according to the expression, $\rho_c \sim \exp(\phi_B/\sqrt{N_D^*})$, where the asterisk on N_D is introduced here to indicate the electrically active dopant concentration. However, assuming ϕ_B to be ~0.2 eV (PtSi), the lower barrier height on the p-poly-Si contact will act to lower ρ_c.

5.1.1 Bipolar Emitter Contacts

Partly for historical reasons, and partly for its relatively low formation temperature, that would not alter the precisely tailored base widths of npn bipolar transistors, PtSi phases have been extensively used as contacts to bipolar devices. This practice continued with the introduction of "poly-Si emitters" which were poly-Si films that overlayed the c-Si contacts. They were used as solid state diffusion sources for forming the emitter concentration profile. This involved the sequence of As ion implantation into the polysilicon followed by a thermal drive-in of As into the underlying c-Si. To better quantify the properties of the PtSi/poly-Si interface,

TABLE III-1 Values of Specific Contact Resistance

Specific contact resistance			Contact structure			Conditions			Source	
High	Low	Units	Metal	Silicon	Doping	T (°C)	Method	Note	Author	Year
0.12		$\mu\Omega$ cm	Al	Si	n-type		e-beam		Chen	1983
0.15		"	Al	Si	p-type	950	RTA	B-1/1-4 $\times 10^{-5}$	Hara	1983
0.15		"	Al	Si	p-type		e-beam		Chen	1983
0.70	0.60	"	Al-2%Cu/MoSi$_2$	Si		500			Yamamoto	1985
0.15		"	Al/Ti	Si					Ting	1982
0.6	0.2	"	CVD TiSi$_2$	Si					Saito	1993
0.10		"	Mo/Ti	Si	n-type				Kim	1985
0.19		"	Mo/Ti	Si	p-type				Kim	1985
0.84		"	PtSi	Si	n-type		furnace		Huang	1988
0.075		"	PtSi	Si	n-type	1050	RTA		Huang	1988
0.08		"	PtSi	Si	n-type				Mallardeau	1989
0.2		"	PtSi	Si	p-type				Mallardeau	1989
0.19	0.15	"	Pd$_2$Si	Si	n-type				Singh	1986
0.31	0.28	"	Pd$_2$Si	Si	p-type				Singh	1986
2.0	0.8	"	Sel CVD W	Si	n-type				Mariya	1983
0.15		"	Si/Al/Ti	Si					Ting	1982
0.5	1.0	"	Ti$_{73}$Co$_{27}$	Si	n-type				Gromov	1995
0.14		"	TiSi$_2$	a-Si	n-type	450			Luiah	1993
0.30		"	TiSi$_2$	a-Si	p-type	450			Luiah	1993
0.09		"	TiSi$_2$	Si	n-type				Mallardeau	1989
1.0		"	TiSi$_2$	Si	p-type				Mallardeau	1989
24.0		"	W/Ti	Si	p-type		furnace		Chen	1981
3.6		"	W/Ti	Si	p-type		laser	$p = 10^{19}$ cm^{-3}	Chen	1981

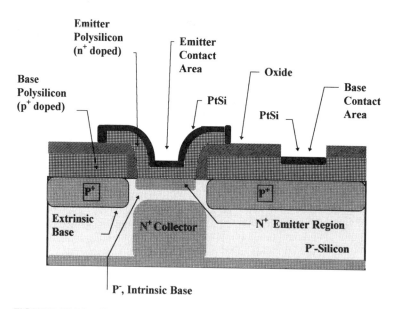

FIGURE III-23 Cross section of bipolar structure including polysilicon base, polysilicon emitter, and PtSi contacts.

Huang et al. (1988) did a detailed study of the contact resistance of PtSi to As-doped poly-Si in which they correlated the value of ρ_c to the fraction of As dopant that was electrically active. Hall measurements determined the carrier concentrations and established correlations between ion implant dose parameters and SIMS determined the As doping level. Postimplant annealing varied the fraction of carriers that were electrically active. Their experimental data is shown in Fig. III-24, where, though convoluted with other contact-physics-related phenomena, the ability to increase ρ_c by charge carrier trapping at grain boundaries is evident through comparison with c-Si data. The As concentration used in this study ranged from 8×10^{19} to 2×10^{21} cm^{-3}. Rapid thermal annealing (RTA) activated a higher fraction of the implanted As, producing ρ_c values as low as 7.5×10^{-8} Ω-cm^2 for samples annealed at 1050°C, as compared to 8.4×10^{-7} Ω-cm^2 for conventional furnace annealing.

5.1.2 Resistor Contact

One of the early patents dealing with the formation of ohmic contacts (Martin, 1969) described a method for defining and making contact to a diffused resistor without the need for enlarging the resistor contact area beyond the width of the body of the resistor. This method, shown schematically in Fig. III-25 through a sequence of structures, utilizes the fact that the diffused region is wider than the masked opening because of the lateral diffusion of the dopant species. Thus the enlarged diffusion region provides for metal contact only to the diffused region, in spite of the lateral etching of the field oxide for defining the contact opening, which exposes more of the diffused area. A metal processing patent assigned to RCA Corp. (RCA, 1967) cited the use of plasma etching to clean a Si surface prior to

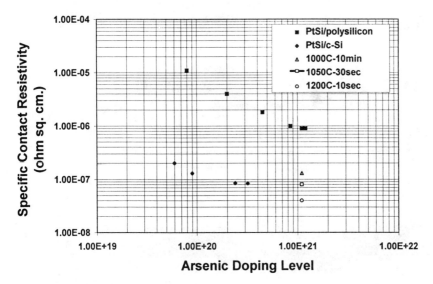

FIGURE III-24 Effect of charge carrier trapping by grain boundaries in polysilicon. PtSi contact resistivity is plotted as a function of As-doping concentration for both single-crystal Si and polysilicon. The increased resistivity of PtSi/polysilicon structures is attributed to charge carrier trapping by grain boundaries. RTA anneals of PtSi/polysilicon samples were as follows: (1) open triangle, 1000°C for 10 min; open square, 1050°C for 30 s; open circle, 1200°C for 10 s. Concentrations are indicated in scientific notation where, e.g., 1E-08 is equivalent to 1×10^{-8} (After Huang et al., 1988)

deposition of Al metal by evaporation. Annealing at 550°C formed the contacts to the source and drain regions of CMOS devices.

5.1.3 Wiring Level Contacts

Cunningham et al. (1970a) described the use of Mo/Au/Mo structures to form, simultaneously, ohmic contacts to Si and the first level of interconnection wiring. Additionally, a second level of metal could be formed using the same metallization. Cunningham et al. (1970b) provided additional refinements to their multilevel metallization scheme in a similar, subsequent patent. Additional details concerning these structures can be found in the original patents.

5.1.4 Cermet Barriers

Beaudouin et al. (1971) patented a metallization structure for contacts employing a Cr-SiO$_2$ cermet as a contacting material to Si and as a diffusion barrier between Si and the Cu/Cr metallization. The cermet also served as an adhesion layer to Cu. Additional material on cermet barriers can be found in Section 6.1.3 on the use of Cr-Cr$_x$O$_y$.

5.1.5 Metal to Metal Contacts

Lowry et al. (1972) proposed using a three-layer structure consisting of Ni-Cr, followed by Ni-Cr-Ag, followed by a conductive layer of Ag, for making ohmic

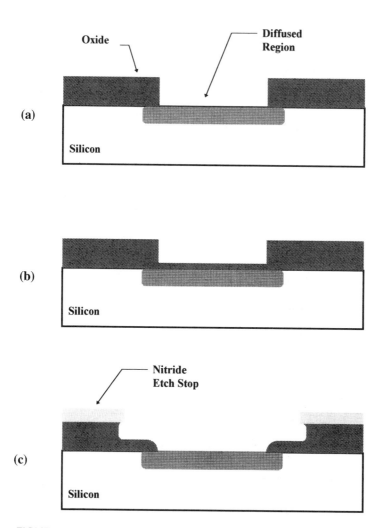

FIGURE III-25 Method of forming a diffused resistor without enlargement of resistor contact area. In (a), a contact is opened in oxide and the exposed Si diffused with an appropriate dopant. In (b), the structure is oxidized to form a thin oxide coating on the diffused region. In (c), a patterned nitride layer masks the field oxide while the oxide over the diffused region is etched, forming a structure for metallization that is confined to the resistor area. (After Martin, 1969)

contacts to Si. They chose materials in the stack so that one or more elements in each layer were common to those in adjoining layers, in the anticipation of providing good adhesion and low resistance contacts from the uppermost conducting layer down to the Si surface. We presently recognize that Ag would not be considered an appropriate choice BEOL applications because of its high susceptibility to corrosion and its degradation of dielectrics like SiO_2. It migrates freely through the glass network causing shifts in gate threshold voltages and promotes growth of dendritic Ag that shorts conductors through intervening dielectrics.

5.1.6 Amorphization of Si

Jaffe et al. (1979) patented a method of contact formation that involved amorphizing the c-Si layer by ion implantation. An Al layer could then be deposited on the a-Si to form the contact structure. The amorphization and doping of the Si could be accomplished in one step by using ion implantation of As at an energy of 180 keV and a dosage of 10^{15} cm^{-2}. An advantage cited for this approach is the reduction of pitting because of the lower accumulation of Si in Al due to the amorphization of Si. It appears likely that the utility of this approach stems from suppression of kinetics of dissolution of Si. From a thermodynamic viewpoint, the higher energy state of amorphous Si versus crystalline Si would favor more dissolution into Al instead of less.

5.1.7 Laser Annealing

Chen et al. (1981) described two ways to improve contact stability to submicron, poly-Si which involved (1) use of W/Ti contact metallization, and (2) use of laser annealing to sinter Al/poly-Si contacts. The R_c values of the W/Ti contacts were comparable to furnace annealed Al/poly-Si but interdiffusion was avoided. Resistance values for the laser annealed Al/poly-Si were the lowest of all and there was essentially no atomic intermixing. For both the furnace annealed Al and W:Ti contacts, ρ_c values of 2.4×10^{-5} Ω-cm^2 were obtained, but for the laser annealed Al contacts, the much lower value of 3.6×10^{-6} Ω-cm^2 was obtained. The P-doping level in the poly-Si film was approximately 10^{19} cm^{-3}.

5.1.8 TiN/Ti Contacts

Ting et al. (1982) investigated the electrical properties of Al/Ti and Si/Al/Ti metallizations on n^+-Si on contacts that were ~ 1 μm^2. They found that the value of ρ_c of these structures was much lower than that of Pd$_2$Si. The measured value of 15 Ω for R_c corresponds to a ρ_c value of 1.5×10^{-7} Ω-cm^2. A summary of their results is given in Table III-1 together with those of other investigators.

A Berger test structure (Berger, 1972) used for these and other contact resistance measurements is shown in Fig. III-26. There are several ways to use this structure to determine contact resistance. Fortunately, for the dc case and where all contacts are physically identical, the method is relatively simple. A reasonably good estimate of R_c can be obtained by measuring the resistance between contacts (a) and (b), R_{ab}, and between contacts (b) and (c), R_{bc}. Then the contact resistance of any one of the contacts is approximately given by:

$$R_c = \frac{R_{bc}L_{ab} - R_{ab}L_{bc}}{2(L_{ab} - L_{bc})} \tag{III-11}$$

L_{ab} is the distance between contacts (a) and (b) and likewise L_{bc} is the distance between (b) and (c).

However, due to the horizontal geometry of the test site, the value of R_c determined from Eq. (III-11) will be somewhat exaggerated relative to a vertical geometry, e.g., a bipolar emitter contact, because of current crowding at the leading edge of the contacts. In the circuit configuration shown in Fig. III-26, the contact

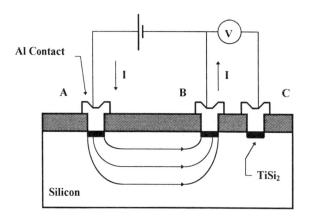

FIGURE III-26 Berger test site for evaluating specific contact resistance. (After Berger, 1972)

labeled C is shown being used to measure the potential of the trailing edge of contact B. This quantity, referred to as the end resistance, R_e provides another measure of the contact resistance but at a different, lower, current density. According to Berger, R_e is given by

$$R_e = \frac{V_{bc}}{I_{ab}} \qquad (I_{bc} = 0) \tag{III-12}$$

where V_{bc} is the voltage measured between contacts (b) and (c) under the condition of no or negligible current flowing between (b) and (c). I_{ab} is the current flowing between (a) and (b). Using transmission line theory, Berger showed that R_e is also given by

$$R_e = \frac{Z}{\sinh(\alpha d)} \tag{III-12}$$

where α is

$$\alpha = \sqrt{\frac{R_s}{\rho_c}} \tag{III-13}$$

and Z, the characteristic impedance, is

$$Z = \left(\frac{1}{w}\right) \sqrt{(R_s \rho_c)} \tag{III-14}$$

Here, R_s is defined in terms of measurable quantities by the expression

$$R_s = \frac{w(R_{ab} - R_{bc})}{(L_{ab} - L_{bc})} \tag{III-15}$$

where w is the length of the contact, i.e., the dimension into the plane of Fig. III-26 (not shown). By using an iterative approach, these expression allow one to calculate the specific contact resistivity, ρ_c. Similar ρ_c values for Al/Si contacts were reported by Hara et al. (1983) using halogen-lamp rapid heating to sinter the contacts. Their ρ_c values were 1.5×10^{-7} Ω-cm^2 for a heat treatment at 950°C and B-implant doses of 4×10^{15} cm^{-2}. In yet another rapid annealing technique, Chen et al. (1983) investigated electron-beam sintering for reducing R_c for VLSI applications. For contacts that were 1 μm^2, and metallized with refractory metals, the observed ρ_c values ranged from 1.2 to 1.5×10^{-7} Ω-cm^{-2} for p^+- and n^+-doped contacts, respectively. Forming gas was used to anneal out any electron beam induced damage in MOS devices. No metal-Si interdiffusion was seen.

A Mo/Ti double-layer contact was used by Kim et al. (1985). The Ti layer was used as the contact to Si to reduce the native oxide and Mo was used as the interconnecting layer. The layers were sputtered in sequence in the same pump-down. The ρ_c values for contacts to n^+ and p^+ Si were 1.0 and 1.9×10^{-7} Ω-cm^2, respectively. The high temperatures needed to cause the onset of electrical degradation of n^+ contacts (650°C) and p^+ contacts (600°C) was evidence of the high thermal stability of the contacts. Rapid thermal annealing (RTA) of Ti and Ti-W metallization on Si was investigated by Mueller et al. (1988) who used Ω_s, ρ_c, surface morphology, I-V properties, and x-ray diffraction measurements. The study concluded that both Ti and Ti-W metallization had good ohmic properties up to 900°C.

The structural and electrical integrity of stacked electrodes consisting of W/TiN/polysilicon(n^+ and p^+)/SiO$_2$/Si which have high conductivity and provide dual work function capability, as required for CMOS circuits, was studied by Pan et al., (1989). The TiN films were thought to be effective barriers to diffusion because there was an absence of any silicide formation, even after annealing at 1000°C for 30 min, and they detected very little dopant diffusion from the polysilicon into the overlying W film using Auger profiling. Moreover, the electrical properties of the 10-nm SiO$_2$ film that underlies the W/TiN/polysilicon stack were quite good. Evidence cited for this conclusion were stable flatband voltages, low interface state densities and tightly distributed breakdown voltages.

Rapid thermal annealing was used by Chung et al. (1990) to evaluate TiW/Ti contact metallizations. They found that RTA had little effect on ρ_c but could convert the Ti in contact with the Si into silicide and, at the same time, formed a thin layer of TiN on the top of the TiW surface. By controlling RTA temperature and time, the interaction of TiW with Si was minimized.

Kubota et al. (1993) described a contact-formation process involving sputtering a Ti$_{>1}$N$_{<1}$ film onto a Si contact and by subsequent annealing simultaneously (1) form Ti silicide by converting the portion of the film in contact with Si and (2) convert the top portion into a stoichiometric TiN layer.

Rapid thermal annealing was used by Uppili et al. (1994), to anneal TiSi$_2$ films used as contacts to emitter poly-Si structures in a double poly-Si bipolar process. They found that the shorter RTA cycle resulted in superior silicide morphology and low ρ_c between TiSi$_2$ and n^+ poly-Si layers. Implanted As increased the polysilicon conductivity upon RTA (Kalnitsky et al., 1994) which they attributed to a reaction between inactive As and lattice vacancies.

Collimated sputtering, which increases coverage of the bottoms of deep submicron contacts, was used by Sekine et al. (1995) to deposit Ti. This, somewhat "directional" sputtering process is shown schematically in Fig. III-27. Here it is apparent that the collimator, interposed between the sputtering target and the sam-

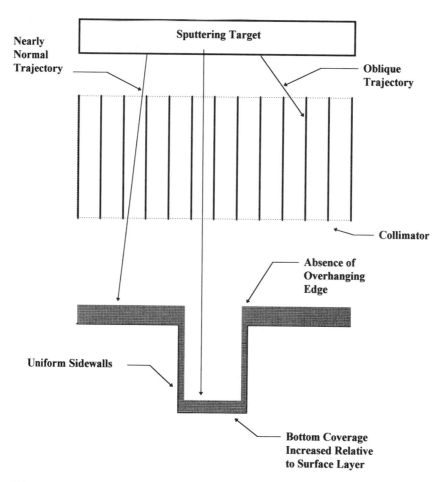

FIGURE III-27 Collimated sputtering process showing influence of collimator on angular distribution of sputtered ion trajectories that reach the substrate. Included is a cross section of a via showing coverage of resulting sputtered film.

ple, plays the role of a directional filter. It only allows ejected Ti atoms with trajectories nearly normal to the target plane to reach the sample. This can substantially improve the thickness of the bottom of a contact relative to the upper surface as indicated by the data in Fig. III-28 where the bottom coverage is expressed as a percentage of the thickness on the upper surface. Collimated Ti, when used in combination with downstream etching using CF_4/O_2 to remove damaged Si layers, generally gave lower values for ρ_c for 0.4-μm contacts having an AR = 4. The magnitude of the improvement was as much as a factor of 2 for rapid thermal processing (RTP) temperatures of 600°C and thin Ti bottom layers ~100 Å thick. For thicker Ti layers of 600 Å and RTP temperatures of 700°C, the improvement was only 5–10%. On the downside, a collimator creates certain problems for integrated circuit manufacturing. These are that the overall deposition rate is substantially slower (which raises costs) and that the collimators are sources of

Semiconductor Contact Technology

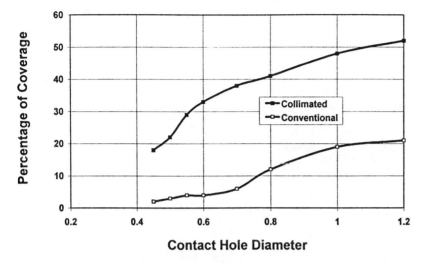

FIGURE III-28 Percent coverage of bottom of contact holes of various diameters relative to top surface. Both collimated and noncollimated processes are shown. (After Sekine et al., 1995)

metal particles which are yield detractors. An extensive discussion on collimation can be found in Chapter VI of this book.

Onuki et al. (1995) used a "switching bias sputtering," i.e., alternating between standard and bias sputtering, to improve step coverage for W metal deposition into contact holes that were 0.3 μm wide and 1.0 μm deep. The authors state that their process results in low values of ρ_c for both n^+- and p^+-Si and does not produce damage during deposition. They obtained values of 20 and 35 Ω for n^+- and p^+-Si, respectively, for a 0.6-μm contact hole.

Chin et al. (1995) described a self-aligned process for a contact diffusion barrier in which N_2 implantation converted a portion of the Ti to TiN which then served as a diffusion barrier. Figure III-9 shows a conducting contact layer such as Ti patterned on a contact structure.

5.2 SELF-ALIGNED STRUCTURES

A self-aligned method was highlighted in an early patent for fabricating transistors from Schottky contacts (Triebwasser, 1980). A MESFET structure, in which the gate is formed by a Schottky diode instead of a MOS field effect structure, was proposed for high-density, integrated circuit applications. A schematic of this structure is shown in Fig. III-29 at an intermediate step prior to completion. It illustrates the novel feature of using a thermally grown oxide to space the souse and drain equally from the gate. Evaporated poly-Si provided the source and drain contacts and any one of several metals or silicides could form the FET gate. Self-alignment is highly advantageous since it avoids the additional area requirements for the overlay tolerances of the masking step. It also tests the creativity of the process designer to provide all necessary processing with one less masking level at his or her disposal. Selected sequential steps for a typical self-aligned structure used for

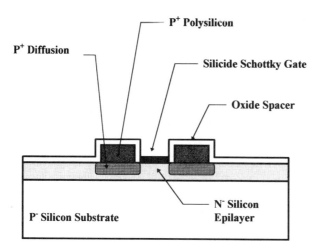

FIGURE III-29 Self-aligned structure used for forming MESFETS for high-density integrated circuits. An oxide grown on polysilicon source and drain pads forms the spacer separating the gate from either the source or the drain. (After Triebwasser, 1980)

present FET fabrication is shown in Fig. III-30. This process shows: (a) growth of the gate oxide defined by a field (or isolation) oxide; (b) masking of the source/drain ion implant by the gate stack to produce a self-aligned structure; (c) growth of side wall spacers on the gate stack by conformal oxide deposition followed by RIE removal; (d) growth of self-aligned silicide (salicide) in the exposed polysilicon or silicon surface by a process of sputtered metal (Ti) deposition, sintering, and strip of unreacted metal. Lavery et al. (1987) described another creative use of the different oxidation properties of doped vs. undoped Si and poly-Si. It provided a way of forming an insulating layer on the poly-Si gate without significant oxidation of the exposed source and drain regions. This is consistent with being a self-aligned process since it avoids the need for a separate contact opening masking step needed when a thick oxide develops on the source and drain. They used "buried" ion implants in the source and drain regions that were also self-aligned to a polysilicon gate contact that served as an implant mask (Fig. III-31a). The proper choice of the energy of the implanted ions, placed their peak concentrations at the desired depth below the Si surface in the source and drain contact regions. The surface regions of the source and drain were thus lightly doped, but a P-diffusion step highly doped the gate region (Fig. III-31b). Since the thermal oxidation rate increases with doping, a thicker oxide grew on the gate poly-Si than on the source and drain when the structure was oxidized thermally. This effectively keeps the source and drain open while electrically isolating the gate from subsequent source and drain metallizations (Fig. III-31c). This ion implant process, in conjunction with other process changes, avoided the need for overlay tolerances and thus produced ROM structures requiring 25% less space. A schematic of the process appears in Fig. III-31. Lin et al. (1994) also patented a method for forming self-aligned contacts for MOSFET structures by using end point detection layers of silicon nitride or silicon oxynitride for opening contacts by plasma etching. This allows for less loss of field oxide in etching, and for incorporating a lightly doped

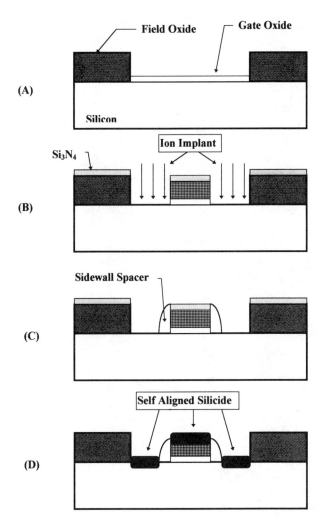

FIGURE III-30 Typical self-aligned process for building FET devices. In (a), a patterned opening to Si is oxidized to form a thin gate oxide. In (b) the gate stack masks the gate area from ion implant of dopants, forming a source and drain that are self-aligned to the gate. In (c), vertical sidewalls are formed on the gate stack by a process of deposition of a conformal oxide film followed by a directional etch of the film from the horizontal surfaces using reactive ion etching. In (d), silicide is self-aligned to the exposed Si in the source and drain contacts and polysilicon in the gate contact by blanket deposition of Ti, sintering to form $TiSi_2$, and removal of the unreacted Ti.

drain spacer etch process. Hodges et al. (1995) patented a process for forming self-aligned lightly doped drain (LDD) structures and low-resistance contacts. A low-resistance ohmic contact between n- and p-type polysilicon occurred because of the formation of an intervening refractory silicide. A $TiSi_2$ layer consumes a portion of both polysilicon layers, forming ohmic contacts with each type of polysilicon, thereby eliminating the rectifying, pn junction that would otherwise exist between the two polysilicon layers. This structure is shown in Fig. III-32 where (a) shows

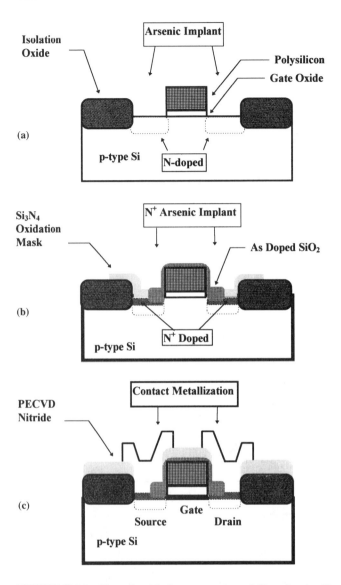

FIGURE III-31 Use of oxidation properties of Si and polysilicon in a self-aligned process. In (a) the gate stack masks the gate area from ion implant of dopants, forming a source and drain that are self-aligned to the gate. In (b), a patterned nitride mask prevents oxidation of portions of the source and drain so that following oxidation, a relatively thick oxide is formed immediately next to the gate stack. This oxide function as a mask for an N^+ arsenic implant to reduce the source and drain doping close to the gate and allows more dopant elsewhere to reduce series resistance in the source and drain regions. In (c), nitride is stripped from the souse and drain regions and contact metallization is applied. (After Lavery et al., 1987)

Semiconductor Contact Technology **199**

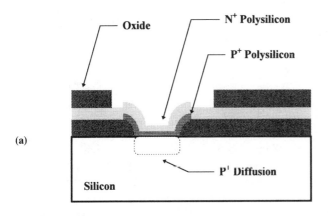

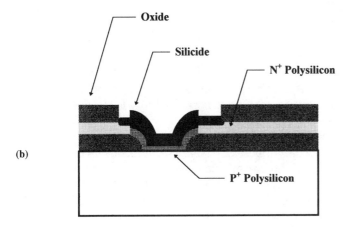

FIGURE III-32 Low resistance contact between *n*- and *p*-type polysilicon formed by an intervening layer of TiSi$_2$. In (a), a structure with overlapped N$^+$ and P$^+$ polysilicon is shown, and in (b), silicide formation that fully consumes the top layer (N$^+$ polysilicon) and a portion of the bottom layer, (P$^+$ polysilicon) is shown. This eliminates the rectifying junction that would otherwise exist, allowing an ohmic contact between the two layers. (After Hodges et al., 1995)

overlapping *n*- and *p*-type polysilicon layers and (b) shows TiSi$_2$ formation that consumes entirely the *n*-doped layer and a portion of the *p*-doped layer, effectively removing the rectifying junction.

5.3 CONTACT PROCESSING

5.3.1 Etching

Contact processing starts with some sort of contact opening step, usually RIE. In this process the Si surface is bombarded with energetic ions once it is exposed after the overlying dielectric film is removed. These are implanted into the silicon lattice and create a wide variety of point defects that are electrically active, functioning primarily as recombination centers for electron-hole pairs. To form well-functioning contacts, it is important, therefore, to minimize induced damage and to anneal any remaining damage where possible. With the use of CMP for planarizing dielectrics and for defining "damascene" or embedded contacts and interconnections, mechanically induced damage can occur in Si or dielectric films as well as ion damage.

With CMP as the one possible exception, present integrated circuit processing avoids causing the overt, mechanical damage investigated by Johansson et al. (1988). These investigators characterized various types of mechanical damage in c-Si using cross-sectional, transmission, electron microscopy. Damage resulted from indentation, controlled scribing, particle impact, grinding with fixed abrasive and polishing with free abrasive, the latter being relevant to scratching by foreign or oversized (agglomerated) slurry particles during CMP. Even in the free abrasive case using 12-μm diamond particles, there was little observable damage and it extended only a few percent of the particle diameter into the substrate. The shallow depth of damage, 50 nm, precluded the authors from determining the precise defect type. The surface deformed plastically and without evidence of spalling that occurred, for example, with fixed abrasives. The inference here is that this extent of damage, though relatively small, is still sufficient to disrupt device and circuit operation. With the increasing use of CMP, problems relating to incidental damage to Si and/or SiO_2 by scratching become of paramount importance for achieving yield targets.

Misra et al. (1989) reported on damage created by reactive-plasma-assisted etching of p^+-n diodes with CF_4-O_2 mixtures. Using p-n junctions as test devices, these authors showed that increased recombination currents and a degraded ideality factor altered the forward I-V characteristics. They used two types of test specimens, those with junctions exposed directly to the RIE and reference diodes with junctions covered by an oxide layer during RIE. These structures behaved altogether differently, the directly RIE-exposed junctions were highly degraded but the covered reference junctions were virtually unaffected. For the exposed junctions, the magnitude of the damage-induced recombination currents increased with increased bias voltages that ranged from 600 to 1000 V during RIE. Thermal annealing at 450°C did not fully recover their initial properties, although junctions experiencing the lowest RIE bias voltage of 600 V recovered substantially more than those etched at 800 and 1000 V. Reverse bias characteristics permanently changed except for the 600 V RIE bias samples that showed partial recovery after annealing indicating that a typical BEOL annealing step can modify, but not eliminate, the defect complexes. Facing similar concerns, Tsukada et al. (1993) studied the physical damage induced in Si by helicon wave plasma etching. These authors exposed Si surfaces to helicon wave O_2 plasma at different helicon wave and bias powers. Their assessment of Si lattice damage involved RBS, TEM, and photoacoustic displacement (PAD). They found that a low damage level occurred by using the combination of high helicon wave and low bias powers, to produce low bias voltages on the sample.

In another example of damage due to contact etching, Awadelkarim et al. (1994) investigated damage in Si substrates, resulting from high-selectivity etching processes. They compared magnetically enhanced reactive ion etching (MERIE) and conventional reactive ion etching (RIE) using two combinations of etchants and substrates. Samples of SiO_2/Si were etched with CHF_3/O_2 and bare Si with CHF_3/Ar. Each etchant-substrate combination was processed in both the MERIE and RIE reactors. All four experimental conditions produced similar types of damage that were cited as: (1) electronic states in the band gap; (2) H permeation into Si; and (3) deactivation of B acceptors. They ascribed the gap states to interstitial-related defects. These arise from Si interstitials, e.g., Si knock-ions generated by the etching process. Simultaneously during etching, H from the plasma dissociated CHF_3 was postulated to permeate into and passivate deep levels in Si. In a controlled experiment, the passivation or deactivation of B acceptors increased with increasing magnetic field for the MERIE etch process using CHF_3/O_2 which was interpreted as evidence for H passivation, since the ion flux incident on the wafer increased with the field. DLTS measurements detected hole traps with activation energies of 0.40 and 0.65 eV for hole emission. They identified the trap at 0.40 eV as the carbon-interstitial oxygen-interstitial defect undergoing a ± 0 charge transition and the trap at 0.65 eV as the Si di-interstitial.

Chien et al. (1986) employed a controlled damage study to investigate the influence of Si defects on Schottky behavior. Removal of the near surface region, consisting of a few tens of nanometers, had no effect on the ϕ_B of the damaged Al/p-Si contact but removal of >100 nm restored ϕ_B to its predamage value.

5.3.2 Contact Cleaning

Sung et al. (1995) analyzed, using TEM and thermal wave spectroscopy (TWS), the damage to Si surfaces caused by dry etching in a Cl_2 plasma in an ECR system and also measured the characteristics of Schottky diodes formed on the damaged Si surfaces. They found several correlations between the electrical measurements and the structural analysis such as an increase in the ideality of diodes from 1.08 to 1.90, a decrease in the breakdown voltage as from 60 to 6 V, an increase in the defect density from 3.6×10^{10} to 1.0×10^{11} cm^{-2} and a decrease in the damage layer thickness from 134 to 91 nm, all occurring as the power increased from 50 to 500 W. Nagomi et al. (1994) observed that suboxides, i.e., Si-rich oxides, formed on the bottoms of contact holes in which the Si surfaces had been damaged by dry etching. Dilute HF could not remove these oxide layers and they also noted that a 1.3-nm-thick oxide, present at the poly-Si/Si interface, increased the contact resistance.

5.3.3 Contact Metrics

Kado et al. (1993) successfully measured the detailed topography of contact holes with an AR = 1.5 using atomic force microscopy (AFM) in which a ZnO whisker probing tip was used in a hopping mode to obtain topographic data under a constant repulsive force at each measuring point.

5.4 LOCAL INTERCONNECTION

5.4.1 Interconnection

Miller et al. (1991) patented a method of producing low-resistance contacts in which patterning of the silicide and underlying adhesion layer defined the interconnection. Lee et al. (1994) later patented a similar concept but applied it to the formation of contacts in source and drain regions. In the later process, Lee et al. (1994) first formed a dielectric layer with an overlying poly-Si conductor and then patterned it to expose the semiconductor substrate. Next, they deposited a blanket silicide layer over the whole structure and again patterned it using an oxide hardmask.

5.4.2 Substrate Contacts

Jerome et al. (1992) patented a method of forming substrate contacts that involved the use of doped poly-Si to form a connection between a channel stop region and the substrate. Substrate contacts provide an electrical connection between the c-Si substrate and an external circuit that maintains the substrate as a constant potential. Substrate contacts are necessary because integrated circuit devices will not function properly unless the substrate is maintained at some predetermined potential. One of the processes described in the patent that produces a silicided substrate contact appears in Fig. III-33.

Chou et al. (1993) described a method for forming transistor contacts that used poly-Si to form buried contacts, gate contacts, and as an implantation mask for the source and drain regions. Buried contacts are contacts to Si that, once formed, are no longer accessible for interconnection by subsequent layers of metallization. They provide circuit connections to the substrate to establish the reference potentials. Moreover, local wiring at the device level improves the density and/or speed of circuits. In the instance cited above, several possible applications stemming from the integration of two patterned levels of doped poly-Si were demonstrated.

6.0 ACTIVE DEVICE CONTACTS

The present use of Schottky diodes in integrated circuits includes discrete circuit elements such as (1) clamps to prevent the emitter currents from reaching saturation levels; (2) protection for diodes in FETs; and, (3) high-frequency circuit elements. Table III-2 contains a summary of the SBD measurements reviewed here. There have been several attempts to utilize these rectifying contacts as one or more of the three ports of field effect or bipolar transistors. For example, in a patent assigned to A. G. Siemens Corp. (Siemens, 1973), Schottky contacts were tried as bipolar transistor elements, providing the function of collector and emitter, with an epitaxial semiconductor region, doped oppositely from the substrate, functioning as the base. The rectifying contacts could consist of Al, Ti, Pt, Rh, Pd, Co, or a silicide thereof. In other examples with a different type of transistor, Drangeld et al. (1971) used Schottky diodes (i.e., Al/Si) as gates in field effect transistors.

6.1 MATERIALS

6.1.1 PtSi

An early SBD process patent cited the use of Pt_2Al as a Schottky barrier, and not one of the three ports of a transistor, but as an independent device for incorporation in a circuit as a diode was by Magdo (1975). The structure (not shown) contained a doped Si region of 10^{18} atoms/cm^3, a deposited layer of Pt that was heated to

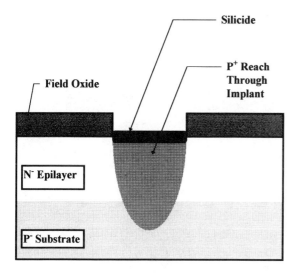

FIGURE III-33 Silicided substrate contact. An implant of boron reaches through the N^- epilayer to reach the P^- substrate.

form PtSi, and a layer of Al on the PtSi which was sintered at 400 to 550°C for 1 h. The latter sintering step reacted Al and PtSi to form the intermetallic, Pt_2Al phase. The patent also cited using PtSi as an ohmic contact but with the doping level above 10^{18} atoms/cm^3. In a subsequent and related patent, Magdo et al. (1976) described a dielectrically isolated Schottky barrier structure. Pockets of Si surrounded by electrically isolating regions of SiO_2 were covered by a second dielectric. This can be a composite layer of Si_3N_4 and SiO_2 or Si_3N_4 which is then opened to expose the Si islands. Pt deposited on the Si was then sintered to form PtSi and overcoated with Al to provide an interconnection. The resulting structure is shown in Fig. III-34.

6.1.2 Al/Si

Observations made by Card (1976) showed that measured values of ϕ_B of Al on Si depended on the details of the surface conditions and the heat treatments. For example, if oxides of 20-Å thickness are present, then the value of ϕ_B on p-type Si can be as high as 0.7 eV and on n-type Si as low as 0.5 eV. Card further showed that heat treatments of at least 300°C are required for reproducible values of ϕ_B of 0.7 and 0.5 eV for n- and p-type Si, respectively. With increasing temperature of sintering (up to 550°C), the value of ϕ_B on n-type Si could reach 0.9 eV and that of p-type could drop to 0.35 eV. The changes in ϕ_B were ascribed to two mechanisms: (1) the removal of positive charges from the oxide and (2) the metallurgical reactions between Al and Si.

Yaspir et al. (1988) used a novel partially ionized beam (PIB) technique to deposit Al on n-Si to form Schottky contacts. Highly uniform characteristics of diodes across a 3-in. wafer were found together with a much reduced pit formation on Si. The energetic Al flux created by the accelerating potentials in the PIB

TABLE III-2 Barrier Height Values

Barrier height			Contact Structure			Conditions			Sources	
High	Low	Units	Metal	Silicon	Doping	T (°C)	Method	Note	Author	year
0.64	0.49	eV	Al	Si	n-type			Ion/Impl't	Ohta	1989
0.70		"	Al	Si	n-type	300			Card	1976
0.90		"	Al	Si	n-type	550			Card	1976
0.50		"	Al	Si	p-type	300			Card	1976
0.35		"	Al	Si	p-type	550			Card	1976
0.62		"	Al	Si(111)	n-type				Hoekelek	1978
0.84	0.72	"	AlCu	Si	-type		quenched		Bhatia	1976
0.65		"	B-NiSi$_2$	Si(111)				>50 nm	Kikachi	1988
0.89	0.61	"	CoSi$_2$	Si(111)	p-type				Fathauer	1988
0.38		"	Gd	Si					Suu	1986
0.39	0.15	"	IrSi	Si	p-type				Tambe	1991
0.68		"	MoSi$_2$			550			Yamamoto	1985
0.76		"	NiAl$_3$	Si(111)	n-type				Hoekelek	1978
0.79	0.78	"	NiSi$_2$	Si(100)				<50 nm	Kikachi	1988
0.55		"	PtSi	poly Si	n-type		Columnar prefered	n = 2.0	Sagra	1991
0.39	0.15	"	PtSi	Si	p-type			n = 1.2	Sagra	1991
0.64		"	Ti$_{73}$Co$_2$,	Si	n-type				Tambe	1991
0.55	0.52	"	TiSi$_2$	a-Si	n-type	450		n = 1.06	Gromov	1995
0.59	0.57	"	TiSi$_2$	a-Si	p-type	450			Liauh	1993
0.52		"	TiSi$_2$	Si	n-type				Mallardeau	1989
0.61		"	TiW	Si	n-type				Babcock	1986

Semiconductor Contact Technology

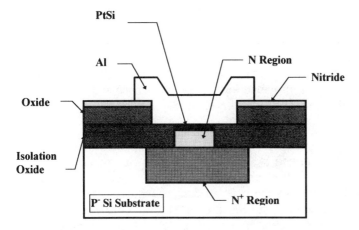

FIGURE III-34 Dielectrically isolated Schottky barrier diode formed with PtSi. Note the PtSi region covers the N region completely, to prevent a parasitic Al/Si contact forming at the periphery. (After Magdo et al., 1976)

deposition system were assumed to help remove native oxide from the Si surface, thus creating a more intimate Al/Si interface.

6.1.3 Cr–CrxOy/Ta

A method of fabricating a low voltage (i.e., ~0.5 eV) Schottky barrier was described by Dalal et al. (1980) which involved evaporating of Ta/Cr-Cr_xO_y/AlCu through a lift-off mask. The Ta/Si interface provided the rectifying contact and the Cr-Cr_xO_y cermet was the diffusion barrier to inhibit the reaction between AlCu and Ta. By forming selected PtSi contacts prior to evaporating the metallization layer, a combination of both low- and high-voltage Schottky contacts, together with ohmic contacts could be formed on a single contact level. In this process, Ta was evaporated first, followed by Cr-Cr_xO_y. The Cr-Cr_xO_y layer was formed by e-beam evaporation of Cr as water was bled into the evaporator. A detailed process for evaporating transition metals to form low-barrier Schottky diodes was described by Dalal et al. (1983); it covered a series of steps to be used for degassing the charge while it is melting, as well as the wafer and evaporator itself, to ensure a deposit of high quality.

6.1.4 Al/NiSi

In a study of the reaction of the Al/NiSi contact structures on n-type, (111) Si, Hoekelek et al. (1978) showed that a NiSi phase was formed in the contact with $\phi_B = 0.62$ eV. After sintering the NiSi in contact with a thin Al overlayer, $NiAl_3$ formed which had a $\phi_B = 0.76$ eV with Si. The electrical characteristics of the $NiAl_3$ layer were stable up to 500°C with no evidence of Al penetration into the substrate.

6.1.5 TaSi$_2$

The application of cosputtered amorphous and polycrystalline TaSi$_2$ for contact metallization was evaluated by Neppl et al. (1984). They found that the atomic transport of Al and Si across the silicide was impeded up to 475°C if the silicide was metal rich. Metal-rich TaSi$_x$ provided low contact resistance and low Schottky barriers to n-type Si with minimal Si consumption.

6.1.6 MoSi$_2$

The value of ϕ_B for MoSi$_2$, formed by sintering a Mo film in contact with Si at 550°C was determined to be 0.68 eV by Yamamoto et al. (1985). They further investigated the thermal stability of the layer in contact with Al-2%Si and found it good up to 500°C. The value of R_c for the structure polysilicon/MoSi$_2$/Al-2%Si was found to be 6.0–7.0 × 10^{-7} ohm cm^2 which matched that of a polysilicon/Al-2%Si contact.

6.1.7 Intermetallics

Intermetallic compounds of Al and transition metals can be formed into stable materials for use as Schottky barrier contacts by a process disclosed by Howard et al. (1982). In their process, Ta and Al can be evaporated sequentially without breaking vacuum in the desired proportions to form an intermetallic compound. Subsequently, photoresist techniques, together with chemical etching, are used to pattern the Ta/Al bilayer. After stripping the photoresist, the bilayer is sintered to form the intermetallic compound.

6.1.8 Selective CVD W

Selective metal deposition could have significant advantages for contact processing. It places the desired metal film directly into the contact and requires no additional processing to remove the excess or overburden of materials elsewhere on the chip as occurs with nonselective, blanket metal depositions. For example, Gargini (1983) used a selective CVD process to deposit W into contact openings to provide a diffusion barrier only on the Al–Si contacts. Besides gaining obvious process efficiency, this showed that selective CVD W also provides an effective diffusion barrier for VLSI applications.

With such strong incentives such as possibly improved yields, reduced processing time and costs, etc., the absence of this potentially high leverage process in manufacturing deserves mention. The Achilles' heel of selective CVD is the apparent vulnerability to incomplete cleaning of contact holes which will prevent metal deposition and create an open contact. While this can happen to some degree with blanket CVD, physically vapor deposited (PVD) nucleation layers such as sputtered TiN will coat all surfaces and provides a reasonably pristine surface for blanket CVD. In effect, for the incompletely cleaned contact, the apparent trade off is an open, nonfunctional, contact for selective CVD vs. a functional, but possibly more resistive one, for blanket CVD. Until the quality of Si surfaces in all contact openings can be guaranteed to be free of nonnucleating films, it is unlikely that selective processes will be used to any great extent in IC manufacturing. Additional processing-related information is contained in a recent study of the selective CVD of tungsten metal on Si followed by ion beam mixing via As implant to form WSi$_2$ diodes (Saraswat et al., 1984).

6.1.9 TiW

Films of TiW alloy, initially developed as diffusion barrier layers, have been studied extensively as Schottky diodes, since in some process sequences they can be used both as diffusion inhibiting layers and as Schottky diodes. This dual use is not entirely free, since to include Schottky and ohmic contacts together, two masking steps are required. The Schottky barrier behavior of TiW on Si(100) has been investigated by Aboelfotoh (1987) who reported that the Ti component of the alloy dominates the interdiffusion with Si as well as the value of ϕ_B. The ϕ_B values for n-Si and p-Si, when combined equaled the indirect band gap for Si and had corresponding temperature dependencies, i.e., a decrease in ϕ_B with increasing temperature. These relationships are indicated in Fig. III-35. Values of ϕ_B were consistent with Fermi level pinning at mid gap.

6.1.10 CoSi$_2$/NiSi$_2$

Advanced molecular beam epitaxial film growth techniques for Co and Ni silicides were utilized by Fathauer et al. (1988) to produce CoSi$_2$ layers on p^+-doped, Si(111). Prior to growing the CoSi$_2$ layer, a fresh Si surface was created by in situ epitaxial growth of p^+ Si on to the Si(111) wafer surface. Measured values of ϕ_B ranged from 0.61–0.89 eV, depending on the thicknesses of the epitaxially grown p^+ layer. Interfaces between single-crystal NiSi$_2$ and Si, of both (100) and (111) orientations, produced by codeposition of Ni and Si on the appropriately oriented

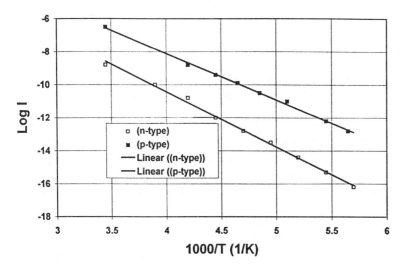

FIGURE III-35 Temperature dependence of the forward current for TiW diodes on n- and p-type Si. The barrier height derived from the data on p-type, (100), Si (solid squares) was 0.54 eV, and for n-type, (100), Si (open squares), the barrier height was 0.68 eV. The sum of the barriers from (a) and (b) is 1.22 eV, which is very close to the Si indirect band gap of 1.20 eV at 0°K. (After Aboelfotoh, 1987)

Si substrate, were analyzed by Kikuchi et al. (1988). For $NiSi_2/Si(100)$, ϕ_B was 0.65 eV, independent of the silicide thickness. For type-B $NiSi_2/Si(111)$, where the term "type-B" refers to a twinned structure, resulting from a 180° rotation about the surface normal to the orientation of the Si(111) substrate, ϕ_B was found to be dependent on $NiSi_2$ thickness, becoming constant at 0.65 eV for silicide thicknesses at or exceeding 50 nm. A value of ϕ_B of 0.78–0.79 eV was estimated by extrapolation to a thickness of $NiSi_2$ of 1–2 atomic layers.

The temperature and pressure dependencies of Schottky barrier heights on Si for a variety of contacts (Ti, W/Ti, PtSi, Pd_2Si, IrSi, $CoSi_2$, $TiSi_2$, Sm, and $NiSi_2$), some of which were epitaxial [$NiSi_2/Si(111)$, types A and B] or faceted [$NiSi_2/Si(100)$] were investigated by P. Werner et al. (1993). In contrast to the "type-B," twinned epitaxial relationship between $NiSi_2$ and Si(111) described above, "type-A" refers to a $NiSi_2$ that is identical in orientation to the underlying Si(111). From the analysis of epitaxial $NiSi_2/Si$ diodes, the authors concluded that there is a direct correlation between interface crystallinity and both the value of ϕ_B and its temperature dependence. Additionally, both pressure and temperature coefficients of the polycrystalline Schottky contacts correlated with the same coefficients for the band gap, indicating that the thermal emission model fully accounted for the observations of ϕ_B.

A novel method of combining both $CoSi_2$ Schottky barriers and TiN diffusion barriers was investigated by Gromov et al. (1995) who utilized a TiCo alloy which was sintered in contact with Si at temperatures of 800–850°C to form $CoSi_2/Si$ contacts (Schottky and ohmic) together with a TiN barrier layer. On n-type Si, the value of ϕ_B and the ideality (n) for an annealed $Ti_{73}Co_{27}/n$-Si contact were 0.64 eV and 1.06, respectively. Ohmic contacts to n-type Si had a value of ρ_c of 5×10^{-7} Ω cm^2, and to p-type Si, 1×10^{-6} Ω-cm^2.

6.2 NOVEL STRUCTURES

6.2.1 Guard Rings

Dreves et al. (1981) disclosed the incorporation of a guard ring as a way to improve Schottky diode performance. Their process employs a peripheral ring or annulus around a diode that follows the diode contour and is directly under a reentrant ledge in the surrounding insulating layers. A diffusion mask of Mo, evaporated into the contact structure, covered the diode area but not the peripheral, guard ring area. Then they diffused the guard ring using a vapor source of the appropriate dopant. Anantha et al. (1987) disclosed a method for making a highly compact, self-aligned guard ring for a Schottky diode. It used the idea of "side wall spacer technology" invented by Pogge (1981). The essential elements of this process are the conformal deposition of virtually any RIE etchable film over a step followed by directional etching to remove the film from the horizontal surfaces. The film can be a metal, semiconductor, polymer, or inorganic dielectric. The only restriction is that the deposited film has to be distinguishable in its etching characteristics from the underlying "step" so that it can be selectively removed. In the case at hand, the films of interest for fabricating the self-aligned guard ring are doped polysilicon and SiO_2. The combination of the steps described above create vertical spacers that adhere to both the vertical side wall and the bottom circumference of the diode opening. The optional SiO_2 film is the first to be deposited and etched, in order to position the diffused zone away from the very edge of the contact opening. The second is the polysilicon film which is processed identically to the oxide, and, upon doping, is the diffusion source for making the guard ring. Ion implantation is typically used to dope the polysilicon with either B or As. The diffusion step takes place by sintering after RIE removal of the remainder of the film. Subsequent deposition of a silicide forming metal into the contact opening lined with a poly-Si sidewall produced a diode structure that is self-aligned with its guard ring as indicated in Fig. III-36.

6.2.2 Hybrid Structures

Rothman et al. (1981) developed a process for combining sputtered diffusion barriers and low-barrier Schottky diodes into an evaporated, lift-off metallization technique for forming contact level connections on Si devices. For example, a blanket layer of sputtered TiW provided a diffusion barrier for the previously silicided, ohmic contacts and high barrier Schottky diodes. In the same TiW deposition step, the unsilicided or bare Si contacts become the low barrier diodes. A lift-off masking structure formed on the TiW surface provided the means for patterning the underlying barrier film. The AlCu conducting metallization was evaporated onto the patterned mask and lifted off to form a desired pattern of Al–Cu over a blanket field of TiW. Reactive ion etching with CF_4 removed the exposed TiW except under the AlCu as the masking layer. For comparative purposes, cross sections of contact structures built by (a) lift-off, (b) combined lift-off and subetch, and (c) subetch are shown in Fig. III-37a–c, respectively. For the lift-off case (a), the barrier layer does not extend fully to the edge of the silicide layer, permitting migration of Al along the side of the barrier down to the silicide, forming Pt_2Al. These metallurgical reactions constitute a well-known precursor to Al penetration. In (b), the sputtered barrier extends over the silicide, offering greater resistance to penetration and the evaporated AlCu allows a greater Cu content than is possible for RIE defined AlCu. For the subetched structures in (c), the sputtered barrier again extends over the

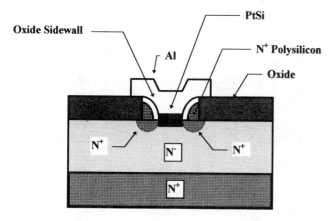

FIGURE III-36 Schottky diode with self-aligned guard ring. The guard ring is formed by the outdiffusion of an n-type, polysilicon sidewall. (After Anantha, et al., 1987)

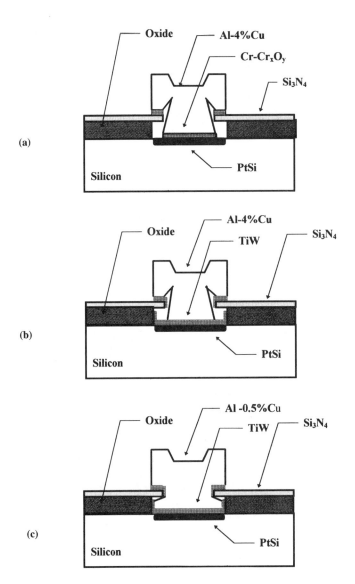

FIGURE III-37 A comparison of contact structures fabricated by lift-off, hybrid, and subetch methods. For the lift-off case, part (a), the evaporated Cr-Cr$_x$O$_y$ barrier does not fully cover the PtSi layer, formed by sintering a sputtered Pt layer. Deep clefts form where the Al metal steps down into the contact hole. For the hybrid case, part (b), the TiW barrier provides coverage fully across the silicide, and part way up the reentrant sidewall. The Al metallization is identical to (a). For the subetch case, part (c), the barrier coverage is the same as (b), but the sputtered Al metal has greater continuity as it steps down into the contact opening.

silicide as in case (b) above, but the cleft in the AlCu at the step from the surface of the insulator into the contact hole is minimal as compared to evaporated AlCu indicated in both (a) and (b). Because the AlCu is RIE etched, only a relatively low Cu content is tolerable, and on this basis the electromigration reliability of the totally subetched structure would be inferior to cases (a) and (b). Fortunately, this limitation was eliminated with the introduction of more highly reliable Ti clad Al-0.5%Cu structures.

6.2.3 Resistor and Diode Structures

Schlupp (1983) patented a more complex structure combining both a Schottky barrier and a resistor. This involved the deposition and patterning of a poly-Si layer over an insulator having a contact opening to the underlying Si. By forming a layer of PtSi in the same shape as the patterned poly-Si, using a salicide process, he formed the combined Schottky diode and resistor structure shown in Fig. III-38. This process can also be used for the simultaneous formation of a silicide field plate around the periphery of the diode to improve the reverse electrical breakdown characteristics. Improvements in Schottky barrier diode characteristics described by Bergeron et al. (1982) linked the anode and cathode of a diode on n-type Si by ion implantation of P. They chose the thicknesses of oxide that overlay the adjoining Si regions to mask the P ion implant, to control the peak implant depth. Proceeding in this fashion, they built guard rings and a high conductivity channel between the anode and cathode while causing only minimal PNP parasitic transistor action.

6.3 PROCESSING EFFECTS

6.3.1 Silicon Damage Effects

Chow et al. 1984 investigated the modification of ϕ_B of Schottky diodes using RIE in NF_3 gas mixtures. They observed that the RIE process caused an increase in ϕ_B for p-Si, and a reduction for n-Si. The introduction of point defects caused by ion bombardment are the likely cause of these effects. In an analogous study, Paz et al. (1984) characterized diodes formed by RF sputtering of TiW onto p-type Si, using such techniques as I-V measurements, deep level transient spectroscopy (DLTS) and electron beam induced current (EBIC) to support their arguments that sputter damage creates a hole trap at $E_v + 0.35$ eV, a strong recombination center.

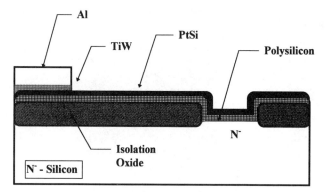

FIGURE III-38 Combined Schottky diode and resistor structure using PtSi formed on polysilicon. (After Schlupp, 1983)

Staayer et al. (1986) studied the influence of sputter damage on the Φ_B value of diodes. As a probe of sputter damage, they used Au sputter deposited onto p-Si. The value of Φ_B of diodes formed using sputtered Au increased over those prepared with evaporated Au, and the increase was dependent mainly on the sputtering voltage. Increases in ϕ_B occurred with sputtering voltages over the range of 0.5 to 1 kV after which they remained fixed. The formation of sputter-induced traps within a thin surface layer (~ 10 nm) of Si fixed the value of ϕ_B.

Ar^+ ion bombardment of Al/Si contacts redistributes residual SiO_x and this reduces spiking of Al into Si by creating a uniformly reacting surface. Thermal annealing at 350°C increased the average value of ϕ_B from 0.49 eV postimplant, to 0.64 eV after annealing, which is close to the undamaged value of 0.69 eV. Alpha particle channeling provided direct evidence that very little damage remained after annealing. Ohta et al. (1989) used DLTS to investigate defects associated with evaporated metal contacts to Schottky diodes. They discovered defect levels at 0.16, 0.14, and 0.12 eV below the conduction band and identified wet chemical etching as the origin of the defects. Supporting evidence came in the form of (1) an increase in the concentration of defects with the amount of Si removed by chemical etching, and (2) a decrease in one of the levels (0.14 eV) with increasing etch rate.

After studying the effects of 150-keV implantations as a function of temperature, Malherbe et al. (1992) determined the influence of Ar^+ ion implantation on the I-V properties of Cr/p-Si(100) diodes. They found that the Ar^+ ion bombardment resulted in higher values of both n and ϕ_B. Drawing upon an earlier explanation given by Fonash et al. (1981) they argued that the surface damage layer reduced band bending near the surface, indicating that this layer stored a net positive charge.

Altman et al. (1975) patented a novel method for improving the reverse leakage characteristics in metal-semiconductor diode contacts that made use of ion bombardment. They extracted ions from an annula shaped radio frequency (RF) diode that was interposed between the substrate and the evaporation source. After a predetermined exposure to ion impingement at the desired bias, movement of the RF diode structure to an off-axis position allowed an in situ deposition of metal. This sputter cleaning step performed in a custom evaporator described above improved the reverse characteristics of Mg, Al, and Pt diodes on p-type Si. Improvements were the replacement of "soft" or gradual increase in leakage current with reverse bias by "hard" characteristics. "Hard" characteristics consisted of a small reverse current that was virtually independent of bias until breakdown occurred and the current increased abruptly.

6.3.2 RTA

Mallardeau et al. (1989) used RTA to form the silicide in $TiSi_2$/Si structures, with n- and p-type Si forming both Schottky diode and ohmic contact structures. The values for Ω_s for $TiSi_2/n^+$-Si and, for comparison, $PtSi/n^+$-Si were 9×10^{-8} and 8×10^{-8} Ω-cm^2, respectively. Since the value of ϕ_B for $TiSi_2$ is much lower than for PtSi, the nearly identical ρ_c values suggest that the dopant, As in this case, is more effectively snow-plowed by PtSi than $TaSi_2$. For p^+-Si, ρ_c values for PtSi (2×10^{-7} Ω-cm^2) are considerably lower than for $TiSi_2$ (10×10^{-7} Ω-cm^2), which corresponds to the lower value of ϕ_B for PtSi. Finally, ϕ_B for TiSi$_2$/n-Si was 0.52 eV and n was 1.08.

6.3.3 Hydrogen Effects

An investigation of the etching of Si by MERIE with HBr as the etchant was reported by Nakagawa et al. (1991). After a surface treatment that removed 400 nm of Si by MERIE with HBr, they formed Ti diodes and used I-V, C-V, and DLTS measurements of these diodes to characterize the damage. They found that permeation of hydrogen was the dominant effect for p-Si, causing deactivation of dopants and an increase in ϕ_B, but for n-Si, no such effects were detected. Annealing at 180°C restored the original properties of the diodes formed on p-Si.

6.3.4 Contact Processing

Bhatia et al. (1976) patented a novel method of increasing ϕ_B. They produced Si-rich structures with improved electrical characteristics by reacting the metal film with a poly-Si layer at an elevated temperature and then cooling quickly (quenching) from the reaction temperature back to room temperature. In the example given, AlCu/poly-Si/Si diodes yielded a ϕ_B varying from 0.72 to 0.84 eV depending on the quench rates. Typical AlCu/Si diodes on n-type Si have values of ϕ_B ranging from 0.68–0.72 eV.

6.4 ELECTRICAL MEASUREMENTS

Difficulties occur in the measurement of ϕ_B for low-barrier diodes because of (1) their relatively low resistance makes it difficult to identify the diode contribution to the overall resistance of the device structure; and (2) the recombination current in the diode can be an appreciable part of the total current, making uncertain the extrapolated value of J_0 needed to evaluate ϕ_B. [See (Eq.) III-4 for an explanation of J_0.] Suu et al. (1986) obtained improved values for low barriers by taking I-V measurements from two front surface contacts of different sizes, where one of the contacts substituted for the usual backside contact. As an example, they measured the value of ϕ_B for Gd/Si contacts and found it to be 0.38 eV.

Tanabe et al. (1991) reported on the spatial nonuniformity of the values of ϕ_B measured laterally across Schottky diodes (PtSi/p-Si, IrSi/p-Si) using internal photoemission. For both silicides, regions with the relatively high ϕ_B value of 0.39 eV coexisted with the anticipated values of 0.24 eV and 0.15–0.17 eV usually found for PtSi and IrSi on p-type Si. Inhomogeneities in interfacial defect densities and therefore in Fermi-level pinning produce the two coexisting regions. However, an additional investigation by Aboelfotoh (1991) cast doubt on the spatial fluctuation of barrier heights across a contact. The author concluded that inhomogeneities must be of the order of the Debye length or less. He drew this conclusion because of the close match between the sum of ϕ_B on n- and p-Si and the band gap. Otherwise, there would be a lowering of the observed values of ϕ_B, making their sum for n- and p-Si consistently less than the band gap.

Sagra et al. (1991) investigated inhomogeneities related to poly-Si grain structures and found that the diode characteristics of PtSi/n-polySi structures were quite different for columnar versus preferentially aligned grains. "Columnar" refers to a fibrous texture where all the grains nucleate on the substrate such that the length of the grain is typically equal to the film thickness but the diameter is much smaller. (Columnar structures are generally undesirable, having poor step coverage, rough morphology, and an overabundance of high-diffusivity paths making them poor diffusion barriers.) Preferentially aligned grains share a common orientation normal to the substrate surface but are randomly oriented in the plane of the film. They tend to form equiaxed crystallites, with diameters that are comparable to the film thickness. For the preferentially aligned polysilicon on Si(100) studied here, the polysilicon grains also exhibited a significant degree of epitaxy with the underlying c-Si. For columnar structures, $n \sim 2.0$ and $\phi_B = 0.55$ eV, but for the preferentially aligned films, $n = 1.2$ and the values of ϕ_B were dependent on the implanted dose.

7.0 CONTACT STUDS FOR ULSI

Increasingly, the contact to the silicon device whether it is a c-Si, poly-Si, or silicide, occurs through use of vertical connections, referred to as *studs* or in this case, *contact studs*. The need to planarize the topography resulting from the device build and to provide flexibility in contact wiring drive this change, despite the additional process complexity it brings. Extensive discussions of the process for forming studs appear in Chapter VI, and therefore only issues and solutions unique to "contact studs" appear here. Some of the pairs of these are: etching contact openings through different insulator thickness and use of etch stop layers, conversion of contacting Si surfaces into silicides, and the use of barriers and borderless, bordered, or self-aligned contacts. Each of these issues modifies the specific processes used in the formation of contact studs. Figure III-39 illustrates barrier issues, the shortcomings of sputtered barriers including sidewall thinning, poor bottom coverage, and breaks at corners. The superiority of the coverage of CVD liners is also indicated.

Dixit et al. (1989) described a method for forming a stable low-resistance ohmic contact. It involved lining the contact hole with an adhesion layer such as Ti, then with a barrier layer such as TiW, and finally filling the opening with either CVD or sputtered W, or sputtered Mo. The Ti adhesion liner makes a low-resistance contact to Si, and the barrier layer prevents encroachment of Si at the Si/SiO$_2$ interface. Moriya et al. (1983) used selective CVD W to form contact studs directly on Si and reported, for contact to n^+ Si, values of ρ_c of 8×10^{-7} to 2×10^{-6} Ω-cm^2. In a typical fabrication sequence, the selected contacts, diffusions, and or gate will be typically selectively silicided, followed by the deposition and planarization of the contact level insulator. However, Saito et al. (1993) used selective deposition of CVD TiSi$_2$ to form low-resistance contacts to Si through the contact hole, using a 0.2-μm hole with AR = 2.5. The value of ρ_c of the selective CVD TiSi$_2$ fell in the range of 2–6 $\times$ 10^{-7} Ω-cm^2. Silane at 720°C was used to preclean the surface prior to the growth of TiSi$_2$ using TiCl$_4$ and silane. Contact stud application usually involves the use of dielectric layers of SiO$_2$, P and B/P doped SiO$_2$, in conjunction with etch stop layers of silicon nitride, Al$_2$O$_3$, undoped poly-Si and MgO as etch stop layers (Kim et al., 1988).

Borderless contact design allows the contact stud to partially intersect the device contact region, providing design flexibility. Bordered contacts on the other

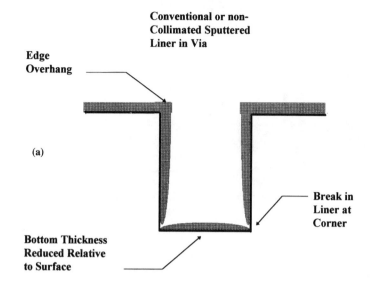

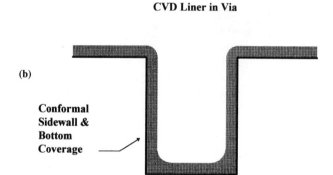

FIGURE III-39 Comparison of liner coverage of vias by conventional sputtering and chemical vapor deposition.

hand require that the contact stud land fully within the device contact opening. This enlarges the contact window dimensions due to the influence of overlay tolerances. A self-aligned contact process simplifies the process by relaxing the lithographic image size, while providing for maximum contacting surface. Figure III-40 illustrates contact level studs built on an FET structure where the device level silicide contacts were self-aligned to the gate source and drain. In Fig. III-40a, the studs are unbordered with respect to the gate, source and drain contacts, whereas in Fig. III-40b, the studs are bordered. Bordering requires larger silicide pads to accommodate the stud and at least a 3σ border for the overlay tolerances. This may become clearer after examining Fig. III-41a,b which shows the top view of borderless and bordered contacts, respectively. Source and drain contacts self-

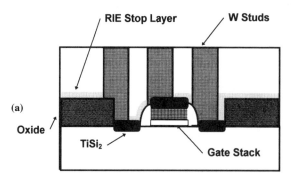

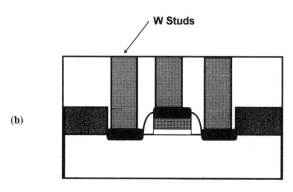

FIGURE III-40 Comparison of borderless and bordered contact studs on a typical FET device structure. The unbordered case, part (a), requires a RIE stop layer in order not to damage the device features. For the bordered case, part (b), it is assumed that the silicide on the contact provides an adequate etch stop.

aligned to the gate sidewall insulation provide for precise gate length and therefore consistent device properties. Achieving such a structure requires an etch stop over the gate stack, including its sidewall and cap insulators. After the required overetch, the barrier layer is removed while retaining the sidewall and cap insulators of the gate. This ensures that neither the source nor drain contact studs short to the gate electrode. A separate etch process is used to open the gate contact. Other variations of this process are possible. Givens et al. (1994) describe an etch process using a high-density plasma to etch 4% PSG in C_2F_6, with a selectivity of 100:1 to PECVD SiN. They followed this by etching SiN in a CH_3F/CO_2 plasma and achieved a 7:1 selectivity over SiO_2, or Si. Gambino et al. (1995) described a contact stud process that involved using at thick silicon nitride cap over the gate electrode and oxidizing the gate sidewall. A thinner blanket nitride film conformally coated the device topography and served as an etch barrier for a C_4F_8/CO RIE process with a selectivity of 15:1. A BPSG film was planarized, and borderless contact hole

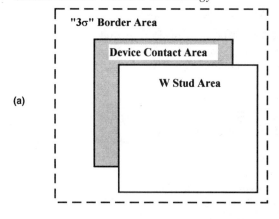

(a)

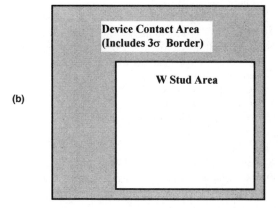

(b)

FIGURE III-41 Top view schematic of borderless and unbordered contacts showing relationship to 3σ overlay tolerance.

openings were made to the source and drain regions. According to the authors, a 256-Meg DRAM technology incorporates this process at a 0.25-μm minimum feature size. The area of the contact studs and the choice of materials and processes will continue to evolve, driven by shrinking devices and need to lower costs.

8.0 CONCLUSIONS

The improvements in contact technology reviewed here have largely taken the form of processes and materials that address the requirements of ever increasing yield and reliability and somewhat less so of electrical performance enhancements. From the earliest structures that employed Al metallizations directly on Si, to the advanced, multilayer processing that so typifies present IC manufacturing, we witness the use of much-improved material systems. These can now provide virtually penetration proof contact structures, highly stable Schottky diodes, and ohmic contacts that consistently approach the theoretical limits. The enhancement of electrical performance is less direct, as the substitution of more conductive materials does not enhance the performance of contacts as it can interconnects, e.g., by the substitution of Cu for Al. However, contacting materials can play a more subtle but still important role by their symbiosis with process design, to make possible self-aligned technologies that improve density, decrease distances, and thereby enhance overall chip performance. Advances in this area will likely be incremental and found more in the areas of improved deposition methods, more consistent material quality and improved understanding of the rather large material set now at the user's disposal as opposed to the sort of quest for altogether new materials (low K dielectrics, Cu) now taking place in the interconnections area.

REFERENCES

Aboelfotoh, M. O., *J. Appl. Phys.*, 61, 2558 (1987).
Aboelfotoh, M. O., *J. Appl. Phys.*, 69 (5), 3351 (1991).
Altman, C., S. G. Chapman, A. Satya, US Pat. 3,924,320, 1975.
Amouroux, C., J. P. Pestie, AVISEM 71, Colloq. Int. Appl. Vide Ind. Semicond. Composants Electron. Microelectron. [C.R.], 3rd, Soc. Fr. Ing. Tech. Vide, Paris, Fr., 249 (1971).
Anantha, N. G., H. S. Bhatia, S. P. Guar, P. Santosh, J. L. Maurer, IV, US Pat. 4,691,435, 1987.
Andrews, J. M., J. C. Phillips, *Phys. Rev. Lett.*, 44, 284 (1975).
Awadelkarim, O. O., P. I., Mikulan, T. J. Gu, T., *J. Appl. Phys.*, 76, 2270 (1994).
Babcock, S. E., K. N. Tu, *J. Appl. Phys.*, 59, 1599 (1986).
Basterfield, J., J. M. Shannon, A. Gill, *Solid-State Electron.*, 75, 290 (1975).
Beaudouin, P. L., R. Glang, J. Riseman, US Pat. 3,559,003, 1971.
Berger, H. H., *Solid State Electron.*, 15, 145 (1972).
Bergeron, D. L., D. J. Fleming, G. B. Stephens, US Pat. 4,357,178, 1982.
Bhatia, H. S., H. C. Calhoun, R. L. Melhado, R. H. Schnitzel, US Pat. 3,987,216, 1976.
Blair, J. C., P. B. Ghate, *J. Vac. Sci. Technol.*, 14, 79 (1977).
Broadbent, E. K., R. F. Irani, A. E. Morgan and P. Maillot, *IEEE Electron Devices*, ED-36, 2440 (1989).
Card, H. C., *IEEE Trans. Electron Devices*, ED-23, 538 (1976).
Card, H. C., K. E. Singer, *Thin Solid Films*, 28, 265 (1975).
Carr, B A., E. Friedland, J. B. Malherbe, *J. Appl. Phys.*, 64, 4775.
Chang, P. H., "Heteroepitaxy on Silicon, Fundamentals, Structures, and Devices" (H. K. Choi, H. Ishiwara, R. Hull, R. J. Nemanich, eds.), *Mater. Res. Soc. Symp. Proc. Vol.*, 116, 471 (1988).
Chen, J. Y., G. Eckhardt, L. D. Hess, in "Semiconductor Silicon" (H. R. Huff, R. J. Kriegler, Y. Takeshida, eds.), *Electrochem. Soc. Proc. Vol.*, PV 81-5, 694 (1981).
Chen, J. Y. T., D. B. Rensch, *IEEE Trans. Electron Devices*, ED-30, 1542 (1983).
Chien, H.-C., S. Ashok, *J. Appl. Phys.*, 60, 2886 (1986).
Chin, M.-R., G. Warren, K. Y. Liso, US Pat. 5,389,575, 1995.
Chino, K., *Solid-State Electronics*, 16, 119 (1973).
Chiou, J.-C., M.-C. Chieh, *J. Electrochem. Soc.*, 141, 2804 (1994).
Choi, J. S., S. H. Paek, Y. S. Hwang, S. H. Choi, D. W. Kim, H. K. Moon, J. K. Chung, W. S. Paek, T. U. Sim, J. G. Lee, *J. Mat. Sci.*, 28, 4878 (1993).
Chou, H. M. J., H. H Chao, US Pat. 5,272,099, 1993.
Chow, T. P., S. Ashok, B. J. Bagila, W. Katz, *J. Electrochem. Soc.*, 131, 156 (1984).
Crowley, A. M., S. M. Sze, *J. Appl. Phys.*, 36, 3212 (1965).
Chung, H. W., A. T. Yao, "Advanced Metallization in Microelectronics" (A. Katz, S. P. Murarka, A. Appelbaum, eds.), *Mater. Res. Soc. Proc. Vol.*, 181, 199 (1990).
Cunningham, J. A., R. S. Clark, IV, GB Pat. 1,203,086, 1970a.
Cunningham, J. A., R. S. Clark, IV, GB Pat. 1,203,087, 1970b.
Dalal, H. M., M. Ghafghaichi, L. A. Kasprazak, H. Wimpfheimer, US Pat. 215,156, 1980.

Dalal, H. M., J. J. Lowney, US Pat. 4,379,832, 1983.
Dixit, P., J. Sliwa, R. K. Klein, C. S. Sander and M. Farnaam, US Pat. 4,884, 123, 1989.
Drangeld, K. E., T. O. Mohr, H. F. Statz, W. von Muench, US Pat. 3,609,477, 1971.
Dreves, R. F., J. F. Fresia, S. U. Kim, J. J. Lajza Jr., US Pat. 4,261,095, 1981.
Eizenberg, M., R. Brener, S. P. Murarka, *J. Appl. Phys.*, 55, 3799 (1984).
Eizenberg, M., R. D. Thompson, K. N. Tu, *J. Appl. Phys.*, 58, 1886 (1985).
Evans, D. R., D. M. Leet, *J. Electrochem. Soc.*, 141, 1867 (1994).
Faith, T. J., C. P. Wu, *Appl. Phys. Lett.*, 45, 470 (1984).
Farahani, M. M., S. Garg, B. T. Moore, *J. Electrochem. Soc.*, 141, 479 (1994).
Farahani, M. M., T. E. Turner, J. J. Barnes, *J. Electrochem. Soc.*, 136, 484 (1989).
Fathauser, R. W., T. L. Lin, P. J. Grunthaner, P. O. Andersson, J. M. Iannelli, D. N. Jamieson, *J. Appl. Phys.*, 64, 4082 (1988).
Fonash, S. J., S. Ashok, R. Singh, *Appl. Phys. Lett.*, 39, 423 (1981).
Freeouf, J. L., *J. Vac. Sci. Technol.*, 18, 910 (1981).
Gambino, J., T. Ohiwa, D. Dobuzinsky, M. Armacost, S. Yoshikawa, B. Cunningham, *VMIC*, 558 1995.
Gargini, P. A., *Ind. Res. Dev.*, 3/25, 141 (1983).
Ghate, P. B., J. C. Blair, C. R. Fuller, G. E. McGuire, *Thin Solid Films*, 53, 117 (1978).
Givens, J., S. Geissler, J. Lee, O. Cain, J. Marks, P. Keswick, O. Cunningham, *J. Vac. Sci. Tech.*, B12, 427 (1994).
Gromov, D. G., A. I. Mochalov, V. P. Pugachevich, *Appl. Phys. A., Mater. Sci. Process.*, A61, 565 (1995).
Grove, A. S., *Physics and Technology of Semiconductor Devices*, John Wiley & Sons, New York, 1967.
Grove, C. L., R. B. Gregory, R. L. Hance, S. W. Sun, N. Kelly, *J. Vac. Sci. Technol.*, A7, 1596 (1989).
Hara, T., N. Ohtsuka, S. Enomoto, T. Hirayama, K. Amemiy, M. Furukawa, *Jpn. J. Appl. Phys.*, 22, Part 2, 683 (1983).
Hirashita, N., M. Kinoshita, T. Ajioka, *J. Electrochem. Soc.*, 135, 3159 (1988).
Ho, P. S., J. E. Lewis, U. Koester, in "VLSI Sci. and Technol." (C. J. Dell'Oca and W. M. Bullis, eds.) *Electrochem. Soc. Proc. Vol.*, PV 82-7, 250 (1982).
Hodges, R. L., R. Sundaresan, EP Pat. 0 632 492 A2, 1995.
Hoekelek, E., G. Y. Robinson, *Thin Solid Films*, 53, 135 (1978).
Howard, J. K., W. D. Rosenberg, J. F. White, US Pat. 4,310,568 (1982).
Huang, H-C. W., R. Cook, D. R. Campbell, P. Ronsheim, W. Rausch, B. Cunningham, *J. Appl. Phys.*, 63, 1111 (1988).
Inoue, M., US Pat. 4,976,839, 1990.
Inoue, T., S. Horiuchi, H. Iwai, H. Shimizu, T. Ishida, *Jpn. J. Appl. Phys.*, 15, 63 (1976).
Jaffe, J. M., J. I. Penton, US Pat. 4,135,292, 1979.
Jerome, R. C., F. Marazita, US Pat. 5,139,966, 1992.
Johansson, S., J.-A. Schweitz, *J. Am. Ceram. Soc.*, 71, 617 (1988).
Kado, H., S.-I. Yamamoto, K. Yokoyama, T. Tohda, Y. Umetani, *J. Appl. Phys.*, 74, 4354 (1993).
Kalnitsky, A., R. MacNaughton, J. Li, *J. Electrochem. Soc.*, 141, 2223 (1984).
Kikuchi, A., T. Ohsima, Y. Shiraki, *J. Appl. Phys.*, 64, 4614 (1988).
Kim, M. J., D. M. Brown, S. S. Cohen, P. Piacente, B. Gorowitz, *IEEE Trans. Electron Devices*, ED-32, 1328 (1985).
Kim, M. J., D. W. Skelly, R. Saia, G. Smith, D. M. Brown, *J. Electrochem. Soc.*, 134, 2603 (1987).
Kim, M. J., B. F. Griffins, D. W. Skelly, US Pat. 4,767,724, 1988.

Kohlhase, A., G. Higelin, ITG-Fachberichte 98, "Grossintegration," Vorträge der Fachtagung vom 16. bis 18. Marz 1987 in Baden-Baden, E. Luder, ed., vde-verlag gmbh, Berlin, Gr., 111, 1987.
Kolawa, E., J. S. Chen, J. S. Reid, P. J. Pokela, M. A. Nicolet, *J. Appl. Phys.*, 70, 1369 (1991).
Kubota, K., E. Hirakawa, A. Hiraki, JP Pat. 92-359885, 1993.
Kubota, K., E. Hirakawa, A. Hiraki, EP Pat 608551 A1, 1994.
Kumar, N., M. G. Fissel, K. Pourrezaei, B. Lee, E. C. Douglas, *Thin Solid Films*, 153, 287 (1987).
La Via, F., V. Privitera, C. Spinella, *Semiconductor Science & Technology*, 7/8, 1196 (1993).
Lane, C. H., *Metall. Trans.*, 1, 713 (1970).
Lavery, J., M. B. Armstrong, H. S. Gamble, *IEEE Trans. Electron Devices*, ED-34, part 1 1039 (1987).
Lau, S. S., W. F. van der Weg, "Thin Films, Interdiffusion and Reactions" (J. M. Poate, K. N. Tu, and J. W. Mayer, eds.), Electrochem. Soc. Series, Wiley-Interscience, New York, p. 433 (1978).
Lee, C.-J., Y.-K. Sung, *J. Electronic Mat.*, 22, 717 (1993).
Lee, K.-H., Yu, C.-H. D., EP Pat. 609014 A2, 1994.
Lee, T. L., L. J. Chen, *J. Appl. Phys.*, 73, 8258 (1993).
Liauh, H. R., M. C. Chen, L. J. Chen, *J. Appl. Phys.*, 74, 2590 (1993).
Lin, J. J., L. S. Tsai, H. W. Chang, C. T. Chiao, US Pat. 5,286,667, 1994.
Lowry, Jr., R. A., B. E. Smith, US Pat. 3,702,787, 1972.
Magdo, I. E., US Pat. 3,900,344, 1975.
Magdo, I. E., S. Magdo, US Pat. 3,956,527, 1976.
Malherbe, J. B., B. de Witt, G. L. P. Berning, *J. Appl. Phys.*, 71 2757 (1992).
Mallardeau, C., Y. Morand, E. Abonneau, *J. Electrochem. Soc.*, 136, 238 (1989).
Martin, R. C., US Pat. 3,468,728, 1969.
McCaldin, J. O., H. Sankur, *Appl. Phys. Lett.*, 20, 171 (1972).
Miller, R. O., C. C. Wei, EP Pat. 310398, 1991.
Misra, D., E. L. Heasell, *J. Electrochem. Soc.*, 136, 234 (1989).
Mitwalsky, A., V. Probst, R. Burnmester, R., in "Semiconductor Silicon" (H. R. Huff, K. Barraclough, J. Chikawa eds.) *Electrochem. Soc. Proc. Vol.*, PV 90-7, 876 (1990).
Moriya, T., S. Shima, S. Hazuki, M. Chiba and M. Kashiwagi, *IEDM 83*, 50 (1983).
Mueller, B. K., T. S. Kalkur, "Reliability of Semiconductor Devices and Interconnection and Multilevel Metallization, Interconnection, and Contact Technologies" (H. S. Rathore, G. C. Schwartz, R. A. Susko, eds.) *Electrochem. Soc. Proc. Vol.* PV 89-6, 289 (1989).
Muller, R. S., T. I. Kamins, *Device Electronics for Integrated Circuits*, John Wiley & Sons, New York, 1977.
Murarka, S. P. *Silicides for VLSI Application*, Academic Press, New York, 99, 1983.
Murarka, S. P., D. B. Fraser, A. K. Sinha, H. J. Levinstein, E. J. Lloyd, R. Liu, D. S. William, S. J. Hillenius, *IEEE Electron Device*, ED-34, 2108 (1987).
Murarka, S. P., *J. Vac. Sci. Tech.*, B4, 1325 (1986).
Murarka, S. P., *Metallization, Theory and Practice for VLSI and ULSI*, Butterworth-Heinemann, Stoneham, Mass., 188, 1993a.
Murarka, S. P., *Metallization, Theory and Practice for VLSI and ULSI*, Butterworth-Heinemann, Stoneham, Mass., 155, 1993b.
Murray, J. L., A. J. McAlister, *Bull. Alloy Phase Diagrams*, 5, 74 (1984).
Nagasawa, E., H. Okabayashi, M. Morimoto, *Jpn. J. Appl. Phys.*, 22, Part 2, 57 (1983).
Naguib, H. M., L. H. Hobbs, *J. Electrochem. Soc.*, 125, 169 (1978).
Nakagawa, O. S., S. Ashok, J. K. Kruger, *J. Appl. Phys.*, 69, 2057 (1991).
Neppl, F., U. Schwabe, "Thin Films and Interfaces II" (J. E. E. Baglin, D. R. Campbell, W. K. Chu, eds.), *Mater. Res. Soc. Symp. Proc. Vol.*, 25, 587 (1984).

Nicollian, E. H., A. K. Sinha, "Thin Films-Interdiffusion and Reactions" (J. M. Poate, K. N. Tu and J. W. Mayer, eds.) Wiley-Interscience, New York, p. 481, 1978.
Nogami, T., S. Takahashi, M. Oami, *Semiconductor Science & Technology*, *11/9*, 2138 (1994).
Norstroem, H., T. Donchev, M. Oestling, C. S. Petersson, 1983, *Phys. Scr.*, *28*, 633 (1983).
Ohdomari, I., T. Takahashi, "Tungsten & Other Adv. Met." (S. Wong, S. Simon, S. Furukawa, eds.), *Mater. Res. Soc. Proc. VLSI-V*, 391 (1990).
Ohta, E., K. Kakishita, H. Y. Lee, T. Sato, M. Sakata, 1989, *J. Appl. Phys.*, *65*, 3928 (1989).
Olowolafe, J. O., C. J. Palmstrom, E. G. Colgan, J. W. Mayer, *J. Appl. Phys.*, *58*, 3440 (1985).
Onuki, J., M. Nihei, *Mater. Trans., J. Inst. Met.*, *36*, 670 (1995).
Ottaviani, G., K. N. Tu, J. W. Mayer, *Phys. Rev. Lett.*, *44*, 284 (1980).
Palmstrom, C. J., J. W. Mayer, B. Cunningham, D. R. Campbell, P. A. Totta, *J. Appl. Phys.*, *58*, 3444 (1985).
Pan, P., J. G. Ryan, M. A. Lavoie, "ULSI Sci. and Technol. 1989" (C. M. Osburn, J. M. Andrews, eds.) *Electrochem. Soc. Proc. Vol.*, PV *89-9*, 104 (1989).
Paz, O., F. D. Auret, J. F. White, *J. Electrochem. Soc.*, *131*, 1712 (1984).
Perera, A. H., J. P. Krusius, *J. Electron. Mater.*, *19*, 1145 (1990).
Pintchovski, F., E. Travis, "Advanced Metallization and Processing for Semiconductor Devices and Circuits" (A. Katz, S. P. Murarka, J. M. E. Harper, Y. I. Nissim, eds.), *Mat. Res. Soc. Proc. Vol. 260*, 777 (1992).
Pogge, B., US Pat. 4,256,514, 1981.
Probst, V., H. Schaber, A. Mitwalsky, H. Kabza, B. Hoffman, K. Maex, L. Vanden hove, *J. Appl. Phys.*, *70*, 693 (1991).
Rand, M. J., J. F. Roberts, *Appl. Phys. Lett.*, *24*, 49 (1974).
RCA Corp (asignee), GP Pat. 1,177,382, 1970.
Reith, T. M., *Appl. Phys. Let.*, *28*, 152 (1976).
Revva, P., A. G. Nassiopoulos, A. Travlos, *J. Appl. Phys.*, *75*, 4533 (1994).
Rosenberg, R., M. J. Sullivan, J. K. Howard, *Thin Films—Interdiffusion and Reactions* (J. M. Poate, K. N. Tu, J. W. Mayer, eds.) Wiley-Interscience, New York, p. 13, 1978.
Rothman, L. B., P. A. Totta, J. F. White, US Pat. 4,272,561 (1981).
Sagara, K., Y. Tamaki, *J. Electrochem. Soc.*, *138*, 616 (1991).
Saito, K., T. Amazawa, Y. Arita, *J. Electrochem. Soc.*, *140*, 513 (1993).
Saraswat, K. C., S. Swirhun, J. P. McVittie, in "VLSI Sci. and Technol." (K. E. Bean, G. Rozgonyi, eds.) *Electrochem. Soc. Proc. Vol. PV 84-7*, 409 (1984).
Schlupp, R. L., WO Pat. 8,301,866, 1983.
Sekine, M., N. Ito, T. Shinmura, *J. Electrochem. Soc.*, *142*, 664 (1995).
Siemens, A.-G., Corp. (asignee), GP Pat. 1,311,839, 1973.
Singh, R. N., D. M. Brown, M. J. Kim, G. A. Smith, *J. Appl. Phys.*, *58*, 4598 (1985).
Singh, R. N., D. W. Skelly, D. W. Brown, *J. Electrochem. Soc.*, *133*, 2390 (1986).
Straayer, A., G. J. A. Hellings, F. M. van Beek, F. van der Maesen, *J. Appl. Phys.*, *59*, 2471 (1986).
Sung, K. T., S. W. Pang, M. W. Cole, N. Pearce, *J. Electrochem. Soc.*, *142*, 206 (1995).
Suu, H. V., F. Pa'szti, G. Mezey, G. Petö, A. Manuaba, M. Fried, J. J. Gyulai, *J. Appl. Phys.*, *59*, 3537 (1986).
Sze, S. M., *Physics of Semiconductor Devices*, Wiley-Interscience, New York, p. 363, 1969a.
Sze, S. M., *Physics of Semiconductor Devices*, Wiley-Interscience, New York, p. 398, 1969b.
Takeyama, M., K. Sasaki, A. Noya, *J. Appl. Phys.*, *73*, 185 (1993).
Tanabe, A., K. Konuma, N. Teranishi, S. Tohyama, K. Masubuchi, *J. Appl. Phys.*, *69*, 850 (1991).
Tang, Y. S., C. D. W. Wilkinson, C. Jeynes, *J. Appl. Phys.*, *72*, 311 (1992).
Thalapaneni, G., US Pat. 5,238,872, 1993.

Ting, C. Y., B. L. Crowder, *J. Electrochem. Soc.*, *129*, 2590 (1982).
Totta, P. A., R. P. Sopher, *IBM J. Res. Dev.*, *13*, 226 (1967).
Triebwasser, S., US Pat. 4,222,164, 1980.
Tsukada, T., H. Nogami, J. Huyashi, K. Kawaguchi, T. Hara, *J. Appl. Phys.*, *74*, 5402 (1993).
Tu, K. N., J. W. Mayer, *Thin Films—Interdiffusion and Reactions* (J. M. Poate, K. N. Tu, and J. W. Mayer, eds.), Wiley-Interscience, New York, p. 359, 1978.
Uppili, S., T. Yamaguchi, S. Alberhasky, *J. Electrochem. Soc.*, *141*, 1663 (1994).
Wang, W.-C., T.-S. Chang, F.-S. Huang, *Solid State Electronics*, *37*, 65 (1994).
Werner, J. H., H. H. Güttler, *J. Appl. Phys.*, *73*, 1315 (1993).
Werner, P., W. Jäger, A. Schüppen, *J. Appl. Phys.*, *74*, 3846 (1993).
Witmer, M., C. Y. Ting, I. Ohdomari, K. N. Tu, *J. Appl. Phys.*, *53*, 6781 (1982).
Witmer, M., C. Y. Ting, K. N. Tu, *J. Appl. Phys.*, *54*, 699 (1983).
Wolters, R. A. M., A. J. M. Nellissen, *Solid State Technol.*, *2/86*, 131 (1986).
Wong, G., US Pat, 5,175,125, 1992.
Wu, C. P., US Pat. 4,525,221, 1985.
Yamamoto, Y., H. Miyanaga, T. Amazawa, T. Sakai, *IEEE Trans. Electron Devices*, *FD-32*, 1231 (1985).
Yapsir, A. S., P. Bai, T. M. Lu, *Appl. Phys. Lett.*, *53*, 905 (1988).
Ye, M., H. Lin, G. Fei, P. Tsien, J. Zhang, S. Yin, *Vacuum*, *39*, 231 (1989).
Yokoyama, N., K. Hinode, Y. Homma, *J. Electrochem. Soc.*, *138*, 190 (1991).
Yu, A. Y. C., *Solid-State Electronics*, *13*, 1189 (1970).
Yu, Y.-C. S., V. F. Drobny, *J. Electrochem. Soc.*, *136*, 2076 (1989).

IV
Interlevel Dielectrics

Geraldine Cogin Schwartz* and K. V. Srikrishnan
IBM Microelectronics
Hopewell Junction, New York

1.0 INTRODUCTION

The reduction of signal propagation delay is one of the driving forces behind the use of multilevel device structures and the principal one responsible for the search for improved materials with which to build them. The dielectric layers, the essential insulating components of a multilevel structure, contribute to the delay (RC) through the capacitance term, ε/d, where ε is the dielectric constant of the insulator and d is its thickness. The total capacitance is due to the capacitance (1) between the first level wires and the Si substrate (2) between wires on the same level and (3) between wires on successive levels. While a thick insulator would reduce the capacitance between levels, there are limits due to processing difficulties such as via hole etching and step coverage and/or hole filling capabilities. Also, thicker interlevel insulators increase the intralevel capacitance, i.e., the cross-talk between conductors in one level. Thus, the insulators best suited for the interlevel dielectric layers are those with low dielectric constants. This chapter will discuss both inorganic and organic insulators; SOGs, also included, bridge the gap between them, since there are both types in use.

Requirements for interlevel dielectric films, in addition to a low dielectric constant, are (1) high breakdown strength; however, since the average field experienced by the dielectric films during operation of the device is quite low this requirement is often exaggerated. Others are (2) low bulk and surface conductivity, (3) low compressive stress (low to minimize wafer warpage and avoid adverse effects on devices and conductors; compressive to prevent cracking), (4) low defect density, (5) good adhesion to underlying layers (metals and dielectrics), (6) surfaces to which photoresist and permanent overlying dielectric and metal films adhere well, (7) low moisture content, (8) high resistance to permeation and absorption of moisture (i.e., high film density) and diffusion of mobile ions, (9) stability to chemicals

---

*Retired.

in the processing and use environment and to thermal excursions, and (10) etchability.

The processes used to deposit the films must satisfy several criteria (1) run-to-run reproducibility, (2) good uniformity within a wafer, (3) wide process window, (4) low contamination due to wafer handling, reactor design, process chemicals, or the conditions used for deposition and etching, (5) compatibility with underlying structures and materials, (6) no radiation damage, particularly important for MOS devices. The equipment used, particularly for manufacturing, must be reliable. System maintenance should be relatively easy and infrequent. It should be capable of high throughput and occupy minimal floor space in the fabrication facility. In addition, in this era of smaller devices and high packing densities, conformal or planarizing films are required as well as void-free filling of the deep, narrow spaces.

No single material or deposition process satisfies every requirement. Compromises are necessary and sometimes complicated sequences of materials and procedures have been devised to try to meet as many demands as possible.

Table IV-1a–c summarize some of the pertinent facts about a number of insulators.

2.0 INORGANIC DIELECTRIC FILMS

2.1 SiO$_2$: INTRODUCTION

Deposited amorphous silica is used most often as the interlevel dielectric in multilevel devices, and deposition of SiO$_2$ will be used to illustrate the apparatus and processes used for depositing many of the inorganic dielectric films. The deposition methods used most widely were discussed in Chapter I. The properties of SiO$_2$ films will be examined in greater detail than will those of the other materials, with attention given to the relationship between the characteristics of the deposition system and the film properties. Some of the techniques to be discussed are rarely used now in manufacturing, but are included for historical perspective and, at times, to indicate the origins of current practices. Other deposition techniques, whose application has never been widespread and thus were not described in Chapter I, will be reviewed briefly together with film properties. This has been done to illustrate new ideas; perhaps future development may make them feasible for manufacturing.

Silicon can be oxidized at high temperatures; the film is often referred to as "thermal oxide." It is used to passivate the silicon substrate after the active and passive devices within it are completed and as the gate in MOS devices. Although it is not used in the interconnections, it is mentioned here because its properties are often used as a standard to which the properties of deposited oxides are compared.

The deposited film may be simply SiO$_2$ (i.e., "undoped," sometimes called USG, undoped silicate glass to distinguish it from a doped oxide, e.g., BSG, borosilicate glass) or the film may have other constituents added intentionally to modify the material properties (dopants such as B, P, As, Ge, and F), incorporated as a result of the deposition method or conditions (e.g., H, Ar), or unintentionally during exposure to various environments (e.g., Na$^+$, H$_2$O). These subjects will be discussed in a later section.

TABLE IV-1a Properties of Inorganic Dielectrics

Material	Refractive index	Dielectric constant ε at 1 MHz	Dielectric breakdown MV/cm
SiO$_2$			
Bulk silica	1.46	3.85	>10
Thermal	1.46	3.9	>10
Sputtered	1.46	3.9	3–7
CVD	1.46	4.1–5^a	8–10
PECVD	1.45–1.47	4.1–5^a	5–10
CVD PSG	~1.45	~4	8.5–11
PECVD PSG	~1.45	4.1–4.3$^{(2)}$	8.5–11
CVD BSG		3.8	
PECVD BSG		3.9	
CVD BPSG		3.8–4.5$^{(1)}$	
PECVD BPSG		~4	
F-doped	<1.46^c	<4.1$^{(3)}$	6–8*, <3†
Nitride			
CVD (Si$_3$N$_4$)	2.01	~7	10
PECVD (SiNxHy)	~2^d	6–9	5
Al$_2$O$_3$	1.6–1.7	7–9	1–3
BN		2.7–7.7	
SiBN	2–1.7^e	6.8–2.9^e	Not sharp BD
SiBNO	~1.8–1.55^f	Minimum: 3.3^f	Not sharp BD

^{a}Depends on deposition conditions.
^{b}Depends on P-content.
^{c}Depends on F-content.
^{d}Depends on H-content and Si/N ratio.
^{e}Depends on B content.
^{f}Depends on [Si] and [O]/[Si].
*From Fukada and Akahori, 1993.
†From Fukada and Akahori, 1995.

The usefulness of SiO$_2$ films for this purpose depends on its physical and chemical properties. When prepared "properly," the films meet the requirement given above for an interlevel insulator.

However, the properties may deviate from ideality depending on the actual stoichiometry, the internal structure, i.e., the bonding, and the presence of other constituents, which depend on the method of deposition, the deposition conditions, and the environments to which the films are exposed.

An SiO$_2$ film that has all the attributes of a "properly" prepared film (described above) needed for a given application, is a "good-quality" film. Nonideal films are of "poorer quality." But, depending on the application, nonideality may be acceptable. For example, films with low *tensile* stress can be used when their thickness is not too great. When device speed is *not* a requirement, a higher dielectric

TABLE IV-1b Properties of Selected Inorganic Insulators

Material	Stress*[1] (MPa)	Thermal stability	Etchability wet/plasma
Sputtered SiO$_2$	C[b]	Stable[c]	BHF[d]/F-containing
CVD SiO$_2$	T	Loses H$_2$O, OH densifies	BHF[e]/F-containing
PECVD SiO$_2$	T or C[f]	Loses H$_2$O, OH Densifies	BHF[e]/F-containing
F-doped SiO$_2$	C or T[g]	Loses HF	BHF/F-containing
Si$_3$N$_4$ (CVD)	High T	Stable[h]	H$_3$PO$_4$/F-containing
SiNxHy (PECVD)	C or T[(1)]	Loses H densifies stress → C	BHF[i] F-containing

*C = compressive; T = tensile (+).
[a]Value usually depends on deposition conditions.
[b]Value depends on [Ar].
[c]May lose Ar at very high temperature.
[d]Etch rate close to that of thermal oxide.
[e]Etch rate depends largely on density, and [OH].
[f]Sign as well as value depends on deposition conditions.
[g]C stress decreases with [F]; may become T.
[h]May lose H at very high temperature.
[i]Rate depends largely on [H].

constant film can be used. The breakdown strength for interlevel dielectrics *need not* be very high, as discussed above. But porosity is never desirable, so that often the term "poor quality" often refers to films that are porous. Nonstoichiometric films which are electrically leaky are "poor quality" films.

The films are characterized, as are all dielectric films, by measuring optical constants such as refractive index and dispersion, physical properties such as stress,

TABLE IV-1c Properties of Selected Inorganic Insulators

Material	Barrier to Na$^+$	Barrier H$_2$O	Use
Thermal SiO$_2$	No	No	Passivate Si
Deposited SiO$_2$	No	No	ID[a]
PSG	Yes	Yes	ID[a]
Si$_3$N$_4$ (CVD)	Yes	Yes	Over-structures formed in Si
SiNxHy (PECVD)	Yes	Yes	ID[a] final passivation

[a]ID: Interlevel dielectric

porosity, and electrical characteristics such as dielectric constant, leakage, and breakdown strength, and chemical properties such as composition, bonding, and etch rates in various etchants. Characterization techniques have been described in Chapter II. It must be reemphasized that a single technique is not adequate for film characterization, although a single property can be used to monitor the reproducibility of a deposition process.

The properties of SiO_2 films depend on the deposition method and the choice of deposition conditions; those deposited by many of the common techniques are reviewed in the following section.

2.2 PHYSICAL VAPOR DEPOSITION

2.2.1 RF Sputtered SiO_2

The basic principles of sputtering and a description of sputtering systems is covered in Chapter I.

RF sputter deposition of SiO_2, despite the superior quality of the films, has never been used widely in device manufacturing because of its high cost. Although the quality of the films produced by the alternate, cheaper processes was (in earlier times) not as good, it was adequate for many of the devices being made. Now even the previous users of sputtered SiO_2 have largely abandoned it for cheaper (faster) processes, which now can compete in quality. Nevertheless work on sputtered oxide has continued to some extent. Many of the lessons learned in developing sputtered SiO_2 have been applied to improving and expanding the usefulness of PECVD processes. Perhaps the most important is the demonstration of the influence of substrate bias on film quality (discussed below) and on step coverage (Kennedy, 1976) and on planarization (Ting et al., 1978) covered in a later section.

SiO_2 sputtered from a dense target of very high purity silica. Sputtering systems were described in Chapter I. The effects of pressure, target-substrate spacing, and temperature on deposition (net accumulation) rates have also been discussed in Chapter I. The absence of chemical interactions makes it easier to design the reactors and to determine the dependencies of the rate, uniformity, and film quality on the deposition parameters.

Increasing the substrate bias, i.e., the energy of ion bombardment, improves film properties. The dielectric constant and dissipation factor are lowered, the resistivity is raised, the etch rate in HF-based solutions is reduced, the pinhole density is decreased and the step coverage is improved (Logan et al., 1970; Maissel et al., 1970; Vossen, 1971; Stephens et al., 1976). The effect on breakdown voltage is not conclusive since both improvement and deterioration (Schreiber and Froschle, 1976) have been reported. The improvement has been shown to be a result of an increase in the "reemission coefficient," R (Jones et al., 1967; Maissel et al., 1970). R is the fraction of the material reaching the substrate which is reemitted due to ion (and possibly neutral) bombardment and to elevated surface temperature (which may result from the increased ion bombardment or external heat sources). R increases as the pressure and input power are decreased and as the substrate bias and temperature are increased. R is related to the binding energies of the deposited species to the surface. Species trapped in nonoptimal sites are more easily removed, leaving behind the more tightly bound species which constitute a high quality film.

It was demonstrated that the reemitted material, collected under conditions of little or no reemission during redeposition, was a very fast-etching, porous film.

SiO_2 films sputtered under optimal conditions are very nearly stoichiometric in composition, contain negligible amounts of Si-OH and no water, have a dielectric constant minimally higher than bulk silica, are dense, and a good moisture barrier. The films have a low compressive stress. The Ar content and its role will be discussed later. Although the presence of O_2 or H_2 in the sputtering gas reduces the deposition rate, benefits of using mixtures of Ar and these gases have been found. Suyama et al. (1987, 1988) reported that although the films were deposited at a low temperature (200°C), they were smoother and had improved electrical characteristics when compared with films deposited in Ar alone. Macchioni (1990) attributed the improvements to decreased deposition rates. Hydrogen addition (in a magnetron system, at a substrate temperature of 200°C) prevented formation of the microvoids seen in films sputtered in Ar alone; the film density was slightly higher than that of thermal oxide. From IR and AES measurements, they concluded that the film structure was not changed by adding H_2 (Serikawa and Yachi, 1984). Apparently there was no (or insignificant) incorporation of H into the film. Previously, Serikawa (1980) had found that H_2 addition enhanced the step coverage of SiO_2 films.

A serious disadvantage of sputtering for SiO_2 deposition is its low rate which translates to low throughout, despite the use of batch systems, and thus higher cost. Several proposals have been advanced for improvement. One is the use of magnetrons (Homma and Tunekawa, 1988). Another is to increase, substantially, the power delivered to the target (Macchioni, 1990). Logan et al. (1990) developed a high rate, low voltage, single wafer reactor. The low voltage operation was achieved by using in-phase target and substrate sheath voltages, 40.68-MHz excitation, and a controlled-area confining wall electrode. Wafer temperature was controlled by gas conduction cooling to <400°C. Rates as high as 1750 A/min were attained under planarizing conditions. A major disadvantage was particulate generation; the size and density of particles were too high for VLSI (and ULSI) applications.

Particulate generation is a problem in all reactors. Poor quality oxide is deposited on the chamber walls and fixturing; this material can flake and become incorporated into the film during deposition. It can become detached when the reactor is cooled and opened to the atmosphere and "rain" onto the wafers. The use of load locks has been tried but is difficult to implement in batch systems. Plasma cleaning has been suggested, but it is not clear that it has been tried; erosion of the target would be a drawback. The usual procedure, which adds to the cost (downtime) is chamber cleaning.

Most sputtering systems are operated at 13.56 MHz, but using a higher frequency has a particular advantage in SiO_2 deposition; damage to FET devices is reduced. Since the damage is probably caused by x-rays generated in the gate by secondary electrons produced at the target (Grosewald et al., 1971), lowering their energy by using the higher frequency reduces the damage (Logan et al., 1977). Another method for damage reduction is insertion of a dc-biased grid between the target and substrate electrodes to absorb the secondary electrons emitted by the target (Hazuki and Moriya, 1987).

Meaudre and Meaudre (1981) studied dc current transport and Meaudre and Meaudre (1984) elucidated the mechanisms for ac conduction in rf sputtered SiO_2.

Interlevel Dielectrics

2.2.2 Reactive Sputtering

Reactive sputtering of SiO_2 has been investigated to some extent, but the availability of dense targets to be used in rf sputtering has reduced interest in this method of film deposition. Fuller and Baird (1963) used DC reactive sputtering of a Si target in an O_2 ambient to form "adhering, continuous" films of SiO_2 at low rates. The measured porosity accounted for the low dielectric constant and high etch rate of the films. Valetta et al. (1966) reported that unless the films were prepared at rates <200–300 A/min and at temperatures >400–500°C, the films were soft, porous, and had poor adhesion. Many of the more recent studies have been part of a series investigating several other oxides for which reactive sputtering may be more appropriate, e.g., Barbee et al. (1984). A "biplanar magnetron" has been used for reactive sputtering at high rates (Rostworowski and Parsons, 1985).

2.2.3 Dual Ion Beam Sputtering

In this technique, a high-purity quartz target was bombarded with an O^+/O_2^+ and Ar^+ beam. The Ar was introduced through a hollow cathode and O_2 was injected into the source chamber (Emiliani and Scaglione, 1985). The deposition rate was maintained at ~60 A/min. The SiO_2 films had the correct stoichiometry, high density, and good mechanical properties. The H-content of the film was ~10% (measured by elastic recoil detection) and the SiOH in the IR spectrum correspondingly high. The films were O-rich (Si:O = 0.45) (Emiliani and Scaglione, 1987).

2.2.4 Deposition by Etching-Enhanced Reactive Sputtering (DEERS)

In this technique, Si targets were sputtered in a planar magnetron system in an O_2/CCl_4 atmosphere in an rf magnetron reactor. The rate rose slowly with small additions of CCL_4, then increased rapidly; at a concentration of 2 at% Cl, the rate reached its maximum value of about 30 times the rate with no CCl_4. The effect of CF_4 was minimal. The mechanism proposed was the formation of a high vapor pressure species at the target; these are sputtered into the plasma, fragment, and impinge on the substrate to react with O-bearing species to form the oxide (Ross et al., 1990). At high concentrations of etch gas, the optical and scratch properties of the film deteriorated. When the gases contained both Cl and F (e.g., CCl_3F) the films contained several at% at Cl and F (Cl < F) but these species were not bonded to Si (Nandra, 1990).

2.3 CHEMICAL VAPOR DEPOSITION (CVD) OF SiO_2

The principles of CVD processing and the apparatus used are covered in Chapter I.

2.3.1 Inorganic Precursors

The most widely used inorganic precursor is SiH_4 but higher hydrides, Si halides, such as $SiCl_4$, or mixed hydride/halides have also been tried. The oxidizing agent is usually N_2O or O_2; CO_2 (Steinmeiner and Bloem, 1964), NO (Rand, 1967) and

O_3 (Maeda and Sato, 1977) and most recently, H_2O_2 (see below) have been used. For application as an interlevel insulator with Al-based metallization, relatively low (400–450°C) deposition temperatures are required. The deposition rate decreases as the temperature is decreased and films deposited at low temperature films are often porous with significant H content (OH and possibly H_2O) and have a high tensile stress. Another *very* serious disadvantage to the use of the inorganic precursors is that the deposited films have poor step coverage/gap fill characteristics. Takahashi et al. (1995) obtained improved step coverage by an oxide deposited using SiH_4 and O_2 at 15 torr, but only at elevated temperatures (≥ 600°C), which reduced the sticking coefficient of the growth species. Gas phase reactions producing particles ("snow," "dust") which result in hazy films with high defect density, and poor dielectric properties are another drawback. Addition of C_2H_4 to SiH_4/O_2 suppressed gas phase particle formation but lowered the deposition rate.

A recent process, called "Flowfill"™, introduced the use of H_2O_2 with SiH_4 and N_2 at 500 m torr; the oxide was deposited on a *cooled* substrate (0°C) and then annealed at 400°C in N_2 for 30 min. The resulting film planarized and filled gaps. SiH_4 and H_2O_2 were introduced separately into the chamber; the reaction rate was moderated by the presence of N_2. The mechanism of gap filling and planarization was postulated to be the formation of $Si(OH)_4$, a very viscous liquid, which condenses into small gaps and, when the thickness increases, is pulled flat by surface tension. The condensate is converted to SiO_2 by condensation polymerization during the annealing step (Matsuura et al., 1994; Dobson et al., 1994; Kiermasz et al., 1995). Another view of the process of gap-fill considered the role of silanol (Si-OH), which has a high surface mobility (Gaillard et al., 1996). The CVD film was encased in SiH_4/N_2O-based PECVD oxides; they were compressive films, deposited at ~300°C, using 380 kHz. The thin lower film was needed to promote adhesion to the underlying metal and act as a moisture barrier during anneal and the thicker upper layer to protect the CVD film during the final furnace bake. The stack was called an Advanced PLanarizing interlayer dielectric (APL); it had a low tensile stress (1×10^9 dyne/cm^2), long term stress stability and only small amounts of OH were detected. The quality of the layers was said to be equivalent, both physically and electrically, to PECVD oxides.

Finally, another drawback to the use of SiH_4 is that, undiluted, it is an explosion hazard and requires safety precautions.

2.3.2 Organic (Organosilane) Precursors

The most commonly used is tetraethoxysilane (or tetraethylorthosilicate), TEOS; tetramethylcyclotetrasiloxane, TMCTS (Fujino et al., 1992b), tetraethylcyclotetrasiloxane, TECTS (Hochberg et al., 1988), octamethylcyclotetrasiloxane, OMCTS (Matsuura et al., 1991), diethylsilane, DES (Huo et al., 1991), hexamethyldisiloxane, HMDSO (Fujino et al., 1992), tetrakis(dimethylamino)silane (Maruyama and Shirai, 1993), and trimethoxysilane, TMS (Suzuki et al., 1996) are among others that have been tried. These precursors are liquids; in the past this required the use of a carrier gas to transport the vapor from the chamber containing the liquid but there are now flow controllers for liquid sources. The advantage cited for the use of precursors other than TEOS is their higher vapor pressures (Galernt, 1990).

There is no safety hazard associated with organic precursors, as there is with SiH_4. In addition, films deposited using organic precursors have much better step coverage/gap fill capabilities than do those prepared using SiH_4; this issue will be discussed fully in a later chapter.

Adams and Capio (1979) as well as Becker et al. (1987) reported deposition of conformal oxide films of excellent quality by the decomposition of TEOS, at ~700°C, in a LP CVD system, using pressures of 0.2 to 0.3 torr, but the high temperature makes the process unsuitable for almost all applications as interlevel insulators. Hochberg and O'Meara (1989) and Patterson and Ozturk (1992) have discussed other organosilane compounds which are suitable for pyrolytic deposition of SiO_2 at lower temperatures. But for the most part, oxidation has replaced pyrolysis. A strong oxidant is needed for low temperature oxidation. Ozone (O_3) is the oxidant of choice, although Pavelscu and Kleps (1990) studied the APCVD TEOS/O_2 system in which rates of 20–100 A/min were obtained.

Maeda and Sato (1977) showed that SiO_2 could be deposited using TEOS and O_3 at temperatures as low as 200°C, although the rate was very low and the films porous.

The pressures commonly used in TEOS/O_3 CVD processes have been <100 torr (60 torr) (Spindler and Neureither, 1989; Ramkumar et al., 1992; 90 torr, Shih et al., 1992), 600 torr (so-called subatmospheric, SACVD) (Lee et al., 1990) and 760 torr, APCVD (Nishimoto et al., 1987; Fujino et al., 1990). Stonnington et al. (1992) reported on the use of *very* low pressures (10^{-1} to 10^{-3} torr). The advantages of low pressure deposition were improved step coverage due to a longer mean free path, less particulate generation, and less background water vapor (which could be incorporated into the film), but the rates were significantly lower than those obtained at the higher pressures. There appears to be no significant difference between APCVD and SACVD nor any advantage of one over the other. It is probable that the small reduction in pressure made adaptation of an existing reactor possible.

The advantage of the higher pressure (SACVD or APCVD) over the lower pressure (60 torr) is improved step coverage/gap fill and increased density so that wet etch rates and film shrinkage upon heat treatment were reduced (Lee et al., 1992); however, there are strong surface related effects (discussed below). Despite this drawback, most of the recent work, described below, has been done in this higher pressure regime.

In the APCVD processes, the deposition rate increases with increasing temperature at low temperature, but decreases with increasing temperature at higher temperatures. The transition to the high temperature regime, which is used in preparing films for VLSI/ULSI applications (~400–450°C), occurs at 300°C according to Kotani et al. (1989), at ~375°C (Maeda and Sato, 1977; Nishimoto et al., 1989), and ~400°C (Nguyen et al., 1990). Fujino et al. (1990b) showed that the rate as well as the transition temperature depended on the concentration of O_3 in the reaction mixture. As the O_3 content of the reaction mixture was increased, the deposition rate increased rapidly, reached a maximum and then decreased slowly (Fujino et al., 1990). In the high temperature range, the wet etch rate decreased (Kotani et al., 1989) but the tensile stress increased (Kotani et al., 1989) as the temperature was increased. Hosada et al. (1992) reported that high temperature and high O_3:TEOS ratios during film deposition, lowered the moisture absorption, and thus, they concluded, increased the film density, but never to a completely acceptable level in terms of subsequent outgassing in vias. They postulated that some of the water in the films was chemically bonded at sites where TEOS oxidation was incomplete; this absorption could be inhibited by annealing the films in steam, but not in O_2 or N_2. Postdeposition exposure to an rf plasma (100 kHz) at 300°C, 0.5 torr for 5 min in NH_3, N_2, Ar, and O_2 reduced the moisture content considerably. The higher the temperature during treatment, the lower the moisture content (Fujino et al., 1993). Reducing the deposition rate, by changing the process conditions, produced films with lower moisture content (Chiang et al., 1992).

Whether the higher or lower pressure processes are used, the deposited films are porous and contain significant amounts of Si—OH, increasing the dielectric constant. The porosity results in moisture absorption which increases the dielectric constant still further. Desorption upon heating is responsible for via poisoning. Absorption/desorption result in stress changes during thermal cycling; the hysteresis has been studied by Ramkumar and Saxena (1992) and Ramkumar et al. (1993). Hysteresis is smaller at higher O_3:TEOS ratios (Kwok et al., 1994). Stadtmueller (1992) related the tensile stress and low density in TEOS-based oxides to the production of byproducts after decomposition of TEOS on a hot surface during film formation.

Another disadvantage of the high pressure TEOS/O_3 CVD oxide is its sensitivity to the underlying material, which influences several properties of the film (Nishimoto et al., 1989; Fujino et al., 1990a,b, 1991a,b, 1992, 1994; Huang et al., 1993; Ong et al., 1993; Kwok et al., 1994). The deposition rate was lower on thermal oxide than on Si (except at very low O_3 concentrations). The wet etch rate of the film decreased with increasing O_3 concentration in the reaction mixture when deposited on Si, on SiH_4-based PECVD oxides, and on N-doped TEOS/N_2O PECVD oxide. The trend was reversed for films deposited on thermal oxide and on PECVD TEOS oxides. Film shrinkage increased with increasing O_3:TEOS ratio when the substrate was TEOS-based PECVD; the trend was reversed on Si substrates. The films were smooth when deposited on Si and on a SiH_4-based PECVD oxide; they were rough on TEOS-based PECVD oxides and the roughness increased when the CVD oxides were deposited using higher O_3/TEOS ratios. The process in which the highest O_3:TEOS ratio was used produced the best films (dense, low shrinkage) when the substrate was Si, but was the most surface sensitive. Stress hysteresis was greater when the base was TEOS-based PECVD oxide than when it was Si or SiH_4-based PECVD oxide.

Ramping down the rf power at the conclusion of the PECVD TEOS-based oxide deposition minimized the surface sensitivity of the SACVD oxide (Huang et al., 1993) In situ exposure of thermal oxide and the TEOS-based PECVD oxide to Ar or N_2 plasmas eliminated the substrate dependence (Fujino et al., 1992; Kwok et al., 1994). Deposition was also a function of the density of the underlying pattern. Although the dimensions of the features may be identical, the thickness of deposit was greater in an area with a less dense pattern. Coating the metal with a TEOS/O_2-PECVD oxide eliminated the difference. (Ahlburn et al., 1991); the PECVD oxide also acted as a moisture barrier between the CVD oxide and the metal film. The surface sensitivity was also eliminated by using P-doped CVD oxide (Ong, 1993). Another process, called pressure-ramp-up (PRU) took advantage of the surface-insensitivity of oxide deposited at 60 torr and the gap-fill properties of films deposited at higher pressures. The initial film was deposited at 60 torr and then the pressure increased to 450 torr for the remainder of the process (Tu et al., 1996).

Fujino et al. (1990, 1991) attributed surface sensitivity to the difficulty of a reaction between hydrophobic TEOS and a hydrophilic surface of SiO_2 as opposed to the ease of reaction with the hydrophobic Si surface. Kwok et al. (1994) refuted this model. Surface analysis showed the presence of F on substrates having surface dependence, but there was no indication of the source. On one surface (PECVD oxide after N_2 plasma treatment) both F and N were detected, and it was postulated that N counteracted the effect of F. The model proposed that the gas mixture containing the film precursors is electronegative due to abundance of O atoms and there is repulsion between the electronegative F-coated surface and the precursors, leading to a lower rate and poorer quality film. Electropositive surfaces (e.g., N, H-coated) attract the precursors and a better quality film is formed. This model explains the efficacy of N_2 and Ar plasma treatments; N on the surface overcomes the effects of F (as stated above) and Ar sputters off the F.

Thick tensile films tend to crack so that the tensile stress in the CVD oxide limits its useful thickness. Where thick films are required, thin layers of the CVD oxide are used in combination with compressive PECVD oxide films to reduce the total stress and thus increase the cracking resistance of a thick film. The porosity requires in situ deposition of a PECVD oxide capping layer oxide or in situ plasma modification. The surface sensitivity dictates modification of the underlying layer or a change in its nature or deposition process. Thus the films deposited using TEOS and O_3 are almost always combined with other films.

A semiselective CVD process using TEOS and O_3 was developed by Homma et al. (1993). It involves capping Al interconnects with TiN and TiW and exposing the surfaces of the metal and the PECVD oxide between the metal stripes to a CF_4 plasma. The thickness of oxide then deposited on the metal was substantially thinner than that deposited in the spaces.

The primary advantage of using TEOS (or other organosilanes) over the inorganic precursors is the improved step coverage. The profiles are rounded and are characterized as having a flowlike appearance. The mechanisms explaining the step coverage are discussed in the chapter on integration.

Nguyen et al. (1990) and Fujino et al. (1990b) discussed the reaction mechanism. They both noted that there were two temperature regimes: (1) the lower temperature one, in which the reaction was thermally controlled (increasing rate with increasing temperature), and (2) the higher temperature one described by Nguyen et al. as a "surface diffusion reaction" region and by Fujino et al., as one in which the high temperature results in fewer adsorbed reactants. At the higher temperature, the higher decomposition rate of O_3 and its loss at the walls may also be a factor. The overall reaction, when the reactants are heated is

$$Si-(O-C_2H_5)_4 + 8O_3 \longrightarrow SiO_2 + 10H_2O + 8CO_2$$

although the intermediate reactions and the precursors in the gas phase and on the surface are not known. Fujino et al. depict the reaction as a chain of reactions in which the Si-containing moiety (the oligomer) is adsorbed in the surface while C_2H_4 and H_2O are desorbed, resulting finally in the deposition of SiO_2 and elimination of the C and H-containing species. They note that the flowlike behavior of the TEOS/O_3 film has been assumed to result from a liquid-like behavior of the oligomers. None of the incompletely reacted TEOS fragments are incorporated into the film; they are probably oxidized to Si-O and Si-OH before incorporation. O_3 may participate in the reaction or act as a catalyst. The higher temperature, although lowering the rate, enhances H and Si-OH elimination and film densification.

2.4 PLASMA-ENHANCED CHEMICAL VAPOR DEPOSITION (PECVD) OF SiO_2

The principles of reactive plasma assisted processing and a description of the reactors used is covered in Chapter I.

PECVD and CVD processes use the same Si-containing precursors, principally SiH_4, and TEOS, although halides such as $SiCl_4$ and SiF_4 (Falcony et al., 1991) and the organic compound TMCTS (Webb et al., 1989) have also been tried. In TEOS-based PECVD deposition, in which the plasma creates reactive species, O_2 can replace the more highly reactive O_3 needed for thermally activated deposition. Except in the high density reactors, discussed below, N_2O is often used when SiH_4 is the precursor. Although lower deposition temperatures are possible in PECVD, since a discharge supplies the energy, produces reactive species and ions to bombard the surface, thereby densifying the film as well as enhancing the surface reactions, it will be seen that higher deposition temperatures result in "good" quality films.

SiH_4- and TEOS-based films will be considered separately, wherever possible, in the following discussion. The advantages of using organic precursors instead of SiH_4 are the same for PECVD as for CVD.

2.4.1 Capacitively Coupled Plasmas; Bell-Jar Reactors

Some of the earliest work was done in bell-jar systems. TEOS was decomposed in an rf oxygen discharge by Alt et al. (1963) who reported that the films were dense. Ing and Davern (1965) found that films deposited at high rates had occluded organic material, were less stable electrically, and had higher dielectric losses, Mukherjee and Evans (1972) noted the presence of OH groups in films deposited below 500°C. Secrist and Mackenzie (1966) used microwave excitation to produce films structurally similar to silica but containing extensive amounts of water.

SiH_4 and N_2O was used by Sterling and Swann (1965) who found that, although films can be formed at low temperature, they contained a large amount of water. Higher deposition temperatures were preferable; the films thus formed were hard, glassy, and adherent. Joyce et al. (1967/1968) also used SiH_4 and N_2O and concluded that the film properties depended on the SiH_4/N_2O ratio and the deposition temperature.

As Sterling and Swann (1965) pointed out, N_2O is preferred over O_2 because of the spontaneous reaction between SiH_4 and O_2, so that, as pointed out by Hess (1987) using O_2 increases the possibility of a homogeneous reaction instead of the desired heterogeneous surface reaction. Gokan et al. (1987) mentioned the low dissociation energy of the N—O bond in N_2O as another advantage of the use of that oxidant. Nevertheless O_2 was used with SiH_4 to prepare a range of SiOx compounds; no mention was made of any homogenous reactions (Pan et al., 1985).

2.4.2 Capacitively Coupled Planar Reactors

The capacitively coupled planar reactor, patented in 1973 by Reinberg, displaced the tubular or bell-jar systems. This reactor or modifications of it were used for batch processing until they were, in turn, largely displaced by single wafer or multiple-station reactors.

SiH_4-Based Oxides

SiH_4 and N_2O and an excitation frequency of 50 kHz were used by Hollahan (1979). He reported the films were compressive, contained some N, and that step coverage of a shallow V-groove by a thin film was conformal. Adams et al. (1981) also used SiH_4/N_2O mixtures in a similar reactor, but 13.56-MHz excitation of the plasma. The deposition temperature was 200–340°C and the pressure 133 Pa. They reported that films deposited over deep straight-sided steps were thin along the vertical walls and contained no N. They emphasized the strong dependence of the film properties on the deposition conditions. They summarized their findings: with increasing deposition temperature, the growth rate and film density increased, the wet etch rate and the H-content decreased, and the film stress was compressive under all deposition conditions.

Gokan et al. (1987) found that the compressive stress decreased as the pressure was increased and the temperature decreased. The opposite trend was reported by van de Ven et al. (1987) who used a multistation reactor with a much higher deposition rate and temperature (400°C); below ~250 Pa the film stress was tensile (~5×10^8 dynes/cm^2) and changed sign at ~250 Pa. The maximum value of the compressive stress was ~1.5×10^9 dynes/cm^2.

Batey and Tierney (1986) reported that excellent film properties were obtained by lowering the SiH_4 and N_2O concentrations with a very large flow of helium to reduce the deposition rate. Chapple-Sokol et al. (1991) also used this highly diluted reaction mixture and found that the deposition rate decreased as the temperature was increased. They postulated that this might be due to reduced reactant residence time at the higher temperatures. At 250°C, H_2O and Si-H were detected in the IR spectrum; as the temperature was increased, the H-concentration (Si-OH only) decreased. The rate increased with increasing input power, reached a maximum and then decreased but the power had little influence on the H-content of the film. The film properties (porosity, density) improved with increasing temperature and power, although temperature has the greater influence in decreasing the Si-OH content. The power played a significant role in incorporation of adsorbed precursors to low energy sites within the growing film.

TEOS-Based Oxides

Kirov et al. (19780 deposited SiO_2 from TEOS and O_2, using a 13.56-MHz discharge. At temperatures below 400°C; the films contained absorbed water. Veprek and Boutard (1991) using a 27-MHz discharge, found that good quality TEOS-based oxides could be prepared at relatively low temperatures if the deposition rates were kept low. The results reported by Hey et al. (1990), for TEOS-based oxides, agreed with those of Gokan (1987), i.e., the stress went from compressive to tensile when films were deposited at pressures of ~170 Pa. Also, the wet etch rate increased monotonically with increasing pressure. They attributed both results to reduced ion bombardment at higher pressure and concluded that the high frequency power had little effect on improving the film properties. The use of TEOS, for deposition of SiO_2 with improved step coverage properties, in a high pressure, single wafer reactor was described in a patent filed in 1991 (Wang et al., 1994). Patrick et al. (1992) showed that the deposition rate and H-content of TEOS-based oxides decreased with increasing temperature; decreasing the $TEOS/O_2$ ratio also, decreased the rate. The rate was independent of power but the stress increased as the power was increased. The dielectric constant increased as the deposition rate increased. Films deposited at a low temperature absorbed moisture at a rapid rate.

Based on the experience gained from sputtering, i.e., higher energy ion bombardment of the substrate (by the use of applied substrate bias) improved the quality of the deposited films, modifying reactors for dual frequency operation was a logical step. Both van de Ven (1990) and Hey (1990) showed that increasing the fraction of low-frequency power made the film stress of a TEOS-based oxide less tensile (more compressive) and lowered the wet etch rate, i.e., the film was denser; therefore it absorbed less atmospheric moisture. These effects illustrated, again, the beneficial effect of higher energy bombardment.

Moisture Absorption

Because of the adverse effects of moisture in the films, effort has been expended on finding ways of ensuring moisture resistance. Harrus et al. (1991) reported that the use of low-frequency during deposition, i.e., by employing a dual frequency reactor, improved the moisture resistance of both TEOS and SiH_4-based oxides. Later results by the same group (van Schravendijk et al., 1992a,b) were that a TEOS-based oxide with a compressive stress $\geq 1 \times 10^9$ dynes/cm^2 and a refractive index ≥ 1.458 would not absorb water "regardless of the deposition conditions" although they also stated that the use of low-frequency power was necessary to produce compressive stress. They used the stress hysteresis (the difference between the film stress as-deposited and at the end of a complete thermal cycle) to demonstrate the influence of stress and refractive index on moisture absorption as shown in Figs. IV-1 and IV-2, since in the absence of moisture, there is no hysteresis. By using the flexibility of a dual frequency plasma, they were able to deposit a film of low compressive stress at 250°C, equivalent to a film deposited at 350°C.

Robles et al. (1992) and Galiano et al. (1992) deposited TEOS-based oxides with a compressive stress of 1×10^9 dynes/cm^2 that did not absorb moisture. There was a negligible influence of deposition rate or the use of a mixed frequency. The higher the deposition temperature, the better the moisture resistance. Thus they concluded that the moisture resistance of TEOS-based oxides was determined by the initial stress and the deposition temperature only.

For SiH_4-based oxides, films with a refractive index ≥ 1.465 do not absorb moisture but such films have a lower compressive stress (van Schravendijk et al., 1992). Incorporation of N or excess Si was required to make low stress films moisture resistant. Blain et al. (1995) agreed with these results, i.e., a high compressive stress does not guarantee a dry, stable film. Very low moisture absorption is related to a high refractive index and a low as-deposited compressive stress. A Si-rich oxide, which met the refractive index and stress criteria did, indeed, have superior moisture resistance. They stated that moisture absorption is controlled by composition, mechanical stress, and density.

Interlevel Dielectrics

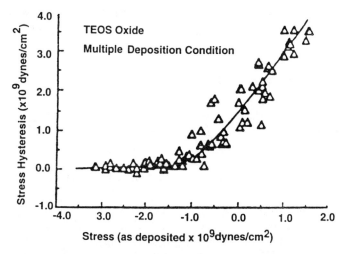

FIGURE IV-1 Stress hysteresis of TEOS-based PECVD SiO_2 films vs. as-deposited stress. (Reprinted with permission from A. S. Van Schravendijk et al.)

Since TEOS- and SiH_4-based oxides behave differently with respect to moisture absorption, it was postulated that there must be a significant difference in their molecular structure (van Schravendijk et al., (1992a).

Step Coverage

As stated before, the step coverage capabilities of PECVD oxides increase when TEOS is used instead of SiH_4 as the Si-containing precursor, but that of the CVD $TEOS/O_3$ film is even better; this is discussed in detail in a later chapter.

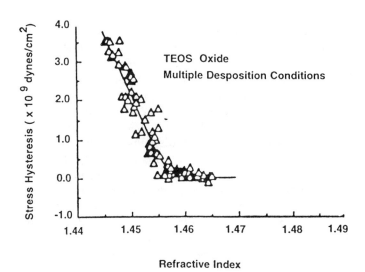

FIGURE IV-2 Stress hysteresis of TEOS-based PECVD SiO_2 films vs. refractive index. (Reprinted with permission from A. S. Van Schravendijk et al.)

In an effort to improve step coverage, a process called TOP-PECVD, which used a pulsed rf plasma, was introduced by Ikeda et al. (1992). Deposition of SiO_2, using TEOS and O_3, was alternated with exposure to an O_2 plasma so that the ion bombardment could dehydrate the film. This cyclical procedure resulted in a thick homogeneous film, more conformal that conventional TEOS/O_2 PECVD oxides but denser than CVD films.

2.4.3 Deposition Models
TEOS-Based Oxide
The explanation of the decrease in deposition rate with increasing temperatures (negative activation energy) has been explained in terms of adsorption/desorption; at higher temperatures the chance for adsorption decreases while that for desorption increases (e.g., Chin and van de Ven, 1988). This model has been discarded by Raupp et al. (1992). They viewed the deposition process as following two paths: oxygen-ion-assisted and O-atom initiated. The rate of the ion-assisted reaction is proportional to the oxygen ion flux with a reactive sticking coefficient close to 1. The rate of the O-atom initiated reaction is nearly independent of temperature and TEOS concentration but was directly proportional to the O-atom concentration. They stated that the evidence favored a model in which the growing film is saturated with TEOS precursors and oxidative attack by O and oxygen ions was the rate-determining step, not precursor adsorption. Thus, they explained that the (apparent) negative activation energy was due to an activated atomic-O surface recombination which decreases the O-atom concentration as the temperature is increased.

SiH_4-Based Oxide
Smith and Alimonda (1993) studied the chemical reactions which occur in the plasma deposition of SiO_2 in undiluted and He-diluted SiH_4/N_2O mixtures; they analyzed the plasma by means of a mass spectrometer connected to the reaction chamber and configured for line-of-sight sampling of the plasma species in the deposition plane. They found that if the rf power was sufficient to generate a supply of O-atoms in excess of that needed to convert all the SiH_4 to SiO_2, the IR spectrum will show Si—OH peaks (a few at%) in addition to the Si—O peaks; but no detectable (<0.5%) HO—H, Si—H, or N—H peaks. O consumes SiH_4 by the generation of various silanols, $SiH_n(OH)_p$. But since O is known to react with the Si—H bond by extracting H to form OH, the silanols must be formed by subsequent reattachment of the OH. The decrease of OH with increasing temperature is evidence for the thermally driven OH elimination. The silanol precursors contribute to film formation and to particle formation (with loss of rate). The importance of silanols in film deposition relative to more direct reactions of SiH_n and O at the surface are not known, although surface reactions must dominate at low pressure when there are not enough gas collisions to form gas-phase precursors. The reaction pathways are shown in Fig. IV-3. They detected no effect of the He dilution on the chemical reactions in the plasma or on the electrical properties of the film. They concluded that excess O was the key to obtaining a good oxide. They measured the quality of the oxides in terms of breakdown fields and electron trapping; these, however, are not of great importance for an interlevel dielectric.

Interlevel Dielectrics 239

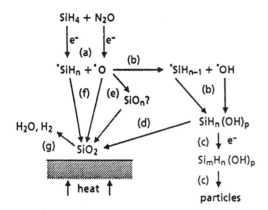

FIGURE IV-3 Reaction pathways for SiH_4-based PECVD SiO_2. (Reprinted with permission from The Electrochemical Society from D. L. Smith and A. S. Alimonda.)

2.4.4 High-Density Plasma

Next in the development cycle were the high-density plasma systems, ECR, RFI (or ICP or TCP), and the helicon. ECR was the first to be introduced commercially. The initial interest in these kinds of systems was fueled by the widely publicized, but mistaken, idea that excellent quality films, capable of filling high aspect ratio gaps, could be deposited at low temperatures, perhaps even at room temperature, and that the low ion energies of such a plasma minimized or eliminated damage. However, as mentioned previously, substrate bias, essential for gap-filling, results in wafer heating and perhaps damage. Pai et al. (1992) discusses this in some detail. Thus, despite the cooled wafer holder, the wafer temperature was high (in some cases ~500°C), although earlier it had been reported that oxides prepared in an identical way had been deposited at low temperature. The systems developed most recently have substrate bias and He-backed chucks for temperature control.

ECR

SiH_4, injected into the reaction chamber, has been used as the Si-source almost exclusively; TEOS is an alternate source used infrequently (Pai et al., 1992). O_2 is the usual oxidant, although N_2O has been tried (Herak and Thomson, 1990; Chau et al., 1991). O_2 is introduced into the plasma chamber usually with Ar.

The deposition rate of the oxide increases with increasing flow of the Si-source and microwave power. The deposition rate decreases with increasing temperature for both SiH_4 and TEOS as the Si-source. This trend is opposite to that seen for oxide deposition in the (capacitively coupled) low density plasma reactors using SiH_4, but the same as that observed for TEOS as the precursor. This implies that the mechanism of oxide deposition from SiH_4 must be different in the low and high density plasmas.

The decrease in rate with increasing temperature has often been explained as due to thermally enhanced desorption of a precursor from the surface on which the film is being grown, although it may be recalled that Raupp et al. (1992) suggested an alternate mechanism involving loss of active O, to explain the negative activation energy in the low plasma density deposition of oxide from TEOS. There has been no suggestion in the literature that Raupp's explanation might also be applicable to oxide deposition from SiH_4 in an ECR reactor.

There is one exception to the decrease in rate with increasing temperature, for SiH_4-based oxide deposition. Herak and Thomson (1990), who used a different ECR configuration, one in which the substrates were placed in the chamber beneath the field coils, i.e., in direct contact with the plasma, or else near the SiH_4 inlet, in the plasma afterglow. They found that when the substrate was located in the afterglow, the deposition rate increased with increasing temperature, suggesting that under these circumstances, there was a thermally activated surface reaction. However, when the substrates were in direct contact with the plasma, the rate decreased with increasing temperature, as has been reported for the divergent-field ECR reactors.

As the substrate bias increases, the deposition rate decreases, except, as discussed below, for very low rf power levels. By analogy to bias sputter deposition, this has been attributed to sputter etching competing with deposition, but this interpretation seems to be open to question.

Lassig (1994) investigated low temperature (~200°C) deposition of oxide in the regime in which very low rf power was applied to the substrate. In the range of 0 to 0.6 watts/cm^2, the rate increased slightly. He attributed this result to an ion-assisted deposition reaction, as had been proposed by Chang et al. (1990) and Raupp et al. (1992) for PECVD and by Chang et al. (1993) for ECR deposition. At low rf power levels there is an excess supply of neutral precursors, possibly silane or one of its radicals, so that the rate is ion-limited. As the ion energy increases, the supply of neutral precursors becomes depleted and the ion-induced rate eventually saturates. As the bias is increased further, the rate falls due to sputter etching. However, Pai et al. (1992) showed that the sputter etch rate was insufficient to account for the rate reduction in the TEOS-based ECR deposition, and that the rise in temperature, due to the ion bombardment, contributed to the rate reduction which they attributed to precursor desorption.

For all values of applied bias, increasing it made the films more like thermal oxide: the wet etch rate decreased (i.e., the film density increased), the position of the Si—O band in the IR spectrum was shifted to higher frequency and the band half-width decreased. Lassig (1994) also concluded that, since for a given ratio of SiH_4 to O_2, increasing the substrate bias resulted in a decrease in H-bonded Si, applied bias increased the oxidation of silane.

The compressive stress was reduced as the substrate bias was increased; the trend opposite to that observed for sputtered oxide films. Films deposited with substrate bias, and optimal reactant ratio, show no hysteresis upon thermal cycling (Chiang et al., 1989). The compressive stress increased with increasing deposition temperature (Chebi et al., 1992) and with increasing O_2:SiH_4 ratio (Lassig, 1994).

Denison et al. (1990) also found that the density of the oxide film decreased with increasing incorporation of Ar. Fukada et al. (1988) reported that a high quality film, equivalent to thermal oxide, can be formed when the ECR position is located close to the substrate; they suggested that the quality is related to highly excited ions transported to the substrate. The results of Herak et al. (1989) are at odds with most of the work reported in the literature, in that they found the properties of their films, deposited at 350°C, were insensitive to the O_2:SiH_4 ratio.

The trends reported here without attribution have been noted by many investigators, some of whom have been cited when discussing other oxide properties, and some of whom have not been cited previously, e.g., S. V. Nguyen and K. Albaugh (1989), Andosca et al. (1992), Fukada et al. (1992), Hemandez et al. (1994).

The influence of deposition conditions on the film composition, particularly the H-content of the films, and the influence of the H-content on film properties, as well as the step coverage and gap-fill characteristics, are discussed in other sections.

Helicon Plasmas

The first report (Charles et al., 1993) of the use of a helicon plasma for deposition of SiO_2 from SiH_4 and O_2 emphasized the usefulness of this kind of reactor for low pressure (2 mtorr) and low temperature (~200°C) deposition of films comparable to those deposited by other techniques at much higher temperature.

O_2 was introduced at the top of the reaction chamber (the helicon source) and SiH_4 via a ring placed above the wafer. Since the wafer was merely loosely clamped to the substrate table kept at 20°C, its temperature was raised by exposure to the plasma to ~200°C. The same power was used for all the experiments. No substrate bias was applied. The deposition rate increased as a function of the SiH_4 flow rate for an O_2:SiH_4 flow rate ratio (R) $\geq$ 3 ([O]/[Si] ratio in the film $\geq$ 1.95). A decrease in deposition rate resulted in a decrease in refractive index, "P" etch rate, XPS line width of both the O 1s and Si 2p peaks, and an increase in the Si—O stretching peak frequency, approaching, but never reaching the values for thermal oxide. The [O]/[Si] ratio in the film increased with increasing R; the film was stoichiometric when R = 10; at this ratio the deposition rate was very low (~25 nm/min).

Nishimoto et al. (1995) also deposited SiO_2; the reactants were SiH_4, O_2, and Ar. A 100 kHz bias could be applied to the substrate in this reactor. The pressure was varied between 8 and 16 mtorr; the substrate temperature was not given. The films had a low compressive stress, low H-content, good water-blocking properties, and wet etch rates comparable to those of thermal oxide. The deposition rate, which had a maximum value of ~100 nm/min, increased with increasing SiH_4 and O_2 flow rates, indicating a diffusion limited reaction. The rate was also increased by increasing the flow of Ar and decreasing the distance between the SiH_4 source ring and the wafer; it was decreased with increasing substrate bias, attributed to sputter etching. The film composition and chemical structure were determined by the value of the O_2/SiH_4 flow rate ratio (R); at R = 1, the refractive index and the wet etch rate were comparable to those of thermal oxide as was the position of the Si—O peak. The concluded that film oxidation and densification were a result of the high density oxygen plasma produced by the helicon source. It is not clear why the value of R at which thermal-oxide-like properties was reached was so different in these two investigations. The only apparent difference in the reactors was the ability to apply substrate bias in the later experiments, but there was no mention of any interaction of bias and R in that paper. The H-content could be reduced by increasing the substrate bias and by increased helicon source power. It was suggested that

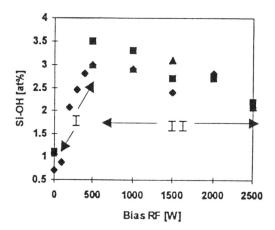

FIGURE IV-4 SiOH content of SiO_2 films deposited in an inductively-coupled plasma reactor vs. rf bias applied to substrate. (Reprinted with permission from P. Weigand et al.)

Ar sputtering increased the film density. Films with low compressive stress were obtained at high helicon power and/or substrate bias and no stress hysteresis (i.e., moisture absorption) was observed; films containing a small amount of Si—H were also stable with respect to moisture absorption.

Thus it appears that "good films" were deposited in a helicon reactor, but comparable films have been deposited in lower density plasma reactors. The advantage appears to be, not in the quality of the films (unless it is confirmed that these good films were deposited at a significantly lower substrate temperature) or in the deposition rate, but in the ability to fill high aspect ratio gaps (as discussed in a later chapter).

Inductively Coupled Plasmas

SiO_2 films were deposited using $SiH_4/O_2/Ar$ in a biased inductively coupled plasma (ICP) reactor. The wafers were placed on an electrostatically clamped holder with He back-side flow for temperature control; the deposition temperature varied between 250 and 350°C. The pressure in the reactor was <5 mtorr. The rates were higher than in the helicon system (180–400 nm/min) but the source power was also higher. Since power density and efficiency were not given, it is difficult to compare the achievable rates in these two kinds of HD reactors.

The deposition rate of the oxide increased with increasing SiH_4 flow and source power, suggesting a species (plasma-activated SiH_4) transport limited process. The sputter etch rate, necessary for gap-fill, increased with increasing substrate bias (Mountsier et al., 1994).

Films deposited using a high O_2/SiH_4 flow rate ratio (~2) did not absorb water and contained only a small concentration of Si—OH; the H-concentration varied between 1.4 to 4 at%, depending on the deposition conditions. The films had a low compressive stress. The Si—O stretching frequency and band width of these films were similar to those of high-temperature LPCVD oxides. Increasing substrate bias and temperature densified the films and reduced the H-content. Films deposited in an O_2-rich region are stoichiometric and have a refractive index of 1.46 and a wet etch rate <2× that of thermal oxide (Nguyen et al., 1995).

SiO_2 films, deposited in the absence of bias and thus, at the lowest temperature, have a lower Si—OH content and less tendency to absorb atmospheric moisture, and were smoother compared with films deposited with bias (Shoda et al., 1996). As the rf bias was increased, the concentration of Si—OH increased monatonically, reaching its maximum value at ~500 W. As the bias was increased further, the Si—OH content decreased slowly, as shown in Fig. IV-4 (Weigand et al., 1996). For biased deposition, the Si—OH concentration decreased monatonically with substrate temperature, which is increased with increasing substrate bias. The explanation offered was that sputtering induced defects in the growing films (Weigand et al., 1996).

The most noteworthy capability of the biased ICP system is its high-aspect-ratio gap-filling (Moutsier et al., 1994; Nguyen et al., 1995) which is discussed in a later chapter.

2.5 OTHER DEPOSITION METHODS

2.5.1 Evaporation

This has never been used to any significant extent for deposition of interlevel insulators. Stoichiometry control is difficult; for example, in a good vacuum, SiO is deposited from an SiO_2 source; when SiO_2 is deposited (by changing the conditions), it is a poor quality film (Pliskin and Lehman, 1965; Pliskin and Castrucci, 1968).

2.5.2 Plasma-Enhanced Evaporation or Activated Reactive Evaporation (ARE) of Thin Oxides

The basic process is evaporation of Si (using an e-gun or a resistance heated source) in the presence of activated oxygen. Bunash (1981) produced a plasma above the source using a dc-biased electrode placed between the source and the substrate. Low-frequency plasma excitation of O_2, magnetically confined in a separate chamber, is introduced into the high-vacuum reaction region. The low temperature of the substrate (30–250°C) and the ability to form SiO_2 patterns by lift-off were emphasized (Chang et al., 1983). Microwave excitation of O_2 was used by Murakami et al. (1985). Another version of this method was described by Lorenz et al. (1991). Silicon was evaporated using an e-gun as O_2 was fed directly into the chamber and activated in an Ar plasma created by a heated filament in a separate cavity connected to the chamber through an orifice. Substrate temperatures were 100–400°C and the deposition rate 120 A/min. The film was stoichiometric, with <1 at% H (from the residual gas); the electrical properties were comparable to those of thermal oxide. The emphasis was on suitability as a gate dielectric although the application of SiO_2 as an interlevel dielectric was mentioned.

2.5.3 Ion Beam Deposition

Minowa et al. (1983) deposited SiO_2 using a reactive ionized cluster beam (RICB); the source was SiO_2 and an adequate supply of O_2 was injected into the chamber.

The substrate temperature was <300°C and the film quality was almost as good as that of thermal oxide; the step coverage was uniform but the deposition rate was 60 A/min in what appears to be a single wafer reactor.

Wong et al. (1985) used what appears to be the same method but they called it nozzle beam deposition, since they say that there was no proof that clusters of oxide were present. Film quality depended on the ionization current and acceleration voltage applied to the beam.

Minowa and Ito (1988) deposited SiO_2 using dual ion beams: an ionized cluster beam (ICB) with an SiO source together with an ionized gas beam (IGB) with an O_2 source. The beams collide in transit or on the substrate surface held at or below 200°C. The properties of the film depended on the partial pressure of O_2, the oxygen ion current density, and the deposition rate. With an adequate supply of O_2 (since the deposition rate depends on the O_2 ion current), deposition rates as high as 2600 A/min can be reached. SiO_2 of a quality almost that of thermal oxide was deposited at a substrate temperature ~200°C and a rate <1200 A/min in a single-wafer system for a 5-in. wafer.

2.5.4 Digital CVD

SiO_2 formation by this method is accomplished by repetitive cycles of (1) deposition of a few monolayers of SiHx, and (2) oxidation. Both SiHx (from He-diluted SiH_4) and oxygen radicals are generated from remote microwave plasma discharges and were alternately irradiated onto a Si wafer. The species are ejected with near-supersonic velocity into a vacuum chamber and fill deep trenches in Si conformally (Nikano et al., 1989, 1990; Sakaue et al., 1993). The oxide grows a little faster on the bottom than on the sidewalls of the trench, it was postulated that the radicals are reflected from the sidewall. The SiOH content decreased with increasing O_2 irradiation; therefore the pulse width of the O_2 was larger (100×) than that of the SiHx. The deposition rate decreased with increasing temperature but the film quality was improved, i.e., the SiOH content was decreased. The deposition rate was 3 A per 5.05 s pulse at 300°C; the area was 1 cm^2. There was no discussion of the possibility of scale-up. Although an interesting concept, the process has no foreseeable application for VLSI processing.

A variation of the method was used by Horiike et al. (1990). Triethylesilane (TES) and H, created in a remote microwave source, produced CxHy radicals which were deposited on a Si wafer and then oxidized, layer-by-layer. The excellent step coverage was attributed to the high viscosity of the radicals. The deposition temperature was 250°C. Both deposition rate and conformality were a function of the H_2 concentration, peaking at about 60% H_2; at high H_2 concentrations, an overhang structure was formed, instead of the filling from bottom to top. The films contained organic species and further work was needed to improve the film quality.

2.5.5 Liquid Phase Oxidation

Trimethylsilane (TMS) was reacted with O atoms, generated in a microwave discharge chamber, on a substrate cooled by liquid nitrogen (Noguchi et al., 1987). Hirose (1988) used the phrase liquefaction CVD. The deposition rate increased with decreasing substrate temperature, reaching a maximum value at about −20°C, decreasing with further decrease in temperature. Above about −20°C, the step

Interlevel Dielectrics

coverage was typical of CVD films, i.e., an overhang profile with sidewall coverage decreasing as the depth increased. Below that temperature, the profile changed and deep narrow grooves were filled completely, reminiscent of liquid flowing into a hole. The proposed mechanism is consistent with that view: as the substrate temperature is lowered, the partially oxidized condensate migrates readily on the Si surface, filling corners; continued growth maintains the initial profile. Thus, oxidation occurred in the liquid phase. The as-deposited film resembles PECVD HMDS. Annealing in O_2 at 300°C for 1 h densified the film ~15%, reduced the Si—OH band, shifted the Si—O—Si band, but had almost no effect on the intensity of the Si—CH_3 band. Although the ability to fill gaps and thereby planarize the surface surfaces is a useful property, the nature of the film appears to make it unsuitable for VLSI application.

2.6 DOPED OXIDES

In this section the effects of the dopants B, P, B/P (Kern et al., 1970, 1977), As (Ashwell and Wright, 1985), and Ge (Chien et al., 1984; Thakur et al., 1994) will be discussed. The doped oxides, called, e.g., BSG, PSG, BPSG, etc. are not mixed oxides, but are silicate glasses as shown by IR absorption and etch rate measurements (Kern and Fisher, 1970). One of the uses of these dopants is to reduce the flow temperature of SiO_2; B is most effective; it greatly reduces the viscosity of the oxide because the smaller atom can enter the interstitial positions easily to weaken the Si—O bond (Malik and Solanski, 1990). The flow temperature reduction improves with increasing dopant concentration, but as the concentrations of dopant increase, the films become unstable to moisture. Another advantage to P and B doping of CVD oxides is stress reduction. Also, since the tendency of B to form boric acid is less than that of P to form phosphoric acid, BSG appears to be preferable to PSG for these applications. Mixtures of dopants are used since flow at lower temperatures can be achieved with reduced sensitivity to moisture. BSG, PSG, and BPSG have other properties (discussed below) besides promoting flow; As (Aswell et al., 1985) and Ge (Chien et al., 1984; Thakur et al., 1994) are used alone or in combination with other dopants only to promote flowage which will be discussed in a later section.

Doped oxides have been deposited using APCVD (Kern), LPCVD (Kern), SACVD, (Robles et al., 1995) PECVD (Law et al., 1989). The growth temperature has been in the range of about 300–800°C. The Si sources have been usually, SiH_4, and TEOS although $SiCl_4$, tetrapropoxysilane, and in the case of BSG, $(CH_3)_3SiBO_3$, (tris(trimethyl)siloxy boron or tris(trimethyl)silyl borate), abbreviated as SiOB has also been used as both a Si and B source. The oxidant has been O_2, O_3 or N_2O. Pyrolysis of TEOS has been another option (Becker and Rohl, 1987).

The most widely used P-dopants have been PH_3, trimethyl phosphite (TMP) and trimethyl phosphate (TMPO) although PCl_3 and $POCl_3$ were used in the earlier days. P-doped oxide was also deposited by reactive sputtering of an undoped Si target in O_2-Ar mixtures to which P vapor was added (Serikawa and Okamotoa, 1985).

The B-sources in common use are B_2H_6, and trimethyl borate (TMB); SiOB has been used more recently. Tri-n-propylborate and BCl_3 have also been mentioned.

2.6.1 PSG

Shioya and Maeda (1986) reported that the preparation method of PSG films deposited from the hydrides affected many of the properties of the PSG films; their results are summarized in Table IV-2. However, it should be noted that the PECVD films were deposited at a lower temperature (300°C) than is customary. The density of all the films increased slightly and their etch rates in HF solution increased markedly with increasing P-concentration although the refractive index did not. The etch rate of PSG in HF solution can be used to measure the P-concentration

TABLE IV-2 Properties of PSG Films vs. %PH_3 in Feed Gas

Property	Plasma	LPCVD	APCVD
Deposition rate	—	—	Incr. monatonic × 6 at 20%
Etch rate in HF	Incr.; highest	Incr.; intermediate	Incr., lowest
RI	—	—	—
Density	Incr. slightly highest	Incr. slightly intermediate	Incr. slightly lowest
Stress	C → T ~ 7%	C → T ~ 7%	T → C; monotonic
Stress after anneal	If C, min ~600° If T, T → C at ~800°	If C, min ~600° If T, T → C at ~800°	T → C; monotonic
IR: P=O	Incr.; intermediate	Incr.; lowest	Incr.; highest
$\varepsilon \sim 3.8-4.5$	Highest incr. slightly	Lowest —	Intermediate —
ε after anneal Almost identical $\varepsilon \sim 3.5-3.9$	—	—	—
Breakdown field	Decr. slightly lowest	Incr. slightly ~same as APCVD	Incr. slightly ~same as LPCVD
Absorption water vapor[a]			
IR: SiOH	Highest	Intermediate	Lowest
IR: P=O	None	None	Some
Etch rate after water vapor	Decr.	Decr.	Decr.

Notes:
—: little or no change
[a]Water vapor treatment: 2 atm, 120°, 2 h.

(Pliskin, 1977). The PECVD films had superior crack resistance; the APCVD films were most susceptible, although P-doping did improve their resistance.

The P in PSG exists as P^{+5} and P^{+3} (Tong et al., 1984; Houskova, 1985; Treichel et al., 1990). Both P—Si (interstitial, i.e., P replacing O in and interspersed between the SiO_2 tetrahedra) and P—O (substitutional, i.e., P substitutes for Si in the SiO_2 tetrahedra) states were detected by ESCA in 8wt% PCVD films and P-implanted thermal oxide (Wu and Saxena, 1985). The populations of these bonding states depend on deposition and annealing conditions. In CVD films, P–Si > P—O, as deposited, but conversion of P—Si to P—O upon annealing has been observed. Wu and Saxena (1985) also postulated that reflow, stress, and etch rates might depend on the relative populations of the states, but did not venture to say just how.

One of the most important effects of P-doping is the inhibition of the movement of cations (usually Na^+) in the presence of an electric field, since Na^+ will move rapidly through an SiO_2 film under the influence of a field and change the threshold voltage of an FET device. PSG "acts as a getter of Na^+ and a barrier against Na^+ drift" (Balk and Eldridge, 1969). Kern and Smeltzer (1985) concluded that the protective mechanism was solely gettering, at least in APCVD BPSG, but whether this is true for all BPSG films has not been confirmed. The protective effects of P-doping increase with increasing P-concentration. However, since the polarizability also increases with increasing P-concentration, too high a concentration will result in instability. Another factor that places an upper limit to the P-concentration is that the films become increasingly hygroscopic with increasing P-content resulting in corrosion of Al-based wiring insulated with PSG (Paulsen and Kirk, 1974; Bhide and Eldridge, 1983). Si—OH bonds and physically absorbed H_2O can be detected in a PSG film, after it has been exposed to water vapor at temperatures of 70–100°C, by using IR spectroscopy. The dielectric constant of CVD PSG, prepared from the hydrides, was reported to be close to that of fused silica (Kern and Heim, 1970). The effect of deposition method and P-concentration on the electrical properties of hydride-based PSG films, as measured by Shioya and Maeda are given in Table IV-2.

However, the dielectric constant of PECVD PSG, prepared using TEOS, O_2, and TMP at a temperature of 390°C (Schwartz, 1991) was lower than that reported by Shioya and Maeda, as might be expected in light of the higher deposition temperature. In this case as well, there was no change in ε as the P-concentration was increased until about 4 wt% P was incorporated in the film and then it increased slightly (from 4.07 to 4.27 at 10 wt% P) (Schwartz, 1991).

Another advantage to the use of PSG is the ability to etch it selectively with respect to the undoped oxide in CF_4/H_2 mixtures. Almost infinite selectivity is possible, depending on the P-content of the PSG and the H_2 concentration in the etch gas (Vender et al., 1993).

2.6.2 BSG

White et al. (1990) reported that for APCVD BSG, incorporation of B (≤4.5 wt% B) by the addition of diborane to SiH_4/O_2 mixtures, produced films stable to moisture, had moderate tensile stress, high breakdown strength, and a lower dielectric constant than the undoped film (3.8 versus 4.1) and these films could be candidates for use as an interlevel dielectric. Both B—O (1380 cm^{-1}) and Si—O bands are found in the IR spectrum. B-doped films tend to have a high silanol content and bind water in their structure, so that upon exposure to the atmosphere, the tensile stress is reduced. Moreover, if the concentration of B in the films is too high, they are unstable with respect to moisture attack, decomposing into boric acid.

APCVD BSG films were also prepared by the addition of either TMB or SiOB to a TEOS/O$_3$ mixture (Fujino et al., 1991), there were no major differences in the film due to the dopant source. Step coverage was very good whether the BSG was deposited using TEOS + O$_3$ and a boron dopant (TMB) or just SiOB + O$_3$. They reported on the effects on a number of properties of changes in temperature and dopant flow rate, for TMB (Fig. IV-5a), for SiOB (Fig. IV-5b) and the effect of O$_3$ concentration for the SiOB + O$_3$ deposition (Fig. IV-5c). The deposition rate of the BSG films was substrate material dependent as was that of the undoped oxide. The leakage current of the TMB/BSG films was very low, without annealing, and

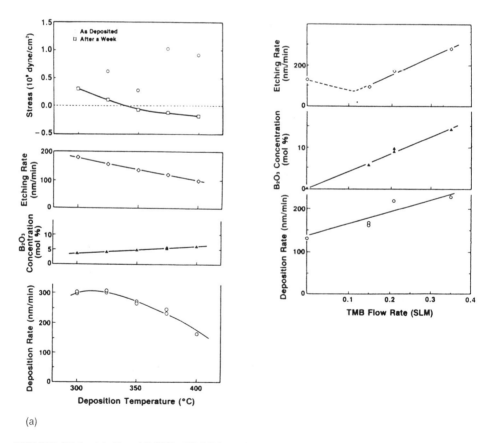

(a)

FIGURE IV-5 (a) For APCVD TEOS-based TMB/BSG films. From top to bottom: (1) Stress, etch rate, B$_2$O$_3$ concentration, deposition rate vs. deposition temperature, (r) etch rate, B$_2$O$_3$ concentration, deposition rate vs. TMB flow rate. (Reprinted with permission from The Electrochemical Society from K. Fujino et al.).

Interlevel Dielectrics 249

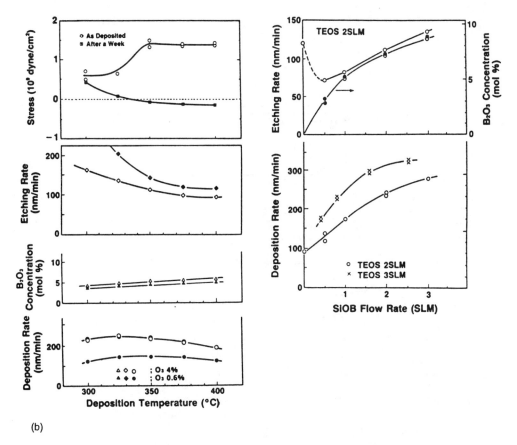

(b)

FIGURE IV-5 (b) For APCVD TEOS-based SiOB/BSG films, from top to bottom: (l) Stress, etch rate, B_2O_3 concentration, deposition rate vs. deposition temperature, (r) etch rate, deposition rate vs. SiOB flow rate. (Reprinted with permission from The Electrochemical Society from K. Fujino et al.)

was independent of the dopant concentration (measured to 9 mol% B); Higher O_3 concentration in the reaction mixture resulted in denser films with the lower leakage. The B—O peak did not change after exposure to the atmosphere, but the Si—OH peak increased. The advantage of SiOB is the existence of the B—O—Si bond leading to an integrated network; Fujino et al. (1991) stated that TMB/BSG is a mixture of SiO_2 and B_2O_3, which contrasts the conclusion of Kern and Fisher (1970). Yuyama et al. (1944) prepared CVD BSG (6 wt% B) using SiOB and O_3. The as-deposited film had a high water content (4.4% H_2O) and $\varepsilon = 4.37$. A postdeposition treatment for 3 min in a N_2 plasma eliminated the water completely and reduced ε to 3.35; neither exposure to an O_2 plasma nor annealing in a furnace at 750°C was as effective in eliminating the water (a surprising result for the high temperature anneal) or lowering ε as much. Their results indicated that the reduction of the moisture, per se, cannot account for the low dielectric constant, since the dielectric constant of an undoped film with equivalent water content was much higher. PECVD BGS was prepared using TEOS, O_2, and TMB. The results were

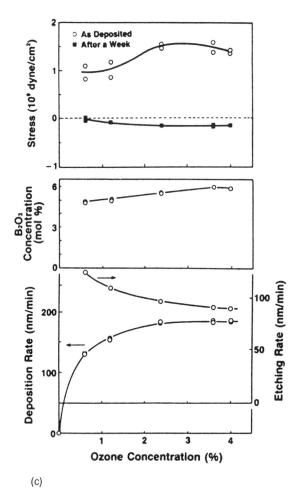

FIGURE IV-5 (c) For APCVD TEOS-based SiOB/BSG films, from top to bottom: stress, B_2O_3 concentration, deposition rate vs. ozone concentration. (Reprinted with permission from The Electrochemical Society from K. Fujino et al.)

similar to those reported by White et al. (1990) i.e., the dielectric constant for a film with B = 3.6 wt% was 3.87 versus 4.03 for the undoped film (Schwartz and Chapple-Sokol, 1992).

2.6.3 BPSG

BPSG is the doped oxide used most extensively as an interlevel dielectric film since it combines the protection against the instability due to ionic contaminants of PSG with the reduced flow temperature of BSG. Table IV-3 is an example of the effect of some of the process variables on the deposition process and film

TABLE IV-3 Effect of Deposition Variables on Properties of SACVD BPSG Films

	Increasing Variables				
	Pressure torr	O_3 Conc. wt%	Boron wt%	Phos. wt%	Anneal °C
Deposition rate	↓	↔	↑	↑	×
% Shrinkage	↔	↓	⇩	↓	↑
WERR	↔	↔	↓	↑	↔
Reflow angle	↔	↓	↓	⇩	↓

Decrease: ↓   Slight decrease: ⇩   Increase: ↑
No effect: ↔   Not tested: ×

properties for SACVD BPSG (Robles et al., 1995). Rojas et al. (1992) compared BPSG films deposited using APCVD, LPCVD, and PECVD using liquid and gaseous sources. They concluded that films deposited using a higher-temperature TEOS-based LPCVD process had superior chemical and physical properties.

The value of ε of BPSG films prepared by CVD from the hydrides appears to depend on the composition, preparation method, and annealing (densification) conditions (Kern and Heim, 1970; Kern and Schnable, 1982) and were in the range of 3.8 to 4.5 (after fusion). Treichel et al. (1990) reported that for a PECVD film, deposited using SiOB, O_2, and TMP (3.6 wt% B/4.2 wt% P), ε was 3.3. In contrast, for a BPSG film, prepared in the same model of reactor, at the same temperature, but using TEOS, O_2, TMP, and TMB, and having the same composition and IR spectrum, ε was ~4 (Schwartz and Chapple-Sokol, 1992). The result of Treichel et al. (1990) is another example of the use of SiOB as both a Si and B source producing a film with an exceptionally low value of ε. Treichel et al. (1990) also reported that the spectra of the films deposited using TEOS and TMB(+TMP) and SiOB (+TMP) were virtually identical, indicating no gross structural difference. Thus it is difficult to explain the discrepancy between the results of Treichel et al. (1990) and Schwartz and Chapple-Sokol, or the role of the specific reactants. It may be recalled that Fujino et al., (1991), in depositing BSG, found that the B-dopant source (TMB or SiOB) had not major effect on the film properties. However, they did not measure ε.

Other papers of interest, not discussed here, are Levy et al. (1987), Pignatel et al. (1991), Dobkin (1992), and Rojas et al. (1993). A paper by Becker et al. (1986) includes an extensive bibliography.

2.7 F-DOPED SiO$_2$

There have been recent efforts to reduce the dielectric constant of SiO$_2$ films by F-doping, i.e, forming Si—F bonds in the silica network. These films also exhibit enhanced step coverage and gap-filling properties to be discussed in a later chapter.

Earlier work on F-doping was directed toward improving the quality of oxide films deposited at low temperatures or for improving its gap-filling capability. Falcony et al. (1993), using SiF$_4$ additions to SiH$_4$/N$_2$O mixtures, had relied on the affinity of F for H to minimize or eliminate H incorporation in SiO$_2$ films deposited at low-temperature. Homma et al. (1993) used a F-substituted organic precursor and water for improved flow during room temperature CVD of SiO$_2$ by forming precursor oligomers which condense, flow, and then polymerize on the surface. NF$_3$ was added to the reaction mixture in PECVD of SiO$_2$ (Ibbotson, 1990) to suppress sidewall growth for improved gap fill. Incorporation of F into the gate oxide increased the radiation hardness of MOS devices (deSilva et al., 1987; Ahn et al., 1991).

F-incorporation (up to ~5%) into SiO$_2$ occurs during liquid phase deposition (LPD) which was introduced as a process for coating a silica film on the surface of glass at a low temperature (Nagayama et al., 1988). LPD was originally carried out by reacting boric acid with a silica-saturated solution of hydrofluosilicic acid, to precipitate, after a series of reaction steps, F-doped SiO$_2$ (Homma et al., 1990; Homma et al., 1993). F-incorporation lowered the value of ε to 3.7 (Homma et al., 1991). The deposition rate, in the range of ~100–300 Å/hr, increased with increasing boric acid concentration and increasing temperature. Chou and Lee (1994) investigated the initial growth mechanism; oxide growth occurs only when the concentration of HF (a reaction by-product) is less some critical value which depends on the condition of the wafer surface.

The selectivity of LPD (oxide deposition requires the existence of —OH groups on the underlying surface so that no oxide was deposited (e.g., on photoresist or W) with the potential for planarization was the advantage claimed for the process (Homma et al., 1990, 1991, 1993), rather than the reduction of ε.

Yeh et al. (1994) eliminated the boric acid, adding only H$_2$O to the silica-saturated hydrofluorosilicic acid. Increasing the amount of water added increased the growth rate, the SiOH concentration, but decreased the F-concentration and film density (Yeh and Chen, 1995). The LPD oxide is somewhat O-deficient and can be represented as SiO$_2$—xFx; both SiOH and water bands are observed in the FTIR spectra.

Annealing the films increased the density and refractive index, decreased the "P" etch rate, and the Si—F bond intensity. FTIR spectra, showing a shift and broadening of the Si—O—Si peak, indicate partial restructuring of the oxide during anneal (Yeh et al., 1995, Yeh and Chen, 1995).

The current method of choice for depositing F-doped oxide is PECVD. Table IV-4 summarizes the F-sources, the basic reaction mixtures to which the dopant is added, and the type of reactor used for PECVD deposition of F-doped SiO$_2$ with lower ε, for potential use as interlevel dielectrics. Among the earliest reports of F-doped SiO$_2$ prepared using PECVD were those by Usami et al. (1993) in a parallel plate reactor and by Takada and Akahori (1993) in an ECR system.

Interlevel Dielectrics

TABLE IV-4 F-Doped SiO2: Deposition Techniques, F-Sources, Basic Reaction Mixtures

Deposition technique	F-Source	Basic reaction mixture	Reference
PECVD	C_2F_6	$TEOS/O_2$	Usami et al. (1993, 1995)
Dual-frequency PECVD	C_2F_6	$TEOS/O_2$	Carl et al. (1995a,b); Mizuno et al. (1995); Matsuda et al. (1995); Takeishi et al. (1995)
Dual-frequency PECVD	SiF_4	$TEOS/O_2$	Matsuda et al. (1995)
Dual-frequency PECVD	FTES (TEFS)	$TEOS/O2$	Mizuno et al. (1995)
Bias ECR	SiF_4	$SiF_4/O_2/Ar$	Fukada and Akahori (1993)
Bias ECR	SiF_4	$SiH_4/O_2/Ar$	Fukada and Akahori (1995; 1995)
HD Plasma	SiF_4	$SiF_4/O_2/Ar$	Carl et al. (1995a, b); Qian et al. (1995)
APCVD	FTES (TEFS)	$TEOS/O_3$	Yuan et al. (1995)
RPECVD	SF_6	$TEOS/O_2$; SiH_4/N_2O	Yu et al. (1995)

The usual oxidant in PECVD processes is O_2, although Hsieh et al. (1996) used N_2O in place of O_2 in a $TEOS/C_2F_6$-based process. They found reduced dependence of the deposition rate on temperature (implying a shift in the deposition mechanism) and the F-doping was more efficient but films of equal F-content were similar to those deposited using O_2. The differences in film properties which have been attributed to the method of deposition and to the kind of reactor used for PECVD will be discussed below.

The F-source used in the deposition reaction affects the film properties. One important distinction among possible sources is whether the Si—F bond is present in the source molecule (e.g., SiF_4) or whether F must be formed in the plasma (e.g., from C_2F_6) and then react with the Si-source to form the Si—F bond. SiF_4 may be used as the sole Si-source or may be mixed with additional ones such as SiH_4.

Although these films are sometimes referred to as SiOF, this does not reflect the stoichiometry; the F-content is usually <10 at%.

As the flow rate of the F-source compound increases, the F-content of the film increases, saturating at high flow rates. Films with a very high F-content (~12 at%) have been prepared, but they be unstable. Mizuno et al. (1995a) stated that film stability played a major role in determining the lowest possible value of e, i.e. the highest practical F-content so that the stability of *all* SiOF films has become an important issue. There is disagreement on this matter in the literature, probably due to differences in the reactor configuration, choices of reactants and process conditions. Carl et al. (1995a,b) and Passemard et al. (1996) concurred that films containing more than ~3–4 at% F were unstable, absorbing moisture readily and outgassing. However, Saproo et al. (1996) reported that the film stability was a complicated function of the process conditions and did not depend solely upon the F-content; they prepared stable films with F-concentrations as high as 11.5 at% and unstable ones with lower F-content. Stability of low-dielectric-constant SiOF films was related to the use of bias in helicon plasma-CVD (Tamura et al., 1996).

The F is incorporated into the silica network; the existence of an Si—F bond has been demonstrated by XPS (the Si—F bond at 687.3 eV) and a band in the IR spectrum at 930 cm^{-1}. SIMS measurements show that the concentration of F is uniform throughout the film (Usami et al., 1994).

The value for ε of the deposited oxide also decreases with increasing F-dopant flow rate and sometimes is shown to saturate (Yu et al., 1995; Mizuno et al., 1995). A typical result is shown in Fig. IV-6, which is derived from data reported by Matsuda et al. (1995).

The final value of ε has often been found to be about 75% of that of the undoped oxide prepared under the same conditions. This comparison is often omitted, and only the ultimate value is given, so that is difficult to assess the effectiveness of the process. The F-content does not, itself, determine ε. For a given F-content, the value of ε depends on the reactor, reaction mixture, and processing conditions. These factors determine the value of ε for the undoped film to which the F-doped film is compared. The lowest value of ε reported is about 2.3 (Lim, 1995; Lim et al., 1996; Shimogaki et al., 1996), but the film was not very stable.

The reduction in ε in SiOF films is due to two factors. The Si—OH bond, which is responsible for the higher dielectric constant of deposited glasses compared with thermally grown SiO_2, is reduced in concentration or eliminated by the fluorine because fluorine scavenges hydrogen. The other is due to the effect of forming Si—F bonds which reduces the force constant in the IR vibrations of SiO_2.

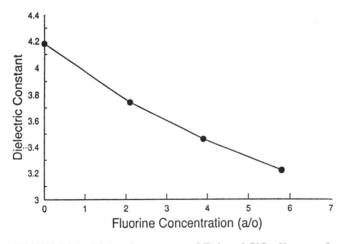

FIGURE IV-6 Dielectric constant of F-doped SiO_2 films vs. fluorine concentration in film.

The positions of the Si—O bands (stretching, bending, and rocking) are shifted to higher wave numbers as the F-content is increased (Nakasaki and Hayasaka, 1994; Usami et al., 1993, 1994; Matsuda et al., 1995; Lim et al., 1995; Shimogaki et al., 1996).

Nakasaki and Hayasaka (1994) attributed the lowering of ε to an increase in the band gap of SiO_2, a reduction in *electronic* polarizability. Lim et al. (1995) and Shimogaki et al. (1996), however, from their analysis of the IR spectrum (using the Kramers-Konig relationship, see, e.g., Chelkowski, 1980) concluded that the decrease in ε was due to a decrease in the *ionic* polarization and the change in the stretching mode was the most important factor.

Annealing at 400°C decreased the value of ε. Usami et al. (1994) who used an H_2 ambient and annealed for 30 minutes reported that the decrease was smallest for the film with the highest F-content. However, Lim et al. (1995) and Shimogaki et al. (1996) reported the opposite trend, i.e, the higher the F-content of the film, the greater the reduction after annealing for 10 minutes (ambient not given).

As stated above, Si—OH is either absent or present in minute quantities in most F-doped oxides. However, Mizuno (1995a,b) reported that there was a significant Si—OH peak in the films deposited using either C_2F_6 or triethoxyfluorosilane (TEFS) in a dual-frequency reactor and this could be reduced only by capping the film with undoped oxide. Higher temperature deposition or annealing were ineffective.

The change in deposition rate with increasing flow rate of the F-source compound depends on the compound and on the base reaction mixture. For example, in a dual frequency reactor using $TEOS/O_2/SiF_4$, the rate decreased and then increased (Matsuda, 1995); in a single frequency reactor using $TEOS/O_2/C_2F_6$, the rate decreased monotonically (Usami et al., 1993, 1994); in a biased ECR reactor, the rate increased linearly in an $SiF_4/O_2/Ar$ mixture (Fukada and Akahori, 1993) but decreased in an $SiF_4/SiH_4/O_2$ mixture (Fukada and Akahori, 1995).

The advantage to using an F-source that has Si—F bonds in its structure is thought to be related to the fact that the Si—F units are probably incorporated directly into the silica network whereas Si—F bonds must be formed by reaction of F with the network when only F is supplied by the dissociation in the plasma. There is experimental evidence to support the conjecture that the latter type of bond is weaker and more reactive than the former. Shapiro et al. (1995) and Matsuda et al. (1995) reported that, for dual-frequency films, uptake of moisture was more immediate and reached a higher level of the C_2F_6 based films than for the SiF_4 based ones and HF evolution upon heating was greater for the former films. Mizuno (1995a,b) reported that, for the same Si—F content, films made using C_2F_6 had a slightly higher value of ε than those prepared using TEFS, perhaps due to moisture adsorption. In addition, HF evolution upon heating mixed-frequency films began at a lower temperature for C_2F_6-based films than for triethyoxyfluorosilane (TEFS or FTES)-based ones. This appears to be a consequence of the fact that there is less unbonded F when SiF_4 or TEFS is used instead of C_2F_6, since there is less dissociation of the Si—F bond in the plasma and, therefore, less F_2 to react with water from the atmosphere (Matsuda et al., 1995). Takeishi et al. (1995, 1996) reported that the value of ε for films prepared in a dual-frequency reactor using $C_2F_6/TEOS/O_2$ increased upon exposure to air but that annealing the film in an N_2O plasma stabilized them. Swope et al. (1996) emphasized that the effect was confined to the surface, that the bulk properties were unaffected. They also found that the plasma treatment could enhance the adhesion characteristics of high F-content films, without changing the moisture stability. Also Fukada and Akahori (1993; 1995) reported NO change in the Si—F concentration after T and H testing of an ECR SiF_4-based film and no F-desorption detected after heat treatment at 1200°C and Qian (1995) reported HD SiF_4/O_2 films (8 at% F) were unaffected by thermal cycling and stable to moisture (50% RH, 5 days).

It appears that high energy ion bombardment of the substrate *may* blur the distinction of the type of F-source. Equivalent results, i.e., good stability toward moisture, were obtained when using C_2F_6/TEOS/O_2 in a dual-frequency reactor in which the low frequency power was very high and an ICP reactor using SiF_4/O_2/Ar. With lower low frequency power, i.e., with less energetic ion bombardment, the C_2F_6-based SiOF films were inferior (Carl, 1995a,b). These results can be explained by ion bombardment densification of the film. In contrast, Mizuno et al. (1995a,b), using a dual frequency reactor, reported that the wet etch rate of their C_2F_6-based, high-power, low-frequency, high-F-content films increased with increasing high energy bombardment. They inferred a lower density due to breaking of the weak Si—F bond. However, the TEFS-based oxide became denser, i.e., the etch rate was lower at higher bombardment energies.

The compressive stress in the film decreases with increasing F-content. It has been suggested that F-incorporation "loosens" the structure so that the stress becomes more tensile and more unstable. However, for the same F-content, increasing the ion bombardment (e.g., by increasing the low frequency power in a dual-frequency reactor) increased the compressive stress of the film (Carl et al., 1995a). They postulated that ion bombardment densifies the film, making the stress more compressive and more stable. The influence of the source of F is again emphasized in the work of Mizuno et al. (1995a,b). They found the same trend in stress with increasing low-frequency power for a TEFS/TEOS-based oxide; the largest change in film stress was for films deposited using the lowest TEFS/TEOS ratios, decrease in tensile stress at low low-frequency power, but this trend was reversed as the power was increased. Only for the films prepared with the lowest C_2F_6 flow rate in the reactant mixture did the stress eventually become and remain compressive, although at the highest low-frequency power there was a trend toward lower compressive stress. This was explained by the weaker F-bonding in the C_2F_6-based oxide. High energy ion bombardment breaks the bond, releases F, so that a more porous film is produced.

Fukada and Akahori (1993) first reported a relatively high breakdown strength, (6–8 MV/cm) for SiOF films deposited using SiF_4 and O_2, but later (1995) reported, for the same kind of film, a very soft breakdown of <3 MV/cm. The ultimate leakage current, however, was lower than that of several other films in current use, as shown in Fig. IV-7. The breakdown strength may be adequate for an interlevel dielectric but more work is needed.

2.7.1 Conclusion

Increasing the F-content of SiO_2 does continue to reduce the value of ε, perhaps down to some limiting value. In addition (as discussed in a later chapter) high F-content SiOF films have better gap-filling capabilities than undoped SiO_2 films.

Interlevel Dielectrics

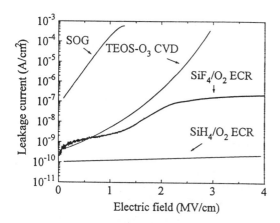

FIGURE IV-7 Leakage current vs. electric field for several dielectric films: SOG, CVD SiO_2(TEOS/O_3), SiOF (ECR, SiF_4/O_2), PECVD SiO_2 (ECR SiH_4/O_2). (Reprinted with permission from T. Fukada and T. Akahorin.)

Both of these attributes make these films good candidates for extending the usefulness of an inorganic dielectric for advanced ULSI devices.

However, there are unresolved issues. When the F-concentration is too high the films become unstable to thermal decomposition of the Si—F bond and to reaction of Si-F with moisture in the environment. The definition of "too high" is uncertain at this time; it seems to range from about >3.5 at% to ~11.5 at% and appears to depend on how the film was deposited. In addition, there is no exact correlation between F-content and ε, i.e., films deposited in different ways may have the same F-content but different values of ε. The reaction of the Si-F bond with moisture results in Si-OH formation, raising the dielectric constant. It also may result in H_2O absorption which, upon thermal desorption may interact with the interconnection wiring and cause blistering and increase corrosion susceptibility. The release of F from the Si-F bond and the possible reaction with the surrounding metal is still another concern that must be addressed.

2.8 CONTAMINANTS: H AND Ar

2.8.1 Hydrogen

H is incorporated into many deposited SiO_2 films. The amount incorporated and the bonding depend on the method of preparation and on the temperature; in general, the higher the temperature, the lower the H-content. H-bonds are formed, as silanols, Si—OH, either H-bonded (IR absorption at 3650 cm^{-1}) or free (3740–3750 cm^{-1}), as absorbed H_2O (at 3300–3400 cm^{-1}) or as Si—H (~2100–2300) (Pliskin, 1973). Incorporation of water may possibly occur during deposition or, more likely by entrapment after the film is formed, within the reactor or after exposure to moisture in the environment. Adams and Douglas (1959) proposed that inclusion of OH groups involves the breaking of Si—O—Si bridges, i.e., Si—O—Si + H_2O = Si—OH HO—Si. Films formed by chemical reactions, i.e., CVD (thermal or plasma-assisted) usually contain H. Bias sputtered SiO_2 films

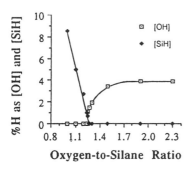

FIGURE IV-8 Percentage of H (as OH and SiH) in SiH_4-based ECR SiO_2 films vs. O_2/SiH_4 ratio in feed gas. (Reprinted with permission from S. Lassig et al.)

contain a negligible amount of Si—OH and no water, but films sputtered at low bias tend to be porous and, therefore, absorb atmospheric moisture readily.

Whatever the source or type of bonding, e.g., as Si—H or as Si—OH, the effect of H incorporation is to increase the dielectric constant of the film, and thus, increase the signal propagation delay. The relationship between OH content and ε has long been recognized. The IR spectra showed that "water" in fused silica was present at Si—OH (Adams and Douglas, 1959). The value of ε increased linearly with increasing OH content (Andeen et al., 1974). This trend was also observed in SiO_2 films deposited using SiH_4 and O_2 in an ECR plasma reactor (Lassig et al., 1993). The dielectric constant of PECVD oxides (using TEOS and O_2 in a capacitively coupled reactor) increased linearly with increasing total H (OH + H_2O) content (Patrick et al., 1992). In the case of films formed in an ECR plasma using SiH_4 and O_2, the incorporation of OH groups depends on the ratio, $R = O_2/SiH_4$. When the ratio is high, the Si is completely oxidized, H is liberated and reacts with the excess O and produces OH and ε increases. In O-deficient plasmas (low R) Si is incompletely oxidized and Si—H bonds are formed, increasing ε. At an intermediate value of R, the bonded H will be at a minimum (Fig. IV-8) as will the value of ε (Fig. IV-9). Increasing the deposition temperature reduced [OH] and

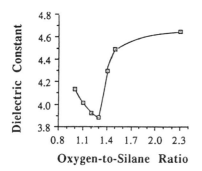

FIGURE IV-9 Dielectric constant of SiH_4-based ECR SiO_2 films vs. O_2/SiH_4 ratio in feed gas. (Reprinted with permission from S. Lassig et al.)

the tendency to absorb moisture upon exposure to a high humidity environment (Patrick et al., 1992). Lassig et al. (1993) attributed the decrease in [OH] to thermally enhanced water desorption during deposition.

The OH-content of porous films is high and such films absorb water from the environment very readily, with a large increase in the dielectric constant. The absorption of water is reversible; heat treatment at 400°C for half an hour desorbs all the water, which is reabsorbed upon subsequent exposure. The OH content, however, is unaffected by this treatment; heating at or above ~600°C is required. Heating at 1000°C transforms deposited films so that they become indistinguishable from thermally grown SiO_2, i.e., free of H.

Stress is also affected by moisture absorption. If the initial stress is tensile, it is decreased (Blech and Cohen, 1982) and may become compressive. If the initial stress is compressive, the stress increases (Gokan et al., 1987). A change of stress upon exposure to humid environments is an indication of moisture pick-up. Another indication is stress hysteresis during thermal cycling (Bhushan et al., 1990; Cramer and Murarka, 1995).

Some of the deleterious effects of desorption of the absorbed moisture are "poisoned" (i.e., high-resistance) vias, metal blistering and corrosion, device instability, as well as the mechanical instability related to stress changes.

2.8.2 Argon

Large amounts of argon can be incorporated into sputtered SiO_2 films (e.g., Ar/$SiO_2 \leq 0.1$); the Ar content increases with increasing substrate bias. Unlike H-incorporation, Ar does not affect the dielectric properties of the films. In fact, sputtered films with a high Ar content most closely resemble thermally grown oxide (Schwartz and Jones, 1970). The room temperature stress of sputtered films is compressive; this is thought to be a consequence of the incorporation of Ar since it does increase with bias as does the Ar content. Although annealing the films at 400°C decreases the Ar content, the stress remains unchanged.

2.9 ETCHING OF SiO_2 FILMS

2.9.1 Wet Etching

SiO_2 films are soluble in HF-based solutions. When used to etch masked SiO_2 films to form through-holes, the HF solution is buffered by the addition of NH_4F (BHF). In a given solution, the etch rate depends on the film "quality" (thermal oxide has the lowest rate). For a given oxide, the rate is determined by the ratio NH_4:HF; the rate decreases as the ratio increases. As discussed in Chapter I, reactive-plasma-assisted etching has replaced wet etching for patterning. BHF may still be used, at times, in brief clean-up steps and in place of "P" etch in evaluating deposited oxides.

2.9.2 Sputter Etching

This technique is rarely used for patterning now. However, in some processes used in fabricating multilevel device structures, in situ sputter etching (cleaning) is used to remove insulating films (e.g., oxides, polymers) from the surface of a metal at the bottom of a via hole, before deposition of the next metal layer (e.g., Bauer, 1994).

2.9.3 Reactive-Plasma-Assisted Etching

The steps in a device fabrication process in which SiO_2 films are etched are (1) forming through holes ("contacts" to the Si substrate and "vias" connecting sequential metal levels), and (2) in etch-back planarization, which will be discussed in a later chapter.

The requirements for a successful process are adequately high rates (for throughput requirements), profile control, uniformity, reproducibility, minimal damage and contamination, and good selectivity to the mask and the substrate.

SiO_2 is etched in F-containing plasmas. Etching in a barrel reactor was the earliest application of dry processing in reactive gases, but it was realized very quickly that it could not meet the requirements of VLSI/ULSI processing. RF sputter etching of SiO_2 by fluoro-chloro-hydrocarbon gases was reported in 1974 by Hosokawa et al. Reactive ion etching (RIE) in capacitively coupled reactors of various designs has since been the method of choice, although, the high density plasmas are coming into use.

2.9.4 Etch Mechanism

The final volatile silicon-containing product is SiF_4 and most of the O is probably desorbed as O_2 (Winters, 1983). According to Winters and Coburn (1979), F atoms (produced from XeF_2 at very low pressure) do not react spontaneously with SiO_2 but do react in the presence of ion bombardment. On the other hand, Flamm et al. (1979) found that F atoms (at a higher pressure) etched SiO_2 at a measurable rate (~100 A/min at room temperature). The difference in results may be due to the difference in atom flux or in the sensitivity of detection. SiO_2 etches slowly and isotropically in a fluorinated plasma in a barrel reactor, in which ion bombardment is negligible but the pressure (reactant supply) is high.

Several experiments demonstrated that RIE of SiO_2 is dominated by an ion-driven mechanism. Schwartz et al. (1979) showed that in a low pressure RIE system, the profile of SiO_2 etched in CF_4, was vertical beneath a nonerodible vertical mask, i.e., there was no undercut and a loading effect, due to depletion of neutral species, was small. When the effect of ion bombardment was minimized, etching became isotropic, the etch rate was reduced significantly, and the loading effect was substantial. The etch rate of SiO_2 in various Freons, had a stronger dependence on the sheath voltage than on the nature or concentration of the chemical species (Fortuno, 1986; Simko and Oehrlein, 1991); the existence of a threshold voltage for RIE of SiO_2 was inherent in the etch model (Fortuno, 1986). Therefore, in the absence of lateral mask erosion, nonundercut vertical profiles would be expected in low pressure RIE systems, irrespective of the specific F-containing etchant.

2.9.5 Etch Gases

The compounds used for etching SiO_2 include CF_4 (one of most widely used in the early days of plasma etching) and other saturated and unsaturated fluorocarbons (e.g., C_2F_6, C_3F_8, C_2F_4), hydrofluorocarbons (e.g., CHF_3), as well as SF_6, NF_3, and chlorofluorocarbons (CFxCl$_4$—x). These gases are often used as mixtures with the reactive gases O$_2$ or H$_2$; inert diluents, such as Ar or He, are sometimes added. The chlorofluorocarbons (CFCs) are ozone-depleting gases and there is international agreement to phase out production of them by the end of the century. The fluorocarbons are members of the perfluorocompound (PFC) family which have a long atmospheric lifetime (and thus a global warning potential). Mohindra et al. (1994) suggested and demonstrated the use of the shorter lifetime PFC alternatives such as the hydrofluorocarbons (HFCs), C$_2$F$_5$H, C$_2$F$_4$H$_2$. CHF$_3$ is unique among the HFCs for its long atmospheric lifetime. Emission controls are now required for the PFCs, CHF$_3$, SF$_6$ and NF$_3$ (Mocella, 1996).

CF$_4$ etches Si more rapidly than it does SiO$_2$, so it cannot be used alone for etching contact holes to a Si substrate. Addition of O$_2$ to CF$_4$ initially increases the concentration of F atoms but ultimately the concentration decreases, possibly due to a decrease in the electron energy; CO or CO$_2$ is also produced. Since etching of Si is largely due to the neutral species (F-atoms) the etch rate of Si follows the same trend (Mogab et al., 1978). The etch rate of SiO$_2$ is increased only slightly, as might be expected, since the basic mechanism is ion-enhancement. This mixture, CF$_4$ + O$_2$, is also unsuitable for contact hole etching. It has been used for interlevel via hole etching; the O$_2$ component serves to tailor the mask profile, as discussed below.

2.9.6 Selectivity

Unlike O$_2$ additions, addition of H$_2$ to CF$_4$ reduces the F atom concentration markedly. Thus, as expected, the etch rate of silicon is reduced significantly; at some concentration of H$_2$ in the feed gas, the silicon etch rate goes to zero. In his paper (1975) and a patent issued later (1976), Heinecke was the first to demonstrate that this gas mixture, as well as fluorine-deficient CF compounds, e.g., C$_3$F$_8$, could be used to obtain selectivity of SiO$_2$ over Si in plasma etching. He also showed that polymer formation would occur under some conditions. Ephrath (1979) extended the work (on CF$_4$/H$_2$ mixtures) to RIE systems. Polymer surfaces are affected in much the same way as Si surfaces. Reduction of the etch rate of Si in CF$_4$/H$_2$ is the result of two processes: reduction of the F-concentration due to scavenging by H$_2$ and deposition on the Si surface a C-containing material (a fluoropolymer) blocking access of the etchant. The H$_2$ concentration at which the etch rate of Si goes to zero has been called the "polymer point" since a visible film can be detected on the Si surface (Ephrath and Petrillo, 1982). The effectiveness of using polymer-forming etchants is enhanced by the use of Si-coated electrodes which scavenge F. The etch rate of SiO$_2$ is affected only slightly in the same range of H$_2$ concentration, probably because the liberation of O removes the adsorbed blocking layer. Only at very high concentrations of H$_2$ does the etch rate of SiO$_2$ decrease significantly. Good selectivity i.e., a high etch rate ratio (ERR) = ER$_{SiO_2}$/ER$_{Si}$, is essential for etching contact holes in SiO$_2$ to the silicon substrate, particularly as the junction depths become smaller. It should be emphasized that not only a high ERR but a low absolute etch rate of Si is needed for a successful process (Jacob, 1976). Coburn (1982) has described a boundary between etching of Si and polymerization on the Si surface (on which SiO$_2$/Si selectivity depends) in terms of the F/C ratio and substrate bias for various CF compounds, showing how the chemical composition of the reactant gas, addition of O_2, H_2, and loading affect that boundary. In addition to the choice of etchant, other system variables such as total flow, total pressure, residence time, rf power, sheath voltage, wafer temperature, and reactor configuration (e.g., RIE, MERIE, HC, ECR) have a strong influence on the selectivity.

2.9.7 Profile Tailoring

If vertical interconnects (studs or plugs) are not used, it is advisable to round and taper the holes which would otherwise, due to the anisotropic nature of RIE of oxides, have straight-walls and would be very difficult to cover adequately with a metal layer. The shape of an etched feature can be controlled by shaping the resist mask during the lithographic processing by various protocols of reflow baking (Huang et al., 1984; Saia and Gorowitz, 1985) or low-dose flood exposure of the resist after the soft-bake step, before exposing the pattern (White and Meyerhofer, 1987). The shape can also be modified, during etching, by controlled mask erosion, selecting a suitable resist to oxide etch rate ratio, or changing the ratio during oxide etching. The change in ratio is accomplished by adding O_2 to the etchant among which have been CF_4, CHF_3, C_2F_6, SF_6, and NF_3 (Viswanathan, 1979; Bondur and Frieser, 1981; Duffy et al., 1983; Light and Bell, 1983; Peccoud et al., 1984; Castellano, 1984; Saia and Gorowitz, 1985; Bogle-Rohwer et al., 1985; Kudoh et al., 1986; Jillie et al., 1987). Nagy (1984), by ion etching, propagated a facet formed in the resist mask (due to the angular dependence of ion etching) to the mask/oxide interface; this results in replicating the facet in the oxide. In another technique, the rf power was pulsed and it was possible to etch the resist isotopically and the oxide anisotropically (Giffen et al., 1989). An "optimized slope multi-tier contact" etch process was developed for a batch reactor (Mautz et al., 1994). It combined an initial anisotropic oxide etch with sequential photoresist removal/oxide etch steps in tandem.

The use of a cantilevered mask resulted in controlled taper of the via independently of the mask profile. The thickness of the organic layer beneath the overhanging nonerodible mask (which controls the via size) determines the taper (Rothman et al., 1981). Changing the ratio of etch gases, CHF_3 and CF_4, in a high pressure reactor tapered the slope independently of the resist shape (Chen and Mathad, 1987). Ohiwa et al. (1992) etched tapered holes at high rates in SiO_2 using CHF_3 by lowering the substrate temperature to about $-50°C$; the results were explained by fluorocarbon polymer effects. A triode reactor has been used to taper oxide profiles by controlled resist erosion and splitting the power between the top and bottom electrodes (Bogle-Rohwer et al., 1985). It was also used to perform an "iso-anisotropic" etch process. In this latter process, the isotropic etch of the oxide, with no resist erosion, was carried out in a "downstream mode" in which the rf power was fed to the side electrode; the anisotropic etch was a standard RIE process (Giffen et al., 1989). Still another approach which is decoupled from the mask shape was to reshape vertical holes in SiO_2, taking advantage of the incident-ion angle-dependency of etching of oxygen ions in an ECR etcher to facet the hole edges (Hashimoto et al., 1990).

2.9.8 Through-Hole Etching

Contact Holes

Etching contact holes in SiO_2 requires the proper profile and high selectivity to Si or a silicide (as well as the mask) or else the use of an etch-stop layer. The high aspect ratio and the variable insulator thickness make contact hole etching a challenging process.

Via Holes

In etching via holes, originally the question of selectivity had been relevant largely in terms of the effect of the various etchants on the photoresist mask, i.e., in profile control, as well as insuring adequate protection during etching. The underlying films may be Al, W, TiW, TiN. In the case of Al, after exposure to a CF_4 plasma, a thin fluorinated film could be detected on its surface by RBS (Chu and Schwartz, 1976). But Al cannot be reactive ion etched (RIE'd) in a fluorinated plasma since AlF_3 is involatile; it is merely sputter etched slowly. The other underlying metals can be RIE'd in these plasmas, but any loss can be accommodated during the deposition of a vertical interconnect, if it is used, or the next interconnection-level metal. Since via holes are often sputter cleaned using Ar or reactive gases, the effect of overetch was not important in this respect.

More recently, as the size of via holes has been reduced substantially, selectivity to the underlying metal has become a major concern; sputtering of the metal which is redeposited on the walls of the via reduces their size still further, making subsequent processing even more difficult. Two relatively recent patents (Arleo et al., 1993; Rhoades et al., 1993) described the addition of N_2 or a N-containing gas to F-based etchants to suppress redeposition by sputtering of the underlying conducting film. Rhoades et al. (1993) postulate that the action of N_2 is to form a volatile compound, such as AlN, TiN, instead of involatile AlF_3 or an organometallic polymer, which would be sputtered and redeposited. Control of sputtering (and microloading) was also achieved in an ECR reactor by increasing the flow rate of C_4F_8 in a mixture of C_4F_8 and O_2. The protection was afforded by fluoropolymer film, deposited only in the bottom of the via (Hisada et al., 1994).

2.9.9 Feature-Size Dependence of Etch Rate

As mentioned earlier, large holes etch more rapidly than smaller ones when holes of different size are etched simultaneously. However, it was found that the rate depends on the *aspect ratio* (*AR*) and not on the absolute width of the opening. The normalized etch rate decreases linearly with increasing AR for $AR \geq 2$ (Chin et al., 1985).

The phenomenon has been given several names: *RIE lag* (Lee and Zhou, 1990, 1991), *aperture effect* (Abechev et al., 1991), *aspect-ratio dependent etching* (*ARDE*) (Gottscho et al., 1992). Fujiwara et al. (1989) and Sato et al. (1991) called it *microloading*. However, Gottscho et al. (1992), in an extensive review called "Microscopic uniformity in plasma etching," emphasize that microloading is a *misnomer* for the effect, since this effect results from microscopic transport within a single feature while microloading and macroloading arise from identical causes. Thus the AR effect can occur without microloading or vice versa.

Because the AR increases as etching proceeds, the etch rate decreases with time. Gottscho et al. point out that, to be sure the phenomenon is AR related, it is necessary to determine the etch rate as a function of time for several feature widths to distinguish it from other time-dependent etch rates which can arise due to polymer build-up or surface poisoning.

The AR effect has been observed in high pressure plasma processes, in low pressure batch RIE, and in low pressure MERIE processes (Jones et al., 1990), as well as in ECR reactors (Fujiwara, 1989; Nojiri, 1989), in etching resists, oxides, and Si, and in a variety of etch gases.

Coburn and Winters (1989), using simple vacuum conductance arguments, concluded that in high AR features, conductance is adequate to allow the outflow of reaction products, but can be expected to limit the flow of reactive species to the bottom of the feature, thus explaining the etch rate dependence of high AR features. Jones et al. considered possible causes for the phenomenon (1) transport limitations in narrow gaps, (2) polymerization, (3) geometric shadowing, (4) ion deflection by surface charges, (5) ion scattering in the plasma. They concluded that ion scattering in the sheath is the principle mechanism, as did Lii et al. (1990). Jones et al. stated that ion-driven processes are more sensitive to RIE lag and that processes in which the sheath width is small with respect to the mean free path in the plasma will exhibit less lag. Such processes would be those using electronegative gases (e.g., SF_6), very low pressures, or low bias, and that the scattering angle could be decreased by using light nonionizing diluents such as He.

Abachev et al. developed equations for the particle flux including particle reflection and adsorption (when the mean free-path was much larger than the width of the feature, i.e., width < 1 μm and pressure < 100 Pa) and concluded that limited ion and radical delivery were the main mechanisms. Both Sato et al. and Fujiwara et al. also concluded decreasing species concentration on the trench bottom was responsible.

Davis (1991) considered the image potential between an ion and the wall of an etched feature for small high AR features. He concluded that low energy ions (<200 eV) are attracted to the walls resulting in loss of directionality and reduced vertical etch rates. This appears to counter the advice of Jones et al. (1990). Another model, based on polymer deposition while etching SiO_2 in selective (to Si) gases was proposed by Joubert et al. (1994) (1) ion scattering reduces the density and energy of ions at the bottom of a high AR trench or hole, (2) oxide etching moves from a regime in which the oxide is etched to one in which a thin fluorocarbon polymer can grow on the oxide surface, slowing its etch rate, and (3) finally, as the AR increases (particularly in a hole) the reduction of the energy flux will result in deposition of polymer on the oxide, and etching may stop.

Gottscho et al. (1992) considered eight mechanism which have been offered to explain ARDE (1) Knudsen transport of neutrals, (2) ion shadowing, (3) neutral shadowing, (4) differential shadowing of insulating microstructures, (5) field curvature near conductive topography, (6) surface diffusion, (7) bulk diffusion, and (8) image force deflection and conclude that only the first four are consistent with AR scaling. "There may be many causes for ARDE; these may be reduced to neutral and ion transport phenomena which in turn are affected by gas phase collisions, surface scattering, and surface charging."

Sato et al. (1991) also observed a larger effect when etching Si with F-containing gas than with Cl-containing gases; which they attributed to enhanced ion-assisted etching with the heavier halogen, which runs counter to the argument of Jones et al. (1990) that ion-dominated etching processes are more susceptible to RIE lag. Both Fujiwara et al. (1989) and Nojiri et al. (1989) report that at very low pressures, in an ECR reactor, the effect is no longer observed, although other investigators have found this not to be true.

Since thick resists exaggerate the AR, finding mask materials or etch gases for which the etch rate of the mask is low, making thinner masks possible, helps to decrease the severity of the problem. Etching at the very low pressures possible in high density plasmas may be a way of avoiding the problem. But Gottscho et al. (1992) state that other factors besides pressure are altered when the reactor geometry is changed, so that pressure reduction alone may not be adequate to account for the improvement in RIE lag.

Several explanations for RIE lag have been proposed but there has been no definitive resolution as to whether there is a single dominant mechanism or whether there are several working in concert.

A phenomenon called "reverse RIE lag," i.e., one in which small holes etch faster than large ones, has been observed in oxide etching (Dohmae et al., 1991; McVittie and Dohmae, 1992). This was attributed to ion reflections off sidewalls which depended on wall angle and ion energy, with reflection effects increasing at lower energies. When reflection is suppressed by using an overhang structure, "standard" lag results.

2.10 SILICON NITRIDE

Because minimization of device delay is of paramount importance, the use of silicon nitride, whose dielectric constant is ~7, is limited. It provides good protection against Na^+ migration and moisture penetration. Thus it is most often used as a protective layer over devices to prevent instability or as a final passivation layer to inhibit moisture permeation, but not as an interlevel dielectric in fast devices. Therefore the present review will be very brief.

A limited thickness of the compound can be formed by thermal nitridation of silicon at high temperatures (e.g., Ito et al., 1980; Nemetz and Tressler, 1983; Delfino et al., 1992, and references therein).

CVD at high temperature is another method used most commonly for the preparation of stoichiometric films containing little or no H. The high temperature required makes CVD films unsuitable as an interlevel dielectric with other than refractory metals. Its use is limited to coating gate electrodes and contacts in Si devices. Milek (1971, 1972) has reviewed the early work on the use of silicon nitride for microelectronic application.

PECVD is carried out at temperatures suitable for use with Al or Cu-based metallization and produces films of acceptable quality for device applications; therefore, it is now used, replacing CVD, when temperature limitations exist. Although stoichiometric (Si_3N_4) films have been deposited, they are most often Si-rich; N-rich films have also been prepared. PECVD nitrides have been referred to as "plasma SiN," but are better characterized as SiNxHy to emphasize the nonstoichiometry, when it exists, but more important, the fact that they usually contain a sizeable amount of H; stable films with H-content as high as 40 at% have been deposited. The stoichiometry (Si/N ratio), the H content and bonding (Si—H, N—H), as well as the chemical (e.g., etchability) and physical (e.g., stress, CTE, electrical properties) characteristics of the film depend on the precursors (e.g., NH_3, N_2 mixtures), deposition conditions (e.g., temperature, excitation frequency, substrate bias), and reactor configuration. The H-content of the films deposited using a given process has been reduced by applying a low frequency bias to the substrate (Martin et al., 1988). While postdeposition annealing is a more effective method, there are temperature limitations for annealing as there are for deposition. Also, the stress in the film is altered significantly when the H-content is reduced.

A comparative study of PECVD nitride films has been published by Kanicki and Voke (1986) and Kanicki and Wagner (1987). Other useful references are Sinha et al. (1978), Claasen et al. (1985, 1987), Tsu et al. (1986), Hirao (1988), Landheer et al. (1991), Parsons et al. (1991), Taylor (1991), Ito et al. (1991), Kikkawa and Endo (1992), Pearce et al. (1992), Nguyen et al. (1992), Stamper and Pennington (1993), Cottler and Chapple-Sokol (1993).

2.11 SILICON OXYNITRIDE

PECVD silicon oxynitrides are used to some extent. The ratio of oxygen to nitrogen in the film determines the properties of the films; these will lie between those of the pure oxide and nitride. The ratio of the source gases will determine the ratio in the films.

2.12 F-DOPED PLASMA SiN

As in the case of SiO_2, F-doping reduces the dielectric constant of silicon nitride; values of ~4–6.5 have been reported (Fujita et al., 1984, 1985, 1988; Chang et al., 1988; Livengood and Hess, 1988). SiF_4, Si_2F_6, NF_3, F_2 have been used as F-sources and the Si-sources have been SiH_4, SiF_4, and SiF_6. It is assumed that Si-F bonds exist, but the IR absorption band is obscured by that of Si—N (Fujita and Sasaki, 1988). The fluorinated nitrides have smaller amounts of H and greater thermal stability with respect to H-loss, higher density, resistivity and breakdown strength, than the conventional PEVCD SiN films. Films with a very high F-content hydrolyzed to SiO_2 on exposure to moisture.

2.13 BORON NITRIDE

This discussion of BN is included here because it appears to have been the inspiration for the development of SiBN films, discussed below. B-rich BN films were reported to have a low dielectric constant, i.e., $\varepsilon < 3$, but were very unstable in moist environments. However, a review of the literature unearthed an exceptional variability of results. The value of ε ranged from 2.7 to 7.7 and there was no clue, from the deposition method, conditions, or choice of precursors to account for the variation. Most recently, Nguyen et al. (1994) prepared stable films with good moisture resistance in a PECVD reactor, using diborane/NH_3 (400°C) and borazine/N_2 (300°C); ε from 4 to 5, depending on deposition conditions. Thus it

Interlevel Dielectrics 267

seems clear that BN would not be likely to replace SiO_2 as the interlevel dielectric, because of its dielectric constant. Other properties, such as its etch characteristics, i.e., anisotropic profiles with excellent selectivity over silicon oxide and nitride, or its ability to act as a polish stop in CMP may be factors influencing its use. It should be noted that some of the previous reports of a very low value of ε may have been due to measurement error. If an *n*-type Si substrate is used for the capacitance measurement; B may be implanted during PECVD deposition, forming a *pn* junction; this causes inaccurate capacitance measurements and thus an inaccurate value of ε.

2.14 FILMS CONTAINING Si, N, AND B

The instability of BN led to the idea of SiBN films. These may be considered Si-doped BN films with better moisture resistance or B-doped SiN films with a lower dielectric constant. The films were prepared by PECVD by using mixtures of the hydrides in an Ar plasma and by varying the ratio of the hydrides, the ratio of the constituents in the films could be varied; N—H, B—H, Si—H, B—N, and Si—N were all detected in the IR spectrum of the films (Maeda and Makino, 1987). Addition of B reduced ε, the refractive index, and the etch rate in HF-based etchants. However, when the dielectric constant was reduced to a value substantially lower than that of SiO_2 by increasing the B-content, the films became hygroscopic, making them unusable.

2.15 FILMS CONTAINING Si, N, B, AND O

O-doping of PECVD SiBN reduces the dielectric constant of the films still further. The films were prepared by the addition of N_2O to the mixture of the hydrides. ECR (Maeda and Arita, 1990a,b) as well as conventional PECVD (Maeda, 1990, 1993) were employed. The minimum value of $\varepsilon = 3.3$ was attained when the ratio $[O]/2[Si] = 1$; increasing the O-content resulted in an increase in ε, instability, and susceptibility to moisture and a decrease in dielectric strength. It was postulated that the B–N groups are replaced by the more polar B—O groups at high O-doping. The reproducibility, stoichiometry control, stability, etc. of the quaternary film require further study, but no further work on this material has been reported.

3.0 SPIN-ON-GLASSES (SOGs)

3.1 INTRODUCTION

Spin-on-glasses (SOG) are solutions of polymers with an Si—O network in their structures. The liquid is spun on the wafer which is then heated to expel the solvents and to cure the material to form an amorphous inorganiclike transparent film.

There are several uses for SOGs in semiconductor processing. They have been used as dopant sources for silicon but that application will not be discussed here. SOGs have been used as O_2-resist barrier layers during RIE of the planarizing organic underlayer in multilevel mask and lift-off stencil fabrication. The O_2-resistance is probably due to the formation of a thin surface layer of SiO_2. SOGs can be etched in the usual plasmas used for Si and Si-compounds so that they can be patterned readily. Currently, the interest in SOGs lies chiefly in their ability to partially planarize and smooth existing topography as well as to fill high aspect ratio spaces between adjacent conductors. They are, therefore, useful as sacrificial layers in etchback planarization; this will be discussed in a later section. Here the emphasis will be on their use as interlevel dielectric layers, usually combined with other inorganic dielectrics, often PECVD SiO_2.

This section will cover the types of SOGs, the processing of the films, their properties, their deficiencies and methods to overcome them, and their interactions with the environment and other materials in the structure. Their use in planarization processes will be discussed in Chapter VI.

3.2 APPLICATION

One of the advantages often cited for the use of SOGs as an interlevel dielectric is that application by spinning is an economical alternative to deposition by CVD, PECVD, or sputtering. However, special equipment must be used to obtain defect-free SOG; when SOGs are applied using standard spin stations filament of the polymer can form ("cotton candy") and adhere to the surface. Shelf life is another important factor; if there is gel formation during storage, the solutions must be discarded to avoid defects. The cure cycles can be lengthy and must be well controlled; often additional water-desorption steps (discussed below) are necessary, adding to the cycle time, increasing the cost. And, as will be shown below, they are usually used with deposited films, so there is still need for that equipment. Care must be exercised to avoid contamination with mobile ions, especially Na^+ so that their formulation and manufacture must be rigidly controlled.

3.3 SOGs FROM ORGANIC PRECURSORS

3.3.1 Introduction

There are three basic types of these SOGs: polysilicates, polysiloxane, and polysilsesquioxanes. The reactions involved in the formation of SOGs are hydrolysis and condensation. The properties of the polymer and the final film are functions of the starting material, the reaction conditions, the nature and constitution of the solvent, so that is has been possible to prepare a variety of materials with somewhat different characteristics.

There are many differences among them but they have some properties in common. After curing, they are all porous to some extent so that they absorb moisture from the environment readily, raising the dielectric constant (Harada et al., 1990). Hirashita et al. (1990) postulated that, upon heating, residual Si—OH groups can react to form water which is released. From whatever source, the moisture is easily desorbed during subsequent metal deposition, leading to the so-called "poisoned" vias and contacts (Chiang et al., 1987), blistering of Al-based wiring (Hirashita et al., 1990), and distorted metal and poor step coverage in vias (Ting et al., 1988). These effects may be minimized or even eliminated by prolonged in situ heating or possibly in situ sputter cleaning before metal deposition (Ting et al., 1988; Wolters and Heesters, 1990). However, the continued evolution of water during sputter cleaning can negate the effect of sputter cleaning when the growth

Interlevel Dielectrics

rate of Al_2O_3 (the source of the high resistance) exceeds the sputter etch rate so that an in situ preheat *before* sputtering may be necessary. Most SOGs shrink when cured so that the film stress is tensile and thus thick films crack, limiting their applicability as interlevel dielectrics. After curing at 900°C the films resemble thermally grown SiO_2 since all the Si—OH and organic groups have been eliminated and the films densified; however such a high temperature is incompatible with anything but refractory materials.

Because of the stress-limited thickness and the proclivity for moisture absorption, when incorporated into the structure as permanent dielectric layers thin films of SOG are almost always combined with thicker layers of a stable oxide. There are several configurations for composite structures, as shown in Fig. IV-10. In the top diagram, SOG is directly in contact with the metal and capped with the more stable oxide to prevent moisture absorption; the middle diagram shows SOG as the outer layer with the stable oxide in contact with the metal to prevent interaction between SOG and the Al-based conductor; the most commonly used structure, with SOG sandwiched between two layers of oxide, is seen in the bottom diagram.

Polysilicates

The starting material is an ester; TEOS is a common reactant but other esters have been used as well (Smolinsky et al., 1989a). All the silicon atoms in the molecule are bonded to oxygen atoms; these, in turn, may be bonded to Si, H, or organic groups (R). Curing at ~425°C eliminates the organic groups but not the Si—OH; thus this SOG has been called an inorganic SOG in some papers. The structure after curing is shown in Fig. IV-11. The presence of the silanol groups increases

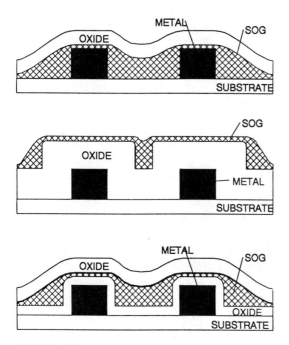

FIGURE IV-10 Composite structures of SiO_2 and SOG: (t) SOG in direct contact with metal, (m) SOG asouter film, (b) SOG sandwiched between two oxide layers.

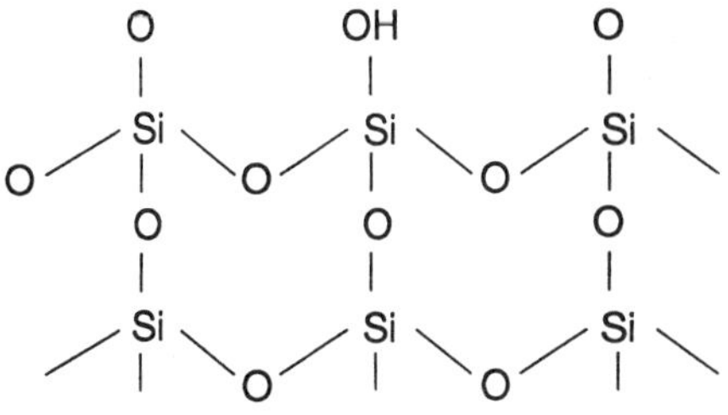

O
OH
O
Si
Si
Si
O
O
O
O
O
O
O
O
Si
Si
Si
O
O

FIGURE IV-11 Structure of a polysilicate film after cure.

the dielectric constant of the film and tendency to absorb atmospheric moisture. Moisture absorption/desorption may cause unreproducible changes in the dielectric constant and possible stress hysteresis upon exposure/storage. In addition water in SOG films has been linked to a mobile charge (H^+) phenomenon affecting the threshold voltage and transconductance of MOS devices (Lifshitz et al., 1989; Lifshitz and Smolinksy, 1989).

Polysilicates films have a pronounced tendency to crack with increasing thickness. Thicker films can be formed by applying several thin coats, curing between successive layers; rapid thermal annealing was found to be very effective (Uoochi et al., 1990). P-doping modifies the network and reduces film stress. However, if $P \gtrsim 4\%$, the films become even more hygroscopic (Chiang and Fraser, 1989). The structure of a P-doped silicate, after cure is shown in Fig. IV-12.

Water is the principal species evolved when a cured polysilicate is heated. Desorption occurred at temperatures below 300°C and was *reversible upon re-exposure to moisture.* Tompkins and Tracy (1989) proposed that the water was not a decomposition product but was evolved from the pores in the film and that heating after curing reduced the number, but not the size distribution, of the pores. Ito et al. (1991) classified the sources of water (1) incomplete removal during curing <450°C, (2) absorption during storage, and (3) absorption during immersion in water. Chemisorbed water could be removed by exposure to a low plasma-density O_2 plasma in a parallel plate reactor with a graphite electrode maintained at a low temperature; the process was named reactive glass stabilization (RGS) by Ito et al.

O O O O O O O
-Si-O-Si-O-P=O O-Si-O-Si-O-Si-O-Si-
O O O O O O O
-Si-O-Si-O-Si-O-Si-O-Si-O-Si-O-Si
O O O O O O O
-P=O

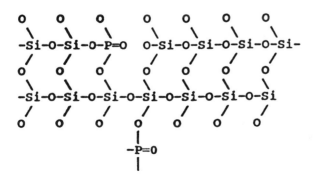

FIGURE IV-12 Structure of a P-doped polysilicate film after cure.

(1990, 1991). Ion bombardment is responsible for eliminating the OH group, forming a dense surface layer. However although this does not inhibit moisture absorption from the air, it does minimize it during immersion in water. The effect is enhanced if the stabilized film is immersed in organic liquids. Treatment of a SOG film, which had been heated to 100°C on a hot plate, with fluoro-alkoxy-silane (FAST) vapor (in N_2) at 25°C, and then curing for 60 min in N_2 at 300°C, reduced the Si—OH content, cracking susceptibility, shrinkage during anneal, leakage current, and moisture absorption/desorption (thus eliminating "poisoned" vias). The density of the film was increased (Homma and Murao, 1993).

Polysiloxanes

In these SOGs, Si atoms are bonded to C atoms as well as to O; usually fewer than 1 Si is bonded to a C atom. After curing at ~425°C, organic groups are still present. The structure is shown in Fig. IV-13. The presence of the organic groups in the matrix provides greater flexibility and results in lower stress, compared to silicate SOGs, making it possible to apply thicker films without cracking. Polysiloxanes absorb/desorb less water than the polysilicates. Nevertheless desorption of water trapped in a polysiloxane film was cited as the cause of metal blistering and void formation in Al interconnections (Hirashita et al., 1990). Outgassing of water during metal deposition results, as noted before, in high resistance vias and contact. Another cause of high resistance is the formation in the vias of a layer consisting of Si, O, F, and C probably formed during RIE to form the hole (Shacham-Diamand and Nachumovsky, 1990). They postulated that the presence of SOG enhanced fluorocarbon polymer deposition and aluminum oxidation. The contamination layer also degrades the adhesion between successive metal layers and delamination may occur. Addition of oxygen to the F-plasma reduced the film formation, but increases the etch rate of the patterning resist. Longer in situ sputter clean times, while time consuming, do yield good results. The organic constituents can be easily oxidized which results in cracking. Since ashing is often necessary (e.g., resist stripping after via etching), this is a serious problem. The RGS process discussed in the last section can be used to improve the crack resistance during ashing (Ito et al., 1990) as well as to lower the water-content of the film and perhaps water absorption as well, minimizing or possibly eliminating "poisoned" vias. They suggested that the thin surface layer prevented the species in the plasma from reaching the inner layers. They used RGS before applying the second stable oxide in a sandwich

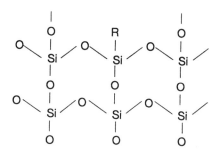

FIGURE IV-13 Structure of a polysiloxane film after cure.

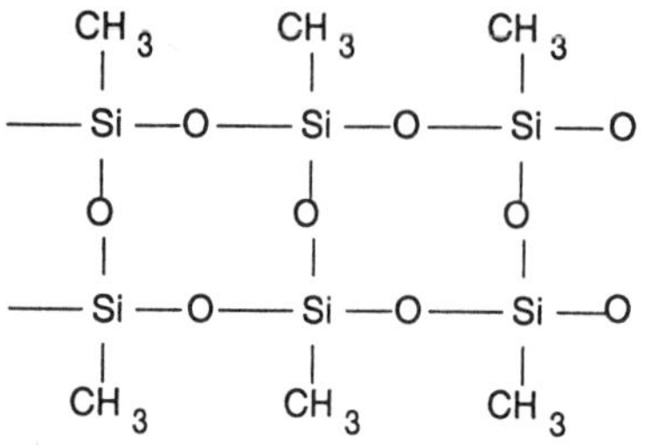

$$\begin{array}{c}
\text{CH}_3 \quad\quad \text{CH}_3 \quad\quad \text{CH}_3 \\
| \quad\quad\quad | \quad\quad\quad | \\
-\text{Si}-\text{O}-\text{Si}-\text{O}-\text{Si}-\text{O} \\
| \quad\quad\quad | \quad\quad\quad | \\
\text{O} \quad\quad\quad \text{O} \quad\quad\quad \text{O} \\
| \quad\quad\quad | \quad\quad\quad | \\
-\text{Si}-\text{O}-\text{Si}-\text{O}-\text{Si}-\text{O} \\
| \quad\quad\quad | \quad\quad\quad | \\
\text{CH}_3 \quad\quad \text{CH}_3 \quad\quad \text{CH}_3
\end{array}$$

FIGURE IV-14 Structure of a simple silsesquioxane, SPS. (Reprinted with permission from The Electrochemical Society from A. Oikawa et al.)

structure as well as after etching via holes to prevent cracking along the walls during resist stripping. Rucker et al. (1990) modified the surface of polysiloxane film chemically by annealing the film in O_2 and cooling in an ambient of N_2 and a silylating compound, thereby reducing the Si—OH bond concentration significantly, eliminating cracking almost completely. Ar^+ implantation into the SOG is reported to increase the density and O_2-plasma resistance and produce high thermal stability (Mizuhara et al., 1995).

Silsesquioxanes

These are ladder siloxanes; the backbone is $(Si_2O_3)n$ with organic groups on every Si atom (Smolinski et al., 1989). Polymethylsilsesquioxane (PMSS), shown in Fig. IV-14 (Oikawa et al., 1990), is one of the simpler examples; the organic group is simply a methyl radical, CH_3. It is stable when heated to almost 500°C in an inert atmosphere, but is oxidized readily below 400°C. Substituting phenyl for methyl groups, to form polyphenylsilsequioxane (PPSQ), improved the oxidation resistance but made the material brittle and prone to cracking (Oikawa et al., 1990). Adachi et al. (1990) reported that the shrinkage of a PPSQ film was nearly zero, its dielectric constant was 3.2 and the breakdown strength was acceptable (4–5 MV/cm). Polysilphenylenesiloxane (SPS), shown in Fig. IV-15 (Oikawa et al., 1990), did not oxidize below 450°C, had a low ($\sim 5 \times 10^8$) tensile stress, and a low dielectric constant ($\varepsilon = 3$). There was very little outgassing when heated, probably because the low silanol content and high film density minimized moisture absorption. Thick

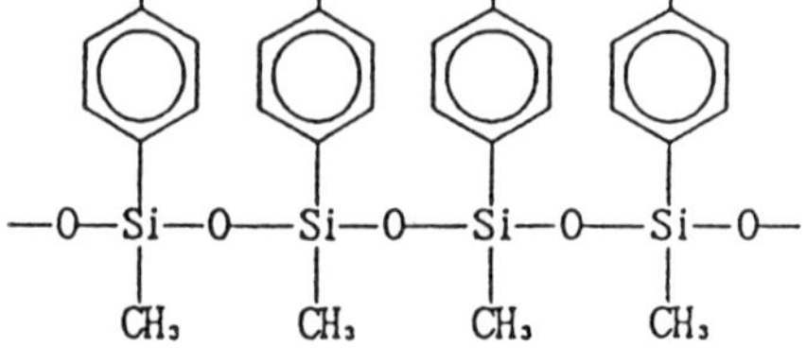

FIGURE IV-15 Structure of polysilphenylyenesiloxane, PMSS. (Reprinted with permission from The Electrochemical Society from A. Oikawa et al.)

films (e.g., 1 μm) withstood cracking when heated to 420°C in air. SPS has excellent planarizing properties, perhaps due to its comparatively low melting point so that it melts into a viscous liquid filling low regions before it hardens. The high organic content of silsesquioxanes which makes them flexible, results in a low etch rate in HF-based solutions and in F-based plasmas.

3.4 SOGs FROM INORGANIC PRECURSORS

A carbon-free spin-on silica film has been receiving increased attention. The material is available commercially as Dow-Corning Flowable Oxide. The precursor is a carbon-free polymer. Pineda et al. (1990) described the material as being "based on a new concept of converting caged $(HSiO_{3/2})n$ structures by eliminating hydrogens and cross linking the caged structures with oxygen," as illustrated in Fig. IV-16.

The polymer is spun on and then the wafer is placed on a hot plate where the film melts at about 200°C and flows prior to conversion to a glass; it does not flow again after conversion. The sequence is shown in Fig. IV-17. The unique melt and flow properties result in good planarization and gap-fill. Pineda (1990) described an optimum cure process which combines a low temperature oven bake followed by exposure to an O_2 plasma. Oxygen is required during curing to ensure formation of stoichiometric SiO_2. They report that the properly cured film is a stable, crack-free oxide, contains no OH bonds, and does not absorb moisture from the atmosphere. The dielectric constant is comparable to that of thermally grown oxide and the breakdown strength was ~8 MV/cm. Ballance et al. (1992) described curing at 800°C in O_2 for premetal application; neither Si—H nor Si—OH are

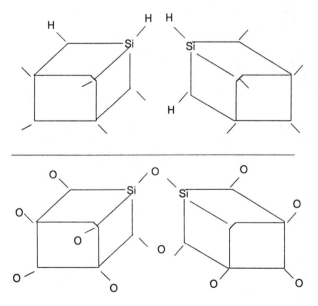

FIGURE IV-16 Reaction to form C-free SOG hydrogen silsesquioxane $[(HSiO_{3/2})n]$. (Reprinted with permission from R. Pineda et al.)

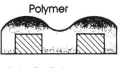  

Spin-On Polymer   Melt and Flow   Convert
(on spin-coater hot plates)   (in quartz tube furnace)

FIGURE IV-17 Process sequence using C-free SOG: left to right: spin-on, melt and flow, and convert to glass. (Reprinted from Dow-Corning production information.)

present and $\varepsilon = 3.8$. For intermetal applications, however, lower temperatures are required and they used an ambient containing steam and ammonia at 425°C, resulting in films with small amounts of Si—OH and ε was high, i.e., 4.7. Ahlburn et al. (1995) reported that the gas ambient, cure temperature, and time affected the extent of curing but did not specify the values of these variables. A full range of mixtures of SiO_2 and polymer, described as oxygen-deficient SiOx, could be formed by changing the cure temperature. The films cured at lower temperatures contained SiH, SiOH, and HOH and are O-deficient, all of which would be expected to increase the value of ε. However, they reported that the lower the temperature, the lower the dielectric constant; this was attributed to the higher proportion of polymer which has lower polarizability. Thus the apparent contradiction: higher cure temperatures increased the dielectric constant, but total conversion to SiO_2 produces a film with $\varepsilon = 3.8–3.9$. The films absorb moisture on standing, and moisture evolution occurs during ashing. Thus the low value of the dielectric constant might possibly be attributed to porosity which decreases with increasing temperature of cure.

The flowable oxide is intended to be used as a "stand-alone" dielectric, i.e., does not require the concurrent use of other oxides. This fact is the basis for the conclusion of Pai and Konitzer (1992) that the use of flowable oxide results in the lowest manufacturing cost/layer, when compared to nonetchback traditional SOG, partial etchback, resist etchback, and dep/etch using PECVD + sputtering.

Other papers describing the use of SOGs in multilevel structures, characterization of the materials, processing alternatives, advantages and disadvantages of the various SOGs, problems encountered in their use, ways of overcoming the problems, etc., are for example, Kojima et al. (1988), Yen and Rao (1988), Chung et al. (1989, 1991), Forester et al. (1989, 1990), Pramanik et al. (1991), Matsuura et al. (1993), Chen et al. (1994), Karim and Evans (1996), and Ohashi et al. (1996).

4.0 ORGANIC DIELECTRIC FILMS

4.1 INTRODUCTION

Interest in organic insulators as dielectrics for semiconductor multilevel interconnection applications has been spurred on by a variety of reasons. During the 1970s and early 1980s, a spin-on organic insulators held the promise of fewer defects, lower cost, and a planar structure, reducing topographical concerns as more than one level of wiring was used. A host of requirements such as adhesion between layers, low stress and high ductility, thermal conductivity, and compatibility with

RIE processes including corrosive gases, etc. prevented significant use of organics in semiconductor applications. However, as a result of continued development in polymer synthesis, many new organic insulators, especially polyimides (PIs), became available in the market under different brand names. It has been claimed that they meet many of the requirements listed above and are, therefore, ready to be used as interlevel dielectrics. Continued improvement in both equipment and processes, resulting in better quality SiO_2 films and higher throughput (and lowered cost) in their deposition, have kept the PIs from making significant gains. More recently, interest in organic insulators has been revived for their potential of having a low dielectric constant (ε), which can reduce on-chip interconnection propagation delays. Polyimides have an ε of 3.0–3.7 which is a significant reduction in ε achievable over plasma SiO_2. Polyimides with fluorine groups have been shown to have ε less than 3; Teflon, another fluorinated polymer, has a value of ε of ~2. Polymer foams with small voids have been synthesized and they also have a low ε values, ~2 (Cha et al., 1996).

Organic insulators are usually polymer chains with a varying number of monomer units in a chain. The arrangement of chains, in a random or linear orientation, the bonding between groups in different chains, and the molecular configuration of the monomers and their arrangement within the chain, determines the mechanical and thermal properties of the polymer as well as any anistropy in the properties of the resulting films. Therefore, there is a wide variation among the polymers in their properties, such as ε, coefficient of thermal expansion (TCE), elastic modulii (E), etc. The spin-on process used for applying the thin insulating films, tended to promote directionality of the structure, which aggravated the anisotropy of the properties of the resulting film; i.e., the properties were different in the in-plane and out-of-plane directions. A list of organic insulators recently developed for use as interlevel dielectrics is shown in Table IV-5. These are referred to by their chemical groups and interested readers should consult the referenced papers for details. Because of their superior thermal stability. PIs, among the different organic insulators, are most commonly considered for semiconductor application. Therefore, this section will be devoted primarily to developing an understanding of the structure and properties of PIs. There will be a brief discussion, at the end of this section, of some of the newer materials developed more recently.

In general, the dielectric requirements are the same for all VLSI/ULSI applications; they include electrical properties such as low leakage and ε, thermal properties such as a high glass transition temperature (T_g) and good thermal stability, and mechanical properties such as toughness and low film stress, interfacial properties such as good adhesion, and low moisture absorption.

4.2 POLYIMIDES

4.2.1 Synthesis and Structure

The most common process for synthesizing PIs is by reacting nearly equivalent amounts of a diamine and a dianhydride in a suitable solvent. The polymer chain is formed by step-wise addition of monomer units, in what is referred to as a condensation process; the diamine and dianhydride form an alternating structure. Since the monomers form a chain with alternating units, the chains that are formed

TABLE IV-5 Organic Dielectric Materials: Commercial Names, Sources, Description

Commercial name	Source	Description
Cyclotene 3022 (BCB)	Dow Chemical	Divinyl siloxane benzocyclobutene. Isotropic, 3-D structure.
FLARE	Allied Chemical	Flourinated poly(Arylethers). Based on perfluoro biphenyl. Linear chain thermoplastic.
Parylene		Poly para-xylelene. Vapor deposited.
Fluorinated PI	DuPont	Rigid rod structures with fluorinated backbone, e.g., 6FCDA or PMDA with TFMB or TFMOB.
Foams	IBM Almaden (Cha et al., 1996)	Triblock copolymer based on propylene oxide dispersed PMDA-3F; PO is thermally decomposed.
Chisso	Chisso Corp.; AMOCO	Polyimide siloxane with SiO_3 bonded to aromatic compounds.
Pyralin	DuPont	Linear chain polyimides such as BPDA-PDA.
PIQ-L100	Hitachi	Directional chain from reacting aromatic diamine or diisocyanate wth aromatic tetra carboxylic acid derivative.
SIM	Occidental Chemical	Soluble, fully imidized polyimide siloxane.

in the beginning of the reaction can have larger numbers of monomer groups, and the ones that are formed toward the end of the reaction are likely to have fewer monomer units (or lower molecular weight). This occurs because access to the chain is limited by diffusion. Additives are used to stop the reaction by tying up the reactive end of the chain. Temperature, stirring, and additives can be used to influence the molecular weight distribution of the polyimide. Edwards (1965) in his patent, discuss methods of making aromatic polyimides and teach the use of offset polymerization, i.e., the use of slightly unequal amounts of each monomer, to control molecular weight and capping the end of the chain with a chain termi-

Interlevel Dielectrics

nating agent. St. Clair et al. (1992), teach in their patent the use of aliphatic polyol solvents to prepare aromatic polyimides. An example of the synthesis reaction is shown in Fig. IV-18, where a pyromellitic dianhydride (PMDA) is reacted with oxydianiline (ODA) to form a PMDA-ODA polyamic acid precursor. This is the precursor, which when cured (i.e., after an imidization reaction has been carried out by heating the solution) becomes polyimide. The solvent used is N-methyl-pyrrolidone 2 (NMP) in which the PMDA, ODA, and the resulting PMDA-ODA are soluble. Also shown is the use of ODA-ester with PMDA which results in a polyamic ester using same NMP as a solvent. It is required that the reactants

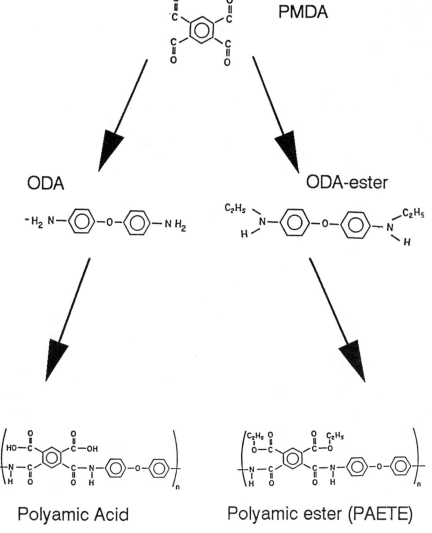

FIGURE IV-18 Synthesis of polyamic acid from PMDA and ODA and polyamic ester from PMDA and ODA-ester.

(monomers) and the resulting polymer are all soluble in the chosen solvent. Among the solvents are NMP, dimethylacetamide (DMAC), dimethylsulfoxide (DMSO), etc.; NMP is preferred for safety reasons. A good discussion on the many choices of anhydrides, amines, and solvents is given in Sroog (1976).

The molecular weight distribution of the chains in the precursor is controlled by the purity of the starting materials, their proportions, the reaction time and temperature, and finally the end cap additives used to stop the chain growth. An excess of dianhydride or diamine results in a lower average molecular weight. The percent of solids in the solvent and the molecular weight distribution of the polymer affects the viscosity of the polyamic solution, which determines the coating thickness. Gel formation is a result of undesirable polymerization and gels are undesirable "defects" which appear as particulates in the coating. The choice of the anhydride and the diamine starting materials determines the chemical structure of the resulting polyimide and many of its properties. Numata et al. (1987) in their patent, teach the formation of low TCE polyimides wherein the chains have a rod like structure and are highly oriented. The low TCE polyimides are made from a tetra-carboxylic anhydride such as PMDA, BPDA, BTDA with diamines such as m-phenyenediamine, diaminodipheny sulfone, etc., which results in a group of polyimides that have TCEs .4 to 30 ppms. Sachdev et al. (1992) in their patent, show the preparation of linear aromatic polyimides that contain dimethylsilane linkages in the dianhydride portions of the molecules and at least one trifluoromethyl group in each of the diamine portions. These polyimides have values of $\varepsilon < 3$ and form nearly transparent films. It is evident that PIs of many varieties can be synthesized and many of their characteristics can be fine tuned.

The PI precursor solution is applied by spin coating (for thin films) and spraying for thicker films. Thermal conversion of polyamic acid to polyimide films (curing) imparts thermal stability and mechanical rigidity to the films. For a given solution viscosity (which is determined by the solids content and molecular weight distribution), the spin speed determines the film thickness. The substrate with the spin coated film is dried at about 100°C to evaporate some of the more volatile solvents and to make the film more vicious and the substrate easier to handle. It is heated to a maximum temperature of usually 350 to 400°C by ramping or in a step-wise manner. The purpose of the final heat treatment is to remove all the solvents completely and complete the imidization reaction. Figure IV-19 shows the imidization for PMDA-ODA acid and its ester version. The completion of the imidization reaction has been studied by IR spectroscopy (Lee and Craig, 1980), weight loss (Numata et al., 1984), wet etch rates (Ginsburg, and Susko, 1984) and microdielectrometry (Day and Senturia, 1984). These studies show that most of the solvent is removed and 70 to 80% of the imidization reaction completed below 200°C. However it takes a temperatures >300°C for 99% completion of the imidization reaction. Incomplete imidization can result as outgassing on subsequent heating; this is unacceptable for semiconductor processing. In the case of polyamic ester, the onset of the imidization reaction is delayed with respect to the loss of solvents, when compared with polyamic acid. When a polyamic precursor is used to fill narrow gaps, voids form when the solvents are evaporated upon heating. If the polymer is only partially imidized, however, it can flow-in and fill the void. If the imidization reaction has progressed too far, the polymer becomes too rigid to flow and incomplete gap-fill results. Therefore, polyamic esters are preferred for

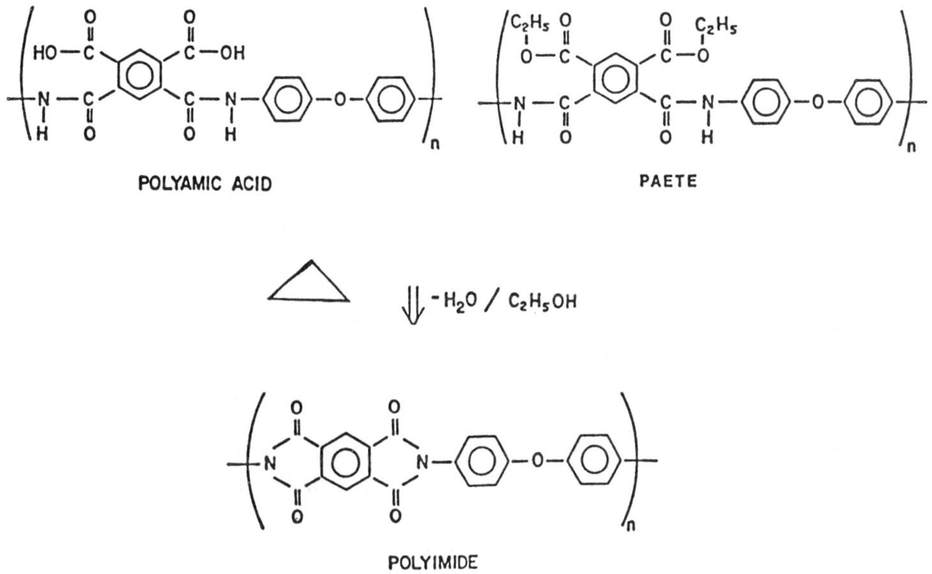

Interlevel Dielectrics 279

FIGURE IV-19 Imidization reaction for PMDA-ODA acid and ester.

filling narrow gaps. This is just an example of fine tuning of the polyimide for a specific requirement.

Another type of PI, with soluble preimidized structures that do not require a curing cycle, has been reported by Wang (1988). Lee (1989) has reported a class of soluble fully imidized polyimidesiloxanes. These PIs will not be discussed further.

4.2.2 Thermal Properties

The thermal stability of cured polyimide is measured using thermogravimetric analysis (TGA), where the weight loss is monitored as the sample is heated at a constant ramp rate or alternatively the sample is held at a constant high temperature and the weight loss monitored as was shown in Fig. II-28a,b. In the first analysis, the ramp rate is typically kept at 10°C/min and the temperature at which the organic insulator starts to fail catastrophically is measured. In the second analysis, the weight loss rate is determined at a selected high temperature. Both measurements are usually carried out in a nitrogen ambient. Complete curing and absence of outgassing during subsequent processing temperature is an important requirement. Mass spectrometric analysis of degradation products of PMDA-ODA by Heacock and Berr (1965) showed that the onset of rapid weight loss due to the gaseous decomposition of polyimide is in the form of CO and CO_2, and small amounts of benzene, H_2, and H_2O, etc.

Another important specification is the glass transition temperature since organic materials do not have well-defined melting points. T_g is defined as the highest temperature reached before the onset of flow, in the absence of a mechanical load. Some polyimides (strongly crosslinked) show no glass transition but merely degrade on heating. Palmese and Gillham (1987) have studied the relationship between T_g and temperature/time used for curing the PI.

Most PIs have thermal coefficients of expansion in the range of 10–40 ppm/°C. By comparison, silicon has a TCE of 3 ppm/°C and silicon dioxide 0.5 ppm/°C. Since the thermal stresses are largely driven by TCE mismatch, this is of great importance. Many of the new PIs have been synthesized to have TCEs less than 10 ppm/°C (Numata et al., 1988; Auman, 1995). The stress due to thermal mismatch during a thermal excursion has been discussed in an earlier chapter. The stress to a first approximation is linearly dependent on the difference between substrate expansion and film expansion; hence there is a strong motivation to match the PI TCE to that of silicon. It might be equally beneficial to match the TCE in the vertical direction to that of the metallization.

Equally important is the knowledge of what happens when the film and the substrate are cycled between room temperature and a high process temperature as is the case in semiconductor processing. In addition to TCE which might be temperature dependent, the behavior of the polymer can change from elastic to plastic at high temperature. As the thermal stress exceeds yield strength, there is stress relaxation. Stress relaxation is usually complete at high temperature. Film stress and thermal cycling behavior of organic thin films are usually studied using a supporting substrate and monitoring the change in the curvature of the substrate. Chen et al. (1988, 1989) have used a quartz cantilever beam, described in Fig. II-20, to study the effect on film stress from thermal cycling, and cycled the PI film from room temperature to 400°C. The deflection of the cantilever free end was measured, and the film stress calculated. Figure IV-20 shows the result of such cycling for BPDA-PDA. As seen in the figure, the polyimide film is totally relaxed and stress free at 400°C, but on cooling the film experience residual stresses. Since stresses are relaxed, catastrophic failure is avoided; however, since stress relaxation is by plastic deformation, thin film insulator may show other problems such as extrusion and creep.

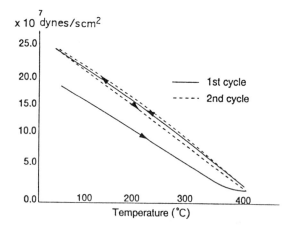

FIGURE IV-20 Changes in film stress due to cycling BPDA-PDA. (Courtesy of Chen, 1988.)

Interlevel Dielectrics 281

Thermal conductivity determines the heat dissipation from the devices and the resistive losses of the interconnection metallurgy and hence the maximum temperature of the chip. Organic insulators typically have thermal conductivities an order of magnitude lower than that of SiO_2. Jin et al. (1996) measured thermal conductivity using the same test structure and test method for both materials, found that PI has a thermal conductivity of 2.4 mW/cm°C, compared with a value of 11.5 mW/cm°C for plasma SiO_2. A low-thermal-conductivity insulator can result in higher device operating temperatures, which can degrade device performance and cause interconnection failures.

4.2.3 Adhesion/Interface Reactions

The basic concepts and methods of measuring adhesion are discussed in Chapter II.

An adhesion strength exceeding 25 g/mm, obtained from the peel test, is considered adequate for PIs; higher values may be required for specific applications. Adhesion tests are used to evaluate the effectiveness of adhesion promotors and surface treatment processes. Greenblatt et al. (1984) studied primary and tertiary amino silanes as adhesion promotors between polyimide and SiO_2. They found that aminopropyltrimethoxysilane upon hydrolysis generates a trihydroxysilane which bonds covalently to the selected PI surface groups. Satsu et al. (1991) using XPS, studied the surface modification of low TCE polyimides that improved adhesion (of metals and PI) as determined in a pressure cooker test. They found that the plasma (O_2, CF_4) treated surfaces had a substantial number of F—C=O functional groups, which appears to improve the adhesion.

Chromium and titanium provide good bonding to PI surfaces and are often used as "glue" layers. Using XPS, the chemical interaction between Cr, Cu, Pd, and Ni over PI was studied and it was shown that metal atoms bonded to the C, N, or O atoms in the PMDA monomer (Chou and Tang, 1984). Burkstrand (1979a,b, 1981), based on his XPS studies, concluded that adhesion of metallic films to PI is directly related to the metallic film electronegativity, with Cu < Ni < Cr. Haight et al. (1988), studying adhesion of Cr and Cu to polyimide by XPS, suggested that based on the energy spectrum, Cr had two types of bonding whereas Cu had a single weak interaction or bonding. LeGoues et al. (1988) showed that Cr—PI formed a well defined interface, but Cu—PI interface was broken by the precipitation of Cu in PI. On annealing, Cu precipitate coarsens, indicative of the poor stability of the Cu—PI interface. Kim et al. (1990) have studied the adhesion/reaction of metal-PI interfaces by use of crossectional TEM and XPS. They found PI/metal was different from metal/PI. When PI is spun on a metallic surface, the polyamic precursor creates a stronger metal-PI bond than when metal is deposited over a cured PI surface.

Faupel et al. (1989a,b) studied the interaction and diffusivity of Cu and Cr in PMDA-ODA, using an isotope of Cu as a tracer. The diffusivity of Cu ranged from 10^{-15} to 10^{-14} cm^2/s in PMDA-ODA in the temperature range of 200 to 300°C; the diffusivity in PI is an order of magnitude higher than that in SiO_2 (McBrayer et al., 1986).

Absorption of moisture probably results in oxidation of the interfaces, thereby causing delamination. The relative ease of moisture absorption and diffusion in PIs has been well documented (Sacher and Susko, 1979, 1981; Denton et al., 1985). The moisture uptake varied linearly with relative humidity. At room temperature, the diffusivity of water in PMDA-ODA is 10^{-9} cm^2/s, which means that it will take only few seconds for water to diffuse through a film 1 μm thick (typically used for an interlevel dielectric) and reach the PI-metal or PI—PI interface. The potential oxidation of the PI-metal interface and consequent loss of adhesion is a matter of concern for ULSI/VLSI processing. Release of moisture trapped within PI or at the PI/metal interface was held responsible for bubble formation in a PI film (Mitchell and Goodner, 1984; Peek and Wolters, 1986).

4.2.4 Mechanical Properties

The mechanical properties include stress-strain behavior at different temperature, Poisson's ratio, TCE, and E. The earlier work on thick polymers has shown the effect of structural anisotropy on the resulting anisotropy of mechanical properties (Gupta and Ward, 1968; Hadley, 1975). Since the spin-on films tends to have a preferred structures in the plane of the film and perpendicular to the film surface, the mechanical properties need to be measured in both the in-plane and out-of plane directions. The measurement of these properties is difficult as the films are very fragile. The multilevel structures are complex; the PI films are interspersed with metal features and the film thicknesses are usually small. As more information about the mechanical properties becomes available, realistic modelling of actual structures using organic dielectrics becomes feasible. Despite the complexity and limited data, finite element methods (van Andel and Gootzen, 1989) are used to obtain approximate stress-strain distributions in the device structures. Thicker free standing films (about 10 μm) are used in a tensile pull tester apparatus (Srikrishnan et al., 1990; Chen, 1995) to measure stress-strain behavior at different temperatures. The stress-strain data for BPDA-PDA and PMDA-ODA are shown in Fig. IV-21. The ductility increases and yield stress decreases appreciably at high temperatures and strength.

Poisson's ratio is determined by measuring the thickness change in response to a force applied in the in-plane direction of the film. The ductility was about 60% for PMDA-ODA at room temperature, and was about 35% for BPDA-ODA. However, BPDA-PDA exhibited higher toughness (area under stress-strain curve) compared to PMDA-ODA, which is desirable. Chen (1995) measured, the out-of-plane modulus for BPDA to be about 1 GPa at 20°C, much lower than the in-plane modulus of 11.7 GPa; this is in contrast to PMDA-ODA which has similar elastic modulas of 4 GPa in both directions. Chen (1995) attributed the PBDA-PDA behavior to the weaker bonding in vertical planes compared with the stronger covalent bonding in the in-plane structure. Auman (1995) measured in-plane mechanical properties of rodlike fluorinated PIs prepared using different monomer groups; their properties covered a wide range of values. Bauer et al. (1988) have measured the change in film stress in response to an applied hydrostatic pressure and concluded that the Poisson's ratio ranged from 0.3 to 0.5.

The TCE is determined by measuring the change in sample length with temperature. The equipment used is called a thermomechanical analyzer (TMA). The values of TCE for organic material range widely with typical polyimides in the range of 30–40 ppm/°C and low TCE polyimides in the range of 0.4–10 ppm/°C.

Interlevel Dielectrics

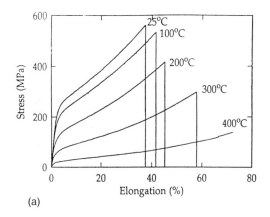

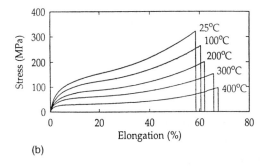

FIGURE IV-21 Stress vs. elongation (strain): (t) BPDA-PDA PI film, (b) PMDA-ODA PI film. (Courtesy of Chen, 1995.)

Tong et al. (1991), using a laser interferometer, measured TCE in the vertical direction (out-of-film plane) and concluded that for PMDA-ODA, the vertical TCE is 2–3 times larger than the in-plane TCE.

The large difference in the TCEs in the in-plane and out-of-plane direction implies that the deformation of PIs that are matched along the plane of the substrate, may be substantially higher in the out-of-plane direction. Since the stress that organic films can support is typically low, the mismatch strain from the metal film in the in-plane direction is accommodated in the first 1- to 2-μm thickness of the polymer film. This is clearly an advantage for thick polymer film applications, as the deformation is localized within the film; however, for multilevel interconnection applications, where the films are about 1 μm, the strain can extend into layers below the interfaces.

4.2.5 Electrical Properties

This section discusses electrical conduction, ε, dissipation factor, the time dependent breakdown (TDDB) of the dielectric film, and the effect of ionic impurities and moisture on these properties. Most published data are on thin PI films and usually in the out-of-plane direction; some recent measurements in the in-plane direction have become available.

Electrical Conduction

Smith et al. (1987) have reviewed electrical conduction in PIs, in particular PMDA-ODA and BTDA-ODA/MPDA using a capacitor structure. Some key observations: The PIs showed a strong current transient when the voltage is impressed, especially at temperature below 100°C. The charging and discharging is fully reversible and the decay current followed a (time)$^{-0.8}$ relationship. The charging and discharging behavior was attributed to the alignment of the weak dipoles in the PI with the externally applied field. The charging and transport current increased with moisture uptake. Above 150°C, the current is primarily transport or leakage. Figure IV-22

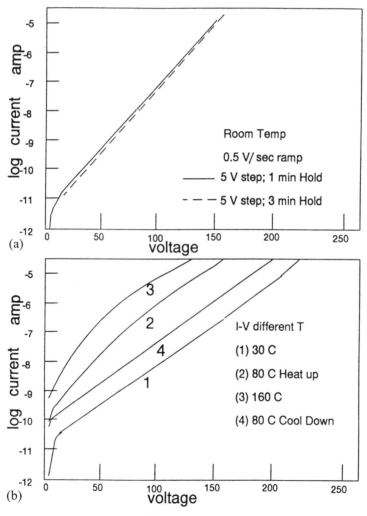

FIGURE IV-22 I-V curves for PI films: (a) film at room temperature, voltage ramped in 5V steps, (b) effect of thermal treatments.

shows I-V behavior of BPDA-PDA at three different temperatures (taken at 0.5 V/cm). The current at a given electric field increases at higher temperature, as one would expect from other insulators. However, when the film was heated from 30 to 80°C, the current at 80°C was larger than when the film was cooled down from 160 to 80°C. A likely explanation is that the film at 80°C on heat-up probably had more moisture than the film that cooled down to 80 from 160°C. Several studies (Sacher, 1979; Sawa et al., 1980; Chang et al., 1982; Rothman, 1980) have proposed that the transport current in polyimide to be primarily due to H^+, H_3O^+, Na^+, etc. Neuhaus et al. (1985) suggested that ionic impurities dominate leakage totally at high temperatures based on a study using Na^+ doped films in the temperature range of 25 to 380°C. They found that below 200°C, the transport current between doped and undoped films was similar.

The rapid diffusion of ionic impurities can cause device instabilities if the PI is used adjacent to a device such as a gate of a field effect transistor (Bergeron et al., 1984; Brown, 1981). Sato et al. (1973) showed that the instability can be avoided by using PIs with low ionic impurities. Beuhler et al. (1989) showed that fluorinated polymers with even high Na^+ did not cause device instability, the assumption being that diffusivity is very low or the Na^+ ions are tied down locally.

Breakdown

Electrical breakdown of an insulator usually occurs when the localized electrical fields exceed a critical value, causing a large current flow which results in excessive heat and melting or vaporizing of the electrode. In the case of inorganic materials, the breakdown invariably occurs at local "weak spots," the shorting is usually irreversible. Rothman (1980) reported that polyimides (PMDA-ODA, Skybond) have a breakdown strength of 5–8 MV/cm comparable to SiO_2, whereas others (Homma et al., 1988) reported a lower breakdown voltage of 3–5 MV/cm. In Rothman's study, the leakage current density corresponding to breakdown field was 2–3 orders of magnitude larger than SiO_2. In most of the studies, the breakdown criterion was not defined explicitly. Figure IV-23 shows a cumulative histogram of the behavior many Al-dot capacitors using PI (BPDA-PDA) or PECVD SiO_2 as the dielectric. The criterion for both PI and PECVD oxide was defined as the voltage at which a breakdown spike was sensed or the current reached 1 μA (~0.5 mA/cm^2). In the case of PI, the current of all capacitors increased monotonically, reaching the predetermined maximum value and in all cases of PECVD oxide a breakdown was noted.

Samuelson (1982) noted that both PIQ and PI2545 had a thickness dependent breakdown characteristic; this was attributed to the existence of pin holes. Samuelson (1982) used statistical breakdown measurements (shorts) at low fields to show that the pinhole probability for a 1.2-μm PI film was 7.5 in a million. This not only attests to the high uniform quality achievable with spin-on PI, but also explains the high current density supported by PI before breakdown, especially if the film is cast from a high purity solution.

Dielectric Constant/Dissipation Factor

Most of the measurements of ε of PI thin films have been made in the range of 10 kHz to 1 Mhz. Depending on the chemical group, the ε of the PI films have a value in the range of 2.5 to 4. Denton et al. (1985) showed that ε of PMDA-ODA

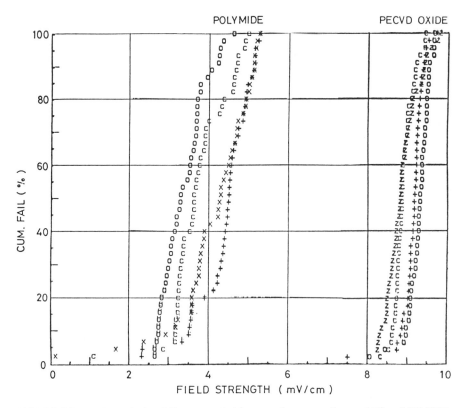

FIGURE IV-23 Cumulative failures vs. field strength: comparison of PI and PECVD.

when dry was 3.5, and increased by as much as 10% when soaked in water. Typical ε values of some of the PIs are listed in Table IV-6. There has been extensive study of ε in many of the new PIs in both in-plane and out-of plane directions. Auman (1995) studied several rodlike fluorinated PIs and showed that their dielectric properties were isotropic, with $\varepsilon < 3$. Ip and Ting (1995) used serpentine metal lines to measure the values of ε, both in-plane and out-of-plane, for a large number of low ε organic insulators and found that in-plane dielectric constants were higher than out-of-plane by 0–14% in their studies. Cha et al. (1996) synthesized polyimide films with nanofoam morphology and showed that ε of PMDA-3F decreased linearly with increasing porosity.

4.2.6 Photosensitive Polyimides

Photosensitive polyimides (PSPI), as the name implies, behave similarly to photoresist in response to exposure and developing to form patterns. In this case, the undeveloped material can be cured to form PI. Both negative (Ahne et al., 1984) and positive (Lazaridis and Chu, 1990) acting PSPIs are available. The presence of photoactive compounds, in addition to solvents, leads to shrinkage of the order of 40 to 50% after curing. This material is more suited to large, sloped-via applications, and is likely of only limited interest for on-chip interconnect applications.

TABLE IV-6 Dielectric Constants of Polymer Films

Polymer	Dielectric constant
BCB	2.7
FLARE	2.6
Parylene-N	2.6–2.8
6FCDA-TFMB	2.5
6FDCA-TFMOB	2.8
PMDA-TFMB	2.6
PMDA-3F (dry)	2.9
Foams (19% poro)	2.3
Chisso Pl	3.3–3.4
PIQ-L100	3.2
BPDA-PDA	2.9
Teflon	1.9
PMDA-ODA	3.3–3.5

4.2.7 Patterning of Polyimides

Reactive Ion Etching

Organic insulators are patterned using RIE systems which were described in Chapter I. Oxygen is used as the primary reactant, with F_2 added at times, for a variety of reasons. Directionality of pattern transfer is due to the directionality of ions; in their absence, (e.g., as in a barrel asher) etching will be isotropic. Heidenreich et al. (1986), using an experimental reactor, studied the effect of RF and microwave plasmas and of ion energies on the anisotropy of etching PIs. They concluded that to preserve anisotropy, the ion assisted etch rate must be dominant; even at 2.6 eV, the ion assisted etch rate was higher than the chemical etch component. Pederson (1982) concluded that aromatic PIs etched at a slower rate than aliphatic ones and that the etch rates are roughly in accord with the susceptibility of polymers to chain scission by high energy radiation. Wrobel et al. (1988) concluded that the etch rates of PIs, polyamides, epoxy, and polyesters correlated strongly with the degree of unsaturation in the polymer structure. Tepermeister and Savin (1991) monitored the etch products, in situ, when etching a PMDA-ODA PI in Ar, O_2, and O_2/F_2 mixtures. In pure Ar and O_2 plasmas, acetylene was a major product. Addition of 2.5% of F_2 to an O_2 plasma increased the etch rate; however increasing the concentration of F_2 tended to passivate the surface and decreased the etch rates. In their patent, Babu et al. (1991) disclosed that pulsing between O_2 and CF_4 plasmas maintained high etch rates. Goldstein and Kalk (1981) showed that in the absence of inorganic masks, the surface finish of the polymers etched in an O_2 plasma was smooth. Iglitto et al. (1992) studied the etching of poly (tetrafluoroethylene) in O_2-CF_4 mixtures in RF plasmas and concluded that polymer radical generation by ion bombardment is a likely step for initiating etching; the maximum etch rate was reached in pure O_2 plasmas, but adding F_2 reduced surface roughness.

Wet Etching

Polyimides can be etched using caustic solutions prior to being fully imidized. After etching, the PI layer is fully cured; this results in a tapered opening. Wet processes are not favored in chip applications because of the need to maintain small dimensions.

If used at all, wet etching is practiced in printed circuit board technology.

4.3 OTHER ORGANIC DIELECTRICS

Mills et al. (1995) reported that benzocyclobutene (BCB) thermosetting polymer has desirable interlevel dielectric properties. Fluorinated poly(arylethers), available commercially as FLARE™ are synthesized using a condensation reaction similar to PI and have been proposed for interlevel dielectric applications (Hendricks et al., 1995). Vacuum deposition of polymers is increasingly being pursued; some of the materials are fluorinated parylene (Wary, 1996), Teflon (Singh and Sharangapani, 1996), and other fluorocarbon films (Takeishi et al., 1996). The interest in these materials is due to their low values of ε. It must still be proven that they satisfy the other requirements for use as interlevel insulators. Those who are interested in these materials are urged to consult the references for further details.

5.0 CONCLUDING REMARKS

Deposited SiO_2, undoped or doped with P, B, or both, has been a reliable component of semiconductor devices since the earliest days of integrated circuit fabrication. Polyimides have also been used, but to a limited extent.

However, reduction of ε, below that of SiO_2, is one of the forces fueling development of new dielectric materials. One important question is: just how low must it be? The answer is tied to the choice, and thus the resistivity (ρ), of the metallization system. Will it be Al alloy or (cladded) Cu? Both ρ and ε are factors in determining the signal delay. Analysis of the components of circuit delay for a given family of chips and the role of circuit design are required before a reasoned decision about the requirement for ε can be made.

Assuming that the need for the lowest possible value of ε is urgent, what materials meet this requirement? Although there have been reports of SiOFs with $\varepsilon = 2.3$, it is not a common result. The very low values of ε (~2) are reported more frequently for fluorinated polymers, which might signal a movement toward organic dielectrics. But there are other requirements that must be satisfied in addition to a low ε.

Most of the chemical and physical properties of SiO_2 are desirable and all are familiar. The deposition systems and basic processes are in place; this means there have been large investments on equipment and development. Thus, although the ultimate value of ε may be higher in the case of SiOF films than in the fluorinated polymers, there may be forces compelling the continued use of an inorganic dielectric. For one, there is the return on the investment already made; for another there is the uncertainty of venturing into the (almost) unknown world of the polymer dielectrics. The price for staying with the inorganic dielectric, however, may be greater than making relatively minor changes involved in doping oxides with F. Major process development may be required to insure film stability and reproducibility, while driving down ε to some "acceptable" value. But are there long-term benefits in trying to push the inorganic insulator technology if the lower limit in the value of ε of a stable SiOF film is too high?

The claims made for low ε inorganic alternatives to oxide, the B-N-based materials, have not stood up to further investigation.

The value of ε in polymer films is lower than that in SiO_2 films. Fluorinating the polymers decreases the value of ε further. The fluorinated polymer films appear to be less sensitive to F-related instability than are the F-doped oxides. But continued material development of polymers is required to overcome deficiencies common to these materials, such as susceptibility to thermal degradation during processing, greater permeability to moisture and metal migration, higher leakage current, higher levels of ionic contamination, anisotropy, high TCE, poor heat conductivity, etc. Some properties, such as low TCE and anisotropy may be linked. Singer (1996) has provided a status review and issues related to the search for low ε dielectric.

One of the attractions of the polymeric materials had been what was perceived as the low cost of spin-coating; but switching to spin-coating of the interlevel dielectric will require investment in the additional equipment such as coating systems, furnaces for curing, etc.

The manufacturers of equipment for oxide deposition and etching have been very active in process and material development, collaborating with the users and probably reducing the cost to them in the long run. But the vendors are working in a market with known customers. Will there be the incentive for commercial formulation of advanced polymers if their use, cost, and chance of success are uncertain? But work must start on new approaches in advance of their being *essential* because of the long time needed to solve the problems inevitably encountered with new materials and processes. Will the challenge therefore have to be met by the individual chip manufacturers? Will they, or can they afford to? Can they afford to ignore it?

The use of aerogels, polymer foams, and air bridges are all low ε possibilities that have been suggested, but there is no evidence that they can be implemented successfully.

Polymer films, both spin-coated and vacuum deposited ones, and PECVD SiOF films fill gaps well; although spun-on films have a better tendency toward local planarization than do oxides, global planarization (as discussed in a later chapter), and thus the use of chemical-mechanical planarization (CMP), is deemed necessary. Development of CMP of organics is behind that of oxides.

The jury is out. There is great reluctance to change. Is it a matter of better the devil you know (if not completely) than the devil about whom you know so little? Whether decisions can be made, without yielding to prejudice, is an interesting question.

REFERENCES

Abachev, M. K., Y. P. Baryshev, V. F. Lukichev, A. A. Orlikossky, K. A. Valiev, *Vacuum*, 42, 129 (1991).
Adachi, H., E. Adachi, Y. Aiba, H. Kandegae, *ECS Ext. Abstr. 302, PV 90-2*, 436 (1990).
Adams, A. C., F. B. Alexander, C. D. Capio, T. E. Smith, *J. Electrochem. Soc.*, 128, 1545 (1981).
Adams, R. V., R. W. Douglas, *J. Glass Technol.*, 43, 147 (1959).
Adams, A. C., C. D. Capio, *J. Electrochem. Soc.*, 126, 1042 (1979).
Ahlburn, B. T., G. A. Brown, T. R. Seha, T. F. Zoes, Y. Yokose, D. S. Ballance, K. A. Scheibert, *1995 DUMIC*, 36 (1995).
Ahlburn, B., R. Nowak, M. Galiano, J. Olsen, *ULSI Science and Technol.* (J. M. Andrews & G. K. Celler, eds.) *ECS PV 91-11*, 617 (1991).
Ahn, J., G. Q. Lo, W. Ting, D. L. Kwong, J. Kuehne, C. W. Magee, *Appl. Phys. Lett.*, 58, 425 (1991).
Ahne, H., H. Kruger, E. Pammer, R. Rubner, *Polyimides*, (K. L. Mittal, ed.), Plenum Press, New York, 1984, p. 905.
Alt, L. L., S. W. Ing, Jr., K. W. Laendle, *J. Electrochem. Soc.*, 110, 465 (1963).
Andeen, C., D. Schuele, J. Fontanella, *J. Appl. Phys.*, 43, 1071 (1974).
Andosca, R. G., W. J. Varhue, E. Adams, *J. Appl. Phys.*, 72, 1126 (1992).
Arleo, P., Henri, J., Hills, G., Wong, J., *US Pat 5,176,790* (1993).
Ashwell, G.W.B., S. J. Wright, *1985 VMIC*, 285 (1985).
Auman, B. C., *1995 DUMIC*, 297 (1995).
Babu, S. V., Joffarth, A. R. Knoll, W. E. Mlynko, J. F. Rembetski, K. D. Mack, *US Pat. 5,053,104* (1991).
Balk, P., J. M. Eldridge, *Proc. of IEEE*, 57, 1568 (1969).
Ballance, D. S., K. A. Scheibert, J. V. Tietz, *1992 VMIC*, 180 (1992).
Barbee, Jr., T. W., D. L. Keith, L. Nagel, W. A. Tiller, *J. Electrochem. Soc.*, 131, 439 (1984).
Batey, J., E. Tierney, *J. Appl. Phys.*, 60, 3136 (1986).
Bauer, H. J., *J. Vac. Sci. Technol.*, B12, 2405 (1994).
Bauer, L., E. Kodak, R. J. Farris, *I Proc. 3rd Int. Conf. on Polyimides*, SPE, Ellenville, NY (University Microfilms International, Ann Arbor, Mich.), p. 249, 1988.
Becker, F. S., S. Rohl, *J. Electrochem. Soc.*, 134, 2923 (1987).
Becker, F. S., D. Pawlik, H. Shafer, G. Staudigl, *J. Vac. Sci. Technol*, B4, 732 (1986).
Becker, F. S., D. Pawlik, H. Anzinger, A. Spitzer, *J. Vac. Sci. Technol.*, B5, 1555 (1987).
Bergeron, D. L., J. P. Kent, K. E. Morrett, *22nd IEEE/IRPS*, 1 (1984).
Beuhler, A. J., M. J. Burgess, D. E. Fjare, J. M. Caudette, R. T. Roginski, "Electronic Packaging Materials Science IV" (R. Jaccodine, K. A. Jackson, E. D. Lillie, R. S. Sundahl, eds.) *Mater. Res. Soc. Symp. Vol.*, 154, 73 (1989).
Bhide, V., J. M. Eldridge, *21st IEEE/IRPS*, 44 (1983).
Bhushan, B., S. P. Murarka, J. Gerlach, *J. Vac. Sci. Technol.*, B8, 1068 (1990).
Blain, S., L. Ouellet, Y. Tremblay, 1995 DUMIC, 111 (1995).
Blech, I., U. Cohen, *J. Appl. Phys.* 53, 4202 (1982).
Bogle-Rohwer, E., D. Gates, L. Hayler, H. Kurasaki, B. Richardson, *Solid State Technol.*, 4/85, 251 (1985).
Bondur, J. A., R. G. Frieser, *Plasma Processing* (R. G. Frieser, C. J. Mogab, eds.) *Electrochem. Soc. Proc. Vol. PV 81-1*, 180 (1981).
Brown, G. A., *Proc. 19th IEEE/IRPS*, 282 (1981).
Bunshah, R. F., *Thin Solid Films*, 80, 255 (1981).
Burkstrand, J. M., *J. Vac. Sci. Technol.*, 16, 864 (1979a).
Burkstrand, J. M., *Phys. Rev. B 20*, 4853 (1979b).
Burkstrand, J. M., *J. Appl. Phys.* 52, 4795 (1981).

Carl, D., D. Mordo, B. Sparks, M. Logan, J. Ritter, *1995 DUMIC*, 234 (1995a).
Carl, D., S. Schuchmann, M. Kilgore, R. Swope, W. van den Hoek, *1995 VMIC*, 97 (1995b).
Castellano, R. N., *Solid State Technol.*, 5/84, 203 (1984).
Cha, H. J., J. Hedrick, R. A. DiPietro, T. Blume, R. Beyers, D. Y. Yoon, *Appl. Phys. Lett.*, 68, 1930 (1996).
Chang, C. P., C. S. Pai, J. J. Hseigh, *J. Appl. Phys.*, 67, 2119 (1990).
Chang, C. Y., J. P. McVittie, J. Li, K. C. Saraswat, S. E. Lassig, J. Dong, *IEDM 93*, 853 (1993).
Chang, H. K., W. M. Shen, J. Yu, *IEEE Conf. on Electrical Insulators and Dielectric Phenomena*, 108 (1982).
Chang, C.-P., D. L. Flamm, D. E. Ibbotson, J. A. Mucha, *J. Vac. Sci. Technol.*, B6, 524 (1988).
Chang, R.P.H., B. Darack, E. Lane, C. C. Chang, D. Allara, E. Ong, *J. Vac. Sci. Technol.*, B1, 935 (1983).
Chapple-Sokol, J. D., W. A. Pliskin, R. A. Conti, E. Tierney, J. Batey, *J. Electrochem. Soc.*, 138, 3723 (1991).
Charles, C., G. Giroult-Matlasowski, R. W. Boswell, A. Goullet, G. Turban, C. Cardinaud, *J. Vac. Sci. Technol.*, A11, 2954 (1993).
Chau, T. T., S. R. Mejia, K. C. Kao, *Can. J. Phys.*, 69, 165 (1991).
Chebi, R., D. Webb, J. Draina, S. Mittal, *ECS, PV 92-6*, 353 (1992).
Chelkowski, A., *Dielectric Physics*, Elsevier Scientific Publishing Co., Amsterdam, 1980.
Chen, L., G. S. Mathad, *US Pat. 4,671,849*, 1987.
Chen, L.-J., S.-T. Hara, J.-L. Lau, 1994 VMIC, 81 (1994).
Chen, S. T., C. H. Yang, F. Faupel, P. S. Ho, *J. Appl. Phys.* 64, 6690 (1988).
Chen, S. T., F. Faupel, P. S. Ho, *Proc. 2nd ASM Int. Conf. on Microelectronic Packaging Technology* (W. T. Shieh, ed.), American Society for Metals, Metals Park, Ohio), p. 345, 1989.
Chen, S. T. in "Low-Dielectric Constant Materials-Synthesis and Applications in Microelectronics" (T.-M. Lu, S. P. Murarka, T.-S. Kuan, C. H. Ting, eds.), *Mat. Res. Soc. Symp. Proc. Vol.*, 381, 141 (1995).
Chiang, C., N. V. Lam, N. Chu, D. Cox, D. Fraser, J. Bozarth, B. Mumford, *1987 VMIC*, 404 (1987).
Chiang, C., D. B. Fraser, *1989 VMIC*, 397 (1989).
Chiang, C., D. B. Fraser, *ULSI/1989* (C. M. Osburn, J. M. Andrews, eds.) *Electrochem. Soc. Proc. Vol. PV 89-9*, 552 (1989).
Chiang, C., K. Yoshioka, N. Cox, J. Ren, D. B. Fraser, J. Sisson, T. O. Curtis, L. Bartholomew, *1992 VMIC*, 115 (1992).
Chien, F. C., R. L. Brown, G. N. Burton, M. B. Vora, *1984 VMIC*, 45 (1984).
Chin, B. L., E. P. van de Ven, *Solid State Technol.*, 4/88, 119 (1988).
Chou, J.-S., S.-C. Lee, *J. Electrochem. Soc.*, 141, 3214 (1994).
Chou, N. J., C. H. Tang, *J. Vac. Sci. Technol. A2*, 751 (1984).
Chu, W.-K., G. C. Schwartz, unpublished (1976).
Chung, H.W.M., S. K. Gupta, T. A. Baldwin, *1989 VMIC*, 373 (1989).
Chung, H., S. Wong, S. Lin, *1991 VMIC*, 376 (1991).
Classen, W.A.P., W.G.J.N. Valkenberg, F.H.P.M. Habrakenm, Y. Tamminga, *J. Electrochem. Soc.*, 130, 2419 (1983).
Claasen, W.A.P., W.G.J.N. Valkenberg, M.F.C. Willemsen, W.M.v.d. Wijgert, *J. Electrochem. Soc.*, 132, 893 (1985).
Claasen, W.A.P., *Plasma Chemistry and Plasma Processing*, 7, 109 (1987).
Coburn, J. W., H. F. Winters, *J. Vac. Sci. Technol.*, 16, 391 (1979).
Coburn, J. W., H. F. Winters, *Appl. Phys. Lett.*, 55, 2730 (1989).
Cottler, T. J., J. Chapple-Sokol, *J. Electrochem. Soc.*, 140, 2071 (1993).

Cramer, J. K., S. P. Murarka, *J. Appl. Phys.*, 77, 3048 (1995).
Davis, R. J., *Appl. Phys. Lett.*, 59, 1717 (1991).
Day, D. R., S. D. Senturia, *Polyimides* (K. L. Mittal, ed.) Plenum Press, New York, p. 249, 1984.
Delfino, M., J. A. Fair, S. Salimian, *Appl. Phys. Lett.*, 60, 341 (1992).
Denison, D. R., C. Chiang, D. B. Fraser, *ULSI/1989* (C. M. Osburn and J. M. Andrews, eds.), *Electrochem. Soc. Proc. Vol.*, PV 89-9, 563 (1989).
Denton, D. D., D. R. Day, D. E. Fiore, S. D. Senturia, E. S. Anolick, D. Schneider, *J. Electronic Material*, 14, 119 (1985).
deSilva, Jr., E. F., Y. Nishioka, T.-P. Ma, *IEEE Trans. on Nuclear Sci.* NS-34, 1190 (1987).
Dobkin, D. M., *J. Electrochem. Soc.*, 139, 2573 (1992).
Dobson, C. D., A. Kiermasz, K. Beekman, R. J. Wilby, *Semiconductor Intl.* 12/94, 85 (1994).
Dohmae, S.-I., J. P. McVittie, J. C. Rey, E.S.G. Shaqfeh, V. K. Singh, in "Patterning Sci. and Technol. II and Interconnection, and Contact Metallization for ULSI" (W. Green, G. J. Hefferon, L. K. White, T. O. Herndon, A. L. Wu, eds.), *Electrochem. Soc. Proc. Vol.*, PV 92-6, 163 (1992).
Duffy, M. T., R. A. Soltis, A. Day, *Electrochem. Soc. Ext. Abstr.* 165, PV 83-1, 265 (1983).
Edwards, *US Patent 3,179,634*, 1965.
Egitto, F. D., L. J. Matienzo, H. B. Schreyer, *JVST A 10*, 3060 (1992).
Emiliani, G., S. Scaglione, *J. Vac. Sci. Technol.*, A5, 1824 (1987).
Ephrath, L. M., *J. Electrochem. Soc.*, 126, 1419 (1979).
Ephrath, L. M., E. J. Petrillo, *J. Electrochem. Soc.*, 129, 3282 (1982).
Falcony, C., A. Ortiz, S. Lopez, J. C. Alonso, S. Muhl, *Thin Solid Films*, 199, 269 (1991).
Faupel, F., C. H. Yang, S. T. Chen, P. S. Ho, *J. Appl. Phys.* 65, 1911 (1989a).
Faupel, F., D. Gupta, B. Silverman, P. S. Ho, *App. Phys. Lett.*, 55, 357 (1989b).
Flamm, D. L., C. J. Mogab, E. R. Sklaver, *J. Appl. Phys.*, 50, 6211 (1979).
Forester, L., A. L. Butler, G. Schets, *1989 VMIC*, 72 (1989).
Forester, L., W. Doedel, Kosinski, W. Heesters, *1990 VMIC*, 28 (1990).
Fortuno, G., *J. Vac. Sci. Technol.*, A4, 744 (1986).
Fujino, K., Y. Nishimoto, T. Tokumasu, K. Maeda, *1990 VMIC*, 187 (1990a).
Fujino, K., Y. Nishimoto, T. Tokumasu, K. Maeda, *J. Electrochem. Soc.*, 137, 2883 (1990b).
Fujino, K., Y. Nishimoto, T. Tokumasu, K. Maeda, *J. Electrochem. Soc.*, 138, 550 (1991a).
Fujino, K., Y. Nishimoto, T. Tokumasu, K. Maeda, *1991 VMIC*, 445 (1991b).
Fujino, K., Y. Nishimoto, T. Tokumasu, K. Maeda, *J. Electrochem. Soc.*, 138, 3019 (1991c).
Fujino, K., Y. Nishimoto, T. Tokumasu, K. Maeda, *J. Electrochem. Soc.*, 138, 3727 (1991d).
Fujino, K., Y. Nishimoto, T. Tokumasu, K. Maeda, *J. Electrochem. Soc.*, 139, 1690 (1992a).
Fujino, K., Y. Nishimoto, T. Tokumasu, K. Maeda, *J. Electrochem. Soc.*, 139, 2282 (1992b).
Fujino, K., Y. Nishimoto, T. Tokumasu, K. Maeda, *Electrochem. Soc. Ext. Abstr.* 279, PV 92-2, 395 (1992c).
Fujino, K., Y. Nishimoto, T. Tokumasu, S. Fisher, K. Maeda, *1993 VMIC*, 96 (1993).
Fujita, S., H. Toyoshima, T. Oshishi, A. Sasaki, *Jpn. J. Appl. Phys.* 23, L144 (1984).
Fujita, S., T. Ohishi, H. Yoyoshima, A. Sasaki, *J. Appl. Phys.*, 57, 426 (1985).
Fujita, S., A. Sasaki, *J. Electrochem. Soc.*, 135, 2566 (1988).
Fujiwara, H., K. Fujimoto, H. Araki, Y. Tobinaga, *SPIE Vol. 1089*, 348 (1989).
Fukada, T., K. Suzuki, S. Takahashi, Y. Mochisuki, M. Ohue, N. Momma, T. Sonobe, *Jpn. J. Appl. Phys.*, 27, L1962 (1988).
Fukada, T., K. Saito, M. Ohue, K. Shima, N. Momma, *IEDM 92*, 285 (1992).
Fukada, T., T. Akahori, *Ext. Abstr. 1993 Intl. Conf. on Solid State Devices and Materials*, 158 (1993).
Fukada, T., T. Akahori, *1995 DUMIC*, 43 (1995).
Fuller, C. R., S. S. Baird, *Electrochem. Soc. Spring Mtg., Ext. Abstr.* 65, 17 (1963).
Gaillard, F., P. Brouquet, A. Kiermasz, K. Beekman, C. Dobson, *1996 DUMIC*, 124 (1996).

Galernt, B., *Semiconductor Intl. 3/90*, 82 (1990).
Galiano, M., E. Yieh, S. Robles, B. C. Nguyen, *1992 VMIC*, 100 (1992).
Giffen, L., J. Wu, R. Lachenbruch, G. Fior, *Solid State Technol.*, *4/89*, 55 (1989).
Ginsburg, R., J. R. Susko, *Polyimides* (K. L. Mittal, ed.) Plenum Press, NY, 1984, p. 573.
Gokan, H., A. Morimoto, M. Murahata, *Thin Solid Films*, *149*, 85 (1987).
Gottscho, R. A., C. W. Jurgensen, D. J. Vitkavage, *J. Vac. Sci. Technol.*, *B10*, 2133 (1992).
Greenblatt, J., C. J. Araps, H. R. Anderson, Jr., *Polyimides* (K. L. Mittal, ed.) Plenum Press, New York, p. 573, 1984.
Grosewald, P., L. V. Gregor, R. Powlus, *Proc. Int. Electron Devices Mtg.*, Washington, D.C., Paper 3.7 (1971).
Gupta, V. B., I. M. Ward, *J. Macromol. Sci. B2*, 89 (1968).
Hadley, D. W., Structure and Properties of Oriented Polymers (I. M. Ward, ed.) Appl. Science Publishers, Barking, UK, 1975, Chap. 9.
Haight, R., R. C. White, B. D. Silverman, P. S. Ho, *J. Vac. Sci. Tech. A6*, 2188 (1988).
Harada, H., I. Kato, T. Takada, K. Inayoshi, *Electrochem. Soc. Ext. Abstr. 188, PV 90-1*, 285 (1990).
Harrus, A. S., B. van Schravendijk, J. Park, E. van de Ven, *Electrochem. Soc. Ext. Abstr. 228, PV 91-2*, 322 (1991).
Hashimoto, S. C., K. Machida, H. Oikawa, *J. Vac. Sci. Technol.*, *B8*, 529 (1990).
Hazuki, Y., T. Moriya, *IEEE Trans. on Electron Devices*, *ED-34*, 628 (1987).
Heacock, I. F., C. E. Berr, *SPE Trans. 5*, 105 (1965).
Heidenreich, J. E., J. R. Paraszczak, M. Moisan, G. Sauve, *Microelectronic Engineering 5*, 363 (1986).
Heinecke, R. A., *Solid State Electronics*, *18*, 1146 (1975).
Heinecke, R. A., *US Pat. 3,940,506*, 1976.
Hemandez, M. J., J. Garrido, J. Piqyueras, *J. Vac. Sci. Technol.*, *B12*, 581 (1994).
Hendricks, N. H., B. Wan, A. Smith, *1995 DUMIC*, 283 (1995).
Herak, T. V., T. T. Chau, D. J. Thomson, S. R. Mejia, D. A. Buchanan, K. C. Kao, *J. Appl. Phys.*, *65*, 2457 (1989).
Herak, T. V., D. J. Thomson, *J. Appl. Phys.*, *67*, 6347 (1990).
Hess, D. W., *J. Vac. Sci. Technol.*, *A2*, 244 (1984).
Hey, H. P. W., B. G. Sluijk, D. G. Hemmes, *Solid State Technol.*, *4/90*, 139 (1990).
Hirao, T., K. Setsune, M. Kitagawa, T. Kamada, K. Wasa, K. Tsukamoto, T. Izumi, *Jpn. J. Appl. Phys.*, *27*, 30 (1988).
Hirashita, N., I. Aikawa, T. Ajioka, M. Kobayakawa, F. Yokoyama, Y. Sakaya, *28th IEEE/IRPS*, 216 (1990).
Hirose, M., *Mat. Sci. and Engr.*, *B1*, 213 (1988).
Hisada, M., S. Nakamura, A. Hosoki, "Proc. 10th Symp. Plasma Processing" (G. S. Mathad and D. W. Hess, eds.) *Electrochem. Soc. Proc. Vol.*, *PV 94-20*, 320 (1994).
Hochberg, A. K., D. L. O'Meara, *J. Electrochem. Soc.*, *136*, 1843 (1989).
Hollahan, J.R., *J. Electrochem. Soc. 126*, 930 (1979).
Homma, Y., S. Tunekawa, *J. Electrochem. Soc.*, *135*, 2557 (1988).
Homma, T., T. Katoh, Y. Yamada, J. Shimizu, Y. Murao, *1990 Symp. VLSI Technol.*, 3 (1990).
Homma, T., T. Katoh, Y. Yamada, J. Shimuzu, Y. Murao, *NEC Res. and Develop.*, *32*, 315 (1991).
Homma, T., R. Yamaguchi, Y. Murao, *J. Electrochem. Soc.*, *140*, 687 (1993a).
Homma, T., Y. Murao, *J. Electrochem. Soc.*, *140*, 2046 (1993b).
Homma, T., T. Katoh, Y. Yamada, Y. Murao, *J. Electrochem. Soc.*, *140*, 2410 (1993c).
Homma, T., M. Suzuki, Y. Murao, *J. Electrochem. Soc.*, *140*, 3591 (1993d).
Homma, T., Y. Murao, R. Yamaguchi, *J. Electrochem. Soc.*, *140*, 3599 (1993e).
Horiike, Y., T. Ichihara, H. Sakaue, *Appl. Surf. Sci.*, *46*, 168 (1990).
Hosada, Y., H. Harada, H. Ashida, K. Watanabe, *1992 VMIC*, 121 (1992).

Hosokawa, N., R. Matsuzaki, T. Asamaki, *Jpn. J. Appl. Phys.*, *8*, Suppl. 2, Pt. 1, 435 (1974).
Houskova, J., K.-K.N. Ho, M. J. Balazs, *Semiconductor Intl.*, *5/85*, 236 (1985).
Hsieh, J., H. teNijenhuis, D. Mordo, R. Swope, W. S. Yoo, S. Schuchmann, F. Nagy, *1996 DUMIC*, 265 (1995).
Huang, I.-W., T. W. Bril, D. Bernard, B. Westland, *Electrochem. Soc. Ext. Abstr.* 396, PV 84-2, 567 (1984).
Huang, J., K. Kwok, D. Witty, K. Donohoe, *J. Electrochem. Soc.*, *140*, 1682 (1993).
Huo, D. T. C., M. F. Yan, P. D. Foo, *J. Vac. Sci. Technol.*, *A9*, 2602 (1991).
Ibbotson, D. E., J. A. Mucha, J. J. Hsieh, D. L. Flamm, *Electrochem. Soc. Ext. Abstr.* 131, PV *90-1*, 192 (1990).
Ikeda, Y., K. Kishimoto, K. Hirose, Y. Numasawa, *IEDM 92*, 289 (1992).
Ing, Jr. S. W., S. W., W. Davern, *J. Electrochem. Soc.*, *112*, 284 (1965).
Ip, F. S., C. Ting, "Low-Dielectric Constant Materials-Synthesis and Applications in Microelectronics" (T.-M. Lu, S. P. Murarka, T.-S. Kuan, C. H. Ting, eds.), *Mat. Res. Soc. Symp. Proc. Vol.*, *381*, 135 (1995).
Ito, S., Y. Homma, E. Sasaki, S'i. Uchimura, H. Morishima, *J. Electrochem. Soc.*, *137*, 1212 (1990).
Ito, S., Y. Homma, E. Sasaki, *J. Vac. Sci. Technol.*, *A9*, 2696 (1991).
Ito, T., T. Nozaki, H. Ishikawa, *J. Electrochem. Soc.*, *127*, 2053 (1980).
Itoh, N., K. Kato, I. Kato, *Electronics and Communications in Japan*, Part 2, *74*, 101 (1991).
Jillie, D., P. Freiberger, T. Blaisdell, J. Multani, *J. Electrochem. Soc.*, *134*, 1988 (1987).
Jin C., L. King, K. Taylor, T. Seha, J. D. Luttmer, 1996 *DUMIC*, 21 (1996).
Jones, H. C., R. Bennett, J. Singh, in Proc. of 8th Symp. on Plasma Processing (G. S. Mathad and D. W. Hess, eds.), *Electrochem. Soc. Proc. Vol PV 90-14*, 45 (1990).
Jones, R. E., C. L. Standley, L. I. Maissel, *J. Appl. Phys.*, *38*, 4656 (1967).
Joyce, R. J., H. F. Sterling, J. H. Alexander, *Thin Solid Films*, *1*, 481 (1967/1968).
Kanicki, J., P. Wagner, in "Silicon Nitride and Silicon Dioxide Thin Insulating Films" (V. J. Kapoor and K. T. Hankins, eds.), *Electrochem. Soc. Proc. Vol. PV 87-10*, 261.
Karim, M. Z., D. R. Evans, *1996 DUMIC*, 63 (1996).
Kennedy, T. N., *J. Vac. Sci. Technol.*, *13*, 1135 (1976).
Kern, W., A. W. Fisher, *RCA Review*, *25*, 715 (1970).
Kern, W., R. C. Heim, *J. Electrochem. Soc.*, *117*, 562 (1970).
Kern, W., R. C. Heim, *J. Electrochem. Soc.*, *117*, 568 (1970).
Kern, W., S. Rosler, *J. Vac. Sci. Technol.*, *14*, 108 (1977).
Kern, W., G. L. Schnable, *RCA Review 37*, 3 (1982).
Kern, W., R. K. Smeltzer, *Solid State Technol.*, *6/85*, 171 (1985).
Kiermasz, A., C. D. Dobson, K. Beekman, A. H. Bar-Ilan, *1995 DUMIC*, 94 (1995).
Kikkawa, T., N. Endo, *J. Appl. Phys.*, *71*, 958 (1992).
Kim, J., S. P. Kowalczyk, Y. H. Kim, N. J. Chou, T. S. Oh, *Proc. Adv. Electronics Packaging Symp.* (A. T. Barfknecht, J. P. Partridge, C. J. Chen, C. Y. Li, eds.), *Mat. Res. Soc. Symp. Proc. Vol. 167*, 137 (1990).
Kirov, K. J., S. S. Georgiev, E. V. Gerova, S. P. Aleksandrova, *Phys. Stat. Sol.*, *48*, 609 (1978).
Kojima, H., T. Iwamore, Y. Sakata, T. Yamashita, Y. Yatsuda, *1988 VMIC*, 390 (1988).
Kortlandt, J., L. Oosting, *Solid State Technol.*, *10/82*, 153 (1982).
Kotani, H., M. Matsuura, A. Fujii, H. Genjou, S. Nagao, *IEDM 89*, 669 (1989).
Kudoh, H., T. Yoshida, M. Fukumoto, T. Ohzone, *J. Electrochem. Soc.*, *133*, 1666 (1986).
Kwok, K., E. Yieh, S. Robles, B. C. Nguyen, *J. Electrochem. Soc.*, *141*, 2172 (1994).
Landheer, D., N. G. Skinner, T. E. Jackman, D. A. Thompson, J. G. Simmons, V. Stevanovic, D. Khatamian, *J. Vac. Sci. Technol.*, *A9*, 2594 (1991).
Lassig, S. E., "Proc. 10th Symp. Plasma Processing" (G. S. Mathad, D. W. Hess, eds.), *ECS PV 94-20*, 546 (1994).

Lassig, S., K. Olsen, W. Patrick, *1993 VMIC*, 122 (1993).
Law, K., J. Wong, C. Leung, J. Olsen, D. Wang, *Solid State Technol.*, 4/89, 60 (1989).
Lazaridis, C. N., J. H. Chu, p. 236 in *Proc. 3rd Du Pont Symp. on High Density Interconnection Technology* (C. C. Schukert, ed.) DuPont, 1990.
Lee, Y. H., Z. H. Zhou, in "Proc. 8th Symp. Plasma Processing" (G. S. Mathad, D. W. Hess, eds.), *Electrochem. Soc. Proc. Vol.*, PV 90-14, 34 (1990).
Lee, Y. H., Z. H. Zhou, *J. Electrochem. Soc.*, *138*, 2439 (1991).
Lee, P., M. Galliano, P. Keswick, J. Wong, B. Shin, D. Wang, *1990 VMIC*, 396 (1990).
Lee, J. G., S. H. Choi, T. C. Ahn, C. G. Hong, P. Lee, K. Law, M. Galiano, P. Keswick, B. Shin, *Semiconductor International*, 5/92, 116 (1992).
Lee, C. J., *Proc. 39th IEEE Electronic Components Conference*, 896 (1989).
Lee, Y. K., J. D. Craig, *ACS Organic Coating Preprints 43*, 451 (1980).
LeGoues, F. K., B. D. Silverman, P. S. Ho, *JVST*, *A6*, 2200 (1988).
Levy, R. A., P. K. Gallagher, F. Schrey, *J. Electrochem. Soc.*, *134*, 430 (1987).
Lifshitz, N., G. Smolinsky, J. M. Andrews, *J. Electrochem. Soc.*, *136*, 1440 (1989).
Lifshitz, N., G. Smolinsky, *J. Electrochem. Soc.*, *136*, 2335 (1989).
Light, R. W., H. B. Bell, *J. Elexctrochem. Soc.*, *130*, 1567 (1983).
Lii, Y.-J., J. Jorne, *J. Electrochem. Soc.*, *137*, 2837 (1990).
Lim, S. W., Y. Shimogaki, Y. Nakano, K. Tada, H. Komiyama, *Abstr. 1995 Intl. Conf. Solid State Devices and Materials*, 163 (1995).
Livengood, R. W., D. W. Hess, *Thin Solid Films*, *162*, 59 (1988).
Logan, J. S., F. S. Maddocks, P. D. Davidse, *IBM J. Res. Dev.*, *14*, 182 (1970).
Logan, J. S., J. M. Keller, R. G. Simmons, *J. Vac. Sci. Technol.*, *14*, 92 (1977).
Logan, J. S., J. Constable, F. Jones, J. E. Lucy, *J. Vac. Sci. Technol.*, *A8*, 1935 (1990).
Lorenz, H., I. Eisele, J. Ramm, J. Edlinger, M. Buhler, *J. Vac. Sci. Technol.*, *B9*, 208 (1991).
Macchioni, C. V., *J. Vac. Sci. Technol.*, *A8*, 1340 (1990).
Maeda, M., T. Makino, *Jpn. J. Appl. Phys.*, *26*, 660 (1987).
Maeda, M., *Jpn. J. Appl. Phys.*, *29*, 1789 (1990).
Maeda, M., Y. Arita, *Jpn. Soc. Appl. Phys.*, 37th Spring Meeting, 630 (1990a).
Maeda, M., Y. Arita, *Jpn. Soc. Appl. Phys.*, 51st Fall Meeting, 663 (1990b).
Maeda, M., *Mat. Res. Soc. Symp. Proc. Vol.*, *284*, 457 (1993).
Maeda, K., J. Sato, *Denki Kagaku*, *45*, 654 (1977).
Maissel, L. I., R. E. Jones, C. L. Standley, *IBM J. Res. Dev.*, *14*, 176 (1970).
Malik, F., R. Solanski, *Thin Solid Films*, 193/194, 1030 (1990).
Martin, R. S., E. P. van de Ven, C. P. Lee, *1988 VMIC*, 286 (1988).
Maruyama, T., T. Shirai, *Appl. Phys. Lett.*, *63*, 611 (1993).
Matsuda, T., M. J. Shapiro, S. V. Nguyen, *1995 DUMIC*, 22 (1995).
Matsuura, M., Y. Hayashide, H. Kotani, H. Abe, *Jpn. J. Appl. Phys.*, *30*, 1530 (1991).
Matsuura, H., Y. Ii, K. Shibata, Y. Hayashide, H. Kotani, *1993 VMIC*, 113 (1993).
Matsuura, M., Y. Hayashide, H. Kotani, T. Nishimura, H. Iuchi, C. D. Dobson, A. Kiermasz, K. Beekmann, R. Wilby, IEDM94, 117 (1994).
Mautz, K., J. Dahm, R. Berglund, "Proc. the 10th Symp. Plasma Processing (G. S. Mathad and D. W. Hess, eds.), *Electrochem. Soc. Proc. Vol.*, PV 94-20, 340 (1994).
McBrayer, I. D., R. M. Swanson and T. W. Sigmon, *J. Electrochem. Soc. 133*, 1242 (1986).
McVittie, J. P., S.-i. Dohmae, in "Proc. the 9th Symp. Plasma Processing" (G. S. Mathad, D. W. Hess, eds.), *Electrochem. Soc. Proc.. Vol.*, PV 92-18, 11 (1992).
Meaudre, R., M. Meaudre, *J. Non-Crystalline Solids*, *46*, 71 (1981).
Meaudre, R., M. Meaudre, *Phys. Rev. B.*, *29*, 7014 (1984).
Milek, J, T., *Silicon Nitride for Microelectronic Applications, Preparation, and Properties*, IFI/Plenum, New York, 1971/1972.
Mills, M., M. Dibbs, S. Martin, P. Townsend, *1995 DUMIC*, 269 (1995).
Minowa, Y., K. Yamanishi, K. Tsukamoto, *J. Vac. Sci. Technol.*, *B1*, 1148 (1983).

Minowa, Y., H. Ito, *J. Vac. Sci. Technol.*, *B6*, 473 (1988).
Mitchell, C., R. Goodner, 1984 VMIC, 130 (1984).
Mizuno, S., A. Varma, H. Tran, P. Lee, B. Nguyen, in "ULSI Science and Technol./1995" (E. M. Middlesworth and H. Massoud, eds.), *Electrochem. Soc. Proc. Vol. PV 95-5*, 354 (1995a).
Mizuno, S., A. Verma, H. Tran, P. Lee, B. Nguyen, *1995 VMIC*, 148 (1995b).
Mocella, M. T., private communication (1996).
Mogab, C. J., A. C. Adams, D. L. Flamm, *J. Appl. Phys.*, *49*, 3796 (1978).
Mohindra, V., H. H. Sawin, M. T. Mocella, J. M. Cook, J. Flanner, O. Turmel, in "Proc. 10th Symp. Plasma Processing" (G. A. Mathad, D. W. Hess, eds.) *Electrochem. Soc. Proc. Vol., PV 94-20*, 300 (1994).
Moutsier, T. W., A. M. Schoepp, E. van de Ven, *Electrochem. Soc. Ext. Abstr. 485, PV 94-2*, 770 (1994).
Mukherjee, S. P., P. E. Evans, *Thin Solid Films*, *14*, 105 (1972).
Murakami, E., S.-i. Kimura, T. Warabisakko, K. Miyake, H. Sunami, *Ext. Abstr. 17th Conf. Solid State Devices & Materials*, 271 (1985).
Nagayama, H., H. Honda, H. Kawahara, *J. Electrochem. Soc.*, *135*, 2013 (1988).
Nagy, A. G., *J. Electrochem. Soc.*, *132*, 689 (1985).
Nakano, M., H. Sakaue, H. Kamamoto, A. Nagata, M. Hirose, Y. Horiike, *Appl. Phys. Lett.*, *57*, 1096 (1990).
Nakao, M., H. Kawamoto, A. Nagata, M, Hirose, Y. Horiike, *Ext. Abstr. 21st Conf. on Solid State Devices and Materials*, Tokyo, p. 49, 1989.
Nakasaki, Y., H. Hayasaka, "Ext. Abstr. 41st Spring Mtg.," *Jpn. Soc. Appl. Phys.*, p. 719 (1994) (in Japanese).
Nandra, S. S., *J. Vac. Sci. Technol.*, *A8*, 3179 (1990).
Nemetz, J. A., R. E. Tressler, *Solid State Technol.*, *2/83*, 79 (1983).
Neuhaus, H., Z. Feit, F. W. Smith, S. D. Senturia, *Proc. 2nd Conf. Polyimides, SPE, Ellenville, NY*, University Microfilms International, Ann Arbor, Mich., p. 152, 1985.
Nguyen, S. V., K. Albaugh, *J. Electrochem. Soc.*, *136*, 2835 (1989).
Nguyen, S. V., D. Dobuzinsky, R. Gleason, M. Gibson, *Electrochem. Soc. Ext. Abstr. 126, PV 92-2*, 209 (1992).
Nguyen, S. V., T. Nguyen, H. Treichel, O. Spindler, *J. Electrochem. Soc.*, *141*, 1633 (1994).
Nguyen, S., G. Freeman, D. Dobuzinsky, K. Kelleher, R. Nowak, T. Sahin, D. Witty, *1995 VMIC*, 69 (1995).
Nguyen, S., D. Dobuzinsky, D. Harmon, R. Gleason, S. Fridmann, *J. Electrochem. Soc.*, *137*, 2209 (1990).
Nishimoto, Y., N. Tokumasu, K. Maeda, *1995 DUMIC*, 15 (1995).
Nishimoto, Y., N. Tokumasu, T. Fukuyama, K. Maeda, *Ext. Abstr. 19th Conf. Solid State Devices & Materials*, Tokyo, 447 (1987).
Nishimoto, Y., N. Tokumasu, T. Fukuyama, K. Maeda, *1989 VMIC*, 382 (1989).
Noguchi, S., H. Okano, Y. Horiike, *Ext. Abstr. 19th Conf. Solid State Devices & Materials*, Tokyo, 451 (1987).
Nojiri, K., E. Iguchi, K. Kawamura, K. Kadota, *Ext. Abstr. 21st Conf. Solid State Devices & Materials*, Tokyo, 153 (1989).
Numata, S., T. Miwa, Y. Misawa, D. Makino, J. Imaizumi, N. Kinjo, *MRS Symp. Proc. 108*, 113 (1988).
Numata, S., K. Fujisaki, N. Kinjo, *Polyimides* (K. L. Mittal, ed.), Plenum Press, New York, 1984.
Numata, S., Fujisaki, N. Kinjo, J. Imaizumi, Y. Mikami, *US Pat. 4,690,999*, 1987.
Ohashi, N., H. Nezu, H. Maruyama, T. Fujiwara, H. Aoki, H. Yamaguchi, N. Owada, *1996 DUMIC*, 86 (1996).

Ohiwa, T., K. Horioka, T. Arikado, I. Hasegawa, H. Okano, *Jpn. J. Appl. Phys.*, *31*, 405 (1992).
Oikawa, A., S.-i. Fusuyama, Y. Yoneda, H. Harada, T. Takada, *J. Electrochem. Soc.*, *137*, 3223 (1990).
Ong, W., A. M. Nguyen, B. C. Nguyen, *Electrochem. Soc. Ext. Abstr. 293*, 1993, *PV 93-1*, 443.
Pai, P.-L., C. G. Konitzer, *1992 VMIC*, 213 (1992).
Pai, C. S., J. F. Miner, P. D. Foo, *J. Electrochem. Soc.*, *139*, 850 (1992).
Palmese, G. R., I. K. Gilham, *J. Appl. Polymer Sci.*, *34*, 1925 (1987).
Pan, P., L. A. Nesbit, R. W. Douse, T. Gleason, *J. Electrochem. Soc.*, 2012 (1985).
Parsons, G. N., J. H. Souk, J. Batey, *J. Appl. Phys.*, *70*, 1553 (1991).
Passemard, G., P. Fugier, P. Noel, F. Pires, O. Demolliens, *1996 DUMIC*, 145 (1996).
Patrick, J., G. C. Schwartz, J. D. Chapple-Sokol, K. Olsen, R. Carruthers, *J. Electrochem. Soc.*, *139*, 2604 (1992).
Patterson, J. D., M. C. Ozturk, *J. Vac. Sci. Technol.*, *B10*, 625 (1992).
Paulsen, W. M., R. W. Kirk, *12th IEEE/IRPS*, 172 (1974).
Pavelscu, C., I. Kleps, *Thin Solid Films*, *190*, L1 (1990).
Pearce, C. W., R. F. Fetcho, M. D. Gross, R. F. Koefer, R. A. Pudliner, *J. Appl. Phys.*, *64*, 1838 (1992).
Peccoud, L., J. Arroyo, P. Lassagne, M. Puech, *Electrochem. Soc. Ext. Abstr. 389*, *PV 84-2*, 554 (1984).
Pederson, L. A., *J. Electrochem. Soc.*, *129*, 205 (1982).
Peek, H. L., R. A. M. Wolters, 1986 VMIC, 165 (1986).
Pignatel, G. U., J. C. Sisson, W. P. Weiner, *J. Electrochem. Soc.*, *138*, 1723 (1991).
Pineda, R., C. Chiang, D. B. Fraser, 1990 VMIC, 180 (1990).
Pliskin, W. A., H. S. Lehman, *J. Electrochem. Soc.*, *112*, 1013 (1965).
Pliskin, W. A., P. P. Castrucci, *Electrochem. Technol.*, *6*, 85 (1968).
Pliskin, W. A., p. 506 *Semiconductor Silicon* (H. R. Huff, R. R. Bergess, eds.) Electrochemical Society, NJ, 1973.
Pliskin, W. A., *J. Vac. Sci. Technol.*, *14*, 1064 (1977).
Pramanik, D., V. Jain, K. Y. Chang, 1991 VMIC, 27 (1991).
Qian, L. Q., H. W. Fry, G. Nobinger, J. T. Pye, M. C. Schmidt, J. Cassillas, M. Lieberman, *1995 DUMIC*, 50 (1995).
Ramkumar, K., S. K. Ghosh, A. N. Saxena, *J. Electrochem. Soc.*, *140*, 2669 (1993).
Ramkumar K. A. N. Saxena, *J. Electrochem. Soc.*, *139*, 1437 (1992).
Rand, M. J., *J. Electrochem. Soc.*, *114*, 274 (1967).
Raupp, G. B., T. Cale, H. P. W. Hey, *J. Vac. Sci. Technol.*, *B10*, 37 (1992).
Reinberg, A. R., *US Pat 3,757,733*, 1973.
Rhoades, P., Halman, M., Kerr, D., *US Pat 5,269,879*, 1993.
Robles, S., M. Galiano, B. C. Nguyen, *Electrochem. Soc. Ext. Abstr. 129*, *PV 92-1*, 215 (1992).
Robles S., K. Russell, M. Galiano, V. Kithcart, V. Siva, B. C. Nguyen, *1995 VMIC*, 122 (1995).
Rojas, S., R. Gomarasca, L. Zanotti, A. Borghesi, A. Sassella, G. Ottaviani, L. Moro, P. Lazzeri, *J. Vac. Sci. Technol.*, *B10*, 633 (1992).
Rojas, S., L. Zanotti, A. Borghesi, A. Sassella, G. U. Pignatel, *J. Vac. Sci. Technol.*, *B11*, 2081 (1993).
Ross, R. C., *J. Vac. Sci. Technol.*, *A8*, 3175 (1990).
Rostworowski, J., R. R. Parsons, *J. Vac. Sci. Technol.*, *A3*, 491 (1985).
Rothman, L. B., J. L. MauerIV, G C. Schwartz, J. S. Logan, in "Plasma Processing," (R. G. Frieser, C. J. Mogab, eds.), *ECS*, *PV 81-1*, 193 (1981).
Rothman, L. B., *J. Electrochem. Soc.*, *127*, 2216 (1980).

Rucker, T., H.-Y. Lin, C. H. Ting, *Electrochem. Soc. Ext. Abstr. 137, PV 90-1*, 283 (1990).
Sachdev, K., J. P. Hummel, R. W. Kwong, R. N. Lang, L. Linehan, H. S. Sachdev. *US Pat 5,115,090*
St. Clair, Anne K., T. L. St. Clair, *US Patent 5,093,453* (1992).
Sacher, E., J. R. Susko, *J. Appl. Polymer Sci.*, 23, 2355 (1979).
Sacher, E., and J. R. Susko, *J. Appl. Polymer Sci.*, 26, 679 (1981).
Saia, R. J., B. Gorowitz, *J. Electrochem. Soc.*, 132, 1954 (1985).
Sakaue, H., T. Nakasako, K. Nakaune, T. Kusuki, A. Miki, Y. Horike, in "Amorphous Insulating Thin Films" (J. Kanicki, R. A. B. Devine, W. L. Warren, M. Matsumura, eds.), *Mat. Res. Soc. Symp. Proc. Vol.*, PV 284, 169 (1993).
Samuelson, G., p. 93 in *Polymer Materials for Electronics*, (E. D. Feit and C. W. Wilkins, Jr., eds.). Amer. Chem. Soc., Washington, D.C., 1982.
Saproo, A., D. R. Denison, J. Lam, *1996 DUMIC*, 239 (1996).
Sato, M., S.-c. Kato, Y. Arita, *Jpn. J. Appl. Phys.*, 30, 1549 (1991).
Sato, K., S. Harada, A. Saiki, T. Kimura, T. Okubo, K. Mukai, *IEEE Trans. Parts, Hybrids and Packaging PHP-9*, 176 (1973).
Satsu, Y., O. Miura, R. Watanabe, K. Miyazaki, *Trans. Inst. Electronics Communication Eng. C II* (Japan) J47C-II(6), 489 (1991).
Sawa, G., K. Iida, S. Nakamura, *IEEE Tran. Electrical Insulators, EI-15*, 112 (1980).
Schreiber, H.-U., E. Froschle, *J. Electrochem. Soc.*, 123, 30 (1976).
Schwartz G. C. and J. D. Chapple-Sokol, (1992) unpublished.
Schwartz, G. C., L. B. Rothman, T. J. Schopen, *J. Electrochem. Soc.*, 126, 464 (1979).
Schwartz, G. C., R. E. Jones, *IBM J. Res. Develop.*, 14, 52 (1970).
Secrist, D. R., J. D. Mackenzie, *J. Electrochem. Soc.*, 113, 914 (1966).
Serikawa, T., *Jpn. J. Appl. Phys.*, 19, L259 (1980).
Serikawa, T., T. Yachi, *J. Electrochem. Soc.*, 131, 210 (1984).
Serikawa, T., A. Okamoto, *J. Vac. Sci. Technol.*, A3, 1988 (1985).
Shacham-Diamand, Y., Y. Nachumovsky, *J. Electrochem. Soc.*, 137, 190 (1990).
Shapiro, M. J., T. Matsuda, S. V. Nguyen, *1995 DUMIC*, 118 (1995).
Shih, Y. C., C. S. Pai, K. G. Steiner, W. G. Wilkins, *1992 VMIC*, 109 (1992).
Shimogaki, Y., S. W. Lin, M. Miyata, Y. Nakano, K. Tada, H. Komiyama, *1996 DUMIC*, 36 (1996).
Shioya, Y., M. Maeda, *J. Electrochem. Soc.*, 133, 1943 (1986).
Shoda, N., P. Weigand, T. Matsuda, S. V. Nguyen, T. E. Jones, M. J. Shapiro, J. Rzuczek, *1996 DUMIC*, 13 (1996).
Simko, J. B., G. S. Oehrlein, *J. Electrochem. Soc.*, 138, 2748 (1991).
Singer, P., *Semiconductor Intl.*, 5/86, 88 (1996).
Singh, R., R. Sharangapani, *1996 DUMIC*, 78 (1996).
Sinha, A. K., H. J. Levinstein, T. E. Smith, G. Quintana, S. E. Haszko, *J. Electrochem. Soc.*, 125, 601 (1978).
Smith, D. L., A. S. Alimonda, *J. Electrochem. Soc.*, 140, 1496 (1993).
Smith, F. W., H. J. Neuhaus, S. D. Senturia, Z. Feit, D. R. Day, T. J. Lewis, *J. Electronic Materials*, 16, 93 (1987).
Smolinsky, G., N. Lifshitz, V. Ryan, *Mat. Res. Soc. Synp. Vol. 154*, 173 (1989).
Spindler, O., B. Neureither. *Thin Solid Films*, 175, 67 (1989).
Srikrishnan, K. V., S. T. Chen, J. P. Yang, p. 268, *Proc. 3rd DuPont Symp. High Density Interconnect Packaging* (C. J. Shuckert, ed.), DuPont, 1990.
Sroog, C. E. *J. of Polymer Science*, 11, 1 (1976).
Stadtmueller, M., *J. Electrochem. Soc.*, 139, 3669 (1992).
Stamper, A. K., S./L. Pennington, *J. Electrochem. Soc.*, 140, 1748 (1993).
Steinmeiner, W., J. Bloem, *J. Electrochem. Soc.*, 111, 206 (1964).
Stephens, A. W., J. L. Vossen, W. Kern, *J. Electrochem. Soc.*, 123, 303 (1976).

Sterling, H. F., R. C. G. Swann, *Solid-State Electronics*, 8, 653 (1965).
Stonnington, K. D., K. Y. Hsieh, L. L. H. King, K. J. Bachman, A. I. Kingon, *J. Vac. Sci. Technol.*, *A10*, 970 (1992).
Suyama, S., A. Okamoto, T. Serikawa, *J. Electrochem. Soc.*, *134*, 2260 (1987).
Suyama, S., A. Okamoto, T. Serikawa, *J. Electrochem. Soc.*, *135*, 3104 (1987).
Suzuki, S., N. Tokumasu, K. Maeda, *1996 DUMIC*, 95 (1996).
Swope, R., W. S. Yoo, J. Ksieh, H. teNijenhuis, *1996 DUMIC*, 295 (1996).
Takahashi, T., Y. Egashira, H. Komiyama, *Appl. Phys. Lett.*, *66*, 2858 (1995).
Takeishi, S., R. Shinorhara, H. Kudoh, A. Tsukune, Y. Satoh, H. Miyazawa, H. Harada, M. Yamada, *1995 DUMIC*, 257 (1995).
Takeishi, S., H. Kudoh, R. Shinohara, A. Tsukune, Y. Satoh, H. Miyazawa, H. Harada, M. Yamada, *J. Electrochem. Soc.*, *143*, 381 (1996).
Takeshi, S., H. Kudo, R. Shinohara, M. Hoshino, S. Fukuyama, J. Yamaguchi, M. Yamada, *1996 DUMIC*, 71 (1996).
Tamura, T., Y. Inoue, M. Satoh, H. Yoshitaka, J. Sakai, *1996 DUMIC*, 231 (1996).
Taylor, J. A., *J. Vac. Sci. Technol.*, *A9*, 2464 (1991).
Tepermeister, I., H. Sawin, JVST *A9*, 790 (1991).
Thakur, R., R. Iyer, B. Benard, S. Fisher, N. Tokumasu, K. Maeda, Y. Nishimoto, 1994 VMIC, 117 (1994).
Ting, C. H., H. Y. Lin, P. L. Pai, T. Rucker, *Electrochem. Soc. Ext. Abstr.* 257, 1988, PV 88-2, 366.
Ting, C. Y., V. J. Vivalda, H. G. Schaefer, *J. Vac. Sci. Technol.*, *15*, 1105 (1978).
Tompkins, H. G., C. Tracy, *J. Electrochem. Soc.*, *136*, 2331 (1989).
Tong, H. M., K. L. Saenger, G. W. Su, in "Proc. SPE Ann. Tech. Meeting, Montreal," *SPE-ANTEC Proc. 1991*, 1727 (1991).
Tong, J. E., K. Schertenleib, R. A. Carpio, *Solid State Technol.*, *1/84*, 161 (1984).
Treichel, H., O. Spindler, R. Braun, T. A. Brooks, R. Nowak, "Proc. 8th Symp. Plasma Processing (G. S. Mathad, D. W. Hess, eds.), Electrochem. Soc. Proc. Vol. PV 90-14, 574 (1990).
Tsu, D. V., G. Lucovsky, M. J. Mantini, *Phys. Rev. B.*, *33*, 7069 (1986).
Tu, T., A. Ku, J. Chu, K. C. Chen, W. Su, T. Chung, *1996 DUMIC*, 311 (1996).
Uoochi, Y., A. Tabuchi, Y. Furumura, *J. Electrochem. Soc.*, *137*, 3923 (1990).
Usami, T., K. Shimokawa, M. Yoshimaru, *Ext. Abstr. 1993 Intl. Conf. on Solid State Devices and Materials*, 1993, p. 161.
Usami, T., K. Shimokawa, M. Yoshimaru, *Jpn. J. Appl. Phys.*, *33*, 408 (1994).
Valetta, R. M. J. A. Perri, J. Riseman, *Electrochem. Technol.*, *4*, 402 (1966).
van Andel, M. A., W. F. M. Gootzen, in "Electronic Packaging Materials Science IV" (R. Jaccodine, K. A. Jackson, E. D. Lillie, R. C. Sundahl, eds.), Mat. Res. Soc. Proc. Vol., *154*, 183 (1989).
van de Ven, E. P., R. S. Martin, M. J. Berman, *1987 VMIC*, 434 (1987).
van de Ven, E. P., J. W. Connick, A. S. Harrus, *1990 VMIC*, 194 (1990).
van Schravendijk, A. S. Harrus, G. Delgado, B. Sparks, C. Roberts, *1992 VMIC*, 372 (1992).
Vender, D., G. S. Gottlieb, G. C. Schwarz, *J. Vac. Sci. Technol.*, *A11*, 279 (1993).
Veprek-Heijman, M. G. J., D. Boutard, *J. Electrochem. Soc.*, *139*, 2042 (1991).
Viswanathan, N. S., *J. Vac. Sci. Technol.*, *16*, 388 (1979).
Vossen, J. L., *J. Vac. Sci. Technol.*, *8*, S12 (1971).
Wang, D. N., J. M. White, K. S. Law, C. Leung, S. P. Umotoy, K. S. Collins, J. A. Adamik, I. Perlov, D. Maydan, *US Pat 5,362,526*, 1994.
Wang, W. W., in "Electronic Packaging Materials Science II" (R. C. Sundahl, R. Jaccodine, K. A. Jackson, eds.), *Mat. Res. Soc. Proc. Vol. 108*, 125 (1988).
Wary, J., R. A. Olson, W. F. Beach, DUMIC 1996, 207 (1996).

Webb, D. A., A. P. Lane, T. E. Tang, in "ULSI Science and Technol./89" (C. M. Osburn, J. M. Andrews, eds.), *Electrochem. Soc. Proc. Vol.*, *PV 89-9*, 571 (1989).
Weigand, P., N. Shoda, T. Matsuda, S. V. Nguyen, T. E. Jones, M. J. Shapiro, R. Ploessl, J. Rzucek, *1996 VMIC*, 75 (1996).
White, L. K., D. Meyerhofer, *J. Electrochem. Soc.*, *134*, 3125 (1987).
White, L. K., J. M. Shaw, W. A. Kurylo, N. Miszkowski, *J. Electrochem. Soc.*, *137*, 1501 (1990).
Winters, H. F., *J. Vac.Sci. Technol.*, *B1*, 927 (1983).
Winters, H. F., J. W. Coburn, *Appl. Phys. Lett.*, *34*, 70 (1979).
Wolters, R. A. M., W. C. J. Heesters, *1990 VMIC*, 447 (1990).
Wong, J., T.-M. Lu, S. Mehta, *J. Vac. Sci. Technol.*, *B3*, 453 (1985).
Wrobel, A. W., B. Lamontagne, M. R. Wertheimer, *Plasma Chemistry and Plasma Processing*, *8*, 315 (1988).
Wu, O. K. T., A. N. Saxena, *J. Electrochem. Soc.*, *132*, 932 (1985).
Yeh, C.-F., C.-L., Chen, G.-H. Lin, *J. Electrochem. Soc.*, *141*, 3177 (1994).
Yeh, C.-F., C.-L., Chen, W. Lur, P.-W. Yen, *Appl. Phys. Lett.*, *66*, 938 (1995).
Yeh, C.-F., C.-L. Chen, *J. Electrochem. Soc.*, *142*, 3579 (1995).
Yu, B..-G.J.-G. Koo, H.-J. Yoo, J-D Kim, 1995 VMIC, 119 (1995).
Yuan, Z., C. Fisher, W. J. Shaffer, L. D. Bartholomew, *1995 VMIC*, 152 (1995).
Yuyama, Y., N. Tokumasu, K. Maeda, S. Fisher, J. Foggiato, J. Park, *1994 VMIC*, 133 (1994).

V
Metallization

Geraldine Cogin Schwartz*
IBM Microelectronics
Hopewell Junction, New York

1.0 INTRODUCTION

Among the factors involved in choosing a metallization system is, as in the case of dielectrics, a need to decrease signal propagation delay. To this end, although Al and its alloys have been the metals of choice for many years, Cu, with its lower resistivity, ρ, is being studied more intensively. The alternative of using thicker Al (alloy) films to reduce the line resistance is of limited value since this results in an increase in the line-to-line capacitance. Figure V-1 (Bohr, 1995) illustrates this point; it shows that the benefit of increased thickness soon becomes negligible. This figure also shows the deleterious effect on the RC delay of the use of higher ρ materials as an underlay for Al and as an encapsulant for Cu as well as the beneficial effect of a low ϵ insulator.

Another factor is the need for higher density and for planarization and gap-fill. The increased interest in CVD metals has been prompted by their hole filling capability; vertical interconnects, called "plugs" or "studs," compatible with vertical vias through the interlevel dielectric, are formed readily by CVD processes. This has been one of the motivations in developing CVD W processes, despite its higher ρ. Innovations in metal patterning make possible the smaller features on the denser chips. The advances in physical vapor deposition methods, driven by this same need for smaller, denser chips, are discussed in Chapter VI.

---

*Retired.

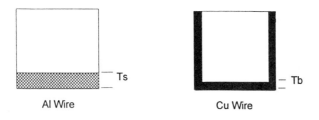

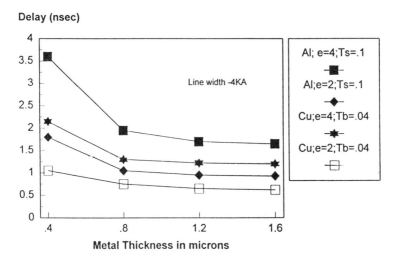

FIGURE V-1 Interconnection delay as a function of metal thickness (AR) for Al and Cu, for a pitch of 0.8 μ and a length of 10 mm; and without an additional layer and for insulators with different values of ϵ. The metal structures are shown at the top. (From Bohr, 1995)

2.0 ALUMINUM

2.1 INTRODUCTION

In the earliest days of integrated circuits (ICs), pure Al was used as the interconnection metallization. Al has a low resistivity (2.65 $\mu\Omega$-cm at 20°C), could be etched in solutions which did not attack the underlying films, adheres well to other metals and to dielectric films; photoresist as well as metals and dielectric films adhere well to it. It is easily evaporated (m.p. 659.7°C) and can be sputtered or deposited by CVD processes. Although pure Al eventually became inadequate, CVD Al has nevertheless been investigated because of its superior hole-filling properties.

Metallization

2.2 CVD OF ALUMINUM

2.2.0 Introduction

The basic advantage of using CVD Al is that good step coverage and void-free filling of small diameter, high AR holes can be realized at low temperature. The use of CVD Al plugs in place of the CVD W plugs used earlier reduces the via resistance significantly.

In this section the focus will be an the deposition processes, i.e., on the precursors, mechanisms, etc. Plug fill and planarization by CVD Al will be discussed in more detail in Chapter VI.

There are several precursors for CVD of Al, one inorganic and a variety of organic ones.

2.2.1 Blanket Deposition

Inorganic Precursor

This method, rarely used, is based on the transport of inorganic Cl$^-$ in an LPCVD reactor. Although AlCl$_3$ is sufficiently volatile, it cannot be reduced to form metallic Al. The method used is the disproportionation reaction:

$$3AlCl \rightarrow 2Al + AlCl_3$$

where the AlCl is formed either by the reaction

$$AlCl_3 (g) + 2Al (l) \rightleftharpoons 3 AlCl$$

in which AlCl$_3$ must vaporized and transported to the LPCVD reactor without condensation or else formed in situ by the reaction:

$$3HCl (in\ Ar\ or\ H_2) + Al (l) \rightarrow AlCl_3 + 3/2\ H_2$$

This is an atypical CVD process since the hot reactant gases decompose on a cool surface. Deposition occurs on both Si and SiO$_2$, and dopant incorporation is possible. It was claimed that the films were pure, adherent, had a random texture, and smoother than those deposited using an organic precursor. Proper activation of the surface is a problem. Exposure to TiCl$_4$ has been shown to be successful (Levy et al., 1985). The activation reaction is the hydrolysis of TiCl$_4$ by the surface hydroxyl groups (Bakardjiev et al., 1976).

Organic Precursors

Introduction. An organic source of AlCl, diethylaluminum chloride (DEAlCl) has been reported (Sasaoka et al., 1989). On Si, the reaction has been given as:

$$4AlCl + Si \rightarrow 4Al + SiCl_4 (g)$$

There are several others; those used most commonly are triisobutyl aluminum, TIBA or TIBAL [(CH$_3$)$_2$CH—CH$_2$]$_3$Al, and dimethylaluminum hydride, DMAH (CH$_3$)$_2$AlH, used almost exclusively for selective deposition. Trimethyl aluminum, TMA [Al$_2$(CH$_3$)$_6$], has been used as a source for pyrolysis (e.g., Biswas, 1983) PECVD (e.g., Kato et al., 1988; Masu et al., 1990) and photolysis (e.g., Higashi and Steigerwald, 1989). Various alanes, whose generic formula is (R)$_3$N:AlH$_3$ have also been Al sources. Some examples are R = methyl (TMAA) (Beach et al., 1989;

Gladfelter et al., 1989), ethyl (TEAA) (Lehmann and Stuke, 1992), and dimethylethyl (DMEAA) (Tsai et al., 1994).

TIBA. This is a volatile compound, explosive in contact with water, mildly toxic, and pyrophoric (but not when diluted with a saturated hydrocarbon). The films are deposited in a hot wall LPCVD reactor. The overall deposition reaction (pyrolysis) at ~250°C is:

$$Al(C_4H_9)_3 \ (l) \rightarrow Al + 3/2H_2 \ (g) + 3i\text{-}C_4H_8 \ (g)$$

The detailed chemical reactions have been discussed by Bent et al., (1989). The surface β-hydride elimination is the rate limiting step:

$$Al(C_4H_9)_3 \rightarrow AlH(C_4H_9)_2 + i\text{-}C_4H_8$$

Activation to prevent random nucleation on both metal and dielectric surfaces has been accomplished by exposure to $TiCl_4$ vapor, (Cooke, et al., 1982; Levy et al., 1984, Piekaar et al., 1989), use of in situ sputtered TiN as a seed layer which also improved surface roughness and eliminated pinholes (Cheung et al., 1990; Lai et al., 1991), and photonucleation, i.e., laser dissociation of TIBA (Tsao and Ehrlich, 1984; Higashi et al., 1987; Mantell, 1988).

One of the major problems is surface roughness, increasing with film thickness, causing lithography problems. In the initial stages of growth, hemispherical islands are formed and these coalesce when their density becomes high enough. It is implied that a higher nucleation density would result in smoother films (Cooke et al., 1982).

Amazawa et al. (1988) introduced a new reactor in which two heaters were used, so that the temperature of the gaseous reactants impinging on the front side of the wafer was higher than that of the substrate; they called this a "super hot wall CVD region." The benefit was production of smooth film because the substrate surface was supersaturated, generating a high density of nuclei. The resistivity reported for the Al films ranged from ~2.8 to ~3.4 $\mu\Omega$-cm. Any C-contamination was found only at the surface.

DMAH. This compound has a higher vapor pressure than TIBA. It is used with H_2 as a carrier gas. Deposition on Si (after exposure to dilute HF with no water rinse) occurred at much lower temperatures than on SiO_2; the rate increased with increasing substrate temperature but eventually fell at the same temperature at which deposition on SiO_2 became appreciable. The difficulty of depositing Al on oxide surfaces was thought to be due to the ease of oxidation of Al on such surfaces. Deposition on TiN would occur after the native oxide was removed (Kawamoto et al., 1990), although, according to (Tsubouchi et al., 1990a,b) deposition on TiN as well as on any other conductive substrate was said to occur (with no mention of any pretreatment) without nucleation on SiO_2.

Dixit et al. (1995) reported deposition of Al on CVD TiN in a single wafer cold wall reactor. At temperatures above 270°C, the reaction was transport limited. Below that temperature, the activation energy was 0.52 eV and had a half order dependence on DMAH concentration. The deposition rate was 4500 A/min at 260°C and small, high AR holes were filled without voids. They capped the CVD Al with PVD AlCu (see below for more details about CVD AlCu).

Deposition on SiO_2 or any other insulating surface in the $DMAH/H_2$ mixture required a short pretreatment with an rf plasma in a system in which the electrodes are so configured that the plasma does not touch the wafer surface. After extinguishing the plasma, deposition continued on all surfaces.

2.2.2 Selective Deposition

TIBA

UV-laser photodeposition predisposes the surface to CVD of Al from TIBA, by forming a catalytic surface for subsequent growth. Thus, on the modified surface CVD film growth can occur at a lower temperature than on the oxide surface. This inhibition occurs because any absorbed Al becomes oxidized and does not promote further absorption. Thus the deposition is selective (Mantell, 1988). By exposing the laser in the interconnection pattern, maskless formation of the interconnections by CVD is possible.

Mechanism. Pyrolysis of TIBA occurs much more readily on Al, Si, and other conducting surfaces than on dielectric surfaces. By means of molecular orbital techniques, Higashi et al. (1990) were able to elucidate the energy for crucial reactions and determine that the reaction was easier when the Al atom is bonded to other Al atoms than when bonded to O atoms. Thus there will be growth on Al and not on oxide. Mantell (1991) examined the β-hydride elimination reaction and concluded that it is suppressed on an oxide surface. Adsorption of more TIBA on that surface is prevented so that nucleation does not occur. However, if the surface temperature becomes too high, selectivity is lost because of homogeneous nucleation (Amazawa et al., 1988).

DMAH

Selective contact fill by Al directly on Si has been reported (Tsubouchi et al., 1990a,b; Masu et al., 1994) but since the need for a barrier between Al and Si has been demonstrated, this makes the accomplishment of doubtful technical importance. Of more significance is the selective deposition on any conducting surface, e.g., TiN, Al. Sugai et al. (1993) found that although the deposition rate of Al on TiN decreased when the temperature was decreased, at the higher temperatures (e.g., 230°C), only islands of Al grew in a hole and capped the top, leaving a void. At 130°C, holes (0.3 μ in diameter, AR = 2.7) were filled solidly with Al. This was explained by the temperature dependence of the sticking coefficient.

Fabricating a direct, low resistance Al to Al vertical interconnect in a small via hole, required removing the oxide on the lower Al surface. This was accomplished by RIE in BCl_3/Ar and an in vacuo transfer to the CVD chamber. The native oxide on the Al surface not only inhibited selective CVD growth but also results in a high via resistance when Al is deposited by any method. The CVD Al was shown to grow from the bottom of the via (Kawano et al., 1994; Ohta et al., 1994).

Maskless patterning has also been performed using DMAH by forming a seed layer by laser-assisted CVD. This has been done by scanning a focused laser beam (Cacouris et al., 1988) or by projection printing (Cacouris et al., 1990; Zhu et al., 1992). Photochemical CVD, using a deuterium lamp as a DUV source was a two-step process; nucleation, and subsequent growth by a thermal process. The nucleation period decreased as the light intensity increased but was independent of the DMAH pressure; after nucleation, the deposition rate became independent of the light intensity but increased with increasing DMAH pressure, continuing to grow in the absence of radiation at about 200°C. It was concluded that during nucleation, photodissociation of adsorbates took place more slowly than adsorption of DMAH on the surface (Kawai and Hanabusa, 1993).

Mechanism. A "surface electrochemical reaction" model has been proposed in which free electrons on the surface catalytically contribute to the reaction. In the case of Si, a surface terminated H atom (supplied by the HF/water treatment) reacts with the methyl (CH_3) radical in DMAH to produce volatile CH_4; the H atom of DMAH remains on the newly deposited Al surface as a new terminated H atom to react repeatedly with the CH_3 radical to deposit Al (Tsubouchi and Masu, 1992).

3.0 ALUMINUM ALLOYS

3.1 INTRODUCTION

Several elements have been added to Al to solve the problems encountered in the use of the pure metal. However, the disadvantage of the use of alloys is increased resistivity; the effect of such commonly used additives as Si, Cu, Ti can be found, e.g., in the CRC Handbook (1983). When using such tables for comparisons, both the measurement temperature and any heat treatments must be taken into account.

Silicon was the first alloying element added to Al (Kulper, 1971). During thermal processing subsequent to deposition, Si diffuses from the junction over which the Al has been deposited. Etch pits are formed and then filled by Al, resulting in what was called junction "spiking." This led to high junction leakage or, in the case of shallow junctions or small contact windows, shorting. To prevent this, Si was added to the Al, in a quantity to satisfy the solid solubility of Si in Al. Alloying prevented penetration. However, upon cooling, Si can grow epitaxially, as Al-doped Si, (i.e., *p*-doped Si) on single crystal surfaces at temperatures well below the eutectic temperature. The growth can be in the form of discrete structures and as a very thin layer superimposed on the discrete structures (Reith and Schick, 1974). This led to high and variable contact resistance and to the insertion of a rectifying layer between the metal and an n^+ contact; it also had deleterious effects on Schottky barrier diodes. Also, Si precipitates on preexisting Si particles, grain boundaries in the Al film or a steps in the dielectric film. Large Si precipitates (nodules) decreased electromigration (current-induced mass transport) lifetime by reducing the cross-sectional area of the conductor locally (Lloyd, 1982). An extended discussion of Al/Si interactions can be found in reviews by Pramanik and Saxena (1983a,b, 1990).

3.2 AlCu

3.2.1 Film Properties

Electromigration in Al conductors, a severe problem affecting the reliability of devices, "exploded" upon the world of IC processing in the sixties. Doping the Al

Metallization 307

with Cu extended the lifetime of conductors (Ames et al., 1973). The subject of electromigration is covered in detail in the chapter on reliability and the difficulties in RIE of Cu-doped Al are treated later in this chapter.

AlCu films have been prepared by many techniques. For example, Cu was evaporated as a separate layer and incorporated into the film by heat treatment. Copper and Al have been coevaporated; however, control of the Cu content was difficult, so that excess Cu was usually incorporated as a safety measure. Excess Cu presented no problem when alloy films were patterned by wet etching, but did when RIE replaced it; this will be discussed below. The proper concentration could be deposited reproducibly by sputtering from an alloy target and this has become the method of choice.

Copper is almost always distributed nonuniformly in a codeposited film; it tends to segregate at the metal/substrate interface as $CuAl_2$ (θ-phase) particles (Denison and Hartsough, 1980; Brukstrand and Hovland, 1983; Schwartz et al., 1986; Hara et al., 1986; Thomas et al., 1986). Segregation occurs in both sputtered and evaporated alloy films. Higher temperature and the use of bias during sputtering increased the pile-up, as seen in the SIMS depth profile of Fig. V-2. Copper appeared to be distributed uniformly after heat treatment; however, high-spatial-resolution profiling showed that there were θ particles, extending throughout the depth of the film but concentrated in small regions, giving the appearance of uniformity for large area mapping (Burkstrand and Hovland, 1983). The θ particles formed at higher temperatures were larger than those formed at lower temperatures, leading to the suggestion that the θ particles forming at the interface grow by depleting adjacent regions of Cu (Thomas et al., 1986). Legoues (1986) found that at lower temperature, θ particles were distributed randomly throughout the lower half of the film, in Al grains and at grain boundaries; at higher temperatures an almost continuous layer of θ particles were formed.

No pile-up was seen when an AlCu film was deposited on an oxide-free layer of Al (Schwartz et al., 1986; Ryan et al., 1988; Lloyd et al., 1990). The Cu distri-

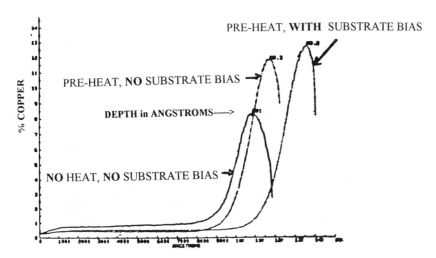

FIGURE V-2 SIMS profile (concentration vs. depth) of Cu in sputtered Al2%Cu: Effect of processing conditions. (From Schwartz et al., 1986)

bution in AlCu/SiO$_2$ and AlCu/Al/SiO$_2$ is shown in Fig. V-3. Schwartz et al. (1986) suggested that nucleation of θ phase was inhibited by diffusion of Cu into the underlying film. If the Al had a native oxide on its surface, the Cu segregated at the interface as it did on any oxide, and on W, Ti, Hf, and TiW.

3.2.2 CVD of AlCu

There have been two approaches to doping CVD Al films. In one, Cu is incorporated into the Al film by diffusion from a sputtered Cu source. A cluster system

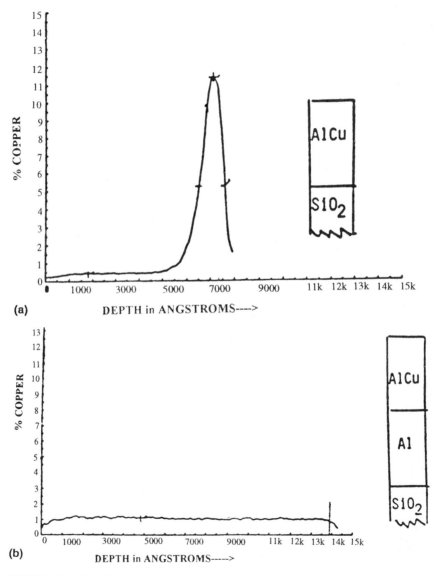

FIGURE V-3 SIMS profile (concentration vs. depth) of Cu in sputtered Al2%Cu deposited on (a) thermally oxidized Si, and on (b) Al/SiO2/Si. (From Schwartz, et al., 1986)

was used which incorporated a single-wafer sputtering module (dc planar cathode, magnetically enhanced) with Cu and AlCu alloy targets and the Al CVD reactor, using TIBA as the precursor. A film can be sputtered before CVD to act as a seed layer. However, if CVD of Al precedes sputtering of the alloying film $TiCl_4$ must be used to insure nucleation of Al on the oxide surface. There was no external source for heating the composite film; the only source was the CVD temperature itself. Cu diffused into the CVD Al film to approximately the solid solubility level at the CVD deposition temperature and was distributed uniformly throughout the CVD and sputtered films. If there is an excess of Cu in the alloy, it stays in the sputtered film. The Cu content of the CVD film can be controlled by heating the final film (Kwakman et al., 1990).

An alternative method is simultaneous CVD of Al and Cu. The details of CVD Cu are discussed in Section 4. DMAH and CpCuTEP (cyclopentadienyl Cu triethylphosphine) were the precursors used along with hydrogen as a carrier gas. The deposition was carried out in a cold wall reactor; the substrate was placed on a graphite susceptor, heated by rf induction. The surface of the deposited films was mirrorlike and the Cu was distributed uniformly throughout the films after annealing at 400°C for 30 min in H_2. No C or P was incorporated into the films. The deposition rate of Al was not affected by CpCuTEP; the Cu concentration was changed by changing the partial pressure of CpCuTEP. The film consisted of Al with $CuAl_2$ (θ-phase) precipitates in both the Al grains and in the grain boundaries; *no* Cu grains were observed. Thus the deposited film was truly an alloy and not a composite of Al and Cu grains. When the precursors were added simultaneously, the Al deposition was no longer selective; it was proposed that Cu was deposited first and acted as a nucleation layer for Al deposition and also to reduce interfacial resistance. Submicron vias, with a sputtered TiN underlayer, could be well-filled with the alloy. Interconnections, both well-filled vias and a wiring level were formed by this method; the interconnection pattern was formed by conventional RIE. The CVD alloy film was more resistive than a sputtered AlCu film of the same dimensions and Cu concentration (Katagiri et al., 1993; Kondoh et al., 1994). An alternate precursor, TMAAl was used and three Cu precursors were investigated (1) $Cu(hfac)_2$, (2) CpCuTEP, and (3) (hfac)CuTEM, using a cold-wall reactor and hydrogen as the carrier gas. The results obtained using (1) were not satisfactory; control of the Cu content was difficult, although the films were pure. Incorporation of Cu when using (2) was greater than when using it without the TMAAl; the authors postulated that a reaction occurred between the precursors or intermediate species. The Cu content was controlled by the temperature of the Cu source. The films were pure and no segregation occurred at the surface or film/substrate interface (1.8 wt% Cu), although small θ phase particles were distributed throughout the film. Resistivities were 2.8–4.0 mΩ-cm and the films were reasonably smooth. Similar results were obtained using (3) as precursor. (Houlding et al., 1992).

3.3 OTHER AL ALLOYS

Electromigration and hillock suppression, plasma etchability, and resistivity have driven the search for alloying elements other than Cu. Ti-doped AlSi (Towner et al., 1986) and Pd-doped AlSi (Koubuchi et al., 1990) have been suggested. Structures other than the simple single-layer film were sought in order to reduce the Cu-content of the AlCu film without sacrifice of electromigration resistance. Incorporation of a layer of a transition metal, such as Cr, Ti, Hf, or Ta between two AlCu films and then annealing the structure to form the intermetallic compound was one solution (Howard and Ho, 1977; Howard et al., 1978). Inverting the sandwich, i.e., placing a sputtered Al(0.5 wt%)Cu film between Ti layers (with an AlCu cap to protect the Ti surface) resulted in a structure more easily etched (due to the lower Cu-concentration) and more resistant to electromigration (Rodbell et al., 1991). Compared with AlCu films. AlSc alloys were harder, contained smaller precipitates after annealing, and had better electromigration resistance (Ogawa and Nishimura, 1991; Nishimura et al., 1993). An Al-Sm alloy was suggested for interconnections because of its low resistivity (after anneal), compared with other Al alloys, and its greater hillock-growth resistance (Joshi et al., 1990). In situ C-doping of PECVD Al suppressed the migration of Al and thus suppressed the growth of grains, hillocks, and spikes and increased the electromigration lifetime but increased the resistivity of the films (Kato et al., 1988a,b).

4.0 COPPER

4.1 INTRODUCTION

4.1.1 Physical Properties

As mentioned earlier, the use of a metal with a lower value of ρ for the interconnections decreases the delay time in ICs. The bulk value of ρ for Cu at 20°C is 1.678 $\mu\Omega$-cm, substantially lower than that of Al and its alloys. Thus Cu is a promising replacement for Al alloys, not only in terms of speed but, as will be discussed in a later chapter, in its higher electromigration and stress-induced voiding resistance.

However, as discussed below, there are a number of problems with the use of Cu which must be solved before Cu can be accepted for use in ICs.

4.1.2 Adhesion

The adhesion between Cu and dielectric layers (e.g., SiO_2, polyimide, etc.). is poor. Improved adhesion has been realized in several ways, such as treatment of the underlying film surface before Cu deposition (Bachmann and Vasile, 1989; Chang, 1987; Ruoff et al., 1988), use of a high substrate temperature during deposition, sputtering instead of evaporation, and adhesion-promoting films (e.g., Cr, Ti, Ta). But unless the improved adhesion is accompanied by inhibition of Cu migration, it is of little value since Cu diffuses rapidly in Si and through dielectric layers. Thus, only the use of adhesion layers which are also good barriers is of any practical value.

4.1.3 Barriers

Copper in Si degrades device performance so that effective diffusion barriers are required (barriers are discussed in Chapter 3).

Copper also diffuses through dielectric layers (McBrayer et al., 1988; Shacham-Diamand et al., 1993). The flux is composed of neutral Cu atoms (thermal diffusion component) and as determined by C-V measurements, positively charged Cu ions

(McBrayer et al. 1988). This causes oxide charges and deep level electron traps and may cause the insulator to become conductive (Gardner et al., 1995). The leakage current increases with increasing temperature and is accelerated in the presence of an electric field; electric field–aided diffusion is exponentially dependent on the electric field. Therefore, electrical measurements as well as physical detection methods (e.g., RBS, SIMS) must be made to determine the effectiveness of any barrier. The stages in the transport of Cu under bias temperature stress was described by Raghavan et al. (1995): (1) injection of Cu ions and diffusion of Cu neutrals into the dielectric, leading to charge build-up which then limits the ionic current (2) thermal diffusion of both species leading to dielectric degradation and increasing leakage, and (3) increased injection of Cu due to neutralization of Cu ions leading to failure. Copper is found at the Si/dielectric interface of a Cu/dielectric/Si capacitor only after the last stage. An example, by Ragahavan et al. (1995), of the leakage current vs. time for one of these capacitors with thermal oxide as the dielectric is shown in Fig. V-4. The source of the oxide had no influence on the results.

If oxynitride and nitride were used as the dielectric, there was no gradual increase in the leakage current until an abrupt breakdown occurred. These findings agree with those of Chiang et al. (1994). It was concluded that the presence of nitride has a significant effect on the transport of Cu through a dielectric. Chiang et al. (1994) attributed the dielectric failure to be due to the thinning of the effective oxide thickness due to Cu diffusion.

Annealing the capacitors (at 150°C) in the absence of an electric field resulted in shorter times to failure under stress, due presumably to the rapid diffusion of

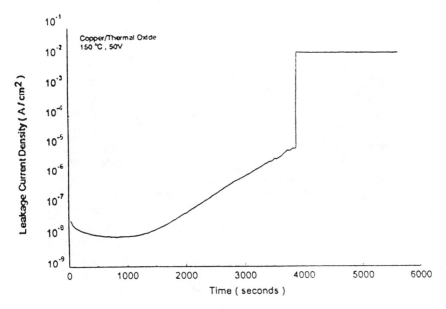

FIGURE V-4 Leakage current vs time for Cu on thermal oxide stressed at 150°C and 50 V. (From Ragahavan et al., 1995)

neutral Cu. Self-healing breakdown was observed at times; this was attributed to local heating from nonuniformly diffused Cu.

Nitride had the best barrier properties, followed by oxynitride. However, the use of these high ϵ materials runs counter to the need to decrease the RC delay.

The effectiveness of metallic barriers has also been investigated. A Ti layer >50 Å was an effective barrier when Cu/Ti/thermal oxide/Si samples were biased at 300°C for 30 h (Shacham-Diamand et al., 1993). When the Ti-Cu layer was annealed in a H_2-containing gas, instead of in an inert or vacuum ambient, Ti was a good barrier to Cu diffusion (in TEOS-based and thermal-oxide, but not PSG) to at least 400°C. (Murarka et al., 1993). Al, either as a thin sputtered layer deposited before the Cu (Nandan et al., 1992; Ding et al., 1994a) or cosputtered with the Cu (Ding et al., 1994b) was shown to inhibit Cu diffusion after annealing. However, only Nandan et al. (1992) who used a very thick Al layer, which resulted in an unacceptable increase in resistivity to (~ 7 $\mu\Omega$-cm), made C-V measurements. Therefore, the effectiveness of using less Al to contain the increase in resistivity has not yet been demonstrated. Gardner et al. (1995) used Mo satisfactorily as a complete encapsulant. One of the advantages of Mo is that it does not diffuse into Cu or react with it.

4.1.4 Oxidation

Finally, the susceptibility of Cu to rapid oxidation during resist stripping and to oxidation and corrosion when exposed to a hostile environment requires that the Cu be passivated. Several techniques have been reported. Doping with Mg (2 at%) reduced the oxidation rate of Cu. After annealing the film in Ar, the resistivity was essentially the same as that of pure copper, presumably due to the formation of a high conductivity phase, Cu_2Mg. Subsequent oxidation formed a thin MgO layer which stopped further oxidation; the protection was greater than for the unannealed film. The films adhered well to SiO_2 and showed good stability against diffusion into SiO_2 (as measured by RBS) and surface roughening (Ding et al., 1994c). Aluminum doping of Cu achieved the same result, but thus far with a greater impact on the resistivity. The mechanism proposed is that the Al diffuses to the surface, forming Al_2O_3, leaving the bulk as pure Cu. However, as mentioned above, increases in resistivity were observed (Nandan et al., 1992; Ding et al., 1994a,b). Copper has also been passivated by forming silicides in a reaction with SiH_4 in a CVD reactor (Hymes et al., 1992) and by ion implantation of B and Al into the surface so that the resistivity remains unchanged while decreasing the oxidation rate by about an order of magnitude (Ding et al., 1992). Coating the surface with a Cu_xGe_y phase by reacting Cu with GeH_4 in a LPCVD reactor passivates it as well (Joshi et al., 1995). Li et al. (1992a,b) Li and Mayer (1993) and Alford et al. (1994) suggested encapsulation of the Cu in refractory metal nitrides, formed by annealing in NH_3, preserving the resistivity of the bulk Cu line but requiring temperatures (550°–750°C) higher than those usually considered compatible with interconnection processing.

4.1.5 Encapsulation

It is clear that, due to the need for encapsulating/passivating layers, the low RC delay expected from the use of low-resistance Cu wiring is not realized. This was

Metallization 313

illustrated in Fig. V-1. It must be pointed out again that, with respect to delay, the choice of dielectric, i.e., ϵ, is as important as the ρ of the metal. Gardner et al. (1995) stated that encapsulated Cu performed better than AlCu but, as seen in Fig. V-1, this may not always be true.

Attempts to reduce the effects of encapsulation appear to introduce other problems. The barrier layer thickness may be reduced, but at the possible expense of unreliable performance. The increased line resistance arising from the use of a barrier layer may be ameliorated by increasing the width of the conductor but reducing the spaces between conductors increases cross talk. To avoid this problem, the wiring pitch must be increased but this comes at the expense of reduced packing density. An alternative might be the use of a nitride-based dielectric as the barrier, but only immediately adjacent to the Cu (to avoid a huge increase in cross talk of the higher ϵ material), with a lower ϵ material forming the bulk of the insulator. But any increase in ϵ is to avoided as is any increase in line resistance.

In addition to its effect on delay time, encapsulation increases process complexity and therefore process cost.

4.2 PHYSICAL VAPOR DEPOSITION OF Cu

Both evaporation and sputtering are feasible methods and were used extensively, particularly in earlier days when hole filling was not of paramount importance.

Park et al. (1991) compared dc planar magnetron sputtering of Cu and Al. They reported that the thickness of metal at the bottom of a hole was greater for Cu and that the sidewall coverage of Cu was continuous. They attributed this to the near normal cosine distribution of the sputtered Cu (Fig. V-5a) which enhances the forward throw of sputtered particles. Sputtered Al showed a preferential emission distribution directed at about 40° from the target normal (Fig. V-5b) which would promote lateral growth at the top, resulting in void formation within the hole as well as thinner deposits on the bottom of the hole.

A so-called low-energy ion bombardment process (using an rf-dc coupled mode of bias sputtering) was used to deposit Cu films on oxide or nitride substrates (Takewaki et al., 1995). The grains in films which were bombarded, during growth, by ions of energy ≥50 V and then annealed at 450°C for 30 min, grew to sizes as large as several hundred micrometers; this probably accounted for the excellent

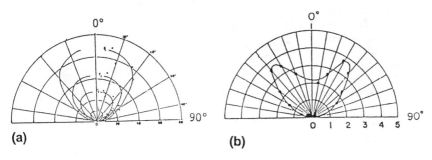

FIGURE V-5 Angular distribution of (a) sputtered Cu and (b) sputtered Al. (From Park et al., 1991)

electromigration resistance of the films. In addition, the crystal orientation of the film was converted from Cu(111) to Cu(200). The ρ of the films was 1.76 $\mu\Omega$-cm (bulk = 1.72 $\mu\Omega$-cm). The drawback for device fabrication was the need to remove part of the dielectric film to couple the dc bias to the substrate surface.

4.3 CVD OF COPPER

4.3.1. Introduction

There are two classes of precursors for CVD Cu, Cu(I) or Cu^{+1}, and Cu(II) or Cu^{+2}.

4.3.2 Cu(I) Precursors

The general formula of the Cu(I) precursor is $XCuL_n$, where X = univalent negative ligand, L = neutral Lewis base ligand, and n = 1 or 2.

A widely used compound is (hfac)Cu(vmts), sold as "Cupra Select" (Norman et al., 1991); in this compound X = hfac (hexafluoroacetylactonate) and L = vmts (vinyltrimethysilane) (Jain et al., 1991; Awaya and Arita, 1993a; Han and Jensen, 1994). A diagram of the compound is shown in Fig. V-6. The deposition reaction is disproportionation:

$$2(hfac)Cu(vmts) \text{ (g)} \rightarrow Cu^0 \text{ (s)} + Cu(hfac)_2 \text{ (g)} + 2 \text{ (vmts) (g)}$$

The reverse reaction is a selective noncorrosive, *isotropic*, dry etch for Cu (Norman et al., 1991).

There are other precursors, with different X and/or L groups, e.g., X may be other β-diketonates, e.g., acetylacetonate (acac), fluoroacetylacetonate (tfac) (Shin et al., 1992) and L may be 1,5-cyclooctadiene (COD) (Jain et al., 1991; Cohen et al., 1992), 2-butyne (Jain et al., 1991), or trimethylphosphine (PMe_3) (Dubois and Zagarski, 1992). Some others have been listed by Jain et al. (1991) and Shin et al. (1992).

4.3.3 Cu(II) Precursors

The most widely used Cu(II) precursor is $Cu(hfac)_2$. A cold wall reactor was often used. Cu can be deposited by heating the precursor in Ar (Temple and Reisman,

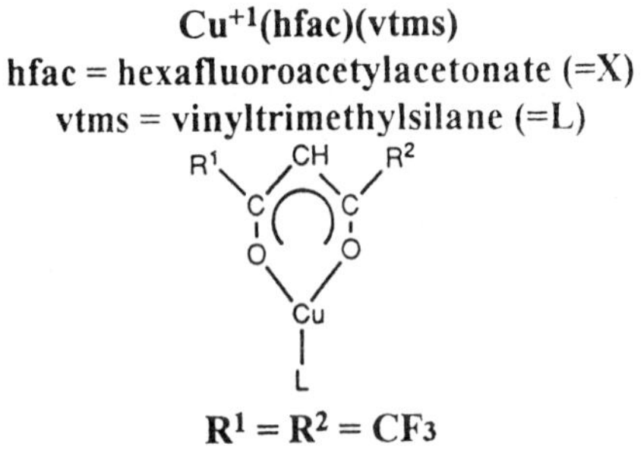

Cu^{+1}(hfac)(vmts)
hfac = hexafluoroacetylacetonate (=X)
vmts = vinyltrimethylsilane (=L)

$R^1 = R^2 = CF_3$

FIGURE V-6 Diagram of structure of Cu(I) precursor.

1989), reduction in H_2 or in an Ar/H_2 mixture (e.g., Awaya and Arita, 1989; Kaloyeros et al., 1990; Lai et al., 1991; Lecohier et al., 1992a), or in an H_2O/He mixture. The growth rate for Cu on Pt-seeded oxide surfaces was the same in a chemically inert carrier gas (He) or a reducing gas (H_2) (Lecohier et al., 1992b).

The reaction with H_2 is

$$Cu(hfac)_2 \text{ (g)} + H_2 \text{ (g)} \rightarrow Cu^0 \text{ (s)} + 2H(hfac) \text{ (g)}$$

Water vapor, added to both Cu(I) (Gelatos et al., 1993) and Cu(II) precursors, increased the deposition rate of Cu (Awaya and Arita, 1991a, 1993b; Lecohier et al., 1992a,b) and improved the surface morphology and resistivity of the deposited film. Addition of water vapor to Cu(hfac)(vtms) at the optimum concentration more than doubled the deposition rate and reduced the nucleation time without affecting the resistivity of the film, although too high a concentration increased the resistivity significantly, perhaps by forming copper oxide (Gelatos et al., 1993). For $Cu(hfac)_2$ the improvement in rate of about a factor of 9 was attributed to the formation of a dihydrate of $Cu(hfac)_2$ which has higher vapor pressure and lower decomposition temperature than the anhydrous compound (Awaya and Arita, 1993b). However, this was refuted by Lecohier et al. (1992b), who observed a dependence of the growth rate on the water vapor content of the reaction mixture when the amount of precursor arriving at the substrate was independent of that of the water vapor. They stated that the concentration of water vapor appears to be rate-determining. Awaya and Arita (1993b) proposed that the CVD mechanism changed when H_2O was used.

Other Cu(II) precursors include Cu acetylacetonate ($Cu(acac)_2$), Cu dipivoylmetanato $Cu(DPM)_2$ (Pelletier et al., 1991; Zama et al., 1992).

4.3.4 Deposition

Deposition of Cu, using any of the precursors and thermally activated reactions most often has been carried out in LPCVD reactors but APCVD has also been used (Temple and Reisman, 1989; Lai et al., 1991). The systems have been hot wall (e.g., Lai et al., 1991, Kaloyeros et al., 1990; Shin et al., 1992; Jain et al., 1993b), warm wall (Jain et al., 1993b; Gelatos et al., 1993), and cold wall (Awaya and Arita, 1989; Cho et al., 1992; Kim et al., 1994). Thermally driven deposition using an Ar-laser resulted in very high purity Cu deposits (Han and Jensen, 1994). Photo-assisted CVD, using Hg-vapor UV light, has also been reported (Zama et al., 1992). PECVD in an rf glow discharge (Oehr and Suhr, 1988; Awaya and Arita, 1991b) and using microwave plasma excitation have also been used. In all these reactors, the precursor was vaporized and flowed into the deposition chamber. An aerosol-assisted, liquid delivery system was used to attain very high deposition rates in a warm wall reactor. The liquid was a solution of Cu(Hfac)(COD) in toluene (Roger et al., 1994). This method does not rely on the delivery of the precursor at the equilibrium vapor pressure at the bubbler temperature. Thus higher precursor feed rates, leading to higher deposition rates, can be obtained.

The resistivity of CVD Cu films deposited using all the methods described above ranged from about 1.7 to 7 $\mu\Omega$-cm; the bulk resistivity is ~ 1.7 $\mu\Omega$-cm. Both blanket and selective deposition can occur. Growth rates as high as $\sim 5000 A/min$

[(hfac)Cu(vtms), Jain et al., 1993] and as low as 900 A/min [Cu(hfac)$_2$ in H$_2$ + H$_2$O, Awaya and Arita, 1993b] have been reported.

Blanket Deposition using (hfac)Cu(vtms)

On Blanket Metal Underlays. In one instance a cold wall reactor was used with a carrier gas (Ar) and a barrier metal, TiW, was the underlay. The wafers were placed face down and heated at the back. Excellent step coverage was obtained; high AR trenches and vias were filled without keyholes. After annealing at 400°C, the resistivity of the films was 2.0 $\mu\Omega$-cm (Cho et al., 1992). In another, a H$_2$ carrier gas was used in a cold wall reactor and the underlay a Ta seed layer. The step coverage and hole fill were temperature dependent. At 260°C, the Cu film was thinner on the edge of a step than at the top and bottom; there was separate island growth and voids were formed at the bottom of a via. Below 210°C, the shape was more fluidlike and below 200°C, vias were filled completely, the films were continuous and the upper surface over the via was planarized, as shown in Fig. V-7 (Awaya and Arita, 1993a, 1995).

No Metal Underlay. There was no substrate dependent nucleation; both Si and SiO$_2$ surfaces were coated when the Cu was deposited in a warm wall reactor, using (hfac)Cu(vtms) without a carrier gas. Excellent step coverage was observed; the film deposited into the deepest corners of an overhang structure; this was con-

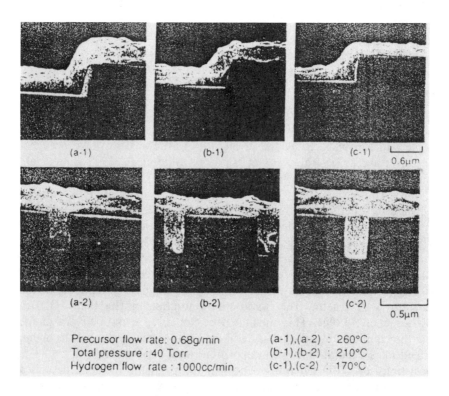

FIGURE V-7 SEMs showing temperature dependence of step coverage and via filling by CVD Cu. (From Awaya and Arita, 1993, 1995)

Metallization 317

sistent with a very low sticking coefficient. Films had very low stress, as indicated by the fact that a thin cantilever (used in the step coverage experiments) was not bent (Jain et al., 1993b), as seen in Fig. V-8.

Selective Deposition

Cu(I) Precursors. Some precursors always result in selective growth, e.g., (hfac(Cu)(PMe)$_3$ and some never do, e.g., (hfac)Cu(2-butyne). Other precursors e.g., (hfac)Cu(vtms) and (hfac)Cu(COD), may or may not, depending on the experimental conditions; Jain et al. (1993) and Dubois and Zegarski (1992) have summarized some of the different results reported in the literature. This indicates the need for careful surface preparation, but there may be other reasons for variability in results e.g., the reactor configuration, wall temperature, use of a carrier gas, etc.

Selectivity might be explained by catalysis of the deposition reaction by conductors so that no Cu could be formed on dielectric surfaces; this is related to the need for electron transfer (Norman et al., 1991).

In a study using (COD)Cu(hfac), the precursor adsorbed on a metal surface, forming a Cu(I)-hfac surface intermediate which then reacted to form Cu. On SiO$_2$, the precursor adsorbed as a very different surface complex. This suggested that the basis for selectivity was the differences in surface reactions and not a lack of precursor adsorption on oxide (Cohen et al., 1992).

In the case of (hfac)Cu(vtms) it was demonstrated that adsorbed H-bonded OH groups on SiO$_2$ surfaces are nucleation sites for film growth, but that isolated OH and strained Si-O-Si groups are not. (Dubois and Zegarski, 1992; Farkas et al., 1994). Selectivity depends on passivation of the OH surface groups. This was done most effectively by in situ introduction (before or during CVD) of gaseous (CH$_3$)$_2$SiCl$_2$; the Si bonds to the O group after the H is removed by the Cl, so that the surface is O-Si(CH$_3$)$_2$ and not OH (Jain et al., 1993b).

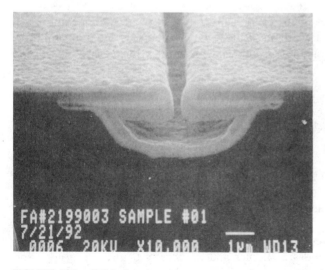

FIGURE V-8 SEM showing step coverage of an overhang structure by CVD Cu. [From Jain et al. (1993).]

The bond strength of Cu-L, and thus the nature of the precursor ligand was shown to affect selectivity. A weak bond (e.g., L = 2-butyne) is readily dissociated on most surfaces with little thermal activation and no electron transfer from the substrate resulting in easy deposition on oxide (Jain et al., 1991; Dubois and Zegarski, 1992). Although (hfac)Cu(vtms) (a high vapor pressure liquid) is stable, it appears to have a very narrow process window. The most stable precursor is (hfac)Cu(PMe$_3$) which should guarantee selectivity; however, it is a solid and difficult to pump. Thus there appears to be no completely satisfactory precursor as yet.

Cu (II) Precursors. Selective deposition appears to be the rule. The same difference in the surface reactions on oxide vs. metal was found to hold for the Cu(hfac)$_2$ as well for (COD)Cu(hfac) (Cohen et al., 1992). Thus selective deposition of Cu can be used to fill via holes etched down to the underlying conductor which can be Al, Cu, AlCu, or silicides (Awaya and Arita, 1989). It can also be used for area-selective deposition on oxide surfaces by forming a thin patterned seeding layer upon which Cu can then grow. The underlay must adhere well to the substrate and must not degrade the morphology, purity, or conductivity of the CVD Cu film (Lecohier et al., 1992). The deposition temperature was found to be the most important parameter in insuring selectivity. Under the deposition conditions investigated, substrate temperatures between 310° and 360°C insured selectivity. When the temperature became too high, Cu was always deposited on oxide, sometimes as a localized deposit or else as a continuous film (Kim et al., 1994). When water vapor was added to increase the growth rate, the selectivity was improved when He was used instead of H$_2$ (Lecohier et al., 1992c).

Selectivity Loss

Loss of selectivity occurs if there is an electrically conducting path through the dielectric (poor quality film, pinholes) or there are metallic residues on its surface (Cohen et al., 1992). A high total pressure and a hot wall system often result in nonselective deposition (Kim et al., 1994). Other factors that can lead to loss of selectivity are heating the precursor or the substrate to too high a temperature, the presence of contaminant hydrocarbons, and residual precursors contaminating the chamber walls when opening the system. Once a film is nucleated, subsequent growth is rapid.

4.4 ELECTROCHEMICAL DEPOSITION OF Cu

4.4.1 Introduction

Deposition of Cu by electrochemical methods may be considered a patterning process as well, since the metal is deposited into holes/grooves etched into a substrate. A more extensive discussion of the wiring processes will be covered in Chapter VI.

Plating has been considered an attractive alternative to CVD because the apparatus and chemicals required are simpler and less costly than those needed for CVD.

4.4.2 Electroless Plating

The reducing agent in this process is formaldehyde (HCHO); a high pH is required for deposition. The reaction is

$$Cu^{+2} + 2HCHO + OH^- \rightarrow Cu^0 + H_2 + 2CHOO^- + 2H_2O$$

The deposition rate of Cu increases with increasing temperature and pH; the rate reaches a plateau above pH ~12.5. No deposition is observed at a pH <11; a pH >12 is needed for stable deposition but too high a rate can degrade the film quality. Jagannathan and Krishnan (1989) showed that replacing formaldehyde by triethanolamine extends the operating range to a pH < 9 and Jagannathan et al. (1991) proposed the use of tetra aza ligands as complexing agents.

Typical formaldehyde-based solutions, with and without Na^+ and K^+ (to minimize incorporation of these ions in the film) were given in a review by Cho et al. (1993) and Shacham-Diamand et al. (1995) who also discussed the mechanisms of electroless plating. The baths include not only the ingredients essential for the reaction shown above, but additives to stabilize the solution, reduce surface tension, retard H_2 incorporation, and otherwise regulate the film properties. Resistivity values of ~2 $\mu\Omega$-cm have been reported. One of the drawbacks to the use of electroless copper is incorporation into the film of the reaction products, H_2 and H_2O, which may degrade the reliability of the film by outgassing during thermal treatment.

4.4.3 Electroplating (Electrodeposition) of Cu

Although electroplating avoids incorporation of H_2, it is used very rarely in chip fabrication; it is used widely on circuit boards and other elements used to package chips. One suggested application to chips was suggested by Contolini et al. (1994) as a way of forming planar Cu interconnects. The process is described in Chapter VI.

5.0 TUNGSTEN

5.1 Introduction

Tungsten is a refractory material (m.p. 3370°C) with a bulk resistivity of 52.8 $\mu\Omega$-cm at 20°C. The possible forms of W are an amorphous phase, α-W, the equilibrium low-resistivity phase, and β-W, the high-resistivity metastable phase.

Films can evaporated but a high substrate temperature is required for low resistivity. Low-resistivity films can be deposited by sputtering, in all types of systems; a high-sputtering pressure minimizes stress (Dori et al., 1990). However, CVD is the method of choice for most applications in IC fabrication.

5.2 CVD TUNGSTEN

5.2.1 Introduction

CVD W is used in current production of ICs chiefly as a "contact plug" and a vertical interconnection between successive wiring levels of multilevel devices. A contact plug connects the active and passive devices in the substrate to the first layer of metallization and to planarize the surface before that film is deposited. It also acts as a barrier, inhibiting interdiffusion and reaction between Si and the metal; this is discussed in Chapter III. Its usefulness for vertical interconnects may be ending as new hole-filling deposition techniques for lower resistivity metals are developed. CVD W has also been used for gate electrodes and *local* interconnections (i.e., very short wires), replacing higher resistivity polysilicon.

CVD W can be deposited selectively, i.e., only on reactive surfaces, or as blanket films, i.e., nonselectively. Although it is toxic, corrosive, highly reactive, and readily hydrolyzed in moist air to form HF, WF_6 is the most common source gas. LPCVD is the usual method of deposition although high-pressure CVD has also been studied. Both hot wall or cold wall reactors have been used.

PECVD has also been used, with various reactants, to deposit blanket W films both in rf reactors (Tang and Hess, 1984; Greene et al., 1988a; Wong and Saraswat, 1988; Kim et al., 1991) and in an ECR system (Akahori et al., 1990). Other methods include laser-induced CVD (e.g., Mogyorosi and Carlsson, 1992), ion-enhanced evaporation (Joshi et al., 1987), and electron beam–induced deposition (e.g., Bell et al., 1994).

5.2.2 Blanket Deposition

The reaction used most frequently is the reduction of WF_6 by H_2.

$$WF_6 + 3H_2 \rightarrow W + 6HF\ (g)$$

SiH_4 reduction of WF_6 has also been used, before H_2 reduction, to deposit a seed layer, as described in a patent by Schmitz et al. (1990). The reaction is

$$WF_6 + 3/2 SiH_4 \rightarrow W + 3/2 SiF_4 + 3H_2$$

Reduction by H_2

This reaction may be preceded by the reduction of WF_6 by SiH_4 or else by coating the surface first with a thin conducting layer, e.g., a sputtered metal film such as TiN or TiW as a seed or a barrier layer. These steps are taken to prevent a reaction with the underlying Si or Al since the reaction between WF_6 and these substrates is favored over reduction by H_2. The seed layer also acts as an adhesion layer. The deposition rate is increased by adding SiH_4 to the reaction mixture (Park et al., 1989). Under *optimal* deposition conditions, a conformal film can be deposited and holes filled without voids; this subject will be discussed at greater length in Chapter VI. The film can be deposited at high temperatures and at high rates, since there are no selectivity restraints, but capping with WSi_x is required to prevent oxidation of the hot W surface as it emerges from the reactor, since in situ cooling reduces throughput unacceptably.

Schmitz and Kang (1993) patented a two-step process in which the first W layer was deposited under conditions (temp. <440°C, press. = 20–100 torr, flow of $WF_6 \geq 0.15$ sccm/cm² of wafer surface) for forming voidless hole-fill, and then by reducing the flow of WF_6 to no more than 0.05, produced a layer with low stress.

Reduction of WF_6 by B_2H_6 (+ H_2) formed α-W with a low resistivity that was attributed to the lower F-content of the film (Hara et al., 1994), although Smith et al. (1994) reported that the films deposited from WF_6/B_2H_6 were high resistivity W_xB_{1-x}.

Mechanisms: $WF_6 + H_2$. The deposition rate of W by the H_2 reduction of WF_6, under the conditions used for practical deposition, varies as the square root of the partial pressure (pp) of H_2 (i.e., rate $\sim P_{H_2}^{1/2}$) and is independent of the pp of WF_6 (i.e., rate $\sim P_{WF_6}^{0}$). The activation energy was found to be 0.71 eV, in the range of 250 to 500°C and pressure range of 0.1–5 Torr. The agreement between this value of the activation energy and that reported for the H_2 surface diffusion on W led to the suggestion that the rate limiting step was H_2 dissociation (Broadbent and Ramiller, 1984). An activation energy of 0.75 eV, reported by McConica and Krishnamanihi (1986) for a smaller temperature range, is in good agreement. Pauleau and Lami (1985) found that a rate limiting step of dissociative adsorption of H_2 on W could not account for the large decrease in the rate of selective deposition of W when the deposition area increased but was instead the surface reaction between fluorine and hydrogen atoms in the adsorbed phase. McConica and Krishnamani also rejected adsorption of H_2 as the rate limiting step. The mechanism they felt was more in keeping with the way the reaction of H_2 actually occurs, i.e., "addition of adsorbed monatomic hydrogen to adsorbed partially fluorinated W." They also pointed out that agreement with rate data does not prove a mechanism and that eliciting the "true" mechanism would require surface analyses. They restated the obvious that "since the reaction is a heterogeneous one, surface cleanliness is critical to film growth and purity." Desorption of HF from the W surface was another suggestion for the possible rate limiting step for the $WF_6 + H_2$ reaction (Broadbent and Ramiller, 1984; McConica and Krishnamanihi, 1986t). However, Bryant (1978) showed that the reaction orders for this limiting reaction are 1/2 with respect to H_2 but 1/6 with respect to WF_6. At low temperatures and very low pp of WF_6, van der Putte (1987) did find that the rate was no longer independent of the pp of WF_6, and that the rate equation was $R = K(P_{H_2}^{1/2})(P_{WF_6}^{1/6})$, suggesting the rate limiting step as desorption of HF in this regime.

5.2.3 Selective Deposition

This is essentially a hole-filling process. In a selective process, nucleation (initiation) occurs only on reactive surfaces; for hole filling, only the bottom of the hole must be reactive. It is more difficult to control than blanket deposition, but has the great advantage of requiring less WF_6 (an expensive reagent) and fewer processing steps since an adhesion layer is not required. Etchback of overfilled holes is less demanding. Since the film grows from the bottom of the hole, there are no seams or keyholes in it. WF_6 can be reduced by Si, Ti, Al, SiH_4, and H_2.

Reducing Agents

Si. The selective deposition of W on Si,

$$2\ WF_6 + 3Si \rightarrow 2W + 3SiF_4\ (g)$$

is a selflimiting reaction. A thin film (~ 100–200A) is formed very rapidly; the limiting thickness is independent of time, temperature (270–450°C), and pressure (Broadbent and Ramiller, 1984; Saraswat et al., 1984; Broadbent and Stacy, 1985; Green and Levy, 1985). The thickness and physical structure depended on the surface preparation, i.e., the W/Si interface. Two models have been proposed to explain the existence of a limited thickness. The first postulates that once a continuous film of W is formed, WF_6 cannot reach the Si surface to react. However, Si reduced W films deposited between 210 and 700°C were found to be porous and discontinuous. The porosity was said to be due to the evolution of the gaseous product, SiF_4; the pores might be remnants of these bubbles (Green et al., 1987). Amorphization of the Si surface by ion implantation decreased the nucleation barrier and allowed the formation of thicker films at lower temperatures (Green et al., 1986). Another proposal was that the presence of a nonvolatile lower fluoride of W was the inhibitor of further reaction between WF_6 and Si. This hypothesis was supported by the presence of F in the W film (Lifshitz, 1987; Park et al., 1988). A pinhole theory was proposed by Broadbent and Ramiller (1984) but was refuted by Green et al. (1987).

Although all the Si-reduced W films are self-limiting, the thickness reached is not always the same. A limiting thickness of 60 Å (equivalent of full density W) was reported to be formed <310°C. The thickness reached a maximum value at 340°C and decreased with increasing temperature, leveling off at about 200 Å. The temperature dependence and the selflimiting behavior were believed to disprove that the reaction is diffusion limited but controlled by factors such as the temperature dependence of nucleation, sticking coefficient, and desorption of intermediates, an atomistic reduction mechanism, and other (unstated) mechanisms; this could also explain the surface sensitivity (Green et al., 1987).

A dependence of the limiting thickness (>400Å) on both the doping condition of Si and the surface preparation has also been reported (Tsao and Busta, 1984).

An advantage to the use of W as a barrier layer between the Si and the interconnections metallization (usually Al-based) is the low and tightly distributed contact resistances, particularly in small contacts (Shibata et al., 1984).

However, what have been called "wormholes" and "tunnels," empty channels with a W particle at the end, are formed at the W/Si interface at the rim of a contact window (Green and Levy, 1985; Stacy et al., 1985; Yang, et al., 1987). Tunnels are shown in Fig. V-9.

Three atoms of Si are consumed for every 2 atoms of W formed; the thickness W deposited is ½ of the thickness of Si consumed (Tsao and Busta, 1984). Since this is an isotropic reaction, there is lateral encroachment at the Si/SiO_2 interface (Moriya, et al., 1983; Stacy et al., 1985), as seen in Fig. V-10. The lateral encroachment distance increases with increasing partial pressure of WF_6 and temperature. Encroachment can be minimized by the use of the proper pressure and substrate temperature, which at the same time improves selectivity (Moriya et al., 1983). Wormholes and encroachment are particularly serious when thick films are deposited. They lead to junction degradation, i.e., high leakage currents.

Several process modifications have been proposed to limit consumption of Si. Addition of SiF_4 to WF_6, by shifting the equilibrium of the $WF_6 + Si \rightleftharpoons W + SiF_4$ reaction, inhibited erosion and encroachment of the Si contacts and resulted in improved performance (Levy et al., 1986). An interlayer, blanket WSi_x, followed by W deposition under selective conditions, prevented excess Si loss and filled contacts without keyhole formation (Hieber and Stolz, 1987). Coating the contact holes with amorphous Si by sputtering at room temperature replaced the Si, which

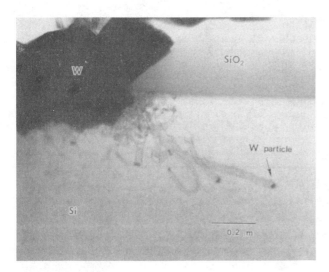

FIGURE V-9 SEM showing "tunnels" into Si after selective CVD of W. (From Stacy et al., 1985)

would have been consumed from the contact holes, with the deposited Si when W was formed by the WF_6 + Si reaction. Therefore, consumption of the diffused layer and encroachment was reduced (Kakiuchi et al., 1987). Selective deposition could be maintained and thick films formed while suppressing Si consumption, tunneling, and encroachment by using SiH_4 as the reducing agent in a range of SiH_4/WF_6 < 1.5 (Kusomoto et al., 1988; Gorczyca et al., 1989). The SiH_4 reduction reaction has the potential for higher deposition rates (e.g., 0.6 μm/min) than other reduction reactions, but when the SiH_4/WF_6 ratio exceeded ~1.6, although the rate was higher, blanket deposition occurred. There is the potential for forming a damage-free surface because there is no interaction with the Si substrate and no HF production. In this case, a cold-wall reactor was used (Tsutumi et al., 1990). Feinerman (1990) found that surface treatment in dilute HF/HNO_3 minimized encroachment, whereas exposure to a CF_4/O_2 plasma maximized it. GeH_4 reduction of WF_6 was

FIGURE V-10 SEM showing encroachment due to selective CVD of W. (From Moriya et al., 1983)

proposed as an alternative reaction. Beta-W containing ~10 a/o Ge was deposited selectively without harmful consumption of Si, tunnels, or encroachment. Contact resistivity to both n^+ and p^+ Si was low, the β-W stable to 600°C and the adhesion of the W to Si was excellent (van der Jeugd, et al., 1990, 1992).

Another phenomenon, "creep-up" in contact holes, has been observed. After W has been formed on the Si surface in a contact hole, the growth continues along the SiO_2 sidewalls, forming a "collar" (Itoh et al., 1985). The collar affords extra protection at the edges of a contact hole where metal coverage by an overlying metal may be thinned.

H_2. To grow thick films selectively, reduction of WF_6 by Si must be followed by reduction by H_2 (Blewer and Wells, 1984). Reduction of WF_6 by Si is favored energetically over the reduction by H_2, so that if H_2 is used at the start, the reaction with Si will still occur so long as the Si surface is exposed. To maintain selectivity, low temperatures, i.e., low growth rates are used.

The introduction of a LPCVD cold wall reactor, in which only the wafer was heated by a lamp or hot plate, allowed *rapid selective* growth of W (by the reduction of WF_6 by H_2) in interlevel vias on Mo interconnection metallization. Mo does not react with WF_6 and so is an excellent substrate for the WF_6 + H_2 reaction. The high rate implied a high deposition temperature. Any metallic impurities were scrupulously removed from the surrounding oxide surface to prevent nucleation (loss of selectivity). This process allowed complete filling of deep vias. If vias of different depths were present, the excess W in the shallow vias could be etched off using a sacrificial resist process (Saia et al., 1987). Such a process would be useful only where resistivity requirements are not stringent since the resistivity of bulk Mo is 5.34 $\mu\Omega$-cm at 20°C. A highly selective deposition of W on Si used alternating cyclic hydrogen reduction of WF_6. Hydrogen reduction was interrupted periodically to vaporize incipient precursor nuclei formed on the oxide surfaces via the disproportionation reaction forming WF_5 from W and WF_6 (Reisman et al., 1990). Selective deposition of W in a polymeric matrix was accomplished using rapid thermal LPCVD and a mixture of $WF_6/H_2/SiH_4$. α-W was always obtained and had a resistivity of ~20 $\mu\Omega$-cm (Bouteville et al., 1991).

Ti. Titanium can reduce WF_6 but the resulting structure consisted of clusters of W on a layer of TiF_3 (Broadbent et al., 1986).

Al. There have been two problems associated with depositing W on an Al surface. One is high interface resistance, due to a surface film of AlF_3, and the other the inability to nucleate W rapidly, reproducibily, uniformly, and then to grow a smooth thick film without loss of selectivity.

The reaction by-product is involatile AlF_3 trapped at the interface, increasing interfacial resistance (Broadbent and Ramiller, 1984; Broadbent and Stacy, 1985). However, Hey et al. (1986) reported that although the via resistance was higher than for an Al-Al contact, it was acceptable. They postulated that the larger contact area of the interface alleviated the problem caused by the AlF_3 insulating film. The W was deposited at a temperature below that used to deposit the Al so that hillock growth was suppressed.

Low via resistance was achieved by controlling two parallel reactions, the reduction of WF_6 by H_2 *and* Al, in a single step in a cold wall reactor described above, in which the deposition temperature was high, while selectivity was maintained. Several explanations have been proposed to explain the effect of the higher deposition temperature on the via resistance. One is that at the higher temperature, AlF_3 was more volatile. Another is that the shortened reaction time allowed less AlF_3 to be formed before the surface was covered with W, i.e., W would cover the Al surface before it could react extensively with WF_6. An additional reason is that at high deposition rates, less or no F is expected to stay at the interface (although the fluorination of Al surfaces occurs readily) (Chow et al., 1987; Wilson et al., 1987; Kang et al., 1988).

Tungsten deposition did not take place on the plasma-oxidized Al surface formed during resist ashing and no F was detected after exposure to WF_6. In the absence of ashing (just native oxide on the Al surface), W layers could be formed reproducibly after cleaning the Al surface by brief etch in hot concentrated HCl. Low levels of F were found at the interface. Since a wet pretreatment was thought to be unacceptable, deposition of a thin (but not too thin) layer of W was sputtered directly on the uncleaned Al surface before exposure to WF_6. These authors reported that accumulation of F was correlated with aluminum oxide, not bare Al. They concluded this because when both W and uncleaned Al surfaces were exposed to WF_6 (the surface condition when the W film was too thin), high concentrations of F were detected at the interface (Ng et al., 1987). Immersion of the Al surface in a dilute HCl solution before CVD W deposition was reported to improve the via resistance, although it was still substantially higher than that of an Al/Al interface (Oshima et al., 1993).

Other processes have been developed to improve the nucleation of W on Al; they involve pretreatment of the Al surface and successive reduction by H_2 and then SiH_4. One consisted of a brief etch in a very dilute HF solution, followed by a dry pretreatment. The most satisfactory was a short exposure to a BCl_3 and then an H_2 plasma or to H_2 at an elevated temperature; this resulted in a uniform W film at the bottom. SiH_4 was used for the final W growth step, ensuring minimal selectivity loss (Hintze et al., 1994; Schulz et al., 1994). Preparing sputtered W and TiN-capped Al surfaces for CVD W deposition required the HF dip and exposure to an NF_3 plasma. In the case of TiN, etching was stopped before the Al surface was exposed. Then the H_2/SiH_4 reduction was carried out (Schulz et al., 1994).

Another process started by removing polymeric residues using an HF/ashing/HNO_3 sequence followed, without further pretreatment, by a two-step W CVD process: H_2 reduction at 350°C, then SiH_4 reduction at 280°C. This sequence resulted in a very low F level at the W/Al interface, with an acceptable via resistance, smooth films, and favorable selectivity (Bae et al., 1994).

Reduction by SiH_4

The growth rate of selective W was proportional to the partial pressure of SiH_4 and decreased slightly with increasing partial pressure of WF_6. The activation energy was slightly negative, indicating competition between adsorption/desorption processes and surface reactions, i.e., adsorption of SiH_4 and competition between film formation on the surface and by-product desorption. The growth rate was inversely proportional to the exposed area indicating a reactant supply limited reaction. The resistivity of the film increased with higher deposition rates and lower deposition temperatures (Colgan and Chapple-Sokol, 1992).

In a patent, Joshi et al. (1993) described a two-step process using a flow ratio of $SiH_4/WF_6 < 1$ for the deposition of an initial layer of W selectively on Si, at temperatures above 500°C, followed by reduction of WF_6 by H_2 at temperatures ≥500°C for the rest of the film.

Mechanisms

$WF_6 + SiH_4$. The apparent activation energy for this reaction varied with both temperature and pressure, suggesting multiple kinetic regimes. The kinetics were modeled to include three competitive reactions of WF_6 absorbed on the surface: (1) reduction by SiH_4, (2) dissociation of SiH_4 and reduction by Si, and (3) reduction by H_2 [formed in (1) and (2)]. The model of Hsieh (1993) is summarized in Fig. V-11.

Since the relative importance of each reaction pathway is highly dependent on process and system conditions it was difficult to establish a universal rule for the process trends. Reactions (1) and (2) have near zero activation energies and for (3), as stated above, $E_{act} \sim 0.7$ eV. A deposition regime controlled by (2) occurs when the ratio SiH_4/WF_6 is too high and loose β-W films and gas phase nucleation can result (Hsieh, 1993). A reaction mechanism in which a fluorinated W surface is reduced by both surface absorbed SiH_x and impinging SiH_4 molecules was proposed by Bolnedi et al. (1994).

Prediction of Selectivity; Loss of Selectivity

Selective deposition is difficult to control. The important factors are pressure, flow rate, and surface preparation/area. According to Carlsson and Bowman (1985) "selective CVD (of W from the WF_6/H_2 reaction) is based on a difference in thermochemical stability between different substrate regions; the higher the difference, the higher the selectivity." Selective W deposition on Si is favored by low temperature and total pressure and a high concentration of WF_6 in the initial stages of deposition. Low temperatures means slow growth and long deposition times but long deposition times, whether due to the rate or the final thickness required, compromise selectivity. As W is formed and growth now occurs on the W surface, the selectivity is less but increases with decreasing WF_6 concentration and increasing temperature. This suggests a two-stage operation should be used. A high temperature increases the probability of reaction with SiO_2 and etching of the oxide may occur. They concluded that temperature was the most important parameter for maximum selectivity and minimum attack of oxide.

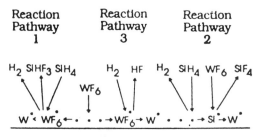

FIGURE V-11 Summary diagram of a CVD-W kinetic model. (From Hsieh, 1993)

Metallization

Loss of selectivity has been linked to an increase in the HF concentration in the reactor; a high flow rate, reducing the residence time of HF on SiO_2 will, therefore, improve selectivity (Pauleau and Lami, 1985; Korner, 1989). Nuclei were formed more readily on nitride than on oxide surfaces and the presence of P (as a surface treatment or as PSG) inhibits nucleation (Bradbury and Kamins, 1986). Metallic contaminants (residues) on the insulator are sites for H_2 reduction of W (Blewer, 1986). Increased numbers of wafers or exposed area degrade selectivity as do improper cleaning, ion implantation, or exposure of the surrounding regions to a plasma (Saraswat et al., 1984). Selectivity loss was found to be caused by the transport of W from the W surfaces to the surrounding oxide. The mechanism proposed was desorption of tungsten subfluorides from the W surface and transport to the other surfaces where they are condensed and disproportionated leaving on the oxide a reactive state of W which catalyzed the reaction between H_2 and WF_6 (Creighton, 1989). A similar model was developed by Desatnik and Thompson (1994), who found that the amount of nucleation on SiO_2 decreased with increasing distance from the metallic surfaces. The model developed was that a short-lived reactive gaseous intermediate diffuses from the metal to the SiO_2 where it reacts to form nuclei in clusters. Nucleation was favored at higher temperatures and increased as the process progressed. However, McConica et al. (1988) developed a model that predicted that *high* temperature, low pp of WF_6, minimal hot metal area, and deep narrow vias enhance selectivity.

Loss of selectivity in the WF_6/SiH_4 process that can occur when $[SiH_4]/[WF_6] < 0.3$ was found to be due to an autocatalytic reaction on the SiO_2 surface, the formation of Si_2H_{2n} from two adjacent SiH_4 molecules on a nucleation site (impurity or adjacent metal area). Formation of this intermediate is retarded by chemisorption of WF_6 on the site but once formed, reacts rapidly to a W-containing species (perhaps W_2F_2, W_2F_6). At monolayer coverage, further reaction with Si_2H_{2n} forms W which acts as nucleation sites. At relatively high SiH_4 densities in the reactor, the Si_2H_{2n} species react with each other to form Si clusters which react with WF_6 to form W, at a reactant ratio >1. If it were possible to remove the W-containing species, selectivity would be maintained (Groenen et al., 1994).

5.2.4 Properties of CVD W

The films formed by the WF_6/H_2 reaction are polycrystalline, with a grain size of ~2000 A, a tensile stress of 7×10^9 dynes/cm^2 and a resistivity of 13 $\mu\Omega$-cm for a film ~1000 A thick, decreasing with increasing thickness. Increasing the deposition temperatures reduced the resistivity to ~8.5 $\mu\Omega$-cm for a 1-mm-thick film. The resistance to both n^+ and p^+ contacts is good and it is a good diffusion barrier between Al and Si up to 450°C. The oxygen content of the film increased with increased deposition temperature or decreased thickness; oxygen was concentrated at grain boundaries. The F content decreased with increasing temperature but was independent of thickness (Green and Levy, 1985; Learn and Foster, 1985).

6.0 PATTERNING OF ALUMINUM AND ALUMINUM ALLOYS
6.1 WET ETCHING

A widely used etchant for Al and its alloys is a mixture of phosphoric, nitric, and acetic acids in water. Others can be found in a listing by Kern and Deckert (1978). Because of the fine dimensions required for VLSI and ULSI interconnections, wet etching for pattern definition has long been discarded.

6.2 SPUTTER ETCHING; ION MILLING

These techniques can be used to pattern any metal and is compatible with the small dimensions, but they are not used widely because of the superior results obtained by using reactive plasmas in manufacturing environments.

6.3 ANODIC OXIDATION OF Al

One of the early attempts to improve dimensional control was the use of anodic conversion of Al (in pure Al or AlCu alloy) to Al_2O_3, i.e., conversion of the unwanted metal to its oxide instead of etching it and depositing a dielectric in the spaces thus formed (Platter and Schwartz, 1974). This process had some advantages. The line width was increased, as compared with wet etching using the same mask (the taper angle was ~60°). The surface was planarized. Monolithic vertical interconnects were easily fabricated (Platter and Schwartz, 1974). However, the value of ϵ for the oxide is high and the manufacturability of the process was doubtful. A better replacement for wet etching was the lift-off process.

6.4 ADDITIVE PROCESSES
6.4.1 Lift-Off

In this process the appropriate patterns (interconnects, vias) are formed in a *disposable* matrix which is dissolved (lifted-off) after deposition of the metal. It was used widely for Al and Al alloy patterning as an improvement over wet etching, for dimensional control and over anodic oxidation for process simplicity.

Among the first of these processes to be used extensively employed a trilayer structure. It consisted of a thick base layer of a hard-baked resist (which accommodated topography differences and withstood elevated metal deposition temperatures and in situ sputter cleaning), an inorganic layer, called a "barrier" layer (which does not etch in an O_2 plasma) and a photoresist film (used to pattern the inorganic layer). In the first version (Franco et al., 1975) the "barrier" layer was a metal but transparent barrier films (for improved overlay capability) soon replaced it. Openings were formed in the "barrier" layer; these corresponded to the metal interconnection pattern (the inverse of a metal etch mask). This layer was used as a mask during RIE in O_2 of the underlying polymer layer. The RIE conditions were such that the inorganic layer was undercut, i.e., an overhang or ledge was formed; the edges of the overhang defined the width of the metal line. *Any* metal may be evaporated into the lift-off stencil. The undercut in the stencil and directional deposition of evaporation kept the edges of the metal from touching the polymer walls,

Metallization

but at times, a thin layer of metal extended from the bottom of the metal line, reaching the base of the stencil, making "clean" lift-off more difficult (Dinklage and Hakey, 1984). The final steps were dissolution of the stencil in a hot solvent which removes the unwanted metal deposited on top of the stencil, followed by rinsing and drying the wafer. The sequence of process steps is shown in Fig. V-12.

A modification substituted a soluble polyimide for the hard-baked resist (Milgram, 1983). Several other closely related processes have been patented. In one the use of sputtered metal was allowed, but the "cusps" extending from the body of the metal at the bottom of the stencil had to be etched off (Sebesta, 1985). Shibata (1986) used a stencil made of a composite of resist and plasma SiN.

A single layer stencil, which did not involve RIE, was developed by Hatzakis et al. (1980). It involved soaking a positive photoresist layer in chlorobenzene to produce differential solubility during development. This resulted in an overhang and an undercut profile. A scheme which relied on the taper angle of the resist but did not need an overhang was described by Batchelder (1982). Separating the metal from the mask required that the metal be etched in resist developer which did reduce the line width.

An image reversal process was the next development. Reversed images have negatively sloped undercut profiles that are ideally suited for lift-off processing. The process steps are: exposing a positive resist, baking it at 100°C, exposing it to blanket illumination, and finally developing it. The top of the mask defined the metal width (Moritz, 1985).

Another single layer stencil was formed by exposing and developing a resist layer, using no profile-shaping procedures. Either evaporated or sputtered metal could be used. After deposition, a second layer of resist was applied; since it is

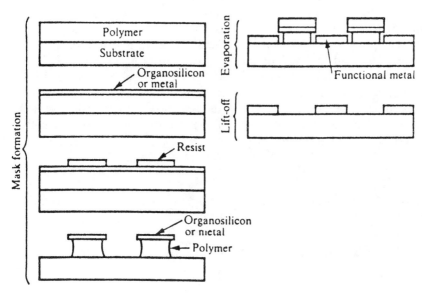

FIGURE V-12 Process sequence for forming a metal pattern by lift-off. (From Fried et al., 1982)

thinnest at sharp edges, a controlled etch-back broke through in that region, exposing the metal which was etched just to clear the sidewalls of the stencil which could then be removed (Pai et al., 1986). In addition to flexibility with respect to the choice of metallization, there was a great improvement in cross-sectional area compared with both wet etching and anodization; the metal taper angle usually varied between 80 and 85° in most of the lift-off processes.

Another advantage of a lift-off process is the ability to expose underlying conductors in a via hole without subsequent attack when patterning the upper metal layer. Highly stressed metals may be used since only limited areas are formed (Fried et al., 1982). One disadvantage of many of the processes is the requirement of line-of-sight evaporation which limits step coverage. The lines are tapered which limits the packing density. The real limit lies in the minimum pitch obtainable. In lift-off patterning, it is not the width of the conductor that limits its usefulness (as it is in subtractive processes) but the space between the lines. The minimum space is determined by the mechanical and adhesive strength of the stencil which defines it. If the stencil is too narrow, it may fall over or collapse or be easily undermined during the premetal processes (Homma et al., 1981). Another problem occurred when using a trilevel stencil (with an evaporated or PECVD oxide barrier layer) to form the vertical interlevel interconnects (studs). Although the process had been used successfully in forming the conductors on a single level, the studs, in the outer regions of an array or in an isolated position, appeared to be corroded i.e., when viewed in the optical microscope, they appeared black. SEM examination, however, showed that this was an optical illusion; the studs were severely tapered and taller than those in the center of an array. The distortion did not depend on the organic material used for the stencil, could not be eliminated by any processing changes, e.g., improved solvent removal, prolonged baking, or elimination of sputter cleaning prior to metal deposition, etc. Since this distortion was not observed when an SiO_2 stencil was used (an impractical solution), outgasing during metal deposition was an obvious culprit. Only along the openings made for conductors or for studs in the middle of an array were there sufficient pathways for removal of the products. The solution was providing a permeable barrier layer, a large area through which the volatile product could escape. The problem was solved when PECVD HMDS was substituted for the usual, but impermeable, barrier materials (Schwartz, 1991).

Although many other versions of lift-off processes have appeared in the literature, both the limitations of the process and the advent of commercially available reactive-plasma-etching reactors designed to etch metals eventually led to the abandonment of lift-off for metal patterning.

6.4.2 Embedment

In this case, the Al or an alloy is deposited into patterns formed in a *permanent* dielectric. This is discussed more fully in Chapter VI.

6.5.0 Reactive Plasma-Assisted Etching of Al and its Alloys

6.5.1 Introduction

Among the earliest reported of plasma-assisted etching of Al was one by Hosokawa et al. (1974), using fluorochlorohydrocarbon gases as etchants. RIE in halogenated gases, such as Cl_2, Br_2, I_2, HCl, HBr, CCl_4 was patented by Harvilchuck et al. (1976) and later described by Schaible et al. (1978). Fluorine-based gases were excluded as etchants since AlF_3 is involatile. Compared with Br-based etchants, the Cl-containing compounds were easier to handle and not as destructive of the existing pump oils and other components of the vacuum systems. Although the vapor pressure of $AlCl_3$ is slightly lower than that of $AlBr_3$ below ~125°C, its vapor pressure is adequate at the use temperatures. Thus, Cl-containing etchants have been used more widely than Br-based ones, despite corrosion and resist-interaction problems, discussed below. I-based plasmas have not been used in any practical process. The toxicity of many of the reactants and products/by-products makes operator safety an important consideration in RIE of Al.

6.5.2 Mechanism of Etching Al in Halogen-Based Etchants

Al + Cl_2 or Br_2: Spontaneous Etching

The native oxide on the Al surface protects it from attack by corrosive species. Once the oxide is removed, Cl_2 and Br_2 etch Al spontaneously, i.e., isotropically. The spontaneous nature of the etch by Cl_2 was demonstrated by Poulsen et al. (1976) who showed that once the etch process has been started, (i.e., the protective oxide removed) the plasma could be extinguished and Al would continue to be etched as long as Cl_2 was kept flowing through the reactor. Plasma beam/mass spectroscopic studies in an ultra-high vacuum were carried out by Smith and Bruce (1982). They found that after the native oxide was removed, the Al etch rate in Cl_x^+ was unaffected by beam bias or plasma power and the rate remained the same even when the plasma was extinguished. They concluded that the etch product was $AlCl_3$ since the appearance potentials in the mass spectrometer of all the $AlCl_x$ fragments were the same as those obtained for sublimed anhydrous $AlCl_3$ crystals. Although at low temperature (33°C) the primary product is the dimer (Al_2Cl_6) and $AlCl_3$ forms at higher temperatures (Winters, 1985), the product is almost universally referred to as $AlCl_3$ when discussing RIE of Al in chlorinated plasmas. Park et al. (1985) observed that clean Al films etched spontaneously in both Cl_2 and Br_2 beams at room temperature without ion bombardment and that the rates were approximately proportional to the halogen molecule pressure and were not enhanced significantly by Ar^+ bombardment. Under bombardment, the rate was approximately the sum of the spontaneous and physical etch rates. Ion bombardment did not change the chemical states of the halogenated Al species significantly. Dissociative chemisorption of the halogen on a clean Al surface seems to be the rate-limiting step, and the sticking coefficient was not changed by ion bombardment. The rates of surface diffusion and reactant desorption were high; the surface coverage of the etch product was very low (~0.1 monolayer). Molecular Cl_2 etches Al four times faster than do Cl atoms, probably due to the enhanced sticking coefficient between the molecule and chlorine bound to the surface. Below 25°C, etching was quenched; this was postulated to be due to the inability of the products and/or contaminants to desorb (Danner and Hess, 1986b). Despite the insensitivity of the Al + Cl_2 reaction to ion bombardment, increased ion energy does increase the etch rate in practical etch systems (e.g., Purdes, 1983). This apparent contradiction is due to the fact that ion bombardment is required to remove surface contaminants (which exist in a RIE system as opposed to an ultra-high vacuum system) and expose the Al surface to the reactants. At pressures at low as 2m Torr, etching of Al in a helicon reactor was via a chemical reaction (Jiwari et al., 1993). Steinbruchel (1986) concluded that in reactive ion beam etching, direct reactive ion etching, i.e., the chemical reaction between the substrate and a reactive ion to form a volatile species, was a component of the etch mechanism.

Native Oxide Removal: The Initiation or Induction Period

It is clear that the native oxide must be removed before the bulk of the film can be etched. The etch rate of the oxide, in a CCl_4 plasma for example, is about two orders of magnitude slower than that of the unoxidized Al (Tokunaga et al., 1981); the etch rate of the oxide is always significantly less than that of Al in any plasma, be it reactive or inert. The oxide film can be sputter-etched by inert ions or by those produced in the etching plasma, but since reoxidation is energetically favored, it can be reformed readily by the residual gases (largely water vapor and air). A hypothesis to account for the improved performance of CCl_4 over Ar in initiating etching was that CCl_3^+ is heavier than Ar^+ and therefore more efficient in sputtering the native oxide (Schaible et al., 1978). But this explanation is untenable since it has been shown that molecular ions are completely dissociated upon impact with the substrate (Steinbruchel, 1984). The improvement is due, in part, to the chemical reduction of Al_2O_3 by radicals formed in the plasma from the parent molecule: CCl_x (from CCl_4) (Poulsen et al., 1976). BCl_x (from BCl_3) (Poulsen et al., 1976; Ingrey et al., 1977), $SiCl_x$ (from $SiCl_4$) (Danner et al., 1987) and BBr_x (from BBr_3) (Keaton and Hess, 1985). Ion bombardment assists the reaction. Initiation time was longer in BBr_3 than in BCl_3, most likely because the etch rate of the oxide was lower. However, another property of the radicals is, perhaps, most important; it is the ability of the radicals to scavenge oxygen and water vapor and so prevent reoxidation. The scavenging efficiency is $BCl_3 > SiCl_4 > CCl_4$ which may be explained by thermodynamic considerations and steric effects (Danner et al., 1987); the scavenging effect of BCl_3 and BBr_3 should be similar (Keaton and Hess, 1985). The initiation period, i.e., the time required to remove the native oxide and start etching the bulk Al, is thus a combination of sputter- and chemical-etching the oxide while preventing reoxidation of the Al. It is generally agreed that the initiation period is shortest in BCl_3 plasmas but that precautions must be taken to reduce moisture contamination in the reactor to minimize the occurrence of nonreproducible total etch times. Vossen (1983) suggested pretreatment by a hydrogen glow discharge for improving variable induction times, but this does not appear to have been used widely. Other means of reducing the moisture in the reactor are: keeping the chamber walls warm at all times, extending the pump-down times, using a cryogenic trap, and installing entrance and exit load locks; this last is now an integral part of modern metal etchers.

Anodization is a method of forming surface oxide films of known thickness (e.g., Young, 1961). Thus a direct correlation between the oxide thickness and initiation time and its influence on total etch time was demonstrated by etching-to-completion Al films which had been anodized to form different thicknesses of oxide (Schwartz et al., 1986).

If there are nonuniformities in the surface oxide, the differential between the etch rates of the oxide and bulk Al can produce random surface roughness and, at times, distinct residues (Chapman and Nowak, 1980).

Metallization

Slow etching of the oxide layer increases the difficulty of RIE of a metal film deposited over a step because oxide etching is directional and the *vertical* oxide thickness on a curved surface is not uniform, as illustrated in Fig. V-13; the steeper the step, the greater the vertical oxide thickness (Maa and Hanlon, 1986). Thus, even after an extensive overetch, a wall of oxide may still remain adjacent to the metal sidewall.

Since grain boundaries contain a higher concentration of oxygen than do the grains, grains etch preferentially to the boundaries as seen in Fig. V-14 (Schaible et al., 1978).

Anisotropic Etching

In other systems (e.g., RIE of Si), it has been shown that directional etching is a result of radiation-enhanced gas-surface chemical reactions where the radiation is energetic directional positive ions (Coburn and Winters, 1979). In the case of Al etching in a Cl- or Br-based plasma, one might postulate that the vertical etch rate is enhanced relative to the lateral rate, to achieve directionality, because the reaction product is $AlCl_x$ or $AlBr_x$ which is further reacted, sputtered, or desorbed by energetic ion bombardment. However, the beam work discussed above disproved this hypothesis, since complete reaction of Al in the halogen gas did not require ion bombardment. The dominance of etching by neutral species is also illustrated by the significant loading effect observed in RIE of Al in CCl_4 (Schaible et al., 1978; Lin et al., 1989; Tsukada, 1991).

It is now accepted that anisotropy is a result of sidewall passivation. Schaible and Schwartz (1979) demonstrated the existence of a polymeric films on the sidewall of an Al structure etched in CCl_4/Ar using an oxide mask. Any polymer formation on the horizontal surface would be removed by the bombarding ions and is one of the reasons for the increase in etch rate with increasing rf power. Smith and Saviano (1982), in mass spectral analyses of Cl_3 plasmas containing various chlorocarbons, concluded that the C_xCl_y plasma products were responsible for sidewall passivation. In BCl_3-based plasmas, a surface-recombinant mechanism,

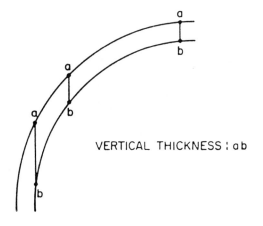

FIGURE V-13 Diagrammatic representation of the variation of the vertical thickness of the surface layer of native Al_2O_3 formed on an Al film deposited over a step. (From Maa and Hanlon, 1986)

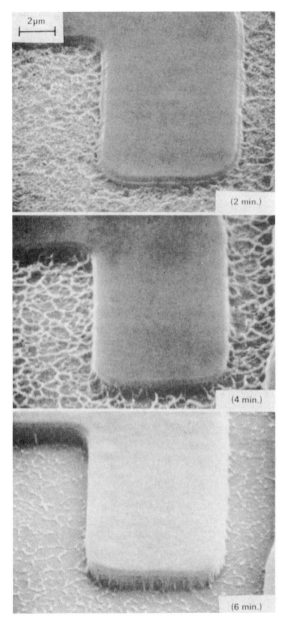

FIGURE V-14 SEMs showing slow etching of oxide concentrated in Al grain boundaries. (From Schaible et al., 1978)

i.e., the removal of Cl_2 at the sidewall by BCl_x, was suggested by Flamm and Donnelly (1981). This was refuted by Schwarzl and Beinvogel (1983) who proposed instead that the protective layer originated in the photoresist mask as suggested earlier by Smith and Saviano (1982).

Etch rates of Al by ion bombardment were reduced in the presence of CCl_4 and CBr_4. These molecules are chemisorbed on a clean Al surface and form a carbidelike layer. Exposure to an ion beam and CCl_4 resulted in the production of chlorinated Al species of lower oxidation states which were easily removed by ion bombardment although the carbidelike species were not. This is consistent with the protective action of the etched sidewall by C-containing species produced in the plasma (Park et al., 1985). The need for sidewall protection has led to the inclusion of polymerizing species in BCl_3/Cl_2-based etchants, such as a fluorocarbon (Iida et al., 1981; Wang et al., 1983), $CHCl_3$ (Bruce and Malafsky, 1982, 1983), and CH_4 (Lutze et al., 1990). In addition to including $CHCl_3$ in the etching gas, Dohmae et al. (1990), used a CHF_3-based encapsulating step before the final overetch step. Fujino and Oku (1992) claimed that in $HBr/BCl_3/Cl_2$ or HBr/BCl_3 plasmas, the sidewall film derived from the photoresist, consisting of C, Br, and Al, was a better passivant than those derived from Cl. Improved profile control was achieved in a TCP reactor when BCl_3/Cl_2 was replaced by HCl/Cl_2 (Yang et al., 1994).

After etching, the sidewall films are usually removed during resist ashing.

There are examples of inorganic sidewall protective films as well. Bollinger et al. (1984) suggested that F-based additives might protect the side walls by forming AlF_3 on them. According to Sato et al (1987) lateral etching is inhibited in $SiCl_4$ RIE of Al by silicon deposition. Sidewall protection was provided by sulfur deposition when etching Al in S_2Cl_2 in an ECR reactor at 20°C; the protective film was removed by sublimation at ~100°C (Tatsumi et al., 1992). Al was etched in Cl_2/N_2 in an ECR reactor, at 60°C with an SiO_2 mask; sidewall protection was provided by a film which consisted of two layers, an outer one Al-O, Al-N, Si-O, Si-N and the inner containing Al with added Cl. The sidewall film was removed in a BCl_3 plasma (Kawamoto et al., 1994). Fu et al. (1991) claim that intrinsically anisotropic etching can be obtained using $SiCl_4$ to etch an Al alloy in MERIE reactor because of the high degree of ionization and low pressure (4 mTorr). Although the low pressure reduces gas scattering so that the etchants impinge in a more normal direction, the high degree of ionization can be useful only in clearing the surface, as discussed above. This suggests that Si-species, from the gas and the oxide mask, may contribute to sidewall passivation, as postulated by Sato (1987). Although biased ECR etching in Cl_2 resulted in vertical profile and the same argument of high ion density and low pressure was invoked to explain the result (Samukawa et al., 1991), it is likely that the resist mask provided the species for sidewall protection.

6.5.3 Practical Etchants

BCl_3 etches Al slowly. This appears to be due to limited dissociation (production of Cl) in the plasma. This view is bolstered by the relative insensitivity of the etch rate to temperature, indicating a gas-phase reaction as the rate-limiting step. CCl_4 etches Al at a faster rate but the rate is temperature sensitive (Schaible et al., 1978; Tokunaga et al., 1981), a characteristic of a surface-reaction rate limit. As expected, addition of Cl_2 accelerated the etch rate (Kurisaki et al., 1982; Horiike et al., 1982); the increase was faster for BCl_3 than for CCl_4 (Danner and Hess, 1986a). Using a mixture of PCl_3 and BCl_3 in a RIE system increased the etch rate over BCl_3 alone; there was no undercut or residue formation and the selectivity to SiO_2 was improved (Nakamura et al., 1981). But this mixture has not been used by others.

Nitrogen was said to be the key to improving the passivation of sidewalls, e.g., using BCl_3 and Cl_2 in a single wafer RIE (Clayton and Besson, 1993) and in a HDP reactor using Cl_2 (Liao et al., 1995).

CCl_4 is no longer used as an etchant because it is a carginogen. This narrows the choices to BCl_3^-, $SiCl_4^-$, and BBr_3^- based etchants, with Cl_2, Br_2, and HBr to increase rates, and polymer-forming gases for sidewall protection, where needed. The combination, $BCl_3 + Cl_2$ is used most frequently. The choice of passivants has become narrower, since $CHCl_3$ is a carcinogen, and the PFCs are now classified as environmental hazards.

6.5.4 Reactors

Although some of the early work was done in the "plasma" mode of reduced ion bombardment, RIE has been the method of choice for many years. Magnetic enhancement of RIE and biased high-density plasma reactors will likely edge out the lower density reactors in time. The high-density reactors, in which high-rate etching can be accomplished at low pressures, are certainly an improvement; the low pressure decreases gas scattering and thus the tendency toward lateral attack and increases the ease of desorption of reaction products, as pointed out by Puttock et al. (1994).

6.5.5 Practical Etching Processes

The issues of etch rates, etch-rate ratios with respect to mask and substrate, profile, and uniformity, are determined by the film itself, e.g., whether Al or an alloy, composition of the alloy, deposition temperature or any heat treatment before etching, oxygen contamination (Eldridge et al., 1987), choice of reactants, their proportions, process parameters such as power, pressure, flow rate (residence time), frequency, masking material, and reactor configuration, which cannot be summarized to any useful extent. The individual papers must be consulted and analyzed. The equipment manufacturers now provide a useful start-up process which they have developed for their own reactors.

6.5.6 Al Alloy Etching

As mentioned earlier, pure Al is almost never used as an interconnect. Si, to prevent junction penetration (Kulper, 1971), and Cu, to improve electromigration resistance (Ames et al., 1973), are added either singly or together. As discussed previously, the deposition conditions, as well as any postdeposition heat treatments affect the film structure which, in turn, influences its etchability (e.g., Maleham, 1984; Hu et al., 1987; Abraham, 1987) Al/Si alloys would be expected to behave like pure Al since Si etches readily in chlorinated plasmas under the influence of ion bombardment, although Maa and O'Neill (1983) found that, in RIE of Al/Si films in $CCl_4/N_2/BCl_3$, the etch rate of Si was much less than that of Al. The possible causes of Si residues, in addition to a rate differential, are oxidation of the Si or the Si precipitates, or inadequate substrate bias to counteract these effects. Cu, on the other hand, is expected to pose a problem since the vapor pressure of the copper halides is very low. An ingenious method of avoiding the problem of RIE of AlCu while forming AlCu conductors, was to deposit Cu using a lift-off stencil and to use RIE to remove the Al regions not coated by Cu (Chiu et al., 1982).

Cu-containing Residues

As indicated above, the Cu-content of Al alloys complicates etching because of Cu-rich residues may be left on the surface after etching. As discussed previously, Cu often concentrates at the metal/oxide interface during deposition, increasing the difficulty of etching. An underlay of Al inhibits the pile-up of Cu, making residue-free etching easier. The residues may be either involatile reaction product or unreacted large $CuAl_2$ (θ-phase) precipitates that had formed within the film (Abraham, 1987; Hu et al., 1989; Suzuki et al., 1992); these act as micromasks, forming conical residues.

In the early days of RIE of Al-Cu alloys, residues were sputter-etched in an inert plasma. Departing from the use of simple sputter-etching, two opposing procedures were used to eliminate the θ-phase particles at the conclusion of RIE. Hu et al. (1989) finished the etch process by lowering the pressure in the reactor, thereby increasing the ion bombardment and, in effect, sputter-etching the particles in the reaction mixture. On the other hand, Suzuki et al. (1992) added a step in which there was significant lateral etching; the residues presumably were undercut and thus etched away.

Elevating the wafer temperature by using a heat-conducting medium between the heated substrate holder and the wafer or enhanced ion bombardment to increase sputtering and heat the wafer, prevent the formation of a Cu-rich residue. Heating the substrate holder without thermal bonding between it and the wafer may be of benefit simply because the gas surrounding the wafer is hotter so that the reaction product may be pumped away before it redeposits on the wafer. By keeping the substrate at 100°C, residue-free etching was accomplished in biased ECR etching in Cl_2 (Samukawa et al., 1991a,b). However, raising the wafer temperature restricts the choice of masking materials and increasing the sputtering component degrades selectivity to both mask and substrate.

Tsukada (1991) reported that magnetron etching was very useful in removing residues. Yet, in $BCl_3/Cl_2/N_2/CF_4$ in a MERIE reactor, residue was observed. There was less residue at high rf powers, slower etch rates (i.e., lower Cl_2 concentration) and high cathode temperatures. There was more residue in the dense areas than in the open ones. Residue formation was minimal at very low pressures, and relatively sparse at very high pressures; residue was densest at intermediate pressures (Mak et al., 1992).

Sparse residues resulted after RIE in a single wafer etcher in HBr/BCl_3 or $HBr/BCl_3/Cl_2$ (Fujino et al., 1992).

Residue-free etching has been reported by a number of workers: RIE in $SiCl_4$ (Sato et al., 1987), ECR using Cl_2 (Samukawa et al., 1991a) BCl_3/Cl_2 (Samukawa et al., 1991b; Bradley et al., 1991), $BCl_3/Cl_2/N_2$ (Marx et al., 1992). Interposing BCl_3 "sputter-etch" cycles at several points during RIE in BCl_3/Cl_2 in a MERIE reactor resulted in residue-free etching at 45°C and 8 Pa (Hattori et al., 1994).

A patent by Webb (1994) disclosed a low temperature (low pressure, high power) process for RIE of AlCu alloys with a Cu content >0.5% Cu. The process was carried out in a single wafer magnetically enhanced, capacitively coupled reactor. Elimination of the etch residue depended on controlling the N_2 content of a BCl_3/Cl_2-based etchant mixture; the higher the Cu-content, the lower the N_2-concentration. However, higher Cu-content alloys required higher temperatures, e.g., for a 2% alloy, the temperature was ~100°C, but whether the temperature referred to that of the wafer or the wafer holder was not specified.

Reacting $AlCl_3$, produced in the chamber in large quantities from a source of Al external to the wafer, with the copper chloride as it was formed on the surface of the film during RIE, prevented the formation of residue by volatilizing Cu during etching. The volatile compound is a copper-aluminum-chloride complex (Bausmith et al., 1990). This process was also used by Narasimhan et al. (1992) in a MERIE reactor and Sato (1994) in a static magnetron triode RIE system (SMTRIE).

Addition of N_2 to a BCl_3/Cl_2 mixture in an ECR reactor suppressed residues (and postetch corrosion), but in a MERIE reactor addition promoted residue formation (and corrosion). It was postulated that BCl_x^+ ($x = 0-3$) in the plasma produced the benefit by enhancing an ion-assisted reaction and/or by physical sputtering); these species were elevated in the ECR and suppressed in the MERIE reactor (Kusumi et al., 1995).

It can be seen that many methods have been proposed for removing or preventing the formation of Cu-rich residues; there are some conflicting reports as to the success of the methods. In order to have a successful process, no residue may remain on the surface at the end of the process. For each user/reactor/etchant system, some method has been evolved which produces a satisfactory product although the experience of others may lead them to disagree with the theory or practice.

It is also possible to remove the residues, post RIE, by rinsing the wafer in HNO_3 (Herndon and Burke, 1977; Nakamura et al., 1981). This treatment also helps passivate the surface, but is rarely done now.

6.5.7 Feature-Size Dependent Etch Rates

A pattern-sensitive clearing effect is seen even when etching pure Al films; it is, perhaps, exaggerated by the formation of Cu-rich residue during AlCu RIE. Overetching to clear the residual metal in narrow spaces may result in drastic overetch of large features and may produce substrate damage. The definition of the various terms used to describe the phenomenon, the distinctions among the terms, and the basic mechanisms proposed have been covered in Chapter I. In many of the reports on etching Al or its alloys, the terms were used somewhat loosely if the guidelines relating to definitions and measurement techniques stated by Gottscho et al. (1992) are taken as definitive. For example, the term ARDE has been used when only a single thickness of metal and mask were etched (space width was translated into aspect ratio) and the etch rate dependence on etch time (for various size spaces) was not measured.

In the case of Al alloy etching, an inhibitor (identified in some cases as redeposited sputtered mask fragments) has been identified as the cause of slow etching in narrow spaces.

In one instance of etching an Al alloy film in $Cl_2/BCl_3/N_2$ in a MERIE reactor, by controlling the process so that it was in a transport rate limited regime, the etch rate difference between tight spaces and open areas could be reversed by changing the BCl_3/N_2 flow rate ratio. The normal lag was observed at a high flow rate ratio. In this case, there was a large sputtering component which was attributed to the BCl_3; a large amount of sputtered material was available to deposit in small spaces. At the same time there was a good supply of reactants to the open areas in which redeposition was minimal. By reducing the flow rate ratio, the sputtering component was decreased so that there was less deposition in the small spaces to inhibit etching. Also, since the residence time was increased, reactant supply to the large areas was reduced. Thus the lag was reversed. Other factors that reduced the normal lag was increasing magnetic field strength, increasing pressure, and decreasing the Cl_2 flow rate. Thus there should be some optimal conditions at which loading is minimized, or perhaps eliminated entirely (Huang and Siegel, 1994).

Ma et al. (1996) examined ARDE in a high density, high etch rate reactor, the decoupled plasma source (DPS) metal etch system. They also found that high etchant concentration enhanced the etch rate in open areas and deposition of resist by-products reduced the etch rate in dense areas. The optimized process conditions (e.g., source power, choice of etchants, and etchant ratios) which balance etch and deposition can minimize microloading at both high and low pressures. They also reported that the breakthrough step played a major role in microloading and could cause poor performance despite good microloading in the main etch.

In an ECR reactor, RIE lag was shown to depend on the substrate rf bias and the mask material, i.e., etching in small spaces was inhibited by the deposition of sputtered mask fragments. The amount of material deposited depended on the distance from the mask, so that narrow spaces receive more of this material. There was less lag when SiO_2 was used as a mask. The angular distribution of ions impinging on the etching surface had no effect. The concept of a limited flow of neutral species into the narrow spaces appeared to be valid only in the case of etching with an SiO_2 mask or etching with a resist mask with zero applied bias. Thus the proposal for reducing RIE lag: lower the rf bias and use an inorganic mask. However, profile control will be exceedingly difficult (Sato et al., 1995). Xie and Kava (1996) and Xie et al. (1996) reported that ARDE in metal etching was more severe in high-density plasmas operating at low pressures. They found that ARDE in a MERIE reactor was controlled by the effective ion to neutral flux ratio and the ion to inhibitor flux ratio. A high ion to inhibitor ratio resulted in a higher etch rate but greater RIE lag; a lower one reduced both the etch rate and lag. At the lowest ratio, reverse lag and a low etch rate may result. They concluded that absolute AR-independent etching may not be realized but may be minimized while keeping a high etch rate. The nature of the inhibitor was not identified.

A simulation code was developed (Aoki and Sasamura, 1996) in which the flux rates (gas particles, ions, and sputtered mask fragments) are calculated as a function of trench AR and then the etch rate is calculated with these flux rates at each AR. The surface reaction parameters are determined by fitting the calculated results to the experimental etch rates. They confirmed that microloading was affected by the sputtered resist fragments. They conclude that there might be less microloading in a RIE system, where the resist sputtering rates are lower, than in the high-density plasma systems.

Microloading was reduced significantly when the HCl/Cl_2 was used instead of BCl_3/Cl_2 in a TCP system. The improvements were said to be due to the change from a more chemical process to a more physical one. The importance of polymer deposition in RIE lag was again emphasized. They had some preliminary results that indicated that replacing photoresist with a carbon mask reduced microloading effects (Yang et al., 1994). Microloading was eliminated, for features with a 0.4-mm space and AR >4, by using an N_2/Cl_2 plasma in an independently biased high-density reactor. The role of N_2 was to enhance sidewall passivation and raise the etch rate by increasing the Cl radical concentration (Liao et al., 1995).

6.5.8 Profile Control

Prevention of undercutting, i.e., producing features with vertical walls and replicating the dimensions printed in the mask, has been the aim of most processes, as discussed above under anisotropic etching. This has involved, for the most part, controlling the concentration of Cl_2 and ensuring adequate sidewall passivation.

Some processes designed to produce positive sidewall tapering (as opposed to undercut) have been developed in order to improve the step coverage by an overlying insulator. The earliest attempts depended on resist erosion: Booth and Heslop (1980) using resist tapered by baking, and Nakamura et al. (1981) using a CCl_4/H_2 mixture, in which the resist and Al were etched at almost equal rates. They were unsuccessful; control of the resist profile and etch nonuniformity were cited as reasons. A later attempt at using the same principle of controlled resist erosion was more successful. Abraham (1986) used a hexode reactor with BCl_3-based etchants and an etch process based on high dc bias/pressure ratios. The faceting of the resist angle resulting, in this case, of tapering independent on feature size (Abraham, 1986).

Another approach has been balancing deposition and etching of a polymeric film on the sidewall. In one instance, the taper angle was determined by the ratio of $CHCl_3$ to Cl_2 in the etch mixture using a MERIE system (Arikado et al., 1986). A similar process used CHF_3/Cl_2/BCl_3 in a hexode reactor; the taper angle was determined by the CHF_3 and Cl_2 flow rates, bias, and resist thickness. By a suitable choice of parameters, angles ranging from 60 to 90° could be obtained (Selamoglu et al., 1991). However, the taper angle decreased with increasing spacing of the aluminum lines.

A process resulting in tapered sidewalls was based on sidewall passivation due to resist *consumption* rather than resist *erosion*. It was carried out in a biased TCP reactor using BCl_3/Cl_2/N_2 and the inclusion of N_2 was the key to forming sloped profiles. Substrate bias, TCP power, and gas flows were determining factors. The dimension of the top of the line was the same as that of the initial mask dimension. In this process, also, the slope angles depended on the space between the lines, decreasing with increasing spacing, with the maximum slope for isolated lines. The deposited polymer was readily removed when the resist was removed (Allen and Rickard, 1994).

6.5.9 Corrosion Control

Prevention and control of post-RIE corrosion of Al and its alloys has presented a major challenge. The presence of Cu in the film and in the residue (if formed) enhances the susceptibility to corrosion because of the galvanic action of dissimilar metals and because the involatility of the CuCl leaves a higher concentration of Cl on the etched surfaces. The higher the copper content in the film, the greater the corrosion susceptibility (Lee et al., 1981; Rotel et al., 1991). The surface oxide, which plays a role in corrosion protection, is though to contain discontinuities in the case of AlCu films (Zahavi et al., 1984).

The reaction product, $AlCl_3$, hydrolyzes in moist air forming HCl. The HCl is a source of Cl^- which destroys the passivating native oxide, reacts with Al to form a soluble compound which hydrolyzes in moist air, forming HCl, which, in turn reacts further. It is the regeneration of HCl that is responsible for the massive corrosion even when the surface concentration is relatively low. Cl has been found to be bonded both to the metal lines and to the C in the sidewall layer as well as the resist. The corrosion product observed on the metal surface or extruding from the sidewalls is hydrated aluminum oxide or aluminum hydroxide. Corrosion results in increased resistance of metal lines and may even cause cracking of the overlying dielectric films (Wada et al., 1987). The use of caps or underlays of other metals (e.g., Ti, TiW) increase the susceptibility to corrosion (Maa, 1990).

Prompt plasma stripping of a resist mask is thought to be helpful in inhibiting post-RIE corrosion, although there is some disagreement as to when this should be done in the post-RIE process cycle, due to the possible acceleration of corrosion by heating the wafer in the presence of Cl. A wide variety of passivation treatments have been proposed, but there is disagreement about the effectiveness of each; in fact the results published in one paper may be contradicted in another. A sample of the proposed post-RIE treatments are (1) low-temperature thermal oxidation of the surface (Lee et al., 1981). (2) rinsing in DI water immediately or keeping the wafers in an inert ambient (e.g., dry N_2) if rinsing must be postponed, (3) rinsing in phosphochromic acid solution (Eldridge et al., 1983), (4) post-RIE exposure to a F-containing plasma, e.g., CHF_3 (Tsukada et al., 1983), CF_4, $CF_4 + O_2$, substituting F for the Cl on the walls, while depositing a passivating polymer film; Fok (1980) suggested following the plasma exposure by rinsing in fuming HNO_3 to remove the fluoride and oxidize the surface and Iida et al. (1981b) by using an NH_3-containing plasma and water rinse, (5) microwave downstream resist strip in O_2/NH_3 (Hwang and Mak, 1992), (6) rinsing in an aqueous alkaline solution and then in water (Iida et al., 1981a), (7) a combination of heat treatment and an organic solvent rinse (Samukawa et al., 1989). Evaluation of corrosion has usually been by visual inspection. However, Brusic et al. (1991, 1993) carried out electrochemical measurements, and concluded on the basis of those experiments, that the most effective post-RIE treatment (of AlCu) was an *immediate* water rinse, followed by resist strip, immersion in phosphochromic acid, then rinsing and drying. CF_4/O_2 treatment prior to the water rinse had a slight but negative influence of corrosion resistance. Brusic and Yang (1996) compared the corrosion susceptibility of AlCu films etched in a RIE system using HCl/Cl_2 and BCl_3/Cl_2 as etchants. They again used electrochemical methods and examined the wafers minutes after being removed from the etcher, which included as ashing station. The wafers exposed to HCl/Cl_2 dissolved more slowly at all potentials. This was explained by the nature of the polymeric films; HCl-treated wafers were more wettable, more cleanable in an O_2 plasma and thus more readily formed a passivating oxide. *Immediate* water rinsing and phosphochromic acid etching were again found to be very effective in further reduction of the corrosion rate. Wafers etched in BCl_3/Cl_2 were more sensitive to any delays before water rinsing. Addition of N_2 and an extended anneal before resist stripping had a beneficial but small effect. Here too, the effects of CF_4 on AlCu corrosion were well-pronounced and negative. These two electrochemical studies appear at variance with many other studies, in which visual inspection was the criterion, in rejecting the benefits of F-containing plasma treatment.

By reducing the substrate temperature during etching in Cl_2 in an ECR reactor, corrosion of AlSiCu was reduced (Aoki et al., 1991). They found that after water rinsing, the Cl concentration was much smaller on wafers etched at $-60°C$ than on wafers etched at $+30°C$. They postulated that the smaller number of Al-Cl bonds left on the surface was due to reduced chemical reaction and diffusion rates at the lower temperature. Adding N_2 to the BCl_3/Cl_2 etch mixture in an ECR reactor also reduced post-RIE corrosion. (Kusumi et al., 1995). Corrosion across a wafer was not uniform but was related to the pattern density. Corrosion was heaviest and the concentration of Cl highest in regions of intermediate spacing. They proposed that there was a limit to the supply of the product $AlCl_3$ in narrow spaces and a depletion of reactant Cl in wide spaces i.e., a microloading effect (Gabriel and Wallach, 1992). Levy (1992) patented a process for etching Al without forming corrosive Cl-containing residues by using a mixture of one or more Br-containing compounds and SF_6. Gebara et al. (1994) pointed out the different types of photoresist absorb Cl differently, so that the choice of resist also has an influence on corrosion.

Polymer build-up in a RIE system when $CHCl_3$ was added to the reaction mixture for better sidewall protection was the cause of increased corrosion susceptibility. The sidewall polymer was contaminated with Cl, either by reaction or absorption. The polymer eventually deposited on the wafers, masking the metal, leaving cluster defects. In addition, it was difficult to remove the deposited polymer which then reacted with moisture and corroded the metal. Reducing the $CHCl_3$, and increasing the Cl_2 concentration, raising the chamber temperature and lowering the pressure, reduced the build-up of polymer (Dang, 1996).

Corrosion pits (voids), in the absence of visible corrosion product, have been observed on the sidewalls of the etched features after RIE in both a batch (hexode) and a single wafer etcher (Daubenspeck and Lee, 1992), and in a TCP reactor (Hill, 1996). Daubenspeck and Lee (1992) stated that the reaction was thermally driven and did not require the presence of moisture. The voids were most numerous at the interface between AlCuSi films and Ti or TiN layers, but existed in the bulk of the film as well. Void formation appears to be due a reaction of the Cl-containing etchants or residues on the sidewalls with the alloy at imperfections in the sidewall passivation either during etching or the stripping process. Daubenspeck and Lee (1992) opt for their occurrence during the stripping operation. Thus to eliminate the voids, the wafers must be heated gradually to remove the residual Cl-contaminant to reduce the probability of a thermally driven reaction, before the resist is stripped. Hill explained that void formation was a case of galvanic corrosion in the neighborhood of θ particles at grain boundaries; the corrosion was caused by rinsing in hot DI water after stripping the resist. Using room temperature water eliminated the problem.

Thus, it is evident that there are many different and sometimes contradictory recipes for corrosion inhibition. Each laboratory or fabrication line most probably has a favorite post-RIE treatment, which satisfy the product requirements.

No mention of corrosion was made in the paper in which HBr was the sole etchant (Aoki et al., 1992). There is essentially nothing in the literature to indicate what might be expected (e.g., is there regeneration of Br^- in a hydrolysis reaction?) apart from the fact that HBr, like HCl, corrodes Al.

6.5.10 Masking

Both inorganic and organic masks are used in RIE of Al and its alloys. In addition to the usual concerns about the etch rate of a mask, and in the case of an organic mask, its thermal stability, there is the chemical interaction of the etch product ($AlCl_3$) with photoresist. The result is excessive global degradation and localized pitting (Spencer, 1983, 1984). This has been explained by the fact that $AlCl_3$ is a Lewis acid which reacts readily with organic materials (Hess, 1982). Another problem is incorporation of $AlCl_3$ into the resist mask; during resist ashing the chloride is converted to the oxide which remains on the surface.

Thus an addition step, such as immersion in phosphochromic acid or a mild alkaline solution, is required. Resist stabilization in a plasma or by uv exposure and hard-bake before RIE was reported to minimize or eliminate the problem (ter Beek, 1985).

The use of a trilevel resist has several advantages; the lithographic resolution is independent of its thickness; loss of masking due to an inadequate etch ratio is no longer an issue. In addition, the walls of the mask are vertical and not distorted by heat. However, during RIE transfer of the pattern in an O_2 plasma into the thick underlay, the Al surface beneath the mask is sputtered and a thick layer of oxidized Al is deposited on the walls of the mask, changing its dimensions and interfering with mask removal (Kinsbron et al., 1982a). There is another residue which is deposited from the hard mask during RIE in O_2; it looks like stalks of grass in the SEM and has, therefore, been named "RIE grass." Both kinds of residue can be removed, before etching, without attack of the Al, by immersion in a mixture of ethylene glycol and buffered HF (Kinsbron et al., 1982b). However, this puts an additional burden on the adhesion of the mask to the substrate, and when the lines are very narrow, the mask may not survive.

6.5.11 Temperature Effects

Elevated wafer-holder temperatures had been reported to be necessary for etching Al-Cu films without leaving residues (Schaible et al., 1978) but this was disputed, e.g., by Chambers (1982) who used as a low-frequency (380 kHz) plasma for etching. Subsequently, it was realized that merely using a heated wafer holder in a low pressure reactor does not heat the wafer (Schwartz and Schaible, 1981) but probably prevents redeposition. However, the low frequency ion bombardment, more energetic than 13.56-MHz bombardment, probably did heat the wafer to assist in volatilization of CuCl and the sputtering would also be more efficient.

The differences between the temperature sensitivity of CCl_4 and BCl_3 etching was discussed earlier.

High wafer temperatures are usually avoided, where possible, to prevent enhancement of the thermally-driven isotropic reaction and to prevent thermal distortion of the resist mask, i.e., flowing and reticulation. When necessary, resists can be stabilized to withstand elevated temperature by either plasma (Ma, 1980) or uv (Haroka and Pacansky, 1981) hardening, followed by a hard-bake.

Cooling the wafer during RIE in a $BCl_3/SiCl_4/Cl_2$ plasma suppressed undercutting; the resulting taper of the profile could be adjusted by changing the temperature (using He between the wafer and the holder for temperature control) (Nakamura et al., 1987).

Thin AlCuSi films were etched in Cl_2 in a biased ECR reactor at low temperatures (down to $-50°C$). Only relative etch rates were given; there were none for the metal itself. SiO_2 was a better mask than resist; the selectivity to SiO_2 was higher and improved as the temperature was reduced. The decrease in etch rate as the spacing decreased (microloading) was smaller as the temperature was reduced from $+40°C$ to $-50°C$. This was explained by postulating that a low vapor pressure precursor was formed on the surface, its thickness independent of the supply of Cl (which is larger in the wider spaces) and its removal dependent on ion bombardment which is also independent of spacing. There was less dependence on line spacing if an SiO_2 mask was used instead of resist and more dependence when HBr was substituted for Cl_2. The change in line dimension, compared with the dimensions of the SiO_2 mask (critical dimension shift) could be controlled by adjusting the temperature. In Cl_2, the line width was decreased at higher temperatures due to lateral etching. At all temperatures there was a positive shift when HBr was used; for Cl_2, zero shift was obtained at 30°C. There was no mention of any residues (Aoki et al., 1992).

The etch rate (in Cl_2, in a biased ECR reactor) was constant as the wafer temperature was reduced from $+50°C$ to $\sim -10°C$; below this temperature the etch rate increased rapidly with decreasing temperature; the rate appears to saturate, at almost double the high-temperature rate, at $\sim -50°C$. The explanation for this phenomenon was based on the model of Bermudez and Glass (1989) which postulated that there are two kinds of adsorption sites for Cl on Al, namely, surface and subsurface sites. Physical adsorption occurs at the surface sites, but the chlorine evaporates before reacting. Chemical reaction occurs at subsurface sites and the product is desorbed thermally. Thermal desorption decreases as the temperature is decreased, but desorption may be induced by ion-bombardment. In addition, there is increased adsorption at surface sites; a cluster model was proposed, i.e., at low temperatures, there are attractive interactions between chlorine molecules on the surface (cluster formation) which decreases the probability of evaporation. Under the influence of ion bombardment two events occur at the surface sites; chemical reaction and desorption. The conclusion is that the increase in etch rate at low temperature is due chiefly to the processes at the surface sites (Uchida et al., 1993).

6.5.12 Non-Cu Alloys

As discussed previously, alternatives to Cu-doped Al have been proposed to increase electromigration lifetime but easy etchability is an important criterion in choosing a new component. Ti, Ta, or Hf, used in some of the multilayer structures, etch readily in chlorinated plasmas, but Cr, which had been suggested, does not. It is claimed that the Pd alloy can be substituted for the Cu alloy; it can be etched in a CCl_4/He plasma with less tendency for undercutting, due to a thin film of palladium oxide on the sidewall. However, no mention was made to the presence or absence of any residues (Koubuchi et al., 1990).

7.0 PATTERNING OF COPPER
7.1 INTRODUCTION

Wet etchants for Cu have been compiled by Kern and Deckert (1978). They have not been seen as suitable for defining the small dimensions now required for VLSI and ULSI devices. However, a new wet etchant for Cu, used to define interconnection test patterns, was reported by Takewaki et al. (1995). It was a mixture of H_2O_2 (to oxidize the Cu), acetic acid (to etch the CuO) and H_2O.

Sputter etching and ion milling are viable techniques but are not now used extensively for patterning.

If it essential that Cu migration into the surrounding dielectrics be prevented, after Cu is etched, it must be encapsulated in an appropriate barrier layer, increasing complexity of processing.

7.2 REACTIVE ION ETCHING OF COPPER

Copper can be reactively ion etched in Cl-containing plasmas at an elevated temperature and with high energy ion bombardment. Rates as high as 5000 A/min were obtained in a CCl_4/Ar plasma using a high rf power, to raise the wafer temperature and a heated electrode to prevent condensation of the etch product. The inability to etch in Cl_2, despite the nonprotective oxide formed on the surface, was postulated to be due to the formation of an inhibiting film of an oxychloride formed from the residual gases in the chamber. Thus the gettering action of CCl_4, and perhaps the ion bombardment, were needed here as well as in etching Al. The elevated wafer temperature required a soluble heat-resistant mask; ashing in the presence of chlorinated residues resulted in the formation of thick black layer on the surface of the copper film. MgO was chosen as the mask. Mo, which etches in the same plasma, was deposited on the Cu surface before deposition of MgO through a lift-off stencil, to insure adhesion of the MgO and prevent interdiffusion. After etching both the MgO and the Mo were removed in solutions which did not attack Cu. The large sputtering component, due to the high energy ion bombardment, resulted in a relatively low etch rate ratio to the underlying SiO_2, but the erosion rate of MgO is very low. In this system, narrow spaces cleared more slowly than did the wider ones. The sidewalls of the Cu were vertical and no corrosion was observed when the etched samples were rinsed in DI water and then exposed to the laboratory environment (Schaible and Schwartz, 1982).

A $SiCl_4/N_2$ mixture was used successfully to etch Cu at temperatures above 220°C. Addition of N_2 had several advantages: (1) it improved the clearing of narrow spaces, presumably by reacting with redeposited organic material to form cyanogen, (2) it formed a protective sidewall film of SiN and the taper angle was controlled by changing the flow of N_2 and (3) it reduced the etch rate of the underlying SiO_2 (Ohno et al., 1989). Howard and Steinbruchel (1991) compared $SiCl_4/N_2$ with $SiCl_4$/Ar mixtures, using a polyimide mask that could be dissolved in a liquid developer after blanket exposure to uv light. They found that etching was initiated at a lower temperature (~190°C versus ~220°C) and reached a higher rate in $SiCl_4/N_2$. They also demonstrated that etching required a balance between adsorption of reactants and desorption (or removal) of products.

High etch rates of Cu (~1 μm/min) were obtained in a Br_2/Ar plasma using a split-cathode MERIE reactor. If the temperature was too low, a copper bromide film formed but the film was not etched. When a resist mask was used for patterning, it could be ashed, apparently with no adverse effects. No etching was observed using I_2 (Rogers et al., 1992).

When etching Cu in a MERIE system at a substrate temperature of 300°C, addition of NH_3 to $SiCl_4/Cl_2/N_2$ resulted in a self-aligned passivation layer on the Cu sidewalls. The samples were rinsed in DI water after etching and were free of Cl. The sidewall layer acted to protect the Cu from oxidation and corrosion. Lateral etching was controlled by the [N]/[Cl] ratio and the sidewall film by [Si]/[Cl]. The sidewall thickness increased with increasing [Si]/[Cl] and was thicken when the NH_3 concentration was increased. The ER of Cu was essentially unaffected by the flow rate of NH_3 but the selectivity to SiO_2 increased for NH_3 flow ratios of >10%. The sidewall film was SiON. The authors suggested that NH_3 removed impurities (Cl) from the sidewall, but it should be noted that the samples were rinsed and this must have played a role in removing Cl (Igarashi et al., 1994; 1995).

Anisotropic RIE of Cu in Cl_2, at lower temperatures (150°C), and at an adequate rate, was reported to result from illuminating the substrate with an IR lamp during etching. Two mechanisms were proposed: weakening of the $CuCl_x$ bond by irradiation or selective heating of the $CuCl_x$ formed on the surface (Hosoi and Ohshita, 1993; Ohshita and Hosoi, 1995). The use of Cl_2 as a single reactant was said to be the key to anisotropic etching of Cu, without the need for sidewall protective films (Miyazaki et al., 1996). The process required precise control of the temperature in a range of 230–270°C. Below 230°C the reaction product(s) cannot be desorbed; above 270°C, voids were created in the sidewalls. Undercutting occurred when the pressure exceeded 2 Pa.

Morita (1994) patented a "low" temperature (200–400°C) dry etch (RIE) of Cu using vaporized acetic acid and water as the reactants.

Although these subtractive processes have been developed, the difficulties in their use have inhibited wide use. In addition, Murarka et al. (1993) pointed out that if high temperature and bias are required for RIE, Cu migration is a cause for concern.

7.3 ADDITIVE PROCESSES

7.3.1 Introduction

The formation of Cu patterns by an additive process may be more feasible than by a subtractive one. In addition to transferring the burden of etching from Cu (which is difficult) to a dielectric (well-characterized processes), the temperatures required are lower, reducing the thermal budget as well as avoiding the problem mentioned by Murarka et al. (1993).

Metallization

7.3.2 Lift-Off

Rogers et al. (1991) described the use of lift-off for forming Cr-encapsulated Cu interconnects. The evaporated metals were a Cr base, then Cu, and then a capping layer of Cr. The stencil was a soluble PI. The metals were clad in PECVD SiON before the permanent PI dielectric layer was spun on. The via plugs were Ti/AlCu/Ti, also formed by lift-off. Cho et al. (1991) illustrated a similar process using CVD W encapsulation of Cu deposited selectively on a W seed layer.

7.3.3 Embedment

In this technique, the metal is deposited into patterns formed in the *permanent* dielectric, usually by CVD or electrochemical plating, although Lakshminarayanan et al. (1994) sputtered Cu into Ti-lined trenches etched into the PECVD oxide Ti.

These additive processes may be combined with blanket removal by CMP, sputter etching, or ion milling in the case of nonselective deposition of overfill with a selective process.

Embedment is discussed more fully in Chapter VI.

7.4 DRY ETCHING

Hampden-Smith and Kodas (1993) have reviewed chemical approaches to dry etching of Cu, but with the possible exception of laser assisted etching which may have the potential for anisotropy, the methods are useful for *blanket* removal, *not patterning*. These include the reversal of the CVD deposition reaction ("Reverse CVD"), oxidation, and the formation of volatile copper(I) halide Lewis base adducts.

8.0 PATTERNING OF TUNGSTEN

8.1 WET ETCHING

There are suitable liquid etchants for W (e.g., Kern and Deckert, 1978) but dry etching, in F-, Cl-, and Br-containing plasmas, is now preferred.

8.2 LIFT-OFF

The high substrate temperatures required for evaporation of low resistivity W are not compatible with the organic stencils used in these processes.

8.3 REACTIVE ION ETCHING

When W is deposited into holes, the excess must be removed, usually using a sacrificial organic overcoat, as discussed in Chapter VI. When etched to form patterns, for gate electrodes or interconnections, dimensional control, i.e., anisotropy, is essential. Etch rate ratios, both to the mask and substrate, are also important factors when choosing a patterning process.

W is etched spontaneously by F only very slowly at room temperature but the rate is enhanced significantly when the surface is bombarded with Ar ions. WF_6 is the only product (Winters, 1985). Mechanisms to explain the ion enhancement of the W + F reaction are (1) chemical sputtering (Winters, 1985) and (2) lattice-damage-induced chemical reactions (Greene et al., 1988b).

WF_6 is less volatile than the oxychloride; therefore O_2 is often added to Cl-based plasmas. Cl_2/O_2 mixtures etched W and a conductive underlayer (e.g., TiN) anisotropically and selectively to an oxide underlay and photoresist mask (Cote et al., 1988). However, W could be etched in a Cl_2 discharge with the addition of a small amount of BCl_3 (Fischl and Hess, 1987). In the absence of ion bombardment, W could be etched by Cl atoms, but not Cl_2 molecules (Fischl et al., 1988).

Anisotropy and selectivity to underlying insulators are inadequate in CF_4 and SF_6 plasmas. Addition of CHF_3 to SF_6 eliminated undercut (Chen et al., 1987). Low temperature ($-10°$ to $-50°C$) etching in a triode reactor using SF_6 was anisotropic and exhibited a reasonably good selectivity to resist. However, when the W film was deposited over a TiN or Ti "glue layer," residues were formed. The mechanism proposed for residue formation was formation of nonvolatile but water soluble TiF_xO_y which covered the entire surface or acted as micromasks during W etching, resulting in "spikes" of unetched W. Following the main etch by RIE in SF_6 at $+55°C$ in a diode and finally in Cl_2 at $5°C$ resulted in clean surfaces (Sridharan et al., 1992).

$CBrF_3/O_2/He$ was found to be more suitable, possibly because a more substantial polymer deposit protects the W sidewalls; the selectivity to both oxide and resist were acceptable (Burba et al., 1986).

Another suitable etchant is $CF_2Cl_2 + CHF_3/O_2$ (Daubenspeck et al., 1989). In a low pressure batch reactor, etch rate and anisotropy depended on the energy of ion bombardment and reactor loading, as well as on power, pressure, and gas composition. In single wafer reactors, in which pressure and power density are higher, multistep processing was required to balance etching and deposition of "polymer" (Daubensleck and Sukanek, 1990).

Addition of C_4F_8 to SF_6 in an ECR reactor eliminated the undercut through sidewall protection; it also decreased RIE lag with the decrease greater for higher C_4F_8 concentrations. At AR = 2.5, the normalized etch depth was increased from 0.72 in the absence of C_4F_8 to 0.82 with $C_4H_8/SF_6 \sim 0.4$. No change in RIE lag with rf power was observed up to 60 W. Application of rf bias increased the etch rate of W but degraded the selectivity to resist significantly. However, anisotropy was achievable in the absence of rf bias and thus unbiased etching was the process of choice (Maruyama et al., 1995).

9.0 STRUCTURE OF METAL FILMS

9.1 HILLOCKS

"Hillocks" is the name given to protrusions on the surface of metal films; they can be formed during deposition but those studied most intensively are a result of postdeposition heat treatments. They can create problems both in processing (interlevel shorts) and in reliability (breakdown during use). It may be necessary to increase the thickness of a dielectric overcoat to insure adequate coverage to prevent these failures. CVD dielectrics conform to the hillocks but sputtered insulators probably afford less protection. Resist may be thinned when applied over hillocks and thus erode prematurely during RIE.

Hillock formation during deposition depends on the deposition rate and the substrate temperature. When the temperature was increased, at a constant deposition rate, the density of the hillocks decreased but their average size increased. As the deposition rate was increased, the hillock density and average sizes decreased. In general, hillocks formed when films were deposited at low temperatures and low rates (Chang and Vook, 1991).

During postdeposition heat treatments, many hillocks are formed on small-grained films; as the grain size is increased, both the number and size decrease (Philofsky et al., 1971). Hillocks formed during heat treatment after deposition may grow to substantial heights (e.g., >2 μm) although whiskers have also been observed. The thickest films had the highest density of hillocks and the biggest ones (Ericson et al., 1991). Some observers reported that hillocks had a well-defined, crystalline appearance. Several classes of hillock were characterized by Santoro (1969); the type depended on the grain structure from which they arose. Edge hillocks grew along grain boundary segments, flat-topped structures arose from small grains, and spirelike ones originated in triple points. Ericson et al. (1991), however, found that annealing hillocks were usually softly rounded, with the well-defined ones arising in very thin films. Hillocks formed on AlCu films were Cu-rich (Philofsky et al., 1971). The hillock density decreased but the size increased as a function of time during isothermal annealing (Chang et al., 1989). Ericson et al. (1991) found that initially hillocks are separated from the film by a grain boundarylike interface along the original film surface, but during prolonged annealing, grain growth of adjacent grains eventually results in integration of the hillock into the film.

One mechanism proposed for hillock formation is stress relief (Paddock and Black, 1968). A high compressive stress is induced, during thermal cycling, due to the thermal mismatch between Al, which has a high TCE, and the substrates (e.g., Si, SiO_2), which have significantly lower coefficients. The low yield strength of Al causes the film to yield. Transport along grain boundaries and possibly along the surface and interface provides the material for hillock growth. To supply material for hillock growth, the film will be thinned elsewhere. The mechanism for surface reconstruction, the mass transport phenomenon responsible for hillock growth, was said to be compressional fatigue, and methods of hillock suppression are those which increase fatigue resistance (Philofsky et al., 1971). An alternative mechanism is a creep phenomenon, a lattice diffusion process; the surface is deformed due to atomic motion in the bulk (Santoro, 1969).

Suppression of hillock growth has often been accomplished in two general ways: by overcoating the films before heat treatment and by doping the Al. Among the specific suggestions are (1) coating with Al_2O_3, grown by anodization (Dell'Oca and Learn, 1971), grown in hot water after immersion of the Al in fuming nitric acid (McMilland and Shipley, 1976) or after sputter-etching the Al surface (Harada et al., 1986); (2) periodically introducing O_2 during Al deposition, i.e., Al/AlO layering (Faith, 1981) or using an Al/Al_2O_3 alloy (Bhatt, 1971); (3) addition of alloying elements, Si (Paddock and Black, 1968t), Cu (Learn, 1976), Ti *and* Si (Gardner et al., 1984), Sn (Sato et al., 1971); and (4) overcoating with, e.g.,

N-doped TiW (Lim, 1982), TiW (Townsend and Vander Plas, 1985; Mak, 1986), TiN (Rocke and Schneegans, 1988), sputtered SiO_2 deposited at low power, i.e., low temperature (Barson and Schwartz, 1970), low temperature CVD (Learn, 1986), refractory metals or silicides (Ho et al., 1987). Alternating layers of Al and $TaSi_x$ remained hillock-free after annealing (Draper et al., 1985). Ar^+ ion implantation, an effect not related to damage (Holland and Alvis, 1987) and interposing a layer of WSi_2 between the Al film and the underlying SiO_2 were also reported to be effective in hillock elimination (Cadien and Losee, 1984).

The similarity between thermally induced hillock formation and electromigration failures is emphasized by the similarity of some of the measures used to minimize both effects (Berenbaum, 1972).

9.2 MICROSTRUCTURE OF METAL FILMS

9.2.1 Evaporated Films

The microstructure of films deposited by vacuum processes is particularly sensitive to the absolute substrate temperature (T, °K) and the melting point of the material (T_m). The basic model is the structure zone model of Movchan and Demchishin (1969), shown in Fig. V-15. Below T_1, i.e., $<0.3\ T_m$ (for metals), this first zone consists of tapered grains; the surface has a domed structure; the diameter of the domes, increases with temperature. For Al, whose melting point is 659.7°C (932.7K), T_1 ~7°C. Levenson (1989) described the structure as columnlike bundles separated by voids. A smooth transition to a second zone occurs at about T_1. In this second zone there is a clearly defined columnar structure with a smooth matt surface. There are well-defined grain boundaries (width ~5A) (Srolovitz et al., 1988) in this structure; the width of the columnal crystallite increases with increasing T. There is no porosity. The slope of the columns is related to the angle of incidence of the incoming vapor. Zone 3 is formed at 0.5 T_m at which temperature

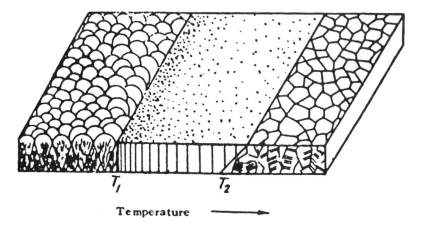

FIGURE V-15 Diagram of temperature dependence of structural zones in a metal film. (From Movchan and Demchishin, 1969)

Metallization

the columnar grains gradually change to relatively large equiaxed (same size in all directions) grains. The surface is rough, with roughness increasing with increasing temperature. The activation energy for grain growth in Al films corresponds to that for grain boundary diffusion in bulk Al (Levenson, 1989). Microhardness, strength, and ductility of the films are determined by the structure features of the zone in which they were formed (Movchan and Demchishin, 1969).

Although the formation of columns depends on the deposition conditions as well as on the material itself, these structures are observed only when the mobility of the atoms is limited (Dirks and Leamy, 1977). Tait et al. (1992) state that their ballistic model confirms that the structure is due to self-shadowing and limited diffusion.

9.2.2 Sputtered Films

A fourth zone (termed T) lying between zones 1 and 2 was identified in sputtered films deposited at low Ar pressures. It consisted of a dense array of poorly defined fibrous grains. Its existence is largely a consequence of the bombarding effect of the ions. Intense energetic ion bombardment during deposition can largely suppress the development of the open structures in zone 1. The formation of voided structures at high Ar pressures was much less at low deposition rates. Also, at high Ar pressures, the columnar grains of zone 2 tended to be faceted. Suppression of open boundaries required an ion flux adequate to resputter 30–60% of the incident flux of incoming species; the required resputtering flux increased with the size of surface irregularities. But resputtering itself is not the major effect; that is postulated to be drive-in of knock-on atoms into the bulk (Thornton, 1986).

9.2.3 CVD Films

Columnar microstructure is seen in refractory metal, nitride, and silicide films, but there have been no reports on structural analysis of CVD films of Al, AlCu, or Cu. The successful application of the zone model to evaporated and sputtered metal films whose melting points cover a wide range, makes it likely that the structure of these CVD films can also be described by the zone model.

9.2.4 Effect of Oblique Incidence

Column formation is more prominent in films produced at an oblique incidence angle. The orientation of columns in films deposited at oblique incidence is always more nearly perpendicular to the substrate than the vapor beam direction (Dirks and Leamy, 1977). This is often expressed by the "tangent rule": $2 \tan \beta = \tan \alpha$, where β is the angle between the columns and the substrate normal and α is the angle between the source direction and the substrate normal. However, it has been reported that the inclination angle of the columns also depends on the deposition parameters, such as temperatures, deposition rate, pressure, and vapor density so that the range of parameters where the tangent rule is obeyed is narrow and very limited (Fujiwara et al., 1988). Nonzero tilting angles are a consequence of atomic shadowing (Mazor et al., 1989). A three-dimensional ballistic deposition model for thin film growth incorporates the mechanisms of shadowing and relaxation. The simulation produces columns which become elliptic as the angle of the vapor flux is increased. A vapor stream normal to the surface of a film is oblique at the edges of steps (Tait et al., 1992, 1993).

In thin films, the grain size and growth depend on substrate temperature, rate of deposition, film thickness, alloy content, and postdeposition annealing temperature. After annealing evaporated films, the θ particles in AlCu films tended to be localized at grain boundaries and triple points. During device processing, films are annealed after patterning; the ultimate grain size was believed to be determined by the film thickness and line width. Columnar and bamboo grain structures were enhanced in annealed submicron lines (Kwok et al., 1985). In a study of sputtered AlCuSi blanket films, it was proposed that the preferred orientation of the grains was affected by the Cu concentration, the mobility of the grain boundary, and the location of Cu precipitates. If the θ-phase is precipitated, the as-deposited film will have a less preferred orientation, and smaller grain growth rate both before and after annealing than a film without θ-phase precipitates. This was thought to be due to the prevention of the coalescence of grains. Excessive heating in the early stages of film growth degrades preferred orientation (Shimamura et al., 1991).

REFERENCES

Abraham, T., *1986 VMIC*, 198 (1986).
Abraham, T., *J. Electrochem. Soc., 134*, 2809 (1987).
Akahori, T., T. Tani, S. Nakayama, p. 209 in *Tungsten and Other Advanced Metals for VLSI/ULSI Applications V* (S. S. Wong, S. Furukawa, eds.), Mat. Res. Soc., Pittsburgh, Penn., 1990.
Alford, T. L., D. Adams, M. Diale, J. Li, S. A. Rafalski, R. L. Spreitzer, S. Q. Hong, S. W. Russell, N. D. Theodore, J. W. Mayer, p. 49 in *Conf. Proc., ULSI-IX*, Mat. Res. Soc. Pittsburgh, Penn., 1994.
Allen, L. R., R. Rickard, *J. Vac. Si. Technol., A12*, 1265 (1994).
Amazawa, T., H. Nakamura, Y. Arita, *IEDM 88*, 442 (1988).
Ames, I., F. M. D'Heurle, R. E. Horstman, US Pat. 3,725,309, 1973.
Aoki, H., E. Ikawa, T. Kikkawa, I. Nishiyama, Y. Teraoka, *Jpn. J. Appl. Phys., 30*, 1567 (1991).
Aoki, H., T. Hashimoto, E. Ikawa, T. Kikkawa, *Jpn. J. Appl. Phys., 31*, 4376 (1992).
Aoki, M., Y. Sasamura, *J. Vac. Sci. Technol., A14*, 398 (1996).
Arikado, T., M. Sekine, H. Okano, Y. Horiike, *IEDM 86*, 54 (1986).
Awaya, N., Y. Arita, *1989 Symp. VLSI Technol.*, 103 (1989).
Awaya, N., Y. Arita, *1991 Symp. VLSI Technol.*, 37 (1991a).
Awaya, N., Y. Arita, *Jpn. J. Appl. Phys., 30*, 1813 (1991b).
Awaya N., Y. Arita, *1993 Symp. VLSI Technol.*, 125 (1993a).
Awaya, N., Y. Arita, *Jpn. J. Appl. Phys., 32*, 3915 (1933b).
Awaya, N., Y. Arita, *Thin Solid Films, 262*, 12 (1995).
Bachman, B. J., M. J. Vasile, *J. Vac. Sci. Technol., A7*, 2709 (1989).
Bae, D. L., Y. S. Kim, Y. W. Park, J. K. Lee, M. Y. Lee, p. 457 in *Advanced Metallization for ULSI Applications in 1993* (D. P. Favreau, Y. Shacham-Diamond, Y. Horike, eds.), Mat. Res. Soc., Pittsburgh, Penn., 1994.
Bakardjiev, I., M. Majdraganova, G. Bliznakov, *J. Non-Crystalline Solids, 20*, 349 (1976).
Barson, F., G. C. Schwartz, *IBM Tech. Discl. Bull., 13, 1122* (1970).
Batchelder, T., *Solid State Technol., 2/82*, 111 (1982).

Bausmith, R. C., W. J. Cote, J. E. Cronin, K. L. Holland, C. W. Kaanta, P.-I.P. Lee, T.M. Wright, USPat. 4,919,750, 4/24/90, 1990.
Beach, D. B., S. E. Blum, F. K. LaGoues, *J. Vac. Sci. Technol., A7*, 3117 (1989).
Bell, D. A., J. L. Falconer, Z., Lu, C. M. McConica, *J. Vac. Sci. Technol., B12*, 2976 (1994).
Bent, B. E., R. G. Nuzzo, L. H. Dubois, *J. Am. Chem. Soc., 111*, 1634 (1989).
Berenbaum, L., *Appl. Phys. Lett., 20*, 434 (1972).
Bermudez, V. M., A. S. Glass. *J. Vac. Sci. Technol., A7*, 1961 (1989).
Bhatt, H. J., *Appl. Phys. Lett., 19*, 3 (1971).
Biswas, D. R., C. Ghosh, R. L. Layman, *J. Electrochem. Soc., 130*, 234 (1983)).
Blewer, R. S., V. A. Wells, *1984 VMIC*, 153 (1984).
Blewer, R. S., *Solid State Technol., 11/86*, 117 (1986).
Bohr, M. T., *IEDM 95*, 241 (1995).
Bollinger, D., S. Iida, O. Matsumoto, *Solid State Technol. 6/84*, 167 (1984).
Bolnedi, S., G. R. Raupp, T. S. Cale, p. 385 in *Advanced Metallization for ULSI Applications in 1993*, (D. P. Favreau, Y. Shacham-Diamond, Y. Horike, eds.), Mat. Res. Soc., Pittsburgh, Penn., 1994.
Booth, R. C., C. J. Heslop, *Thin Solid Films, 65*, 111 (1980).
Bouteville, A., T. Charrier, J. C. Remy, J. Palleau, J. Torres, *J. de Physique IV, Colloque C2, Suppl. au J. de Physique II, Vol. 1*, September 1991, C2, 857 (1991).
Bradbury, D. R., T. I. Kamins, *J. Electrochem. Soc., 133*, 1214 (1986).
Bradley, S., C.-H. Chen, G. Kovall, *1991 VMIC*, 298 (1991).
Broadbent, E. K., C. L. Ramiller, *J. Electrochem. Soc., 131*, 1427 (1984).
Broadbent, E. K., W. T. Stacy, *Solid State Technol., 12/85*, 51 (1985).
Broadbent, E. K., A. E. Morgan, J. M. DeBlasi, P. van der Putte, B. Coulman, B. J. Burrow, D. K. Sadana, A. Reader, *J. Electrochem. Soc., 133*, 1715 (1986).
Bruce, R. H., G. P. Malafsky, in "Proc 3rd Symp. Plasma Processing" (J. Dielman, R. G. Frieser, G. S. Mathad, eds.), *Electrochem. Soc. Proc. Vol., PV 82-6*, 336 (1982).
Bruce, R. H., G. P. Malafsky, *J. Electrochem. Soc., 130*, 1369 (1983).
Brusic, V., G. S. Frankel, C.-K. Hu, M. M. Pletchaty, B. M. Rush, *Corrosion, 47*, 35 (1991).
Brusic, V., G. S. Frankel, C.-K. Hu, M. M. Plechaty, G. C. Schwartz, *IBM J. Res. Devel., 37*, 173 (1993).
Brusic, V., C. H. Yang, in "Proc. 11th Intl. Symp. Plasma Processing" (G. S. Mathad, M. Meyyappan, eds.), *Electrochem. Soc. Proc. Vol., 96-12*, 308 (1996).
Bryant, W. A., *J. Electrochem. Soc., 125*, 1534 (1978).
Burba, M. E., E. Degenkolb, S. Henck, M. Tabasky, E. D. Jungbluth, R. Wilson, *J. Electrochem. Soc., 133*, 2133 (1986).
Burkstrand, J. M., C. T. Hovland, *J. Vac. Sci. Technol., A1*, 449 (1983).
Cacouris, T., G. Scelsi, P. Shaw, R. Scarmozzino, R. M. Osgood, *Appl. Phys. Lett., 52*, 1865 (1988).
Cacouris, T., R. Sacarmozzino, R. M. Osgood, Jr., *1990 VMIC*, 268 (1990).
Cadien, K. C., D. L. Losee, *J. Vac. Sci. Technol., B2*, 82 (1984).
Carlsson, J.-O., M. Bowman, *J. Vac. Sci. Technol., A3*, 2298 (1985).
Chambers, A. A., *Solid State Technol., 9/82*, 93 (1982).
Chang, P. C., K. K. Chaa, in "Proc. 5th Symp. Plasma Processing (G. S. Mathad, G. C. Schwartz, G. Smolinsky, eds.) *Electrochem. Soc. Proc. Vol. PV 85-1*, 12 (1985).
Chang, C.-A., *Appl. Phys. Lett., 51*, 1236 (1987).
Chang, C. Y., R. W. Vook, *J. Mater. Res. 4*, 1172 (1989).
Chang, C. Y., R. W. Vook, *J. Vac. Sci. Technol., A9*, 559 (1991).
Chapman, B., M. Nowak, *Semiconductor International, 11/80*, 139 (1980).
Chen, C.-H., L. C. Watson, D. W. Schlosser, p. 357 in *Tungsten and Other Refractory Metals for VLSI Applications II* (E. K. Broadbent, ed.), Mat. Res. Soc., Pittsburgh, Penn., 1987.

Cheung, K. P., C. J. Case, R. Liu, R. J. Schutz, R. S. Wagner, L. F. Tz. Kwakman, D. Huibregtse, H. W. Piekaar, E. H. A. Granneman, *1990 VMIC*, 303 (1990).
Chiang, C., S.-M. Tzeng, G. Raghavan, R. Villasol, G. Bai, M. Bohr, H. Fujimoto, D. B. Fraser, *1994 VMIC*, 414 (1994).
Chiu, G. T., T. R. Joseph, G. M. Ozols, US Pat 4,335,506, 6/22/83, 1982.
Cho, J. S. H., H.-K. Kang, M. A. Beiley, S. S. Wong, C. H. Ting, 1991 Symp. on VLSI Technol., *Digest of Technical Papers*, 39 (1991).
Cho, J. S. H., H.-K. Kang, I. Asano, S. S. Wong, *IEDM 92*, 297 (1992).
Cho, J. S. H., H.-K. Kang, S. S. Wong, Y. Shacham, Diamand, *MRS. Bull., 6*, 31 (1993).
Chow, R., S. Kang, R. H. Wilson, B. Gorowitz, A. G. Williams. *1987 VMIC*, 208 (1987).
Clayton, F. R., S. A. Besson, *Solid State Technol., 7/93*, 93 (1993).
Coburn, J. W., H. Winters, *J. Appl. Phys., 50*, 3189 (1979).
Cohen, S. L., M. Liehr, S. Kasi, *J. Vac. Sci. Technol. A10*, 863 (1992).
Colgan, E. G., J. D. Chapple-Sokol, *J. Vac. Sci. Technol., B10*, 1156 (1992).
Contolini, R. J., A. F. Bernhardt, S. T. Mayer, *J. Electrochem. Soc., 141*, 2503 (1994).
Cooke, M. J., R. A. Heinecke, R. C. Stern, *Solid State Technol., 12/92*, 62 (1982).
Cote, W. J., K. L. Holland, T. M. Wright, US Pat. 4,786,360, 1988.
CRC *Handbook of Electrical Resisitivites of Binary Metallic Alloys* (K. Schroder, ed.), 1983.
Creighton, J. R., *J. Electrochem. Soc., 136*, 271 (1989).
Dang, K., in "Proc. 11th Intl. Symp. Plasma Processing" (G. S. Mathad and M. Meyyappan, eds.), *Electrochem. Soc. Proc. Vol., 96-12*, 296 (1996).
Danner, D. A., D. W. Hess, *J. Electrochem. Soc., 133*, 151 (1986a).
Danner, D. A., D. W. Hess, *J. Electrochem. Soc., 133*, 940 (1986b).
Danner, D. A., M. Dalvie, D. W. Hess, *J. Electrochem. Soc., 134*, 669 (1987).
Daubenspeck, T. H., H. K. Lee, in "Proc. 9th Symp. Plasma Processing" (G. S. Mathad, D. W. Hess, eds.), PV *92-18*, 381 (1992).
Daubenspeck, T. H., E. J. White, P. C. Sukanek, *J. Electrochem. Soc., 136*, 2973 (1989).
Daubenspeck, T. H., P. C. Sukanek, *J. Vac. Sci. Technol., B8*, 586 (1990).
Dell'Oca, C. J., A. J. Learn, *Thin Solid Films, 8*, R47 (1971).
Denison, D. R., L. D. Hartsough, *J. Vac. Sci. Technol., 17*, 388 (1980).
Desatnik, N., B. E. Thompson, *J. Electrochem. Soc., 141*, 3532 (1994).
Ding, P. J., W. A. Lanford, S. Hymes, S. P. Murarka, in *Advanced Metallization and Processing for Semiconductor Devices and Circuits—II* (A. Katz, S. P. Murarka, Y. I. Nissam, J. M. E. Harper, eds.), *Mat. Res. Soc. Symp. Proc. Vol. 260*, 757 (1992).
Ding, P. J., W. Wang, W. A. Lanford, S. Hymes, S. P. Murarka, *Appl. Phys. Lett., 65*, 1778 (1994a).
Ding, P. J., W. A. Lanford, S. Hymes, S. P. Murarka, *J. Appl. Phys., 75*, 3627 (1994b).
Ding, P. J., W. A. Lanford, S. Hymes, S. P. Murarka, *Appl. Phys. Lett., 64*, 2897 (1994c).
Dinklage, J. D., M. C. Hakey, *IBM Tech. Discl. Bull., 26*, 363 (1984).
Dirks, A. G., H. J. Leamy, *Thin Solid Films, 47*, 219 (1977).
Dixit, G. A., A. Paranjpe, Q.-Z. Hong, L. M. Ting, J. D. Luttmer, R. H. Haveman, D. Paul, A. Morrison, K. Littau, M. Eizenberg, A. K. Sinha, *IEDM 95*, 1001 (1995).
Dohmae, S.-i., S. Mayumi, S. Ueda, *1990 VMIC*, 275 (1990).
Dori, L., A. Megdanis, S. B. Brodsky, S. A. Cohen, *Thin Solid Films, 194*, 501 (1990).
Draper, B. L., T. A. Hill, H. B. Bell, *1985 VMIC*, 90 (1985).
Dubois, L. H., B. R. Zegarski, *J. Electrochem. Soc., 139*, 3295 (1992).
Eldridge, J. M., M. H. Lee, G. C. Schwartz, US Pat. 4,368,220, 1983.
Eldridge, J. M., G. Olive, B. J. Luther, J. O. Moore, S. P. Holland, *J. Electrochem. Soc., 134*, 1025 (1987).
Ericson, F., N. Kristensen, J.-A. Schweitz, *J. Vac. Sci. Technol., B9*, 58 (1991).
Faith, T. J., *J. Appl. Phys., 52*, 4630 (1981).
Farkas, J., M. J. Hampden-Smith, T. T. Kodas, *J. Electrochem. Soc., 141*, 3539 (1994).

Feinerman, A. D., *J. Electrochem. Soc., 137*, 3683 (1990).
Fischl, D. S., D. W. Hess, *J. Electrochem. Soc., 134*, 2265 (1987).
Fischl, D. S., G. W. Rodrigues, D. W. Hess, *J. Electrochem. Soc., 135*, 2016 (1988).
Flamm, D. L., V. M. Donnelly, *Plasma Chemistry & Plasma Processing, 1*, 317 (1981).
Fok, T. Y., *Electrochem. Soc. Ext. Abstr. 115, PV 80-1*, 301 (1980).
Franco, J. R., J. Havas, H. A. Levine, US Pat. 3,873,361 3/25/75, 1975.
Fried, L. J., J. Havas, J. S. Lechaton, J. S. Logan, G. Paal, P. A. Totta, *IBM J. Res. Dev., 26*, 363 (1982).
Fu, C. Y., R. Hsu, V. Malba, in "Low Energy Ion Beam and Plasma Modification of Materials" (J. M. E. Harper, K. Miyake, J. R. McNeil, S. M. Gorbatkin, eds.), *Mat. Res. Soc. Proc. Vol., 223*, 385 (1991).
Fujino, K., T. Oku, *J. Electrochem. Soc., 139*, 2585 (1992).
Fujiwara, H., K. Hara, M. Kamiya, T. Hashimoto, K. Okamoto, *Thin Solid Films, 163*, 387 (1988).
Gabriel, C., R. Wallach, in "Proc. 9th Symp. Plasma Processing" (G. S. Mathad, D. W. Hess, eds.), *Electrochem. Soc. Proc. Vol. PV 92-18*, 367, 421 (1992).
Gardner, D. S., R. B. Beyers, T. L. Michalka, K. C. Saraswsat, T. W. Barbee, Jr., J. D. Meindl, *IEDM 84*, 114 (1984).
Gardner, D. S., J. Onuki, K. Kudoo, Y. Misawa, Q. T. Vu, *Thin Solid Films, 262*, 104 (1995).
Gebara, G., K. Mautz, P. Wootton, in "Proc. 10th Symp. Plasma Processing" (G. S. Mathad, D. W. Hess, eds.), *Electrochem. Soc. Proc. Vol. PV 94-20*, 421 (1994).
Gelatos, A. V., R. Marsh, M. Kottke, C. J. Mogab, *Appl. Phys. Lett., 63*, 2842 (1993).
Gladfelter, W. L., D. C. Boyd, K. F. Jensen, *Chemistry of Materials, 1*, 339 (1989).
Gorczyca, T. B., L. R. Douglas, B. Gorowitz, R. H. Wilson, *J. Electrochem. Soc., 136*, 2765 (1989).
Gottscho, R. A., C. W. Jurgensen, D. J. Vitkvage, *J. Vac. Sci. Technol., B10*, 2133 (1982).
Green, M. L., R. A. Levy, *J. Electrochem. Soc., 132*, 1243 (1985).
Green, M. L., Y. S. Ali, B. A. Davidson, L. C. Feldman, S. Nakahara, in "Thin Films—Interfaces and Phenomena" (R. J. Nemanichm, P. S. Ho, S. S. Lau, eds.), *Mat. Res. Soc. Symp. Proc. Vol. 54*, 723 (1986).
Green, M. L., Y. S. Ali, T. Boone, B. A. Davidson, L. C. Feldman, S. Nakahar, in "Multilevel Metallization, Interconnection, and Contact Technologies" (L. B. Rothman, T. Herndon, eds.), *Electrochem. Soc. Proc. Vol., 87-4*, 1 (1987).
Greene, W. M., W. G. Oldham, D. W. Hess, *Appl. Phys. Lett., 52*, 1133 (1988a).
Greene, W. M., W. G. Oldham, D. W. Hess, *J. Vac. Sci. Technol., B6*, 1570 (1988b).
Groenen, P. A. C., O. F. Tekcan, J. G. A. Holscher, H. H. Brongersma, *J. Vac. Sci. Technol., A12*, 737 (1994).
Hampden-Smith, M. J., T. T. Kodas, *MRS Bull., 6/93*, 39 (1993).
Han, J., K. F. Jensen, *J. Appl. Phys., 75*, 2240 (1994).
Hara, T., N. Ohtsuka, T. Takeda, T. Yoshimi, *J. Electrochem. Soc., 133*, 1489 (1986).
Hara, T., T. Ohba, H. Yagi, H. Tsuchikawan, p. 353 in *Advanced Metallization for ULSI Applications in 1993* (D. P. Favreau, Y. Shacham-Diamond, Y. Horike, eds.), Mat. Res. Soc., Pittsburgh, Penn., 1994.
Harada, H., S. Harada, Y. Hirata, T. Noguchi, H. Mochizuki, *IEDM 86*, 46 (1986).
Haroka, H., J. Pacansky, *J. Vac. Sci. Technol., 19*, 1132 (1981).
Harvilchuck, J. M., J. S. Logan, W. C. Metzger, P. M. Schaible, US Pat. 3,994,793, 1976).
Hattori, K., M. Hori, M. Aoyama, *J. Electrochem. Soc., 141*, 2825 (1994).
Hatzakis, M., B. J. Canavello, J. M. Shaw, *IBM J. Res. Dev., 24*, 452 (1980).
Herndon, T. O., R. L. Burke, *Interface '77*, Kodak Microelectronics Symp., Monterey, Calif., 1977.
Hess, D. W., *Plasma Chemistry and Plasma Processing, 2*, 141 (1982).
Hey, H. P. W., A. K. Sinha, S. D. Steenwyk, V. V. S. Ranam, J. L. Yeh, *IEDM 86*, 50 (1986).

Hieber, K., M. Stolz, *VMIC*, 216 (1987).
Higashi, G. S., G. E. Blonder, C. G. Fleming, V. R. McCray, V. M. Donnelly, *J. Vac. Sci. Technol., B5*, 1441 (1987).
Higashi, G. S., M. L. Steigerwald, *Appl. Phys. Lett., 54*, 81 (1989).
Higashi, G. S., K. Raghavachari, M. L., Steigerwald, *J. Vac. Sci. Technol., B8*, 103 (1990).
Hill, R., *J. Vac. Sci. Technol., B14*, 547 (1996).
Hintze, B., S. E. Schulz, T. Gessner, p. 449 in *Advanced Metallization for ULSI Applications in 1993* (D. P. Favreau, Y. Shacham-Diamond, Y. Horike, eds.) Mat. Res. Soc., Pittsburgh, Penn., 1994.
Ho, V. Q., H. J. Nentwich, H. M. Naquib, US Pat. 4,680,854, 1987.
Holland, O. W., J. R. Alvis, *J. Electrochem. Soc., 134*, 2017 (1987).
Homma, Y., A. Yajima, S. Harada, *IEDM 81*, 570 (1981).
Horiike, Y., T. Yamazaki, M. Shibagaki, T. Kurisaki, *Jpn. J. Appl. Phys., 21*, 1412 (1982).
Hosoi, N., Y. Ohshita, *Appl. Phys. Lett., 63*, 2703 (1993).
Hosokawa, N., R. Matsuzaki, T. Asamaki, *Jpn. J. Appl. Phys., 13*, 435 (1974).
Houlding, V. H., H. Maxwell, Jr., S. M. Cochiere, D. L. Farrington, R. S. Rai, J. M. Tartaglia, in *Advanced Metallization and Processing for Semiconductor Devices and Circuits*—II (A. Katz, S. P. Murarka, Y. I. Nissim, J. M. E. Harpre, eds.), *Mat. Res. Soc. Symp. Proc. Vol. 260*, 119 (1992).
Howard, J. K., P. S.-C. Ho, US Pat. 4,017,890, 1977.
Howard, B. J., Ch. Steinbruchel, *Appl. Phys. Lett., 59*, 914.
Howard, J. K., J. F. White, P. S. Ho, *J. Appl. Phys. 49*, 4083 (1978).
Hsieh, J. J., *J. Vac. Sci. Technol., A11*, 3040 (1993).
Hu, C. K., M. B. Small, M. Schadt, in *Science and Technology of Microfabrication* (R. E. Howard, E. L. Hu, S. Namba, S. Pang, eds.), *Mat. Res. Soc. Proc. Vol. 76*, 191 (1987).
Hu, C. K., B. Canney, D. J. Pearson, M. B. Small, *J. Vac. Sci. Technol., A7*, 682 (1989).
Huang, Z., M. Siegel, in "Proc. 10th Symp. Plasma Processing" (G. S. Mathad and D. W. Hess, eds.), *Electrochem. Soc. Proc. Vol PV 94-20*, 380 (1994).
Hwang, J., S. Mak, in "Proc. 9th Symp. Plasma Processing" (G. S. Mathad and D. W. Hess, eds.), *Electrochem. Soc. Proc. Vol. PV 92-18*, 374 (1992).
Hymes, S., S. P. Murarka, C. Shepard, W. A. Lanford, *J. Appl. Phys., 71*, 4623 (1992).
Igarashi, Y., T. Yamanobe, T. Ito, *Ext. Abstr. of the 1994 Int. Conf. on Solid State Devices and Materials*, 943 (1994).
Igarashi, Y., T. Yamanobe, T. Ito, *J. Electrochem. Soc., 142*, L36 (1995).
Iida, S., Ueki, K., H. Komatsu, T. Mitzutani, US Pat. 4,267,013, 1981a.
Iida, S., Ueki, K., H. Komatsu, K. Hirobe, US Pat. 4,308,089, 1981b.
Ingrey, S. I. J., H. J. Wentwich, R. G. Pousen, US Pat. 4,030,967, 1977.
Itoh, H., R. Nakata, T. Moriya, *IEDM 85*, 606 (1985).
Jagannathan, R., M. Krishnan, US Pat 4,818,286, 1989.
Jagannathan, R., R. F. Knarr, M. Krishnan, G. P. Wandy, US Pat. 5,059,243, 1991.
Jain, A., H. K. Shin, K. M. Chi, M. J. Hampden-Smith, T. T. Kodas, J. Farkas, M. F. Paffett, J. D. Farr, *SPIE 1596*, 23 (1991).
Jain, A., K.-M. Chi, T. T. Kodas, M. J. Hampden-Smith, *J. Electrochem. Soc., 140*, 1434 (1993a).
Jain, A., T. T. Kodas, R. Jairath, M. J. Hampden-Smith, *J. Vac. Sci. Technol., B11*, 2107 (1993b).
Jiwari, J., I. Hiroaki, A. Narai, H. Sakaue, H. Shindo, T. Shoji, Y. Horiike, *Jpn. J. Appl. Phys., 32*, 3019 (1993).
Joshi, A., D. Gardner, H. S. Hu, J. Mardinly, T. G. Nieh, *J. Vac. Sci. Technol., A8*, 1480 (1990).
Joshi, R. V., T. N. Nguyen, J. Floro, Y. H. Kim, F. D'Herule, J. Angeilello, *Symp. VLSI Technol.* 59 (1987).

Joshi, R. V., C.-S. Oh, D. Moy, US Pat. 5,202,287, 1993.
Joshi, R. V., M. J. Tejwani, K. V. Srikrishnan, US Pat. 5,420,069, 1995.
Kakiuchi, T., H. Yamamoto, T. Fujita, *1987 Symp. VLSI Technol.*, 73 (1987).
Kaloyeros, A. E., A. Fenf, J. Garhart, K. C. Brooks, S. K. Ghosh, A. N. Saxena, F. Luehrs, *J. Electronic Materials, 19*, 271 (1990).
Kang, S., R. Chow, R. H. Wilso, B. Gorowitz, A. G. Williams, *J. Electronic Materials, 19*, 213 (1988).
Katagiri, T., E. Kondoh, N. Takeyasu, T. Nakano, H. Yamamoto, T. Ohta, *Jpn. J. Appl. Phys., 32*, L1078 (1993).
Kato, T., T. Ito, H. Ishikawa, *IEDM 88*, 458 (1988a).
Kato, T., T. Ito, M. Maeda, *J. Electrochem. Soc., 135*, 455 (1988b).
Kawai, T., Hanabusa, *Jpn. J. Appl. Phys., 32*, 4690 (1993).
Kawamoto, H., H. Sakaue, S. Takehiro, Y. Horiike, *Jpn. J. Appl. Phys. 29.* 2657 (1990).
Kawamoto, H., H. Miyamoto, E. Ikawa, in "Proc. of 10th Symp. on Plasma Processing" (G. S. Mathad, D. W. Hess, eds.), *Electrochem. Soc. Proc. Vol. PV 94-20*, 398 (1994).
Kawano, Y., E. Kondoh, N. Takeyasu, H. Yamamoto, T. Ohta, p. 317, *Advanced Metallization for ULSI Applications in 1993* (D. P. Favreau, Y. Shacham-Diamond, Y. Horiike, eds.) Mat. Res. Soc., Pittsburgh, PA, 1994.
Keaton, A. L., D. W. Hess, *J. Vac. Sci. Technol., A3*, 962 (1985).
Kern, W., C. A. Deckert, table XIV in *Thin Film Processes* (J. L. Vossen, and W. Kern, eds.), Academic Press, New York, 1978.
Kern, W., C. A. Deckert, p. 317 in *Thin Film Processes* (J. L. Vossen, W. Kern, eds.), Academic Press, New York, 1978).
Kim, D.-H., R. H. Wenyorf, W. N. Gill, *J. Vac. Sci. Technol., A12*, 153 (1994).
Kim, Y. T., S.-K. Min, J. S. Hong, C.-K. Kim *Jpn. J. Appl. Phys., 30*, 820 (1991).
Kinsbron, E., W. E. Willenbrock, H. J. Levenstein, in "VLSI Sci. and Technol." (C. J. Dell-Oca and M. W. Bullis, eds.), *Electrochem. Soc. Proc. Vol. PV 82-7*, 116 (1982a).
Kinsbron, E., H. J. Levenstein, W. E. Willenbrock, US Pat. 4,343,677, 1982b.
Kondoh, E., Y. Kawano, N. Takeyasum T. Ohta, *J. Electrochem. Soc., 141*, 3494 (1994).
Korner, H., *Thin Solid Films, 175*, 55 (1989).
Koubuchi, Y., J. Onuki, S. Fukada, M. Suwa, *IEEE Trans. on Electron Devices, 37*, 947 (1990).
Kulper, L. L., US Pat. 3,609,470, 1971.
Kurisaki, T., Y. Horiike, T. yamazaki, US Pat. 4,341,593, 7/27/82, 1982.
Kusumi, Y., N. Fujiwara, J. Matsumoto, M. Yoneda, *Jpn. J. Appl. Phys., 34*, 2147 (1995).
Kwakman, L. F. Tz, D. Huibretgtse, H. W. Piekaar, E. H. A. Granneman, K. P. Cheung, C. J. Case, W.Y-C. Lai, R. Liu, R. J. Schutz, R. S. Wagner, *1990 VMIC*, 282 (1990).
Kwok, T., C. Y. Ting, J.-U. Han, *1985 VMIC*, 83 (1985).
Lai, W. G., Y. Xie, G. L. Griffin, *J. Electrochem. Soc., 138*, 3499 (1991).
Learn, A. L., *J. Vac. Sci. Technol., B4*, 774 (1986).
Learn, A. J., D. W. Foster, *J. Appl. Phys., 58*, 2001 (1985).
Lecohier, B., J.-M. Philipoz, H. van den Burgh, *J. Vac. Sci. Technol., B10*, 262 (1992a).
Lecohier, B., B. Calpini, J.-M., Philipoz, T. Stumm, H. van den Bergh, *Appl. Phys. Lett., 60*, 3114 (1992b).
Lecohier, B., B. Calpini, J.-M. Philipoz, H. van den Bergh, *J. Appl. Phys., 72*, 2022 (1992c).
Lee, W.-Y., J. M. Eldridge, G. C. Schwartz, *J. Appl. Phys., 52*, 2994 (1981).
LeGoues, F., unpublished (1986).
Lehmann, O., M. Stuke, *Appl. Phys. Lett. 61*, 2027 (1992).
Levenson, L. L., *Appl. Phys. Lett., 55*, 2617 (1989).
Levy, R. A., M. L. Green, P. K. Gallagher, *J. Electrochem. Soc., 131*, 2175 (1984).
Levy, R. A., P. K. Gallagher, R. Contolini, F. Schrey, *J. Electrochem. Soc., 132*, 457 (1985).
Levy, R. A., M. L. Green, P. K. Gallagher, Y. S. Ali, *J. Electrochem. Soc., 133*, 1905 (1986).

Levy, K. B., US Pat 5,126,008, 06/30/92 (1992).
Li, J., J. W. Strane, S. W. Russell, P. Chapman, Y. Shachman-Diamand, J. W. Mayer, *Advanced Metallization and Processing for Semiconductor Devices and Circuits—II* (A. Katz, S. P. Murarka, Y. I. Nissim, J. M. E. Harper, eds.) *Mat. Res. Soc. Symp. Proc. Vol., 260*, 605 (1992a).
Li, J., J. W. Mayer, Y. Shacham-Diamand, E. G. Colgan, *Appl. Phys. Lett., 60*, 1983 (1992b).
Li, J., J. W. Mayer, *MRS Bulletin 6/93*, 52 (1993).
Liao, K. Y., C. T. Chiao, A. Chen, M. Lo, H. T. Chu, H. W. Cheng, *1995 VMIC*, 617 (1995).
Lifshitz, N., *Appl. Phys. Lett., 51*, 967 (1987).
Lim, S. C. P., *Semiconductor Intl., 4/82*, 135 (1982).
Lin, J.-H., W. Ray, M. D. Waelch, in "ULSI Science and Technol." (C. M. Osburn and J. M. Andrews, eds.), *Electrochem. Soc. Proc. Vol. PV 89-9*, 445 (1989).
Lloyd, J. R., *Thin Solid Films, 91*, 175 (1982).
Lloyd, J. R., K. A. Perry, G. C. Schwartz, K. V. Srikrishnan, F. E. Turene, J. M. Yang, *IBM Tech. Discl. Bull., 33, 3B*, 477 (1990).
Lutze, J. W., A. H. Perera, J. P. Krusius, *J. Electrochem. Soc., 137*, 249 (1990).
Ma, H.-L., *1980 IEDM*, Washington, DC, 574 (1980).
Ma, D. X., T. R. Webb, Z. Zhao, Z. Huang, D. Tajima, P. K. Lowenhardt, in "Proc. 11th Int. Symp. Plasma Processing " (G. S. Mathad, M. Meyyappan, eds.), *Electrochem. Soc. Proc. Vol. 96-12*, 250 (1996).
Maa, J.-S., J. J. O'Neill, *J. Vac. Sci. Technol., A1*, 636 (1983).
Maa, J.-S., B. Hanlon, *J. Vac. Sci. Technol., B4*, 822 (1986).
Maa, J.-S., H. Gossemberger, R. J. Paff, *J. Vac. Sci. Technol., B8*, 1052 (1990).
Mak, S., *1986 VMIC*, 65 (1986).
Mak, S., S. Arias, C. S. Rhoades, in "Proc. 9th Symp. Plasma Processing", (G. S. Mathad and D. W. Hess, eds.), *Electrochem. Soc. Proc. Vol. PV 92-18*, 340 (1992).
Maleham, J., *Vacuum, 34*, 437 (1984).
Mantell, D. A., *Appl. Phys. Lett., 53*, 1387 (1988).
Mantell, D. A., *J. Vac. Sci. Technol., A9*, 1045 (1991).
Maruyama, T., N. Fujiwara, K. Shiozawa, M. Yoneda, *J. Vac. Sci. Technol., A13*, 810 (1995).
Marx, W. F., D. X. Ma, C.-H., Chen, *J. Vac. Sci. Technol., A10*, 1232 (1992).
Masu, K., K. Tsubouchi, N. Shigeeda, T. Matano, N. Mikoshiba, *Appl. Phys. Lett., 56*, 1543 (1990).
Masu, K., H. Matsuhashi, M. Yokoyama, K. Tsubouchi, p. 301 in *Advanced Metallization for ULSI applications in 1993* (D. P. Favreau, Y. Shacham-Diamand, Y. Horiike, eds.), Mat. Res. Soc., Pittsburgh, PA, 1993.
Mazor, A., B. G. Bukiet, D. J. Srolovitz, *J. Vac. Sci. Technol., A7*, 1386 (1989).
McBrayer, J. D., R. M. Swanson, T. W. Sigman, *J. Electrochem. Soc., 133*, 1242 (1988).
McConica, C. M., K. Krishnamani, *J. Electrochem. Soc., 133*, 2542 (1986).
McConica, C. M., J. K. Hunter, K. Tan, M. D. Szczepaniak, p. 125 in *Tungsten and Other Refractory Metals for VLSI Applications—III* (V. A. Wells, eds.), Mat. Res. Soc., Pittsburgh, Penn., 1988.
McMillan, L. D., R. E. Shipley, US Pat. 3,986,897, 1976.
Milgram, A. A., *J. Vac. Sci. Technol., B1*, 490 (1983).
Miyazaki, H., K. Takeda, N. Sakuma, K. Hinode, K. Kusukawa, T. Furusawa, Y. Homma, S. Kondo, *1996 VMIC*, 498 (1996).
Mogyorosi, J.-O. Carlsson, *J. Vac. Sci. Technol., A10*, 3131 (1992).
Morita, K., US Pat 5,336,363, 1994.
Moritz, H., *IEEE Trans. on Electron Devices, ED-32*, 672 (1985).
Moriya, T., K. Yamada, T. Shibata, H. Iizuka, M. Kashiwagu, *1983 Symp. on VLSI Technol.*, 96 (1983).
Movchan, B. A., A. V. Demchishin, *Fiz. Metal. Mettaloved, 28*, 83 (1969).

Murarka, S. P., R.J. Gutmann, A. E. Kaloyeros, W. A. Lanford, *Thin Solid Films,* 236, 257 (1993).
Nakamura, M., M. Itoga, Y. Ban, in "Proc. Symp. Plasma Etching and Deposition" (R. Frieser and C. J. Mogab, eds.), *Electrochem. Soc. Proc. Vol. 81-1,* 225 (1981).
Nakamura, M. T. Kurimoto, H. Yano, K. Yanagida, *Electrochem. Soc. Ext. Abstr.* 728, PV 87-2, 1042 (1987).
Nandan, R., S. P. Murarka, A. Pant, C. Shepard, W. A. Lanford, in *Advanced Metallization and Processing for Semiconductor Devices and Circuits II,* (A. Katz, S. P. Murarka, Y. I. Nerram, J. M. E. Harper, eds.) *Mat. Res. Soc. Symp. Proc. Vol. 260,* 929 (1992).
Narasimhan, M., J. Sasserath, E. Ghanbari, *J. Vac. Sci. Technol.,* A10, 1100 (1992).
Ng, S. L., S. J. Rosner, S. S. Laderman, T. I. Kamins, p. 93 in *Tungsten and Other Refractory Metals for VLSI Applications—II* (E. K. Broadbent, ed.), Mat. Res. Soc., Pittsburgh, Penn., 1987.
Nishimura, H., S. Ogawa, T. Yamada, *IEDM 93,* 281 (1993).
Norman, J. A. T., B. A. Muratore, P. N. Dyer, D. A. Roberts, A. K. Hochberg, *1991 VMIC,* 123 (1991).
Oehr, C., H. Suhr, *Appl. Phys.* A 45, 151 (1988).
Ogawa, S.-i., H. Nishimura, *IEDM 91,* 277 (1991).
Ohno, K., M. Sato, Y. Arita, *Jpn. J. Appl. Phys.,* 28, L1070 (1989).
Ohshita, Y., N. Hosoi, *Thin Solid Films,* 262, 67 (1995).
Ohta, T., N. Takeysau, E. Kondoh, Y. Kawano, H. Yamamoto, *1994 VMIC,* 329 (1994).
Oshima, H., M. Katayama, K. Onoda, T. Hattori, H. Suzuki, Y. Tokuda, Y. Inoue, *J. Electrochem. Soc.,* 140, 801 (1993).
Paddock, A. D., J. R. Black, Electrochem. Soc. Ext. Abstr. 98, May Meeting, 247, 1968.
Pai, P.-L., Y. Shacham-Diamand, W. G. Oldham, *1986 VMIC,* 209 (1986).
Park, S., L. C. Rathbun, T. N. Rhodin, *J. Vac. Sci. Technol.,* A3, 791 (1985).
Park, H. L., C. D. Park, J. S. Chun, *Thin Solid Films,* 166, 37 (1988).
Park, H. L., S. S. Voon, C. O. Park, J. S. Chun, *Thin Solid Films,* 181, 85 (1989).
Park, Y. H., A. H. Chung, M. A. Ward, *VMIC,* 295 (1991).
Pauleau, Y., Ph. Lami, *J. Electrochem. Soc.,* 132, 2779 (1985).
Pelletier, J., R. Pantel, J. C. Oberlin, Y. Pauleau, P. Gouy-Pailler, *J. Appl. Phys.,* 70, 3862 (1991).
Philofsky, E., K. Ravi, E. Hall, J. Black, *9th IEEE/IRPS,* 120 (1971).
Piekaar, H. W., L. F. Tz. Kwakman, E. H. A. Granneman, *VMIC,* 122 (1989).
Platter, V., G. C. Schwartz, US Pat. 3,827,949, 9/6/74, 1974.
Platter, V., G. S. Schwartz, *IBM Tech. Discl. Bull.,* 17, 1605 (1974).
Poulsen, R. G., H. Nentwich, S. Ingrey, *Proc. Int. Electron Devices Mtg.,* Washington, D. C., December 1976, p. 205, 1976.
Pramanik, D., A. N. Saxena, *Solid State Technol.,* 1/83, 127 (1983a).
Pramanik, D., A. N. Saxena, *Solid State Technol.,* 3/83, 131 (1983b).
Pramanik, D., A. N. Saxena, *Solid State Technol.,* 3/90, 73 (1990).
Purdes, A. J., *J. Vac. Sci. Technol.,* A1, 712 (1983).
Puttock, M., A. Iacopi, G. Powell, M. Clausen, P. Bennett, in "Proc. 10th Symp. Plasma Processing" (G. S. Mathad, D. W. Hess, eds.), *Electrochem. Soc. Proc. Vol. PV 94-20,* 368 (1994).
Raghavan, G., C. Chiang, P. B. Anders, S. M., Tzeng, R. Villasol, G. Bai, M. Bohr, D. R. Fraser, *Thin Solid Films,* 262, 168 (1995).
Reiseman, A., D. R. Shin, G. W. Jones, *J. Electrochem. Soc.,* 137, 722 (1990).
Reith, T. M., J. D. Schick, *Appl. Phys. Lett.,* 25, 524 (1974).
Rocke, M., M. Schneegans, *J. Vac. Sci. Technol.,* B6, 1113 (1988).
Rodbell, K. P., P. A. Totta, J. F. White, US Pat. 5,071,714, 1991).

Roger, C., T. S. Corbitt, M. J. Hampden-Smith, T. T. Kodas, *Appl. Phys. Lett.*, *65*, 1021 (1994).
Rogers, B., S. Bothra, M. Kellam, M. Ray, *1991 VMIC*, 137 (1991).
Rogers, B., S. Bothra, S. Bobbio, D. Temple, *1992 VMIC*, 239 (1992).
Rotel, M., J. Zahavi, H. C. N. Huang, P. A. Totta, G. C. Schwartz, in "Corrosion of Electronic Materials and Devices" (J. D. Sinclair, ed.), *Electrochem. Soc. Proc. Vol. PV 91-2*, 387 (1991).
Ruoff, A. L., E. J. Kramer, C.-Y. Li, *IBM J. Res. Develop.*, *32*, 620 (1988).
Ryan, J. G., N. P. Marmillion, J. P. Fineman, D. P. Bouldin, *Thin Solid Films*, *166*, 55 (1988).
Samukawa, S., M. Stark, S. Nakamura, R. Chow in "ULSI Science and Technol. 1989" (C. M. Osburn, J. M. Andrews, eds.), *Electrochem. Soc. Proc. Vol. 89-9*, 463 (1989).
Samukawa, S., T. Tomohiko, E. Wani, *Appl. Phys. Lett.*, *58*, 896 (1991a).
Samukawa, S., T. Tomohiko, E. Wani, *J. Vac. Sci. Technol.*, *B9*, 1471 (1991b).
Santoro, C. J., *J. Electrochem. Soc.*, *116*, 361 (1969).
Saraswat, K. C., S. Swirhun, J. P. McVittie, *VLSI Sci. and Technol./1984* (K. E. Bean, G. A. Rozgonyi eds.) *Electrochem. Soc. Proc. Vol. PV 84-7*, 409 (1984).
Sato, K., T. Oi, H. Matsumaru, T. Okubo, T. Nishimura, *Met. Trans.*, *2*, 691 (1971).
Sato, M., H. Nakamura, A. Yoshikawa, Y. Arita, *Jpn. J. Appl. Phys.*, *26*, 1568 (1987).
Sato, M., *SPIE Vol. 2091*, 220 (1994).
Sato, T., N. Fujiwara, M. Yoneda, *Jpn. J. Appl. Phys.*, *34*, 21421 (1995).
Schaible, P. M., W. C. Metzger, J. P. Anderson, *J. Vac. Sci. Technol.*, *15*, 334 (1978).
Schaible, P. M., G. C. Schwartz, *J. Vac. Sci. Technol.*, *16*, 377 (1979).
Schaible, P. M., G. C. Schwartz, US Pat. 4,352,716, 1982.
Schmitz, J. E. J., A. J. M. vas Dijk, R. C. Filwanger, US Pat. 4,892,843, 1990.
Schmitz, J. J., Kang, S. G., US Pat. 5,272,112, 1993.
Schulz, S. E., B. Hintze, C. Wurzbacher, W. Gruenewald, T. Gessner, p. 465 in *Advanced Metallization for ULSI Applications in 1993*, (D. P. Favreau, Y. Shacham-Diamand, Y. Horike, eds.), Mat. Res. Soc., Pittsburgh, Penn., 1994.
Schwartz, G. C., P. M. Schaible, *J. Electrochem. Soc.*, *130*, 1777 (1983).
Schwartz, G. C., P. M. Schaible, in "Proc. Symp. Plasma Etching and Deposition" (R. G. Frieser and C. J. Mogab, eds.), *Electrochem. Soc. Proc. Vol. 81-1*, 133 (1981).
Schwartz, G. C., J. M. Yang, M. Chow, P. M. Schaible, R. K. Lewis, F. Legoues, unpublished (1986).
Schwartz, G. C., P. M. Schaible, G. R. Larsen, unpublished (1986).
Schwartz, G. C., *J. Electrochem. Soc.*, *138*, 621 (1991).
Schwarzl, S., W. Beinvogel, in "Proc of 4th Symp. on Plasma Processing" (G. S. Mathad, G. C. Schwartz, G. Smolinsky, eds.), *Electrochem. Soc. Proc. Vol. 83-10*, 310 (1983).
Sebesta, E. H., US Pat 4,497,684, 2/5/85, 1985.
Selamoglu, N., C. N. Bredbenner, T. A. Giniecki, *J. Vac. Sci. Technol.*, *B9*, 2530 (1991).
Shacham-Diamand, Y., A. Dedhia, D. Hoffstetter, W. G. Oldham, *J. Electrochem. Soc*, *140*, 2427 (1993).
Shacham-Diamand, Y., V. Dubin, M. Angal, *Thin Solid Films*, *262*, 93 (1995).
Shibata, T., T. Moriya, K. Kurosawa, T. Mitsuno, K. Okmura, Y. Horiike, K. Yamada, M. Muromachi, *IEDM 84*, 75 (1984).
Shibata, H., US Pat 4,576,678, 3/18/86 (1986).
Shimamura, H., A. Yajima, Y. Yoneoka, S. Kobayashi, *J. Vac. Sci. Technol.*, *A9*, 595 (1991).
Shin, H. K., K. M. Chi, M. J. Hampden-Smith, T. T. Kodas, J. D. Farr, M. Paffett, *Chem. Mater.*, *4*, 788 (1992).
Smith, P. M., J. G. Fleming, R. D., Lujan, E. Rohert-Osman, J. S. Reid, A. K. Hochberg, D. A. Roberts, p. 345 in *Advanced Metallization for ULSI Applications in 1993* (D. P. Favreau, Y. Shacham-Diamand, Y. Horike, eds.) Mat. Res. Soc., Pittsburgh, Penn., 1994.
Smith, D. L., R. H. Bruce, *J. Electrochem. Soc.*, *129*, 2045 (1982).

Smith, D. L., P. G. Saviano, *J. Vac. Sci. Technol., 21*, 768 (1982).
Spencer, J. E., in "Proc. 4th Symp. Plasma Processing" (G. S. Mathad, G. C. Schwartz, G. Smolinsky, eds.), *Electrochem Soc. Proc. Vol. PV 83-10*, 321 (1983).
Spencer, J. E., *Solid State Technol., 4/84*, 203 (1984).
Sridharan, U. C., D. Hartman, R. Wright, J. Carter, M. Kent, in "Proc. 9th Symp. Plasma Processing" (G. S. Mathad, D. W. Hess, eds.), *Electrochem. Soc. Proc. Vol. PV 92-18*, 409 (1992).
Srolovitz, D. J., A. Mazor, B. G. Bukiet, *J. Vac. Sci. Technol., A6*, 2371 (1988).
Stacy, W. T., E. K. Broadbent, M. H. Norcott, *J. Electrochem. Soc., 132*, 444 (1985).
Steinbruchel, C.H., *J. Vac. Sci. Technol., B2*, (1984).
Steinbruchel, C. H., *J. Appl. Phys., 59*, 4151 (1986).
Stoll, R. W., R. H. Wilson, in "Multilevel Interconnection and Contact Technologies" (L. B. Rothman, T. Herndon, eds.), *Electrochem. Soc. Proc. Vol. 87-4*, 232 (1987).
Sugai, K., T. Shinzawa, S. Kishida, H. Okabayashi, *Electrochem. Soc. Ext. Abstr. 315, PV 93-1*, 482 (1993).
Suzuki, T., H. Kitagawa, K. Yamada, M. Nagoshi, *J. Vac. Sci. Technol., B10*, 596 (1992).
Tait, R. N., T. Smy, M. J. Brett, *J. Vac. Sci. Technol., A10*, 1518 (1992).
Tait, R. N., T. Smy, M. J. Brett, *Thin Solid Films, 226*, 196 (1993).
Takewaki, T., H. Yamada, T. Shibata, T. Ohmi, T. Nitta, *Materials Chemistry and Physics, 41*, 182 (1995).
Tang, C. C., D. W. Hess, *Appl. Phys. Lett., 45*, 633 (1984).
Tatsumi, T., T. Nagayama, S. Kadomura, in "Proc. 9th Symp. Plama Processing" (G. S. Mathad and D. W. Hess, eds.), *Electrochem. Soc. Proc. Vol. 92-18*, 283 (1992).
Temple, D., A. Reisman, *J. Electrochem. Soc., 136*, 3525 (1989).
ter Beek, M., *Electrochem. Soc. Ext. Abstr. 314, PV 85-2*, 475 (1985).
Thomas, M. E., T. K. Keyser, E. K. W. Goo, *J. Appl. Phys., 59*, 3768 (1986).
Thornton, J. A., *J. Vac. Sci. Technol., A4*, 3059 (1986).
Tokunaga, K., F. C. Redeker, D. A. Danner, D. W. Hess, *J. Electrochem. Soc., 128*, 851 (1981).
Towner, J. M., A. G. Dirks, T. Tien, *24th IEEE/IRPS Symp.*, 7 (1986).
Townsend, P. H., H. A. Vander Plas, *1985 VMIC*, 76 (1985).
Tsai, M. H., S. C. Sun, S. Y. Yang, H. T. Chiu, *1994 VMIC*, 362 (1994).
Tsao, K. Y., D. J. Ehrlich, *Appl. Phys. Lett., 45*, 617 (1984).
Tsao, K. Y., H. H. Busta, *J. Electrochem. Soc., 131*, 2702 (1984).
Tsubouchi, K., K. Masu, N. Shigeeda, T. Matano, Y. Hiura, N. Mikoshiba, S. Matsumoto, T. Asaba, T. Mauri, T. Kajikawa, *1990 Symp. on VLSI Technol.*, 5 (1990a).
Tsubouchi, K., K. Masu, N. Shigeeda, T. Matano, Y. Hiura, N. Mikoshiba, *Appl. Phys. Lett., 57*, 1221 (1990b).
Tsubouchi, K., K. Masu, *J. Vac. Sci. Technol., A10*, 856 (1992).
Tsukada, T., H. Takei, E. Wani, E. Ukai, *Electrochem. Soc. Ext. Abstr. 201, PV 83-1*, 327 (1983).
Tsukada, T., *Jpn. J. Appl. Phys., 30*, 2956 (1991).
Tsutsumi, T., H. Kotani, J. Komori, S. Nagao, *IEEE Trans. on Electron Devices, 37*, 569 (1990).
Uchida, T., H. Aoki, M. Hane, S. Hasegawa, E. Ikawa, *Jpn. J. Appl. Phys., 32*, 6095 (1993).
van der Jeugd, C. A., G. J. Leusink, G. C. A. M. Janssen, S. Radelaar, *Appl. Phys. Lett., 57*, 354 (1990).
van der Jeugd, C. A., G. S. Leusink, T. G. M. Oosterlaken, P. F. A. Alkemade, L. K. Nanver, E. J. G. Goudena, G. C. A. M. Janssen, S. Radelaar, *J. Electrochem. Soc., 139*, 3615 (1992).
van der Putte, P., p. 77 in *Tungsten and Other Refractory Metals for VLSI Applications—II* (E. K. Broadbent, ed.), Mat. Res. Soc., Pittsburgh, Penn., 1987.

van Maaren, A. J. P., R. L. Krans, E. de Haas, W. C. Sinke, *J. Vac. Sci. Technol.*, *B9*, 89 (1991).
Vossen, J. L., US Pat 4,372,806, 2/8/83 (1983).
Wang, D. N., F. D. Egitto, D. Maydan, US Pat. 4,412,885, 1983.
Webb, E. A., US Pat. 5,350,488, 1994.
Wilson, R. H., R. W. Stoll, M. A. Calacone, *1985 VMIC*, 343 (1985).
Wilson, R. H., B. Gorowitz, A. G. Williams, R. Chow, S. Kang, *J. Electrochem. Soc.*, *134*, 1867 (1987).
Winters, H. F., *J. Vac. Sci. Technol.*, *B3*, 9 (1985).
Wong, M., K. C. Saraswat, *IEEE Electron Device Lett.*, *9*, 582 (1988).
Xie, R. J., J. D. Kava, in "Proc. 11th Intl. Symp. Plasma Processing" (G. S. Mathad, M. Meyyappan, eds.), *Electrochem. Soc. Proc. Vol.*, *96-12*, 242 (1996).
Xie, R. J., J. D. Kava, M. Siegel, *J. Vac. Sci. Technol.*, *A14*, 1067 (1996).
Yang, C., J. Multani, D. Paine, J. Bravman, *VMIC*, 200 (1987).
Yang, C. H., V. Grewal, J. Laney, J. Yang, *VMIC*, 510 (1994).
Young, L., *Anodic Oxide Films*, Academic Press, New York, 1961.
Zahavi, J., M. Rotel, H.-C. W. Huang, P. A. Totta, *Electrochem. Soc. Ext. Abstr. 253, PV 84-2*, 354 (1984).
Zama, H., T. Miyake, T. Hattori, S. Oda, *Jpn. J. Appl. Phys.*, *31*, L588 (1992).
Zhu, N., T. Cacouris, R. Scarmozzino, R. M. Osgood, Jr., *J. Vac. Sci. Technol.*, *B10*, 1167 (1992).

VI
Chip Integration

Geraldine Cogin Schwartz* and K. V. Srikrishnan
IBM Microelectronics
Hopewell Junction, New York

There are two aspects of chip integration: One is concerned with the problems created by topography (Part I) and the other the compatibility of materials and processing (Part II).

The issue of topography encompasses not only the elimination of surface irregularity, i.e., planarization, but also step coverage and gap-fill (filling the spaces between adjacent conductors with dielectric films and filling via holes in the dielectrics with metal films).

Compatibility issues, to be covered in Part II, are (1) the interactions of the various components of the device structure with each other and their response to the processing conditions and (2) the effects of the choice of processing methods and conditions on the components and on the final device, i.e., its structure, performance, and reliability.

PART 1: TOPOGRAPHY, STEP COVERAGE, AND PLANARIZATION

1.0 OVERVIEW

Topography is a natural result of building a device structure and using multiple levels of wiring; these processes require depositing or growing films and patterning them repeatedly, thus creating steps in the different layers. Coverage of these steps by films depends on the deposition process involved and good step coverage is almost always a required attribute to meet reliability and yield objectives. Planarization processes are developed to mitigate the adverse effect of the creation of

---

*Retired.

topography and reduce the step-coverage requirements. But these issues are interconnected in ways that a clear understanding of one is vital for solving needs of the other. These issues and a variety of processes used in film deposition and their characteristics are discussed in the following sections.

2.0 TOPOGRAPHY
2.1 INTRODUCTION

After the structures within the semiconductor substrate have been completed, the surface is coated by an insulator to isolate the devices from the overlying structures. In the case of silicon devices, this layer is thermally grown SiO_2. Connections must then be made to the devices at the appropriate locations; therefore, so-called contact holes must be etched through the insulator. This introduces topography, i.e., a nonplanar surface. A metal film may be deposited into this hole or, more recently, the hole may be filled by a vertical interconnect. This illustrates facets of the problems of topography, i.e., step coverage (how well does the metal layer cover the sides of the contact hole?) and gap-fill (how well does the metal fill the contact hole?).

2.2 CONSEQUENCES
2.2.1 Structural

Vertical interconnects can result in a planar surface, eliminating the topography due to the contact hole. But the problem reemerges when the first-level conductor is formed. And, as each successive level of the metal-insulator structure is built, the steps become deeper, since the successive layers are usually thicker. Metal films are thickened to decrease line resistance and increase electromigration resistance; insulator films are thickened to reduce capacitance and to cover the thicker metal layers adequately. The trend toward increased numbers of levels is accompanied by a decrease in device and interconnection dimensions as well as an increase in packing density. And as deep steps become steeper to accommodate the increased density, the difficulties of step coverage and hole-fill increase. In addition, residues formed during RIE are less tractable if the sidewalls are steep, adding to RIE-selectivity problems. Thickness variations in the interlevel dielectric result in variation of the interlevel capacitance.

2.2.2 Lithographic

Another deleterious consequence of topography is the loss of line-width control in lithography. The thickness of a photoresist film changes as the film is spun-on over a step (Widman and Binder, 1974; Lin et al., 1984; Thompson, 1994). This is illustrated in Fig. VI-1a. The smallest resist thickness, t_a, at the top of the step t_a:

$$t_a \sim t_n \left[1 - \frac{1}{2} \left(\frac{s}{t_n} \right)^2 \right]$$

Chip Integration

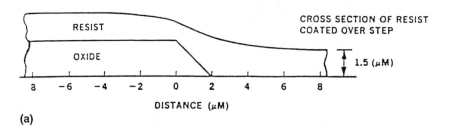

(a)

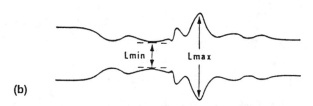

(b)

FIGURE VI-1 (a) Diagram of the thickness change at a step when a film is spun on a surface. (b) Corresponding change in resist dimensions at the step. (From Lin et al., 1984)

where s = step height
t_n = nominal thickness of resist

and the thickest resist, t_b, at the bottom of the step t_b:

$$t_b = t_a + s$$

The corresponding change in pattern dimensions as seen in Fig. VI-1b (Lin et al., 1984). The variation increases as the step height increases; any variation in dimension becomes more serious as the dimensions decrease. As seen in Fig. VI-2, the use of an antireflection (AR) coating reduces, but does not eliminate, the variation (Lin et al., 1984; Horn, 1991).

Thick resist films are more planar than thinner ones (White, 1983; Pampalone et al., 1986). However, since Fresnel diffraction limits the resolution of the minimal pitch (resolution = $3[\lambda(d/2)]^{1/2}$), increasing the resist thickness (d) degrades the lithographic performance.

The use of multilevel resist masks is a technique for achieving planarity and acuity. Figure VI-3a shows the starting structure in the fabrication of a trilevel mask after patterning the thin resist film. The resist pattern is used as a mask during RIE of the intermediate (barrier) layer (e.g., silicon oxide, nitride or SOG) which does not etch in O_2. The patterned barrier layer then acts as the mask during RIE in O_2 of the organic film, forming the ultimate thick mask for etching the substrate. However, during RIE of the organic underlay in O_2, there is redeposition on the sidewall of the resist, during etching by sputtering of the inorganic barrier layer and during the overetch period by sputtering of the substrate (Kinsbron et al., 1982).

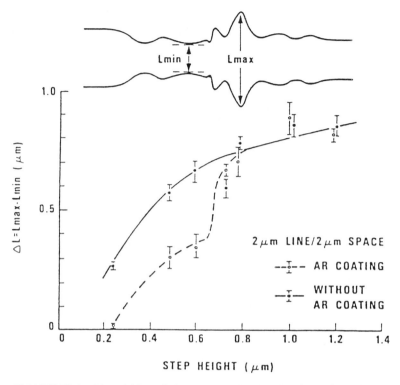

FIGURE VI-2 Linewidth variation vs. step height showing effect of an antireflection (AR) coating. (From Lin et al., 1984)

The coating on the sidewalls of a mask formed by RIE in O_2 (with an overetch of 100%) on an AlCu substrate is shown in the SEM of Fig. VI-3b, but such sidewall films have been found on many other kinds of substrates. This film must be removed before etching the underlay, without shifting or otherwise distorting the organic mask, in order to preserve the dimensions of the initial mask. Thus it is clear that the use of such masks increases both the number of layers and processing steps and requires RIE equipment in addition to the usual ones used for lithography. The patterning process becomes more complex and therefore, more costly.

There are also line-width variations caused by reflections off topographic features within the resist layer although it may be planarized. This is due to standing waves in the resist image (Lin, 1983), as shown in Fig. VI-4.

Another lithographic issue is the depth-of-focus (DOF). To improve the resolution of the resist image, the DOF must be decreased. However, variation in surface height can result in exceeding the in-focus condition for submicrometer imaging; out-of focus images lose fidelity, especially for closely pitched features (Moreau, 1988; Bothra et al., 1995).

Processes have been developed to modify existing steep steps, to improve step coverage and hole-fill, and to planarize, usually partially, but in some cases, globally. These will be discussed in detail below.

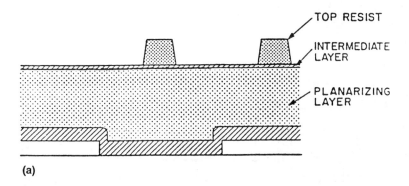

FIGURE VI-3 (a) Starting structure for a typical trilevel resist. (b) SEM of the sidewall coating of resist formed, during RIE in O_2, of the thick organic (planarizing) layer.

2.3 ORGANIZATION

The sections on planarization are organized to consider first the "traditional" aspects of planarization; these include step coverage and gap-fill as well as surface leveling. In the second part, we cover mechanical polishing (CMP), which tacitly assumes that the underlying structures have been adequately protected using the methods in the earlier sections. In both parts, dielectrics will be covered before metals.

3.0 SPIN-ON FILMS
3.1 INTRODUCTION

All the spin-on films used in device fabrication have planarizing properties, but they differ in their effectiveness. The films may be temporary layers or permanent

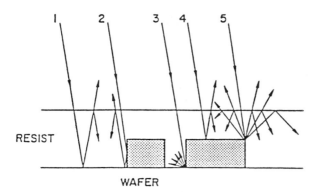

FIGURE VI-4 Diagram showing reflections off topographical features. (From Thompson, 1994)

dielectrics although some materials are used either way. Included in the materials of interest are photoresists, PIs, SOGs, as well as resins such as polystyrene, Accuflo (Applied Materials), Futurex (Futurex Co.), benzocyclobutene (BCB) (Dow Chemical), fluorinated poly(arylethers) (FLARE) (Allied Signal), etc. PIs and SOGs were discussed in Chapter IV.

3.2 PLANARIZATION

The degree of planarization of spin-on films is defined as $1 - (t_s/t)$ as shown in Fig. VI-5; the slope of the edge of the coated structure as θ. The extent of planarization after coating a structure consisting of a set of rectangular features is shown in Fig. VI-6. Figure VI-6 (top) illustrates partial planarization and smoothing of the edges of each feature. Figure VI-6 (middle) shows complete *local* planarization (planarization of a group or groups of closely spaced features). In Fig. VI-6 (bottom) there is complete, or *global* planarization, i.e., planarization independent of size and pitch, but this is much more difficult to achieve.

During spinning, centrifugal and capillary forces drive the flow (gravitational forces are usually not important); viscous forces resist flow. When the centrifugal and viscous forces are equal, the film is uniform; if centrifugal forces dominates, it is conformal in the wide spaces. Local planarization occurs when capillary forces dominate. Baking can improve planarity by decreasing solution viscosity, allowing increased flow; but this is counteracted by solvent evaporation which increases the

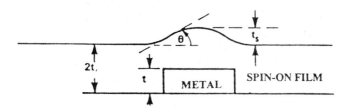

FIGURE VI-5 Definition of degree of planarization.

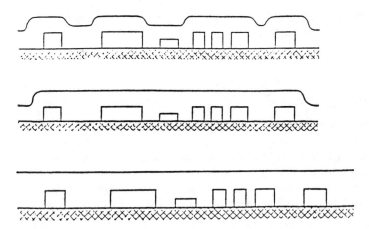

FIGURE VI-6 Characterization of planarization: (top) partial planarization and edge smoothing, (middle) "local" planarization, (bottom) "global" planarization.

concentration and hence the viscosity. Heating also enhances cross-linking of polymer units and shrinkage; both factors oppose flow. As the film dries, solids may be redistributed, leading to planarity. Wide gaps are filled after spinning only if the film can flow long distances. Details can be found in Stillwagon (1987a,b, 1988, 1992), Sukanek (1989), Bornside et al. (1991), and Gu et al. (1995).

3.2.1 Improvements

The planarizing capability of a given spin-on solution depends on the chemical composition of the solute, its molecular weight and concentration, all of which determine its viscosity. White (1983) showed that the use of low-viscosity materials resulted in better planarization of isolated features; however, in a real circuit structure, the situation is more complex. He also stated that the use of high-viscosity materials, coupled with high spin speed, minimized the thickness variation within the pattern. If the coating is to be used as an interlevel dielectric, uniformity in thickness is important to minimize capacitance variation. If the film is used as a sacrificial layer in etchback, nonuniformities in the layer can be transferred into the underlying dielectric, resulting again in capacitance variation. As noted before, if the film is a resist, the thickness variation will be reflected in the loss of dimension control. Crapella and Gualandris (1988) suggested the use of two materials with different viscosities: partially fill steps using a low viscosity material, finish coating with one of high viscosity, and finally, bake. Other material properties can effect planarization. For example, the use of low-molecular-weight polymers was shown by Gokan (1988) to result in better planarization. According to Pampalone et al. (1986) the best planarizing resins are those in which planarization and subsequent polymerization occurred independently. This requires the use of low-melting-point resins. Processing conditions also influence the final result. For example, flow during cure may override the effect of viscosity. Schlitz et al. (1987) reported that destroying the photoactive compound in a resist by exposure to uv light before final bake improved the flow properties of the material. Planarization usually improves as the film thickness is increased (White, 1983; Pampalone et al., 1986). Applying multiple coats (curing between them) is often more effective than using a single coat of the same final thickness (Chao and Wang, 1984; Chen et al., 1988; Kojima et al., 1988).

3.2.2 Effect of Topography

The geometry of the underlying structure, i.e., the pitch and distribution of the features, is very important. Wilson and Piacente (1986) showed that the relative densities of the features affect the film thickness. Since equal volumes of material are deposited in each unit area and solution fills all available space, the film surface is highest where the circuit density is highest. White (1985a,b; White and Miszkowski, 1985) have modeled the effect of topography on film thickness.

3.2.3 Concluding Remarks

In general, small features are planarized more easily than large ones and densely packed features (tight pitch) are planarized better than sparse (or isolated) ones (Rothman, 1980; Chen et al., 1988; Kojima et al., 1988). Narrow, high AR gaps are filled and planarized readily; filling holes is easier than filling trenches.

4.0 STEP COVERAGE BY DEPOSITED FILMS

4.1 CHARACTERIZATION

Step coverage by deposited films, such as those formed by evaporation, sputtering, and blanket CVD and PECVD processes, is very different from that of spin-on films. Geometric shadowing as well as the deposition conditions now play a dominant role (Levin and Evans-Lutterodt, 1983). The sides of a feature limit the acceptance angle of the incoming species so that, in most cases, the deposit along the sidewall decreases in thickness with increasing depth. Step coverage is often characterized by the ratio: $\text{thickness}_{\text{inside corner}}/\text{thickness}_{\text{horizontal surface}}$. If the sides are closely spaced, the films deposited on the top edges may converge, creating a void within the deposit.

When the mobility of the incoming species is large, i.e., the sticking coefficient is low, so that shadowing effects become negligible and/or the mechanism of film formation ensures it, equal thicknesses can be formed on all surfaces; such conformal coverage, shown in Fig. VI-7a (top), is very reliable. Although voids are not formed when the coverage is conformal, a seam may exist at the center of the space where the films growing out from each side merge. If conditions are optimal, W deposited by LPCVD from WF_6 and H_2 and SiO_2 deposited by HPCVD from TEOS plus O_3 exhibit conformal step coverage. The conformality of the SiO_2 film has been attributed to the "liquidlike" flow of the precursor.

More usual are the contours shown in Fig. VI-7a (middle, bottom). Figure VI-7a (middle) illustrates the coverage when the film precursor is mobile and the conditions are such that they have a long mean free path. SiO_2 deposited by PECVD using TEOS and O_2 is one example. Figure VI-7a (bottom) is an example of the coverage when the mean free path is short and/or there is little or no precursor

Chip Integration

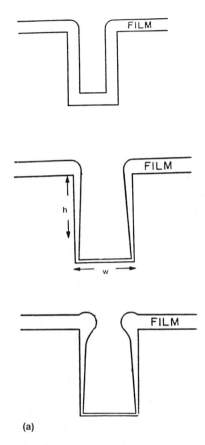

FIGURE VI-7(a) Typical step coverage profiles of deposited films: (top) conformal; (middle) nonconformal: long mfp, mobile precursor; (bottom) nonconformal: short mfp, little or no mobility.

mobility; reduced mobility apparently accounts for the SiO_2 profile when the film is deposited using SiH_4 as the precursor or in sputter deposition of Al.

Step coverage degrades and voids are formed more readily as the sidewall angle becomes steeper and the AR increases; this is illustrated in Fig. VI-7b.

4.2 CVD OXIDES

4.2.1 Models

The step coverage of CVD SiO_2 was modeled by, e.g., Cheng et al. (1989) and by Rey et al., (1991), and compared with experimental results. Several possible mechanisms were considered: (1) direct deposition, (2) reemission of deposited material resulting in indirect deposition, and (3) surface diffusion. A special test structure, shown in Fig. VI-8, was used to isolate the mechanisms. Tapering of the film along the upper edges of the hole was not observed experimentally in SiO_2 deposition,

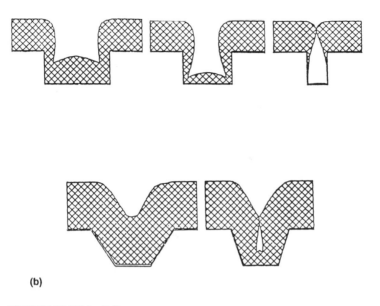

(b)

FIGURE VI-7(b) Effect on step coverage of (top) aspect ratio and (bottom) wall angle.

and therefore it is concluded that surface diffusion is negligible in this case. Reemission of deposited material is determined by the reaction probability or sticking coefficient (sc) of the impinging species; a low sc is key to good step coverage.

When the precursor flux is low, it is valid to use a single value of sc to simulate step coverage, at a given temperature, for any shape of feature (Cheng et al., 1990). The value of sc decreases with increasing temperature and depends on the nature of the Si-source: it is high for SiH_4 (leading to "bread-loafing" or overhang formation) and low for organic precursors. The lower the value of sc, the steeper the sidewall angle that can be filled without void formation (Rey et al., 1990).

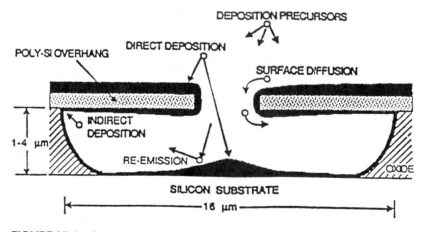

FIGURE VI-8 Overhang test structure used to isolate the mechanisms of CVD step coverage, showing possible mechanisms. (From Rey et al., 1991)

IslamRaja et al. (1993) found that more accurate simulation of the entire profile evolution required the use of two rate-limiting precursors: an intermediate species, sc ~ 1, formed by gas phase reactions, and one with low sc (formed from the source gas). These react separately with the surface to form the film. The increase in step coverage and deposition rate with increasing pressure is due to the change in the ratio of the partial pressures of the two precursors with pressure.

Profile evolution for processes ranging from vacuum to atmospheric conditions was simulated by Ikegawa and Kobayashi (1989); the physical model of the deposition process is shown in Fig. VI-9. Movement in the gas phase was characterized by the Knudsen number (Kn). For $Kn > 10$, the flow is collisionless (PVD, LPCVD); $10^{-2} < Kn < 10$ transitional (APCVD). Other parameters investigated were reaction probability or sticking coefficient (sc) and AR/feature geometry. The step coverage degraded as the value of sc increased. Coverage is better over trenches than holes and depends on the position in the hole as well. As observed experimentally, the simulation predicted that as the value of AR increased, step coverage degraded.

Other simulations/models for CVD processes can be found in the literature, e.g., Watanabe and Komiyama (1990), Cale and Raupp (1990), Wulu et al. (1991), Schroder et al. (1991), Cale et al. (1992), Fujino et al. (1993), Sorita et al. (1993), Coronell and Jensen (1994).

4.2.2 Improvements

Improved planarization of films deposited using TEOS and O_3 was accomplished by treating Ti or Ti-alloy capped Al or W interconnects with a CF_4 plasma before oxide deposition. These metals reduce the adsorption of the precursor oligomers and assist their flow into the spaces between them, resulting in substantial step height reduction (Suzuki et al., 192).

The use of F-substituted Si-sources improved step coverage and gap-fill. The monomer hydrolyzed to yield the oligomer and alcohol, both of which were ad-

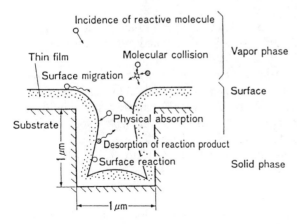

FIGURE VI-9 Species and processes involved in physical model for thin film deposition used in simulation by the Monte Carlo method. (From Ikegawa and Kobayashi, 1989)

sorbed on the surface. The mechanism proposed was that the oligimer flows readily, covering the features, and then polymerizes (Homma et al., 1993).

Both a high rate at low temperature and improved step coverage and gap fill were the advantages given for changing the precursor from TEOS to tetra-isocyanate-silane (TICS) (Taniguchi and Sugiura, 1994); however, the oxide contained silanol groups and water.

Planarized, gap-filling SiO_2, formed using a CVD process called "FLOW-FILL"™, was discussed in Chapter IV.

4.3 CAPACITIVELY COUPLED PECVD OXIDES

The profiles of films deposited in these reactors are similar to those deposited by thermally activated CVD (Ross and Vossen, 1984) (except for those deposited using TEOS and O_3 which are usually conformal, as noted above).

4.3.1 Models

PECVD in a capacitively coupled reactor was studied by Chang et al. (1992) who used the overhang structure described above to study the role of ion bombardment. They concluded that the ion and neutral (LPCVD) components behave independently. The ion flux was responsible for the high rate and directionality of the deposit. Ion bombardment increased sc on the exposed surfaces; the neutral component was nearly isotropic with a low (~0.15) sc.

Cale et al. (1992) simulated film conformality for $TEOS/O_2$ based PECVD SiO_2, also treating the deposition as a combination of an almost directional and an almost isotropic component. They predicted that both conformality and rate decreased as the temperature increased and as the pressure increased (more collisions in the sheath). Conformality increases as the input rf power increases since the fraction of ions increases.

4.3.2 Improvements

PECVD processes have been modified in an attempt to improve the conformality or gap-fill.

By making the reaction mixture oxygen-rich, the profile became more conformal; in an oxygen-lean plasma, the process is ion-dominated, more directional but with poor sidewall-film quality (Hsieh et al., 1989).

Addition of NH_3 (Olmer and Daverse, 1990) or NF_3 (Ibbotson et al., 1990) made the deposition more directional by suppressing growth along the sidewall.

Using hexamethylcyclotrisilazan (HMCTSZ) as the precursor resulted in planar films and void-free gap-fill; it was postulated that low molecular species were formed in the plasma, were adsorbed on the surface where they were able to migrate to the depressions in the surface and then polymerize (Nishio et al., 1994). The film, however, resembles an SOG film.

F-Doped Oxide

PECVD F-doped SiO_2 films fill gaps better than do the undoped ones. Deposition of these films was discussed in detail in Chapter IV.

Chip Integration

Figure VI-10 (Usami et al., 1994) shows the improvement of gap-fill with increasing F-doping, but also the degradation of filling when the F-concentration becomes too high.

The mechanism for gap-fill has been postulated to be plasma-assisted etching of the depositing film by the F-species in the plasma (Matsuda et al., 1995).

However, as discussed in Chapter IV, the low ε may be more of a driving force in the development of F-doped oxides while questions of stability, leakage, reliability, etc., may ultimately determine the acceptance of them.

5.0 IN SITU PLANARIZATION/GAP-FILL OF DIELECTRIC FILMS

5.1 INTRODUCTION

In situ processing refers either to the use of reactors in which deposition and contour modification occur simultaneously or else to the use of integrated reactors in which the films are deposited in one chamber and transferred in vacuo to a second chamber for modification.

5.2 BIAS SPUTTER DEPOSITION

5.2.1 Planarization

By using a high substrate bias during sputter deposition of SiO_2 over metal conductors, partially planarized layers can be produced. The extent of planarization increases with increasing substrate bias which increases the extent of resputtering (sputter etching); this, however, decreases the net accumulation rate significantly. The process can be carried out using a fixed bias ratio (Ting et al., 1978; Singh et al., 1987) or a lower bias can be used to deposit the initial film and a higher bias for planarization (Mogami et al., 1985; Singh et al., 1987). In a reasonable length of time, only the narrow lines can be planarized by this technique, limiting its usefulness. Faceting of the edges of the metal steps can occur; in the case of Al, faceting should be minimal because the native oxide which covers all Al surfaces has a very low sputtering rate. The mechanism proposed by Ting et al. (1978) is

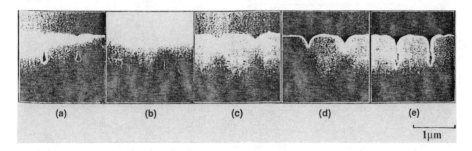

FIGURE VI-10 Gap-fill by PECVD SiOF films; each successive SEM shows the effect of increasing F-doping. Final SEM shows the deleterious effect of too much doping. (From Usami et al., 1994)

based on the fact that the sputter removal rate depends on the angle of incidence of the impinging ions (Fig. VI-11a), which is determined by the taper angle of the feature to be covered (Fig. VI-11b). At low bias, the topography of the oxide coat replicates that of the metal (Fig. VI-11c). At a higher substrate bias, a stable facet angle is formed at the corners of the oxide film; the angle is determined by the equilibrium point at which deposition and etch rates are equal. The higher the substrate bias, the flatter the taper angle. Oxide continues to be deposited on the flat portion of the feature and the oxide cone is buried. When the cone is buried completely, the surface is planarized (Fig. VI-11d). Less oxide is needed to planarize the surface if the resputtering rate is increased. However, since the net accumulation rate is lowered if the input power is held constant, it will take longer to deposit the same total amount of oxide. If, after deposition of the film, the bias is raised so that no further net deposition occurs, the protrusions of oxide will eventually be removed by the inward movement of the sloped surfaces (which etch at a higher rate than the horizontal ones) until the metal and oxide are coplanar (Fig. VI-11f). But this is an extremely slow process.

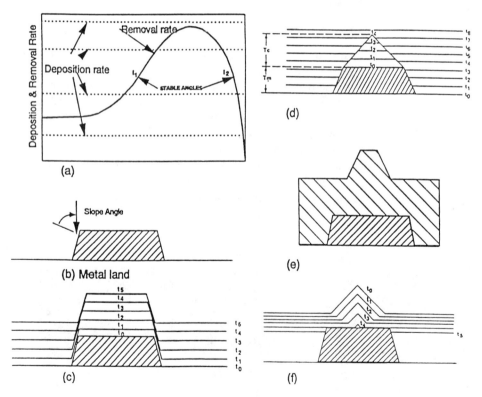

FIGURE VI-11 "Planar" sputtered SiO_2: (a) deposition/removal rate vs. slope angle, (b) metal line to be coated, (c) typical (nonplanar) oxide layer contours, (d) planarized narrow metal land, (e) nonplanarized wide line, (f) oxide contours during formation of coplanar structure by sputter etching after deposition. (From Ting et al., 1978)

5.2.2 Gap-Fill

The ability to fill deep narrow spaces between adjacent conductors also increases as the substrate bias is increased and the model is based on the same phenomenon used to explain planarization (Logan et al., 1989). As the oxide is deposited into a gap with a side-wall angle of A, stable facet angles are formed at both the outside corner (discussed by Ting et al., 1978) and the inside corner (ϕ). Oxide continues to be deposited on the sidewall (at a slow rate due to the restricted acceptance angle) until it intersects ϕ. At some depth, the oxide films deposited at the stable angles at each inner corner intersect at the center. This is shown in Fig. VI-12. Since the deposition rate at the bottom is the slowest one, a void is formed and is eventually pinched off. The critical AR (D/W) that can be filled without void formation is given by:

$$\left(\frac{D}{W}\right)_{max} = \frac{0.5 \tan \phi}{1 - (\tan \phi / \tan A)}$$

The size of the void, when formed, increases with increasing AR and sidewall angle A so that vertical-walled gaps are most difficult to fill. The higher the resputtering rate, the steeper the angle ϕ and the higher AR of the gap that can be filled without a void. But the use of very high resputtering rates will result not only in very low deposition rates but in beveling of the corners of the metal unless they are protected with an adequate thickness of material with an extremely low sputtering rate.

Although more favorable results are obtained when 40.68 MHz excitation is used instead of 13.56 MHz, the gap-filling capability of bias sputtered films is limited and often inadequate for the deep, narrow spaces between adjacent conductors which exist in today's multilevel structures.

5.3 BIAS-PECVD IN A CAPACITIVELY COUPLED REACTOR

5.3.1 Sidewall Tapering

Tapering the sidewalls of PECVD films depends on the angle-dependent sputter etch rate. In one version, by applying a large negative bias to the substrate, deposition and etching occur at the same time; one of the reactants etches the film as it grows. SiH_4/N_2O mixtures, in which N_2O was the etchant (Smith and Purdes, 1985) and $SiCl_4/O_2$ in which $SiCl_4$ was the etchant (Sato and Arita, 1986) have been used. McInerney and Avanzino (1987), on the other hand, used cycles of deposition in SiH_4/N_2O and in vacuo transfer to a sputtering chamber to achieve this result.

5.3.2 Planarization and Gap-Fill

By increasing the substrate bias, McInerney was able to planarize the oxide surface.
More recently, however, the emphasis has shifted from using substrate bias for surface planarization to its use in increasing the gap-fill capabilities of PECVD processes. PECVD SiO_2 films, prepared using SiH_4 as the Si-source have a "bread-

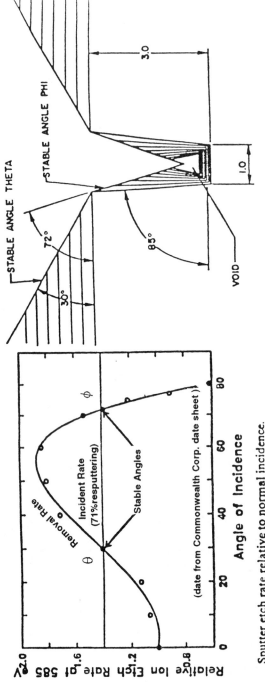

FIGURE VI-12 (right) Diagram showing hole filling by sputter deposition with reemission of 71% with a sidewall angle of 85°; (left) definition of θ and ϕ. (From Logan et al., 1989)

$$(D/W)max = 0.5 \times \tan \varphi / 1 - (\tan \varphi / \tan A)$$

Chip Integration 379

loafed" profile, making gap filling difficult. Using TEOS minimizes this difficulty and TEOS/O$_2$ is therefore preferred for deposition.

A variation of McInerney's process is the use of TEOS instead of SiH$_4$ for deposition. A typical process sequence is shown in Fig. VI-13, in which the final oxide is etched back by RIE to the required thickness (Bader et al., 1990).

A simplification is to deposit only the required thickness as the final step (Schwartz and Johns, 1992). The resulting structure may then be planarized by an etchback process or by CMP (to be discussed below).

It becomes more difficult to fill gaps without forming voids or low density regions (which occur before the appearance of physical voids) if the AR, sidewall angle, or depth of the gap increase (Schwartz and Johns, 1992). The term low density was chosen because such regions etch rapidly in BHF; however there is no acceleration of the RIE rate, so that misalignment of the via pattern will not result in overetching in the gap. The number of cycles required for void-free fill depends on the characteristics of the gap. Their duration as well as their number can be tailored, to some extent, to fill very steep, deep, and narrow gaps. However, throughput concerns, rather than physical limits, will probably determine the ultimate process capability.

A more complex process involves PECVD (TEOS/O$_2$) and low temperature (~400°C) CVD (TEOS/O$_3$) for deposition and reactive and inert plasmas for etchback (Mehta and Sharma, 1989; Pennington et al., 1989; Spindler and Neureither, 1989; Shih et al., 1992). A typical process sequence is shown in Fig. VI-14 with a final etchback or CMP process for polarization.

Another approach, which is reported to result in a greater degree of planarization as well as good gap-fill, uses a sacrificial film of PECVD B$_2$O$_3$ which flows during deposition (Marks et al., 1989; Pennington and Hallock, 1990). This process is illustrated in Fig. VI-15.

In order to minimize particulate contamination, reactive plasma cleaning of the chambers is often carried out after processing each wafer, which adds to the overall process time.

5.4 BIASED HIGH-DENSITY PECVD

The most promising techniques for filling deep, narrow holes with SiO$_2$ are independently biased PECVD in an ECR (Machida and Oikawa, 1986; Denison et al., 1989), induction-coupled (van den Hoeck and Mountsier, 1994), or helicon (Nishimoto et al., 1995) reactor, all of which produce similar results. Deposition and etching occur simultaneously. In these systems, SiH$_4$ has been used almost universally as the source gas, despite the superior step coverage of TEOS-based oxides. As discussed in a Chapter I, despite the high ratio of ion to neutral species and the low operating pressure minimizing collisions, bias is necessary to ensure gap-fill. If the bias is too low, the oxide at the top corner of the gap grows faster than sputtering can remove it so that the gap is closed at the top, leaving a void (Patrick, 1991; van den Hoek, and Mountsier, 1994; Nishimoto et al., 1995). However, if the bias is too high, the corners of the metal become faceted (Tisier et al., 1991). As the AR of the gap increases, the process window becomes smaller.

The simulations for gap-fill and planarization of bias-ECR describe the other bias high-density PECVD processes as well.

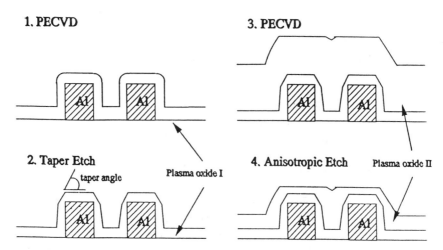

FIGURE VI-13 Integrated process for forming an intermetal dielectric layer using TEOS-based PECVD oxide films and etchback. (From Bader et al., 1990)

Labun (1994) modeled the process as simultaneous PECVD (D) and sputter etching (E). In the simulation, the ratio E/D and the sticking coefficient sc were varied. High values of E/D, which result in low values of sc, improved gap-fill, and planarization. The simulation of Lassig (1995) was more complex. The components of the process, treated independently, are (1) a small LPCVD (low-pressure CVD) component, independent of ion bombardment, (2) IID (directional ion-induced deposition), the dominant component whose directionality is a result of high applied bias which narrows the angular distribution of the incoming ions, (3) angle-dependent sputter etching which prevents void formation by faceting the oxide at the top of the gap so that gaps fill from the bottom to the top, (4) direct line-of-sight-redeposition of sputtered material from the bottom of the gap to the sidewalls (~10–15%) which increases with bias and is responsible for the ability to fill a gap with an overhang or with a reentrant sidewall, and (5) deposition of sputtered material backscattered from the gas phase (a negligible contribution). The simulation replicated the experimental results in which the evolution of the profile was delineated by periodic deposition of thin Si-rich oxide layers which were made visible by BHF etching.

It is possible to planarize small features by continued deposition at high bias but, as in the case of bias sputtering, it is an inefficient planarization technique.

5.4.1 Hollow Cathode

A hollow cathode, low frequency, low pressure, capacitively coupled reactor for PECVD of SiO_2 resulted in good gap-fill, but at much lower deposition rates than the other methods discussed above (Gross and Horowitz, 1993) so that the usefulness of this system for device production is questionable.

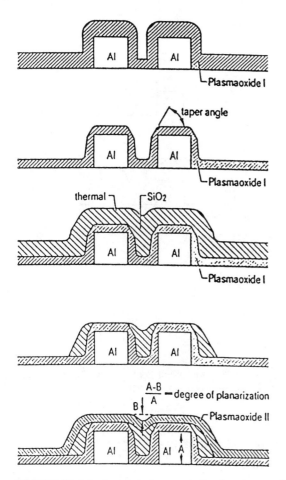

FIGURE VI-14 Integrated process for forming an intermetal dielectric layer using TEOS-based PECVD and thermal CVD oxide films and etchback. (From Spindler and Neureither, 1989)

6.0 ETCHBACK OF INTERLEVEL INORGANIC DIELECTRIC (ILD) FILMS

6.1 INTRODUCTION

There are various options for planarizing an ILD using etchback processes. An inert or a reactive plasma may be used (although in the earliest practice of this technique wet etching was employed). The principles of inert- and reactive-plasma etching have been discussed previously. The dielectric film is most often etched with a sacrificial overcoat although Kotani et al. (1983) sputter-etched and Thomas (1992) ion-milled uncoated films; planarization or partial planarization was effected because of the angular dependence of the etch rate on the angle of ion incidence. When a sacrificial layer is used, it may be a blanket film or may be patterned (selective masking).

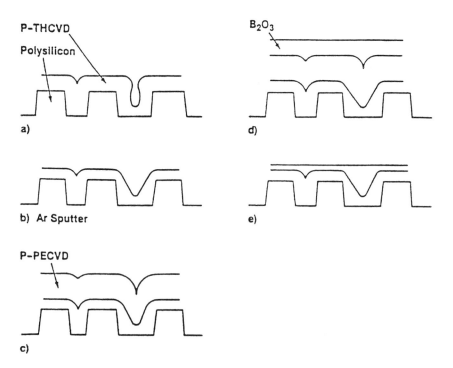

FIGURE VI-15 In situ gap fill and planarization using sacrificial B_2O_3 with P-doped thermal CVD and PECVD oxide films. (From Pennington and Hallock, 1990)

6.2 ETCHBACK WITH A BLANKET SACRIFICIAL OVERCOAT

6.2.1 General Concept

The general concept of etchback planarization of a dielectric film using a sacrificial overcoat is an idealized description of the process. The sacrificial layer is assumed to be planar; it is etched until the very tops of the underlying film are uncovered and then *both films are etched at equal rates, i.e., the etch rate ratio* (ERR = $\text{rate}_{\text{sacrificial layer}}/\text{rate}_{\text{ILD}}$) *is unity*. The process may start with an ERR = 1 or a high etch-rate process may be used for the first step to improve throughput. Etching may be stopped within the ILD and a subsequent masked via etch used to reach the underlying metal surfaces. Or etchback may be continued to expose the metal surfaces (Geiger and Sharma, 1986).

If the metal features are not all in the same plane, two options are available. Expose the highest surface and etch vias to the lower ones, or etch until the lowest metal surfaces are exposed. Complete planarity is not possible in this case.

6.2.2 Etch-Rate Ratio

The requirement that the etch rates of both films be the same when both are exposed to the plasma simultaneously is satisfied more easily when inert plasmas are used. There is no chemical component in such plasmas and thus, sputtering under the right conditions is often nearly nonselective (Homma et al., 1979). In ion milling;

equality of etch rates can be obtained by adjusting the angle of incidence (Hou et al., 1983; Johnson et al, 1983; Kubo, 1986). But the etch rates are too *low* to make this a practical process for chip fabrication.

Etch rates in reactive plasma are suitable for manufacturing needs but such plasmas are usually *selective*. Therefore, mixtures of etchant species, tailored to the specific films and etch conditions; must be formulated. However, the chemical component of a reactive plasma leads to another complication (not a factor in inert plasmas), the so-called "local loading effect." Although the etch parameters remain constant, the local etch rate of each material (and thus the local etch rate ratio) may change as the relative areas change in case the underlying film is not exposed uniformly. The process parameters must therefore be varied to compensate for the pattern factor.

6.2.3 Process Optimization

Optimization of the etch parameters is not a trivial matter. The starting point for process development is usually the determination of the ERR from the etch rates of the blanket films of each material as a function of such variables, such as the choice of the etchant mixture and the ratio of its components, rf power, frequency, pressure, flow rate, and electrode spacing. The values of ERR quoted in describing etchback processes are obtained in this way. However, the answers derived from such a study are often inadequate due to the loading effect and an ERR less than one may be required to produce planarity (Riley et al., 1987, 1988, 1990). Optimization of the process may be by trial and error or, more efficiently, by statistical design (Ray and Marcoux, 1985; Riley et al., 1987, 1988, 1990). Another approach to determining loading effects was the use of a series of masks with different pattern factors (deBruin and Laarhoven, 1988). Multiple-step processes may be required to accommodate the loading effect; optical emission signals have been used to track the process and signal either an end-point or the need for a change in the etching conditions, usually a change in the input gases or their flow rates (e.g., Koyama and Thomas, 1985; Fritzsche et al., 1986; deBruin and van Laarhoven, 1988). Koyama and Thomas (1985) introduced a two-step process in which a thin oxide was deposited first to prevent void formation, coated with a thin resist, and then etched back. This was followed by a thicker oxide and resist and a final etchback. Here too, the ERR was varied during the etchback.

6.2.4 Processes

The processes chosen vary widely. They depend on the dielectric films to be planarized and the nature and planarity of the sacrificial layer, on the initial choice of the reactant gases (which may depend on the common practice in the lab/fab), reactor configuration, pattern factors, and the criterion for the final result.

Although the specific gas mixtures vary widely, they always include a F-containing species, the principal etchant of inorganic films, although a marginal etchant of organics. Examples are the fluorocarbons, SF_6, NF_3. In the case of the fluorocarbons, another choice is the C/F ratio which determines the specific compound used. H is added to control polymerizing reactions; it may be either as H_2 or as part of the molecule. If an organic sacrificial layer is used, O_2 may be added to enhance its etch rate and moderate polymer formation. On the other hand, at some point in the process, the O_2 flow rate may be reduced to counteract the release of O-containing species if the inorganic film being etched is SiO_2, since they accelerate the etch rate of an organic layer. An alternative approach was to cover the oxide layer with plasma nitride to prevent the release of such species (Mayumi et al., 1988). The mixture may contain an inert diluent as well.

The assumption that a single coat of a sacrificial film is planar is not a valid one and the nonplanarity has been dealt with in several ways. Fujii et al. (1988) used two coats of resist to improve the planarity of the sacrificial layer; a multistep process, using a single layer, was devised by Gimpleson and Russo (1984). Rabinzohn et al. (1990) suggested that the etch conditions be tuned to compensate for the circuit density which affects the planarity of the sacrificial layer, as well as the etch products which modify the etch mixture. They also state that it is possible to compensate for a nonplanar sacrificial layer by etching with an ERR less than unity, as Toi and Choi (1985) had suggested earlier.

6.3 SELECTIVE MASKING

Another approach, requiring extra photolithographic processing, is selective masking, in which recessed regions are protected during the etch of the elevated ones. In the earliest use of this concept, a photoresist mask was used to cover the recessed regions of a nonplanar deposited oxide and the exposed, elevated areas etched in BHF (Feng, 1976). The same concept was extended when RIE processing was introduced. The wide recessed regions were filled by a lithographically defined mask, called a "dummy" pattern. The more nearly planar surface was then covered either by polyimide (Mitsuhashi, 1986) or another coat of resist (Sheldon et al., 1988) before RIE etchback. Schlitz and Pons (1986) also used a two-layer structure. After the recessed areas were covered by a lithographically defined mask, the resist was cured to flow it to level the surface before applying and curing a second resist layer. The nearly planar surface was now ready for etchback. The process sequence is illustrated in Fig. VI-16.

The process described by Grivna and Goodner (1994), resembled that of Feng. After the recessed regions were protected by a resist mask, the resist and exposed dielectric film were etched isotropically in a downstream etcher using a timed etch (whose ERR was adjusted for optimal planarity) to reach the desired dielectric thickness. If, after a final resist ashing step, oxide trenches or peaks persisted, a second (thin) smoothing overcoat was applied and a short nonselective RIE was carried out. A "global" planarity of 230 Å (1σ) was claimed for this process.

Daubenspeck et al. (1991) studied the design considerations associated with the use of the underlying mask in such a two layer planarization scheme; they called this mask a "block mask" or a pattern density compensation mask.

6.4 ORGANIC SACRIFICIAL LAYERS

The most commonly used sacrificial layers are the organic resins, such as photoresists (Adams and Capio, 1981; Bartush, 1985, 1987; Jang et al., 1987), polyimides (Sato et al., 1973; Misawa et al., 1987), and more recently, resins formulated spe-

Chip Integration

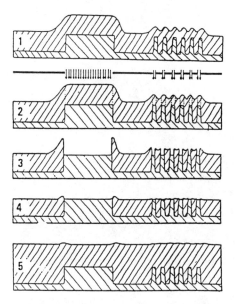

FIGURE VI-16 Process flow to prepare for oxide planarization etchback, using selective masking by a sacrificial layer: (1) spin on resist layer; (2) use mask to cover recessed areas, expose; (3) develop resist; (4) flow resist; (5) apply, cure second resist layer. (From Schlitz and Pons, 1986)

cifically to have enhanced flow properties and, hence, better planarizing capabilities (Neureither et al., 1994).

6.5 ETCHBACK PROCESSES USING SOG AS THE SACRIFICIAL LAYER

A typical process sequence (Elkins et al., 1986) is shown in Fig. VI-17. It can be seen that SOG remains as a permanent part of the structure, in the spaces between adjacent conductors.

One advantage to the use of SOG in planarizing oxide layers, as opposed to the use of an organic resin was thought to lie in the similarity between oxide and SOG which would lessen the severity of the loading effect (Molnar, 1989). However, an area ratio effect had been observed (Hausamann and Mokrisch, 1988) when an O-free gas was used in the etchback process. As the area of oxide exposed was increased, the etch rate of SOG increased more than that of the oxide. In this case it was proposed that the ERR was controlled by polymer deposition which occurred at a higher rate on the SOG surface. The exposed oxide liberated O-containing species as it was etched, thereby accelerating the etch rate of the SOG.

Many etchback processes using silicate or siloxane SOGs have been developed, using the same principles illustrated in the diagram above, but differing in their details (e.g., Naguib et al., 1987; Parekh et al., 1987; Kawai et al., 1988; Nishida et al., 1989; Whitwer et al., 1990; Kishimoto et al., 1992; Bacchetta et al., 1993).

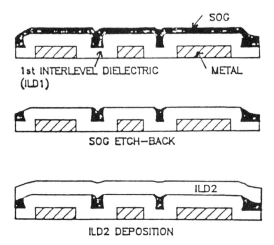

FIGURE VI-17 A typical sequence for an etchback process using SOG as a sacrificial layer. (From Elkins et al., 1986)

A number of process and materials modifications have been proposed to improve the results. Among these are (1) change the etchants and etch mixtures (Chu et al., 1986; Schlitz, 1986; Bui et al., 1987; Bogle-Rohwer and Nulty, 1990); (2) alter the P-content (Elkins et al., 1986) or the Si-content (Weling and Jain, 1992; Weling et al., 1993) of the oxide; (3) modify the cure cycle of the SOG or use a nondirectional etch instead of RIE (Schlitz, 1986); and (4) adjust the pressure (Moriomoto and Grant, 1988) or the electrode spacing (Hausamann and Mokrisch, 1988), and use a composite of an SOG and an organic film (Forester et al., 1994), as shown in Fig. VI-18.

More recently, efforts have been directed toward increasing the planarizing capabilities of SOGs by modifying the structure of the material. Park et al. (1995) has illustrated the improved results obtained using a flowable SOG, i.e., an SOG which flowed during curing, at 100–200°C. The flow of an SOG, and thus its planarizing capabilities, can also be enhanced if solvent vaporization is suppressed during coating (Ohashi et al., 1994; Yen et al., 1995). For very severe topographies, Chen (1995) proposed a two-step SOG etchback process, in which a highly selective etch was used to etch part of the SOG layer and one of low selectivity used to finish the process in order to achieve good planarity and preserve the underlying TiN.

6.6 OTHER ETCHBACK PROCESSES

There have been examples in which a portion of an overthick inorganic film, which has a more sloped contour than a thinner one, is sacrificed (Mercier et al., 1985).

Deposited inorganic layers have also been used as sacrificial layers. Geiger and Sharma (1986) used a sacrificial layer of a PECVD oxide. This was then covered by another sacrificial layer (resist) and the composite was etched down to the metal surface, leaving a locally planarized surface with PECVD in the intermetal gaps.

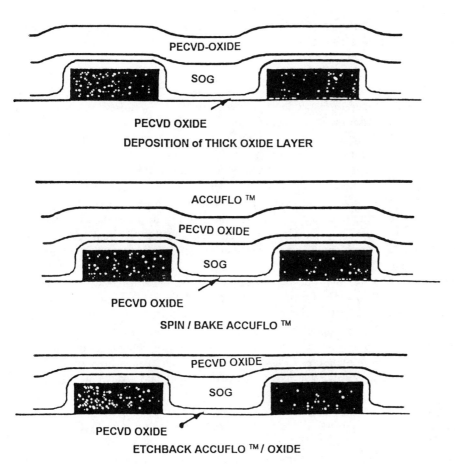

FIGURE VI-18 A Schematic of an etchback process using both SOG and an organic film. (From Forester et al., 1994)

Sum et al. (1987) used a thick film of SiN over a thin conformal coat of SiON so that when the SiN was etched to the SiON surface, a nearly planar surface resulted.

Composite films, such as polystyrene + polyimide (Lii and Ng, 1990) and SOG + organic resin (Parekh et al., 1990; Forester et al., 1994; Lee et al., 1995) have been used to enhance the planarizing/etchback properties of the individual films.

Finally, nonplanarizing layers which have desirable etch properties have been chosen (Hazuki et al., 1982). In this case, SiN was chosen because its etch rate in a CF_4/H_2 plasma was suppressed in grooves.

7.0 CONTOUR MODIFICATION OF A BLANKET FILM/ PATTERNED STRUCTURE

7.1 FLOWAGE OF BLANKET OXIDE

This process requires a high temperature, either in a furnace or using rapid thermal annealing, to smooth the surface and round the corners of CVD oxide films after deposition. Thus it can be used only when the underlying conductors are refractory metals and the devices are stable at the high temperatures. Contour modification improves the step coverage of the next level of metal.

The oxides are doped with P, B, Ge, or As, either alone or in combination, to reduce the temperature required for flow by reducing the viscosity of the glass. Despite doping, temperatures needed for flow are well above those compatible with Al (alloy) or Cu interconnection metallization. Reflow is affected by the viscosity of the oxide, which depends on the dopant type and its concentration. Increasing the dopant concentration reduces the flow temperature as well as the final taper angle. B is more effective than P, because it is more effective in lowering the viscosity. The B—O bond is more stable than the P—O bond and has a lesser tendency to form an acid and remains in the oxide. B atoms are small and enter interstitial positions and break the Si—O bond, making flow easier. The extent of flow is also dependent on the temperature and time but the practical limits are set by the allowable thermal budget and throughput compatibility of the process. Ion implants enhance flow by causing surface damage. Rinsing and delay between deposition and flow inhibit flow. Steam is the best ambient for flowage, although O_2 is satisfactory; flow is least pronounced in N_2 and Ar ambients (Tsai et al. 1990; Malik and Solanksi, 1990). Rapid thermal processing instead of thermal annealing allows a reduction in the dopant concentration for achieving the same final contour.

A patent disclosing a process in which a high reactant velocity at the substrate surface results in simultaneous LPCVD deposition and fusion flow of BPSG at temperatures of 800–875°C was issued to Monkowski and Logan in 1992. The process is reported to accomplish fusion tapering of topography, void-free trench filling and partial to complete planarization. Favorable surface tension effects cause viscoelastic flow resulting in filling from the bottom of the trench (Monkowski et al., 1991; Kern and Hartman, 1991).

The physical basis for the model of glass flow (PSG, BPSG) developed by Leon (1988) is the deformation of the film so as to decrease the surface free energy; the kinetics of the process is assumed to be controlled by surface diffusion. Umimoto et al., (1991), White et al., (1992), and Thallikar et al. (1995) described the flow of BPSG and PSG on the basis of viscous flow with the surface tension as the driving force.

7.2 TECHNIQUES FOR MODIFYING ETCHED FEATURES

7.2 Spacers

Spacer technology combines the deposition of a conformal insulator with anisotropic RIE. The oxide remaining is called a sidewall or spacer and smoothes the edges of existing steps. The shape of the spacer is a complex function of the etch process, but practical processes have been developed by Yao et al. (1985) and by Brader et al. (1986). However, since the effective space between adjacent conductors is reduced, thereby increasing the effective AR, this technique has only very limited application.

7.2.2 Reactive Facet Tapering (RFT) of an Oxide Step

Abraham (1987) and Iyer (1994) have described processes for tapering the edges of a deposited oxide in which the rate of formation of a facet, due to the angular dependence of the sputter yield, is enhanced by the use of a reactive gas.

7.2.3 Edge Cutting Process for a Metal Step

Mayumi et al. (1987) proposed a method for eliminating the sharp edges of an Al conductor. A polymer film was formed on the edges of the Al by exposure to a CF_4 plasma after RIE of the metal using a resist mask. The mask was etched slightly and the exposed Al etched in a wet etchant, and finally, the polymer film was stripped.

8.0 STEP COVERAGE, HOLE-FILL, PLANARIZATION OF METALS

When devices were large and packing density low, the only issue related to topography was coverage by the metal of the steps in the insulator. Good step coverage prevented open circuit failure and early wear-out. But as devices became smaller and packing density greater, not only was there a need for improved step coverage but, beyond that, the need to fill small holes (contacts and vias). In addition planarization during deposition, to simplify processing, became another goal.

The early work, using evaporation, will be covered first, followed by the use of sputtering in a variety of configurations, beam techniques, CVD, flowage, and finally, the first embedment methods.

9.0 EVAPORATION

Evaporation was used almost exclusively in the early years of device fabrication. As is the case for all deposited films, shadowing leads to poor step coverage with thinning or, in the worst case, cracks in the film in the inside corner, although crack-free deposits can be obtained by using elevated substrate temperature ($\geq 300°C$). Another effect of shadowing, in the case of metal deposition, is reduced density of the sidewall deposit. Step coverage and hole-fill are degraded as the holes get narrower and steeper as shown in the case of deposited dielectric films. Blech's model of evaporation (1970) showed that step coverage is also sensitive to the radius of curvature of the step as shown in Fig. VI-19. A small radius results in a "crack" at the bottom corner but if the top is rounded, the crack is not as deep. Increasing the radius of curvature improves the coverage.

Step coverage is also dependent on the configuration of the evaporator. The geometry of the source was characterized by Bader et al. (1987) by what they called the target aspect ratio (TAR) = source (target) to substrate distance/source diameter, illustrated in Fig. VI-20.

An extended source, with a wide angle of incidence, is characterized by a low value of TAR and a directional source by a high value. Figure VI-21 illustrates the effect of both TAR and hole diameter on step coverage/hole-fill.

Increasing TAR degrades the coverage of the sides of the trench but leads to increased film thickness at the bottom; this is useful for lift-off patterning. And, as

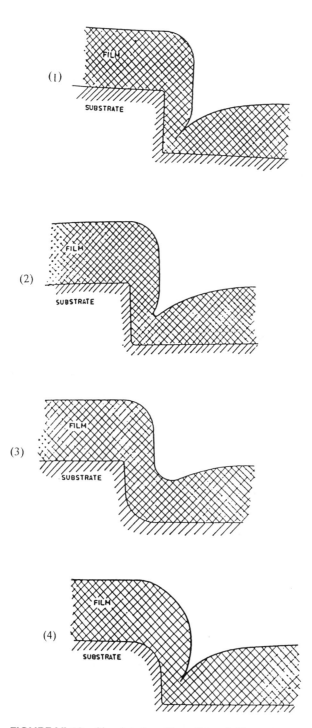

FIGURE VI-19 Simulated profiles of metal films evaporated over a step: (1) sharp step (2); (3) increasing curvature on both top and bottom of the step; (4) rounded at top but sharp at bottom of step. (From Blech, 1970)

Chip Integration

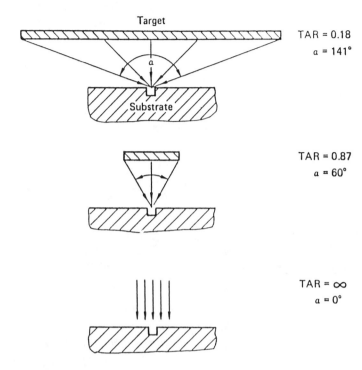

FIGURE VI-20 Illustration of the effect of the target aspect ratio TAR [ratio of the distance of the substrate from the target (source) to the diameter of the target] on α, the angle of incidence of depositing species with respect to the normal to surface. (From Bader, 1987)

the hole size is decreased, the tendency for void formation is increased, particularly at low TAR.

Tait et al. (1993) simulated evaporation at low temperatures (limited surface diffusion) using a ballistic deposition model and compared the results with evaporated films. The decreased density of the deposition and columnar microstructure were predicted and observed, when the surface of the substrate was tilted away from the normal direction, as occurs when there is a step in the substrate.

9.0.1 Elevated Temperature

The reduction of dimensions, requiring more accurate pattern transfer, as well the trend toward multilevel structures emphasizes the need for planarity. The use of high temperature (~450°C) to provide mobility to the incoming Al has had some limited success; 2-μm wide features were planarized but windows with AR > 1 had voids (Georgiou et al., 1989).

This same result was obtained by Smy et al. (1992) who simulated the effect of temperature on the hole-fill by evaporated Al (Fig. VI-22).

9.1 EVAPORATION/SPUTTER ETCH CYCLES

The step coverage of evaporated films can be improved by using alternate cycles of deposition and sputter etching (Lardon et al., 1986; Bader and Lardon, 1986;

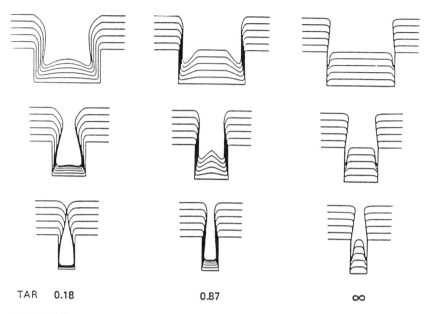

FIGURE VI-21 Simulation of the influence of TAR on the profile evolution during evaporation into grooves with aspect ratios of 0.5, 1.0, 2.0 (top to bottom). (From Bader, 1987)

Bader et al., 1987). A computer simulation of the effect of sputtering on step coverage is shown in Fig. VI-23.

The processes cannot be done simultaneously; evaporation is a high vacuum process ($\sim 2 \times 10^{-3}$ Pa) whereas sputter etching requires a pressure of ~ 0.5 Pa. Evaporation, using a directional source is required for filling and sputter etching for redistribution (discussed more fully below) to improve sidewall coverage. The effectiveness of the process is improved by increasing the number of cycles. The less directional the source, the greater the attack of the substrate by sputter etching in very narrow holes as seen in Fig. VI-24.

The procedure is a slow one and there is a limit to the hole-filling capabilities. For example, in a simulation, using 20 cycles and resputtering 30% of the Al at a

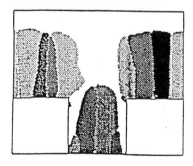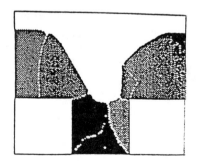

FIGURE VI-22 Simulation: effect of temperature on via fill by evaporated metal: (left) 250°C; (right) 500°C. (From Smy et al., 1992)

Chip Integration

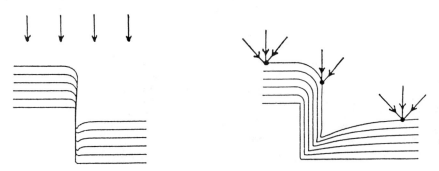

FIGURE VI-23 Simulation of step coverage: (left) evaporation; (right) sputtering. (From Lardon et al., 1986)

bias of 1000V, buried cavities are produced in vertical groves of AR $\gtrsim$ 1.5 and their size increases as AR is increased further, as seen in Fig. VI-25.

10.0 SPUTTER DEPOSITION OF METALS

Sputtering in a capacitively coupled rf plasma system eventually replaced evaporation, almost completely, both for control of the composition of alloys and the ability to apply substrate bias *during* deposition, rather than cyclically, as discussed above.

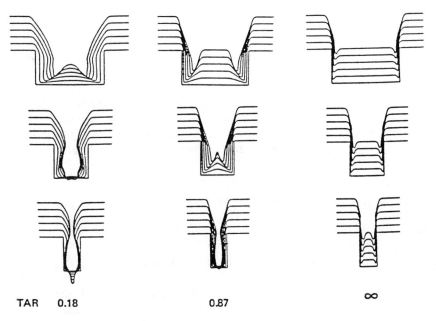

TAR   0.18   0.87 ∞

FIGURE VI-24 Simulation of the influence of TAR on profile evolution over grooves with aspect ratios of 0.5, 1.0, 2.0 (top to bottom) when 33% of the deposited Al is resputtered by Ar^+ of -1000 eV energy. (From Bader et al., 1987)

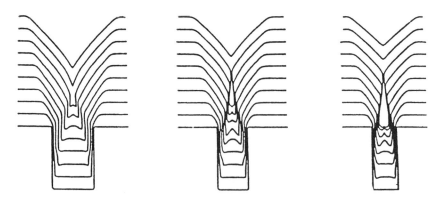

FIGURE VI-25 Simulation of coating shape for 20 cycles of evaporation/sputtering (30% resputtering with −1000 V substrate bias); aspect ratios 1.5, 2.0, 2.5 (left to right). (From Bader and Lardon, 1986a)

A sputtering target is an extended source; the pressure is higher during sputtering than in evaporation so that there is more scattering. Therefore the step coverage will resemble that obtained in low TAR evaporation *unless* bias is applied to the substrate.

10.1 BIAS SPUTTERING

In *bias* sputtering, material *deposited* at the bottom of the step is *resputtered* at small angles, *redepositing* on the sidewalls (Vossen, 1971; Vossen, et al., 1974). Material is deposited on the walls, instead of being sputtered away, because, when the incident ion energy is low (as it is when the substrate is biased), the angular distribution of the sputtered material is "undercosine" (Chapman, 1980; Morgan, 1989), i.e., the material is directed more to the side than perpendicular to the surface, as illustrated in Fig. VI-26.

Since sputter etching of the sidewalls is negligible (ion impingement near the glancing angle), the material stays on the walls. As Vossen has pointed out, the arrival rate of material cannot be too high to ensure that the resputtering keep pace with it. Substrate heating, due to ion bombardment (or an external heat source), enhances the effect of the ion-induced redistribution.

10.2 SIMULATIONS OF BIAS SPUTTERING

Bader and Lardon (1985) have simulated the effects on step coverage/hole-fill. Figure VI-27a compares sputter deposition with and without bias; Fig. VI-27b, with and without redeposition; and Fig. VI-27c shows the effect of lowering the bias potential.

Bias improves the step coverage, adding redeposition results in planarity, and, since at lower energies the stable facet angle is shallower, faster planarization occurs at low energy. However, as seen in Fig. VI-28 (Bader and Lardon, 1986b) although *planarization* (which is better at high AR than at low) is *superior at low energy*, the *onset of void formation occurs at higher AR at the higher energy*.

Chip Integration

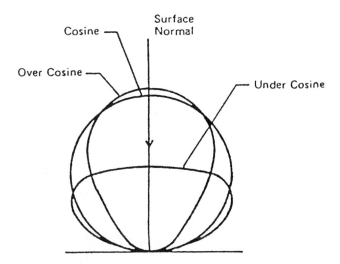

FIGURE VI-26 Possible angular distributions of sputtered material. (From Chapman, 1980).

The simulations agree with experimental observations where the comparison can be made.

Smy et al. (1990) and Dew et al. (1991) used a ballistic deposition model (SIMBAD) to predict step coverage and the structure of films deposited over topographical features. The model includes film resputtering, redeposition, ion reflection, and ion-induced diffusion. The angular dependence of the sputter yield, the ion reflection probability, the distribution of resputtered material, and the angular distribution of material arriving at the substrate from the target are all parameters included for the detailed simulation.

The simulations showed columnar microstructure and variation of film density, with density decreasing as the AR decreased and the wall angle increased. They also showed the improvement of step coverage (approximately doubling the sidewall coverage) and surface planarity with increasing bias (Fig. VI-29) both in the simulation and in experimental results of sputtered W. It should be noted that as bias was applied, the net accumulation rate decreased, indicating the importance of resputtering. Also seen was a raised peak at the center of the bottom of the hole.

As the AR of the hole was increased, substrate attack resulted (Fig. VI-30); this was also seen in the simulations of Fig. VI-26.

10.3 EXPLANATIONS (MODELS) OF BIAS SPUTTERING

There are many examples of improvement of step coverage, hole-fill, and planarization using bias sputtering, although, most often, not for the most aggressive dimensions. And there are several explanations (models) to account for the results.

10.3.1 Resputtering

The first is resputtering, but the issue is whether the bias should be low or high. Homma and Tsunekawa (1985), Mogami et al. (1985) and Lin et al. (1988) all

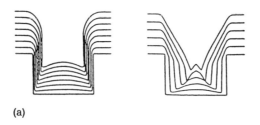

(a)

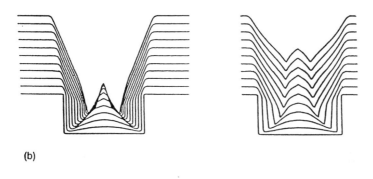

(b)

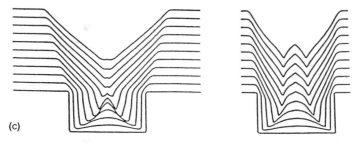

(c)

FIGURE VI-27 (a) Simulation of step coverage/hole-fill by sputtering using an extended source: (left) no substrate bias, (right) substrate bias = −1000 V. (b) Simulation of influence of redeposition during sputtering using an extended source and −1000 V substrate bias and 30% resputtering: (left) no redeposition, (right) with redeposition. (c) Simulation of the effect of ion energy on profile evolution using an extended source and 30% resputtering: (left) 300 V (profile angle ~45°), (right) 1000 V (profile angle ~70°C). (From Bader and Lardon, 1985)

Chip Integration

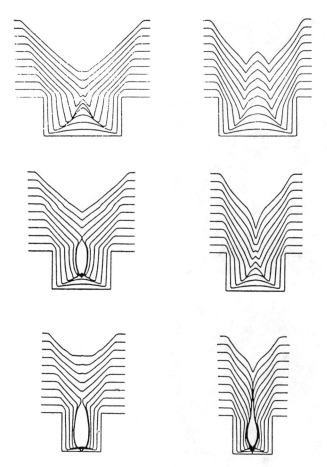

FIGURE VI-28 Simulation of the influence of bias potential (30% resputtering) on profile evolution on grooves with aspect ratios of 0.5, 0.67, 1.0 (top to bottom): (left) substrate bias potential −300 V, (right) substrate bias potential −1000 V. (From Bader and Lardon, 1986b)

emphasized the use of high bias. The role of resputtering in hole-fill and subsequent planarization is seen in the SEMs of Fig. VI-31 for AR = 1.

The net accumulation rate decreased with increasing substrate bias, indicating resputtering was the operative mechanism. Any effect of increased temperature was discounted. Mogami et al. (1985) pointed out that when the bias was too high, regions in the underlying material were etched and subsequently filled with metal.

10.3.2 Low Bias

Homma and coworkers (1993), in contrast to their earlier work and the model of Bader, stated that a *low* bias was required for good fill (invoking "undercosine" redeposition) since at high energies the angular distribution was "overcosine" so that there was outward ejection of the resputtered particles.

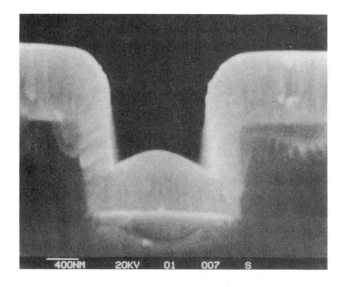
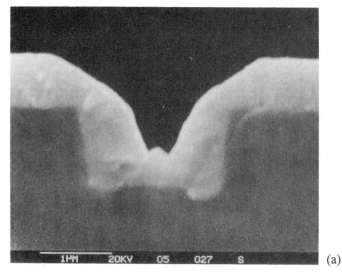
(a)

FIGURE VI-29 (a) SEMs showing influence of substrate bias on W films deposited over 0.55 aspect ratio grooves: (top) 0 rf bias, (bottom) −338 V rf bias. (b) Simulated film deposited over 0.55 aspect ratio groove: (top) no ion bombardment, (bottom) ion bombardment. (From Dew et al., 1991)

10.3.3 Enhanced Mobility

Other models *discount* resputtering because substrate biasing did not reduce the deposition rate. Smith et al. (1982), Smith (1984), attributed the improvement in step coverage to ion bombardment enhanced surface mobility (diffusion), or redistribution of material (Park et al., 1985). "Forward scattering" of surface atoms at

Chip Integration

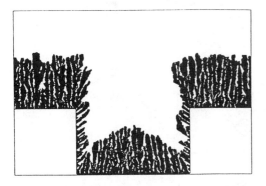

(b)

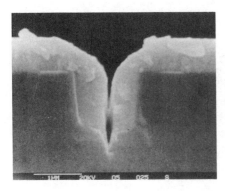 

FIGURE VI-30 Comparison of simulation and experimental result of a film deposited over a 0.9 aspect ratio groove: (left) SEM: −338 V substrate bias, (right) simulation with ion bombardment. (From Dew et al., 1991)

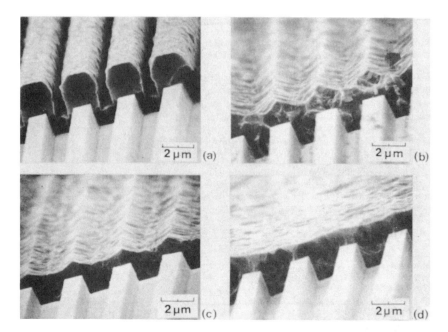

FIGURE VI-31 SEMs showing effect of resputtering on step coverage of Al films deposited over grooves of aspect ratio ~1: (a) 0% resputtering, (b) 40% resputtering, (c) 50% resputtering, (d) 70% resputtering. (From Homma and Tsunekawa, 1985)

the edge of the step by the incoming ions was proposed by Skelly and Gruenke (1986) who also stated that the effect of temperature could not be clearly separated. Demary et al. (1987) stated that, in a high growth-rate process, the primary role of bias is to provide effective rapid coupling of a front-side heat source to the film growth process to enhance mass transport.

10.3.4 High-Temperature Bias Sputtering

Introduction

High-temperature bias sputtering, i.e., the use of substrate heat (from an external source) in addition to bias, improves the results. In most cases, the high-temperature sputtering of the conductor metal is preceded by the lower-temperature deposition of a thin underlay; this may be the conductor metal itself (Park et al., 1991; Talieh et al., 1993) or a film (e.g., Ti, TiW, Ti/TiN), which acts as a barrier against junction penetration and/or as a liner to improve wetting/flow (Gupta et al., 1987; Lin et al., 1988; Hariu et al., 1989; Chiang et al., 1994). Nishimura et al. (1995) found that filling vias with AlCuSi by sputtering at high temperature depended on the thickness of the underlying Ti layer. They postulated that the wettability of the metal film on the sidewall was affected by the uniformity of the reaction between the Ti underlayer and the AlCuSi film.

Collimated sputtering (see below) is sometimes used to deposit this thin layer (Talieh et al., 1993; Xu et al., 1994). The sample is transferred between deposition chambers without a break in vacuum; multichamber ("cluster tool") systems

(Chapter I) are now usually used for this integrated processing. The wafer temperature during high-temperature sputtering has ranged from ~400 to ~550°C (Ono et al., 1990; Park et al., 1991; Nishimura et al., 1992; Xu et al., 1994). Taguchi et al. (1992) reported that diffusion of oxygen from the sidewalls of the via through the PSG ILD onto the surface of the Ti underlayer impeded the flow of AlSi sputtered at 500°C; SiN coating of the via walls acted as a diffusion barrier and improved the fill.

Mechanisms

Several mechanisms have been proposed to account for the improvement due to increased substrate temperature. Hoffman et al. (1987) concluded that thermally enhanced surface self diffusion explains the results, eliminating viscous flow (temperature too low), evaporation/recondensation (vapor pressure too low) and volume diffusion (only insignificant contribution). Kamoshida and Nakamura, (1987) demonstrated, as shown in Fig. VI-32, that planarization occurs through plastic deformation of Al films, i.e., through viscous flow. Since the temperatures *are* too low for plastic deformation, they invoke a decreased deformation resistance, a result of enhanced vacancy generation due to ion bombardment. Smy et al. (1992, 1994) have simulated step coverage, film density, and gap-fill. The use of bias reduces the minimum temperature needed for fill, and increased temperature enhances the effect of bias, increasing the range of adatom surface diffusion so that material is transported from the top of a film into a hole and bulk diffusion of vacancies out of the hole (Fig. VI-33). Figure VI-34 simulates the progression of hole fill on a hot substrate with time.

10.4 OTHER METHODS

10.4.1 AlGe Deposition

Hole filling by sputtering was accomplished at a lower temperature (200–300°C) by using AlGe which flows at this temperature (Kikuta et al., 1993).

10.4.2 Electron Bias Sputtering

A variation of high temperature bias sputtering is "electron bias sputtering" (Onuki and Mehei, 1993) in which Al is deposited using cycles of sputtering, with and without electron irradiation, produced by a pulsed positive bias applied to the substrate. The irradiation increases the temperature of Al, resulting in flow into the

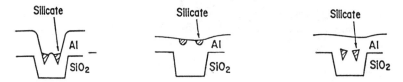

FIGURE VI-32 Model of planarization mechanism in the flowage bias sputter method by the use of a marker motion experiment: (left) a silicate SOG in the grooves of a deposited Al film, (middle) after exposure to Ar plasma with substrate at −1600 V for 5 min, (right) expected result from a surface diffusion model. (From Kamoshida and Nakamura, 1987)

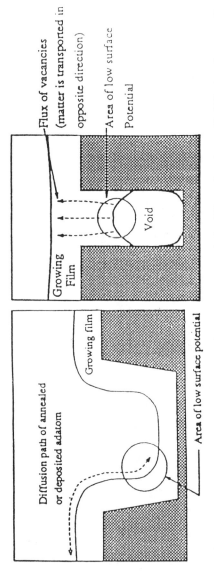

FIGURE VI-33 A model of the effect of temperature on hole filling showing self diffusion is a result of (left) surface diffusion and (right) bulk diffusion of vacancies. (From Smy et al., 1994)

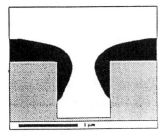  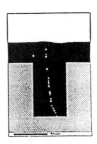

FIGURE VI-34 Simulation of film growth on a hot substrate (left to right) increasing time. (From Smy et al., 1994)

holes. There is a restricted optimum duty cycle, the flow zone, which depends on the substrate temperature. If the duty cycle is too short, there is no flow; if it is too long, the depositing Al aggregates instead of forming a smooth film.

10.4.3 Selective Sputtering

An interesting suggestion for achieving planarization (although its low rate makes it impractical) is ion-assisted *selective* deposition of Al into via holes, an application of what has been called the sputter yield amplification effect. Under low energy ion bombardment, Al can be selectively deposited onto an Al or SiO_2 surface while growth on a W-covered SiO_2 surface is prohibited because W is sputtered away during Al deposition. This effect arises from the fact that the sputtering yield is determined by the ion/substrate momentum transfer efficiency and the energy of the incoming ions so that, during bias-sputter deposition, heavy atoms but not light ones will be etched from a surface consisting of both kinds of material (Barklund et al., 1993).

10.4.5 Elevated Temperature, No Bias

There is one example of the use of high temperature (500–550°C) sputter deposition without substrate bias (Ono et al., 1990). At temperatures >500°C, the liquidity of the metal suddenly increases, perhaps due to a local decrease of the eutectic temperature at grain boundaries, permitting rapid movement. This process was used for hole fill by AlSi but no flow region existed for Al.

10.5 CONCLUDING REMARKS

Despite the improvements realized by the use of high temperature, usually with bias sputtering, its application may be limited because of incompatibility with the materials and performance of other components of the structure.

11.0 DIRECTIONAL SPUTTERING

Advances in hole filling using physical vapor deposition, beyond the capabilities of sputtering at elevated temperature, have been achieved by methods of directional sputtering.

11.1 COLLIMATED SPUTTERING

In sputtering, a large fraction of the impinging particles do not arrive at normal incidence because the atoms are emitted from a sputtering target in a cosine distribution and the pressures are such that gas scattering is significant. To obtain line-of-sight deposition to fill high AR holes, a collimator (an array of directional filters) was placed between the sputtering target and the substrate (Rossnagel et al., 1991; Cheng et al., 1995), as illustrated in Fig. VI-35.

Since the angular distribution of the impinging species is reduced, coverage of the bottom of a hole is increased significantly. This is due to the elimination of the off-normal particles which would lead to pinch-off at the to of the hole and to an increased flux at the bottom of the hole. However, the net accumulation rate is reduced, but according to Bang et al. (1994), the deposition rate loss at the bottom of the hole is not as severe as measurements on a flat surface suggest. A natural consequence of directionality is a minimal, but more uniform, coverage of the sides of the hole; this prevents/delays closure at the top of a hole as seen in Fig. VI-36 showing the initial stages in hole filling with Cu. As the fill proceeded, the microstructure of the film on the sidewalls became more columnar, but the fill was homogeneous if the substrate temperature was increased (200–300°C) or a low bias was imposed.

The higher the aspect ratio of the collimator, the more directional the deposit. Liu et al. (1993) and Dew (1993, 1994) have compared experimental results with computer simulations of collimated sputtering over topographical features and conclude that the deposit on the sidewall is of lower density than that on the top surfaces but the discontinuity at the bottom corners of a hole is reduced when a collimator is used. The thickness of the deposited film is not uniform; there are oscillations in the film thickness which correspond to the collimator pitch. Increased operating pressure would improve uniformity but the increased scattering would broaden the flux distribution, thus reducing the benefits of collimation. In addition, the holes in the collimator become plugged reducing the deposition rate even further and so that the collimator may eventually require cleaning or replacement.

11.2 LONG TARGET-TO-SUBSTRATE SPUTTERING

An alternative to collimated sputtering is sputtering using a long target-to-substrate (T/S)-distance (also called long-throw-distance), a low (~1 mtorr) pressure, a smaller cathode (to mimic a point source) and possibly, the use of a light sputtering gas (Turner et al., 1993; Broughten et al., 1995; Wagner, 1995). These system modifications result in a narrow angular distribution of the sputtered species and, as expected, the fill characteristics and reduction of net deposition rate are the same as those obtained with collimated sputtering. For both collimated and long T/S sputtering, deposition in holes at off-center positions is skewed, as shown in Fig. VI-37, because the angular distribution of the flux is centered at a nonzero angle.

Another application of collimated or long T/S sputtering is the deposition of "seed" or "liner" layers at low temperature before filling a hole with the conductor metal (Talieh et al., 1993; Broughton et al., 1995). The term "coherent sputtering" has been used for "collimated sputtering" (Xu et al., 1994). The directionality of the deposit makes this method suitable for lift-off patterning of a metal as well.

Chip Integration

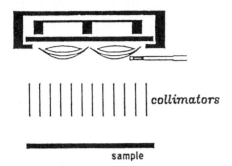

FIGURE VI-35 Cross section of a magnetron sputtering system with (grounded) collimators. (From Rossnagel et al., 1991)

11.3 Hollow Cathode-Enhanced Sputtering

Hollow cathode enhanced magnetron sputtering has been suggested as another alternative to collimation. The hollow cathode is the source of additional electrons which are injected into the plasma in the vicinity of the magnetron cathode and

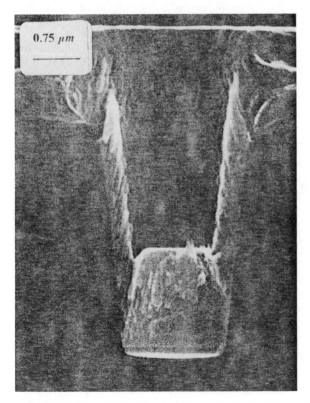

FIGURE VI-36 SEM of the initial stage of hole filling with Cu using collimated sputtering. (From Rossnagel et al., 1991)

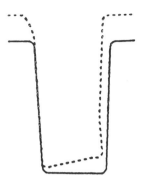

FIGURE VI-37 Diagram of asymmetrical deposition in a hole located away from the center of the substrate in long target-to-substrate (T/S) sputter deposition. (From Wagner, 1995)

are coupled directly into the magnetron discharge. The result is an increase in plasma density so that it is possible to sustain the plasma at lower pressures (<1 mtorr) than otherwise possible. The angular distribution of the sputtered atoms is not a cosine distribution but is "forward peaked" and becomes increasingly so with increased ion energy. At the low pressure there is no rate reduction due to scattering, making long T/S sputtering possible, thereby improving the uniformity of the deposit. The net rates are higher than those achieved using collimated sputtering. If the pressure is increased (>1 mtorr) voiding and incomplete fill of high AR holes results (Cuomo and Rossnagel, 1986; Turner et al., 1993; Joshi, 1995).

11.4 SELF-SPUTTERING

Self-sputtering in a planar magnetron at very low pressures (<0.1 mtorr) has also been suggested as a way of depositing thin films at the bottom of deep holes at high rates. Self-sputtering is the sputtering of metals using ions of the same element in the absence of an inert gas, eliminating diffusion of the sputtered atoms thus resulting in line-of-sight deposition. Self-sputtering can be initiated only when a critical plasma density is reached and the self-sputtering yield must be >1. Efficient ionization of the sputtered material must occur and the ions must be redirected to the target (Posadowski and Radzimski, 1993). Self-sputtering has been demonstrated for Cu and Ag, but not Al because of its low yield and negative effects of O_2 in the chamber. Asamaki et al. (1994) used the technique to deposit thin films of Cu and demonstrated that the thickness at the center of the bottom of a narrow deep trench is the same as that on the top surface. The deposition rate is high making possible large T/S distances. However, the target lifetime is limited, even with improved target design so that the usefulness of the technique in production is questionable.

12.0 HIGH-DENSITY PLASMAS

12.1 INTRODUCTION

High-density plasmas were developed to overcome the limited capability of the rf sputtering systems, discussed above, to fill high aspect ratio holes.

12.2 ECR

An ECR plasma reactor was used to fill high AR holes with Cu using an *evaporated* source (Holber et al., 1993). The system configuration (Fig. VI-38) was somewhat different from the commercially available reactors described in Chapter I. The essential features are: (1) the quartz window through which the microwaves enter the plasma region is out of line-of-sight of the plasma to prevent coating of the window and (2) the substrate is out of line of sight of the resistively heated Cu source. The evaporated Cu atoms are ionized in the ECR source; the flux at the substrate may be ~100% ionized (although this high degree of ionization may not be needed) and no buffer or carrier gas is used. The substrate temperature is controlled by the use of a helium-backed electrostatic chuck which can be biased by an rf or dc supply. The substrate temperature (which was usually ~150°C) appears to have no effect on the filling capability although it may affect the microstructure of the film.

The most important deposition parameter determining whether a hole of a given AR can be filled is the resputtering rate at the surface. If the substrate bias is too low, voids are formed; if too high, the net accumulation of Cu on the top surface is very small. The best results were obtained using a high/low bias process in which a hole of AR ~4.2 (depth ~3 mm) was filled solidly, as seen in the SEM cross section of Fig. VI-39. The hint of porosity in the grain structure as highlighted by the etch used in preparing the specimen.

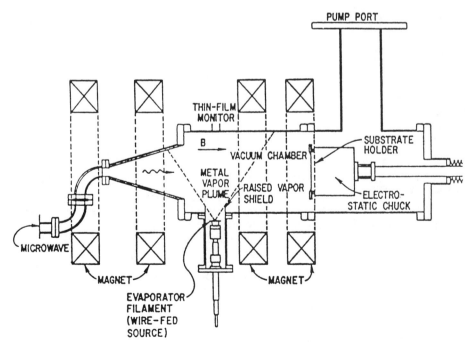

FIGURE VI-38 Diagram of a system used for Cu deposition by an ECR plasma. (From Holber et al., 1993)

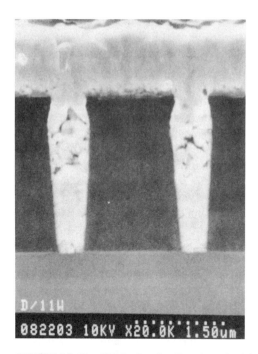

FIGURE VI-39 SEM of a Cu film deposited into a hole of aspect ratio ~4.2 using a two-step (-230 V/-100 V) process in the ECR reactor shown in Fig. VI-38. (From Holber et al., 1993)

This system can also be used to line high AR features conformally with a thin film before the main deposition.

Hole filling with Al, using a sputtering target, and a moderately high substrate temperature of 300°C (to assist by a flow effect) appears to be equally feasible (Ono et al., 1994).

12.2 INDUCTIVE PLASMA

An alternative to the technique described above is magnetron sputter deposition using an RFI plasma to produce high levels of metal ionization (Rossnagel and Hopwood, 1993; Cheng et al., 1995; Hamaguchi and Rossnagel, 1995). In the absenceof the additional ionizing plasma, the system is, of course, a conventional sputtering system incapable of filling high AR trenches/holes without voids. The advantages cited are (1) the use of a sputtering target instead of an evaporation source is more compatible with alloy deposition, and (2) an RFI system is simpler than an ECR system (as was seen in Chapter I). The essentials of the reactor are shown in Fig. VI-40.

A large fraction of the sputtered atoms are ionized in the RFI plasma and accelerated across the sheath of the negatively biased sample holder to be deposited at normal incidence. The processes which occur during deposition are illustrated in Fig. VI-41.

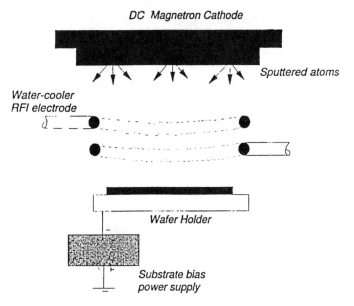

FIGURE VI-40 Essentials of the reactor used for magnetron sputter deposition using an rf inductively coupled plasma (RFI). (From Cheng et al., 1995)

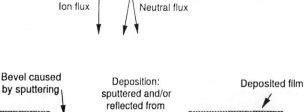
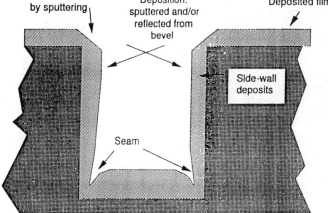

FIGURE VI-41 Diagram showing the processes occurring during deposition in the reactor shown in Fig. VI-40. (From Cheng et al., 1995)

Directional deposition results in good coverage of the bottom of a hole with a dense film but on the sidewalls there is a low-density, columnar deposit because of the grazing incidence of the incoming particles. Figure VI-42a shows the growth of a Cu film.

As the substrate bias is increased, increased reflection and sputtering help fill the corners and deposit film on the sidewalls; elevated substrate temeprature (~250°C) also helps to fill the corners. Bevel formation also occurs as result of sputtering and this has two opposing effects: (1) elimination of overhang formation which increases acceptance angle for the arriving particles, and (2) formation of a lateral build-up by material sputtered from the bevel, which shadows the trench, resulting in void formation. This is illustrated in Fig. VI-42b. However, under the proper conditions, a trench can be filled completely as was demonstrated for dual damascene structures using both AlCu and Cu.

Increasing the rf power increased the relative ionization of the sputtered atoms, increasing the deposition rate, directionality, and bevel angle. Increasing the Ar pressure decreased the overhang thus improving the fill. A decrease in magnetron power lowers the density of sputtered particles in the plasma, increasing the relative ionization and hence results in more directional sputtering.

The mechanisms for highly ionized magnetron sputtering have been discussed by Hopwood and Qian (1995).

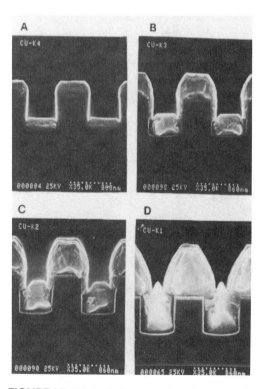

FIGURE VI-42(a) SEMs showing time evolution (increasing time left to right). Films deposited in the reactor shown in Fig. VI-40. (From Cheng et al., 1995)

Chip Integration

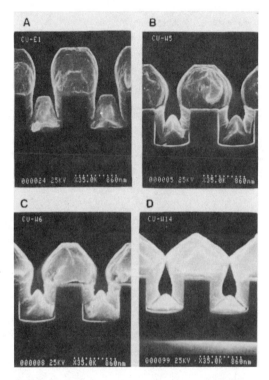

FIGURE VI-42(b) SEMs showing effect of substrate bias (increasing bias left to right). Films deposited in the reactor shown in Fig. VI-40. (From Cheng et al., 1995)

13.0 BEAM TECHNIQUES
13.1 IONIZED BEAMS

Ionized Cluster Beam (ICB) Also Called Nozzle Jet Beam

The material to be deposited is heated in a crucible, *evaporated* through a nozzle, undergoes adiabatic expansion, and upon cooling, forms clusters of up to ~1000 atoms which are partially ionized by electron impact. The substrate can be biased to accelerate the clusters and minimize their spread. A diagram of such a system is shown in Fig. VI-43. The self-ion bombardment of the growing film cleans the surface of sorbed species. The apparatus has been used, most frequently for Al deposition (Younger, 1984, 1985; Takagi, 1986; Ramanarayanan et al., 1986, 1987; Harper, 1987; Yamada and Takagi, 1987, Knauer and Poeschel, 1988; McEachern et al., 1991; Ito et al., 1991). This technique has also been used for SiO_2 deposition, using high purity silica grains in the crucible (Wong et al., 1985).

Partially Ionized Beam (PIB)

The system has capabilities similar to those of ICB systems, but there is no nozzle at the mouth of the crucible, as seen in Fig. VI-44. An ionization grid and an extraction electrode are placed above the crucible. Biasing the substrate is also possible (Mei et al., 1987; 1988; Yapsir et al., 1989; Bai et al., 1990; Gittleman et al., 1990).

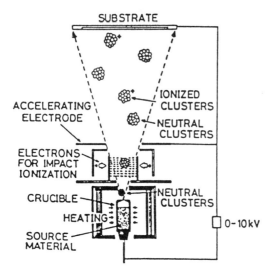

FIGURE VI-43 Diagram of an ionized cluster beam (ICB) system. (From Younger, 1985)

A comparison of ionized beam and sputter deposition conditions reveals that there are several factors which account for the greater directionality of the beam techniques. Since beam systems operate at much lower pressures than sputtering systems, the depositing atoms have a much longer mean-free path and thus suffer fewer collisions. The source-to-substrate distance is greater in beam systems. These factors contribute to the directionality. Also, in sputtering, single atoms are deposited whereas in beam deposition clusters of many atoms are deposited. It has been demonstrated that the larger the cluster size, the better the hole filling capability.

Energetic Cluster Impact (ECI)

In this system (Haberland et al., 1994) atoms are *sputtered* into a rare gas and then aggregate to form the large clusters (~1000–3000 atoms) which become charged (20–80%) by the many charged or energetic particles in the discharge; additional

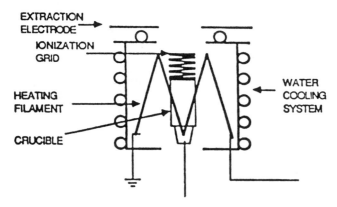

FIGURE VI-44 Diagram of a partially ionized beam (PIB) system. (From Mei et al.)

electron impact is not needed. By means of charge separation, beams which are 100% ionized are available. Excellent fill is obtained when the Cu beam (negatively charged) is 100% ionized and the kinetic energy of the arriving particles is ~10 eV/atom. The substrate temperature is held at ~225°C and the deposition rate is several hundred Å/min. Good filling is attributed to the fact that the clusters arrive in a directed beam, reducing self-shadowing and each impact is accompanied by local heating producing annealing and enhanced diffusion. The effect of ion energy and substrate temperature are illustrated in Fig. VI-45.

13.2 NEUTRAL BEAMS

13.2.1 Molecular Beam Deposition (MBD) Plus Annealing

The MBD system is comparable to a molecular beam epitaxi (MBE) system in which the direction of the Al vapor stream is controlled by the use of an aperture and very low pressure ($\sim 10^{-8}$ torr) during *evaporation* (Mukai and Ozawa, 1994). After deposition, the sample is transferred, in vacuo, to a high vacuum ($<10^{-9}$ torr) annealing chamber. Figure VI-46 shows both hole-fill after evaporation, with Al deposited fully on the sidewall and bottom of the hole without an overhang, and the final structure in which the hole is filled completely and the surface planarized after annealing at 400°C. The mass transport, via surface diffusion, is attributed to the surface tension forces generated by the three-dimensional geometry of the heated Al. the high vacuum during deposition and annealing prevents the formation of an oxide skin which would inhibit surface diffusion.

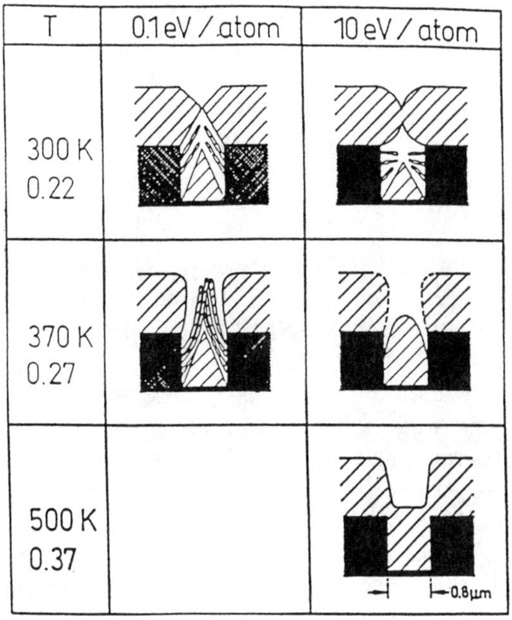

FIGURE VI-45 Effect of temperature and ion energy on hole fill in energetic cluster impact (ECI) deposition. (From Haberband et al., 1994)

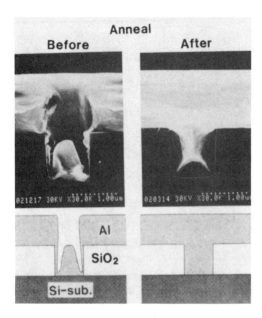

FIGURE VI-46 Molecular beam deposition (MBD): (top) SEMs showing hole fill before and after anneal at 400°C; (bottom) diagrammatic representation of the SEMs above. (From Mukai and Ozawa, 1994)

14.0 FLOWAGE OF METAL FILMS

Postdeposition heat treatment is still another way of filling via holes in the interlevel dielectric layer and of planarizing the surface. There have been two approaches: laser reflow and thermal annealing (1) in the deposition system, (2) in a furnace, (3) using a rapid thermal processor, (4) by the use of high pressure in addition to heat.

14.1 LASER REFLOW

Flashlamp-pumped dye lasers (λ = 504 nm) and pulsed eximer lasers of shorter wavelength (XeCl, λ = 308 nm; KrF, λ = 248 nm; ArF λ = 193 nm) have been used. A schematic diagram of the commercially available XMR Model 7100 pulsed eximer laser system is shown in Fig. VI-47.

The important parameters of laser operation are the fluence, pulse width, and uniformity. A dye laser has a wide pulse width and heats the metal nearly uniformly. An eximer laser has a much shorter pulse which induces a large, brief temperature differential across the metal. A recoil impulse may be generated by evaporation and this deforms the molten surface but mass distribution far outweighs mass loss by evaporation (Marella et al., 1990a). For thinner Au films, the process margin is the same for both lasers, but for thicker films the process margin is about twice as large for the dye as for the eximer laser (Marella, 1990b). The laser pulse melts the metal film deposited over the vias momentarily. If the metal surface is

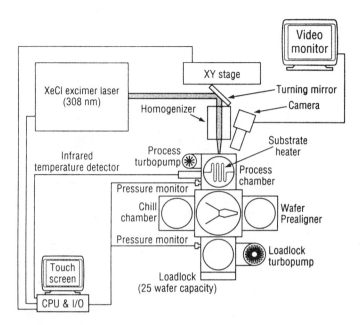

FIGURE VI-47 Diagram of a commercially available system for laser reflow processes. (Courtesy of XMAR, Inc.)

clean, hole-fill and planarization are rapid because clean metals have low viscosity and high surface tension (Tuckerman and Weisberg, 1986a,b). The pulses should be very short, sufficient to melt the metal but short enough to minimize other thermal reactions. The process is feasible since metals have high thermal diffusivity and the underlying insulators are good thermal barriers so that damage to underlying structures can be avoided, or at least minimized (Tuckerman and Weisberg, 1986a,b; Marella et al., 1990; Ong et al., 1991).

The progression of events occurring as the energy density of the laser beam increases is (1) melting (2) planarization (3) via fill (4) spotty ablation, and finally (5) systematic ablation; a higher fluence is required to fill holes completely than to planarize the surface (trapping voids) (Liu et al., 1989). The process window for laser processing is, therefore, bounded at the lower end by the fluence required to planarize or to fill and on the other, by the fluence at which ablation (damage) starts; it is wider for planarization than for complete fill. Using an eximer laser, the damage is localized, confined to the edges of the irradiated spot and is attributed to evaporative recoil effects; using the dye laser, damage is distributed over the entire irradiated spot, but the mechanism is not understood (Marella, 1989, 1990b).

The process window is widened by improving the initial step coverage (Yu et al., 1991a). A pattern density sensitivity exists. It requires more energy to fill isolated vias than closely spaced arrays. At low energy densities an Al film will melt in the neighborhood of a hole but not in open areas (Liu et al., 1989). The minimum fluence required for the planarization of Al increases as the via size increases, i.e., there is a larger process window for smaller vias (Baseman et al.,

1990) who also noted that since the total volume of metal available to reflow over a given feature after it has been patterned would probably reduce the extent of planarization.

There is no significant effect of the AR on the minimum fluence for *planarization*, but at a fixed substrate temperature a higher fluence is required to *fill* higher AR features. If the AR is very high, the radial, inward surface tension forces act to close the hole before fill is complete, although in low AR vias, the same surface tension forces are responsible for complete fill. This is a result of the difference in the fraction of the via surface exposed to the incident laser beam. For the low AR hole, most of the via surface is exposed and the entire surface of the via coating is temporarily melted to some depth; for the high AR hole, the metal at the top of the via is readily melted and flows to pinch off, leaving a void. If conditions are right, the void may collapse and the via will fill (Marella, 1990a,b; Pease et al., 1991).

Increasing the substrate temperature decreases the fluence needed for fill because the higher initial temperature reduces the energy required to reach the melting temperature, increases the absorption of the film, and prolongs the period during which the metal is molten. Although an increased initial substrate temperature also decreases the threshold for damage, it still results in a wider process window because the effect is much smaller for damage than for planarization (Baseman et al., 1990; Carey et al., 1990; Woratschek et al., 1990; Wand and Ong, 1992). The fluence required for planarization and for the onset of ablation increases with increasing film thickness but the process window is almost unaffected (Yu et al., 1992).

A number of film properties affect the efficiency of the process. When the reflectivity of the film is very high (e.g., Al), a higher fluence is required: Cu and Au, with lower reflectivities, are planarized more easily than Al. The use of an antireflection coating reduces the fluence requirement for Al planarization and, in addition, prevents oxidation of the surface which impedes the flow. Si has been used to cap Al, since Al_2O_3 impedes flow but SiO_2 does not because it is reduced rapidly by the molten Al (Tuckerman and Weisberg, 1986). However, the use of the low wavelength ArF laser and a very short pulse, producing very rapid heating, may make a Si cap unnecessary (Mukai et al., 1987). Other caps suggested for Al are, e.g., Cu (Mukai et al., 1988), Ti (Liu et al., 1989), etc. (Ong et al., 1991).

Al alloy films are more readily planarized than pure Al. This is attributed to the fact that the period during which the metal remains molten after a given laser pulse is shorter for Al than for the alloys AlCu, AlSi, AlCuSi (Yu et al., 1992; Woratschek et al., 1989a,b). AlTiSi films have better flow characteristics than AlCu, but AlTiSi films are smoother than AlCuSi after planarization. The wider process window for AlCuSi, compared with AlCu, has been attributed to the slightly lower melting point of AlCuSi, and perhaps, to differences in melt times.

When filling closely spaced submicron holes, AlSi films can be planarized as readily as the ternary alloys, but in the large planarized regions, spot ablation, resulting in deep holes, has been observed after laser exposure. The absence of spot ablation in the case of AlSiTi was explained by the fact that Si precipitates are smaller in the ternary alloy (Woratschek et al., 1989a,b). The underlying film (barrier layer or liner) could affect wetting and, therefore, reflow and filling (Ong et al., 1991), as well as the duration of melt times (Carey et al., 1990) although

Yu et al. (1990) found no effect of the barrier material on the process window for fill.

An attempt was made to planarize Cu deposited over PI (Baseman and Turene, 1990) using an eximer laser. Planarization was accomplished readily but the defects produced in the PI made the process unacceptable.

Mukai et al. (1991) developed a process for producing plugs which involves patterning a thin metal layer to cover a hole and then irradiating it with an eximer laser to form the plug. The mass transport was attributed to surface tension forces created by the three-dimensional geometry of the molten metal on the sidewall.

14.2 THERMAL ANNEALING

A rapid thermal processor, in which tungsten-halogen lamps were used to illuminate only the metallized side of the wafer, planarized AlCuSi films sandwiched between a TiN/Ti barrier layer and a Ti layer, which prevented oxidation of the Al surface (and also acted as an antireflection coating) (Lee et al., 1993). The sample placed on a larger "dummy" wafer (isolated thermally from the sample) to keep the substrate temperature as low as possible (estimated at $<300°C$). The temperature was ramped up, kept at the maximum temperature (t_{dummy} = 620–640°C) for 10 s, and then cooled. The total process time was 120 s. When the "dummy" temperature exceeded 650°C, the surface morphology was poor and local ablation occurred.

Thermal annealing, after vacuum transfer from the deposition chamber (module) to an annealing chamber, has been used to fill high AR holes with Al or Al alloys. In one version, (Xu et al., 1994) the process started with collimated sputtering of a thin Ti "wetting" layer. Al was then sputtered at ~50°C; the wafer was transferred to a UHV flow chamber and heated at ~500–540°C. A two-step process, in which half the final thickness was deposited and annealed in each step was more effective than a single anneal for a thicker film.

The use of high pressure (700 atm) lowered the temperature needed for reflow of Al and AlCuSi films to about 430°C in a process called "FORCEFILL"™ (Dixit et al., 1994; Shterenfeld-Lavre et al., 1995; Barth et al., 1995). The high-pressure chamber, part of an integrated system, is shown in Fig. VI-48. The barrier layer, TiN, was sputtered in a so-called "HI-FILL" chamber which employed a large target-to-substrate distance instead of collimation. The conductor metal was sputtered in a conventional sputtering system. It is essential that the thickness of the layer be such that a continuous bridge is formed over the hole. The metal was then forced into the hole by applying a high pressure of argon. The process sequence is shown in Fig. VI-49a and the final filled via in Fig. VI-49b.

Hirose et al. (1994) simulated Al reflow into submicron holes using a surface melting model. Mass transport, driven by surface tension, occurs only in a very thin reflow layer whose thickness depends on the temperature. Evolution of the surface topography was predicted from the initial Al step coverage and window curvature and agreed well with experimental observations. They compared laser melting, in which the entire film melts, with the lower temperature surface reflow process. The voids in high AR holes were smaller (or even absent) after laser melting than after the lower temperature reflow.

Gardner and Fraser (1995) filled holes with sputtered Cu (on a "wetting" layer) using a procedure similar to that of Xu et al., (1994); they emphasized the need

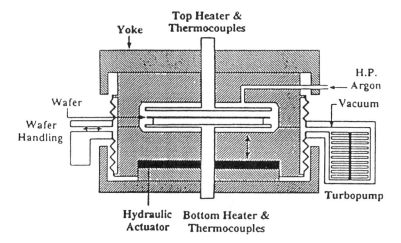

FIGURE VI-48 Diagram of the reactor used for "FORCEFILL"™ technology. (From Dixit et al., 1994)

for an ultra-high-vacuum sputtering system and very high purity targets and gases. After a vacuum transfer the films were annealed at 350–450°C. By minimizing surface contamination, the film flowed readily at lower temperatures by surface diffusion (E_{act} = 0.8 eV), not by volume diffusion (E_{act} = 2.19 eV). The reflow time is a function of the initial profile, film thickness, AR, and the trench dimensions. If a void is formed, there is no continuous surface for diffusion; grain boundary or volume diffusion (higher E_{act} processes) is required, prolonging the fill time. The process is more suitable for submicron features than for larger ones; for example, 0.25-μm trenches could be filled but not those 0.35 μm wide.

The low throughput of this process led Abe et al. (1995) to attempt an air transfer to a furnace. Holes were filled successfully after heating at 450–500°C for 30 min in 800 mtorr of H_2. Gardner and Fraser found that good fill required a good wetting layer, whereas Abe et al. (1995) found agglomeration necessary.

Friedrich et al. (1996) used simulation and experimental observation to study the Cu reflow process. Microstructural factors such as grain boundary grooving and faceting were found to account for trench-to-trench variability in filling rates, probably by affecting the surface diffusion rates. Thus they concluded that reflow times must be sufficiently long or the temperature sufficiently high to accommodate the microstructural configurations. They also suggested that rate variations might be controlled by tailoring the grain structure of the Cu film by adjusting the deposition conditions or by optimizing the wetting characteristics of the underlayers or, perhaps, by incorporating nucleation sites.

14.3 MODIFICATION OF UNDERLAY

Park et al. (1994) improved the wetting characteristics of an underlying TiN barrier layer by ECR sputter etching, using a low bias voltage, so that a conformal Al film

Chip Integration

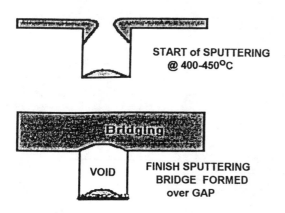

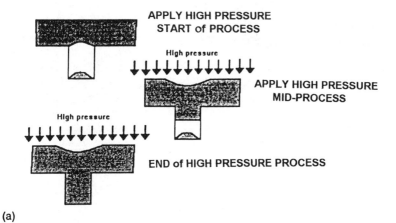

(a)

FIGURE VI-49(a) Diagrammatic representation in the steps of the "FORCEFILL"™ process. (From Shterenfeld et al., 1995)

could be deposited on the sidewall. TiN was used instead of Ti to retard the formation of Al_xTi but it had poor wetting characteristics so that discontinuous islands of Al were formed. Sputter-etching amorphized the surface layer which became Ti-rich, annihilated the surface grains and left the walls of the holes smooth. In addition, ion bombardment creates preferred adsorption sites for the incoming Al atoms. Thus, the wetting characteristics of TiN became equivalent to those of Ti.

15.0 CVD METALS

One of the attractive features of this method of deposition is its superior step coverage and hole fill capability.

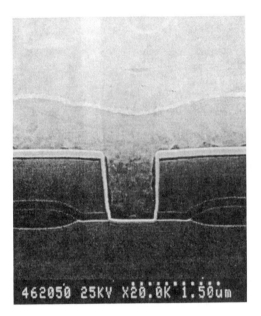

FIGURE VI-49(b)   SEM of a contact hole filled with high pressure Al. (From Barth et al., 1995)

15.1 CVD OF W

15.1.1 Introduction

CVD W has been studied more extensively and is used more widely than the other CVD metals. The first widespread use of CVD W was as a "contact plug" (to planarize the substrate, after completion of the processes of fabricating and passivating the active and passive devices in the substrate) before the first metallization layer was deposited. The next application was as a vertical interconnection between successive wiring levels in a multilevel structure.

15.1.2 Selective CVD

Selective deposition of W was discussed in Chapter V; it is intrinsically a hole filling technique, and will not be discussed further in this section.

15.1.3 Blanket CVD

Blanket CVD W films can be deposited with 100% step coverage, filling holes without void formation, when the reaction of $WF_6 + H_2$ is carried out under optimal conditions. The higher the AR of the hole, the more difficult it is to get good step coverage; step coverage is higher in trenches than in holes of the same dimensions. Although there are no voids, seams which are more vulnerable to etchants than the bulk W, may exist. All the experimental observations, both LPCVD (e.g., McConica and Churchill, 1988; Chatterjee and McConica, 1990; Schmitz and Hasper, 1993) and HPCVD, i.e., 80 torr versus 0.5–10 torr. Clark et al. (1991) showed that optimization of the deposition conditions for plug applications included lowering the temperature and the partial pressure of H_2 and raising that of WF_6. They also found that reducing the spacing between the gas inlet (showerhead) and the wafer (i.e., reducing the reactor volume) also improved step coverage.

Blumenthal and Smith (1988) examined, in detail, the effect of temperature on step coverage. In the higher temperature (higher deposition rate) regime in which the reaction rate is controlled by mass transport and therefore nonactivated, the step coverage was nonconformal and depended strongly on the size of the groove. In the lower temperature range, the surface reaction-controlled region (activated), 100% step coverage could be attained for grooves larger than 1 μm.

Schmitz and Hasper (1993) emphasized that when the step coverage is less than 100%, the existence of a void (and its size) has a greater impact on etchback complications and reliability than does the value of the step coverage, i.e., the ratio of film thickness at the top of the hole to that at the bottom.

When SiH_4 was used to reduce WF_6, to increase the deposition rate in a cold-wall LPCVD reactor, sizeable voids were formed (Schmitz et al., 1987, 1988; Lee et al., 1989); H_2 reduction, in the same reactor, did produce void-free fill. Better hole-fill was obtained when SiH_2F_2 was used instead of SiH_4, but H_2 reduction resulted in the best fill (Goto et al., 1991).

Adding H_2 to SiH_4/WF_6 mixture (under conditions of equal growth rate) degraded the step coverage dramatically (Schmitz et al., 1988) although a two-step process, in which the initial reduction with SiH_4 was followed by H_2, resulted in void-free fill (Lee et al., 1989).

Step-Coverage/Hole-Fill Models

A model was developed by McConica and Churchill (1988) for predicting the step coverage during blanket CVD W deposition by the WF_6/H_2 reaction. When diffusion and reaction occur simultaneously in a deep hole, concentration gradients of the reactant species are established along the length of the pore. This occurs because in LPCVD processes, transport in small holes is dominated by Knudsen diffusion which is slower than bulk diffusion. This limited transport, combined with the decrease in the radius of the pore as deposition proceeds, restricts access of the reactant. The deposition rate at any location in the hole is dominated by the local reactant concentration but the concentration gradient within the hole does not itself result in a thickness gradient. The nonuniform step coverage results from a limiting W thickness which results when the WF_6 concentration reaches zero and all growth stops. The time growth stops is a function of the hole characteristics, the reaction rate, and the mass of the diffusing species. The predictions of the model agree with the experimental results described above. McConica and Churchill (1988) point out that, although the reaction rate is zero order in WF_6 (the diffusion limited species), the step coverage is *not*.

Schmitz and Hasper (1993) simulated the hole filling process, basing it, as did McConica and Churchill (1988), on the balance between diffusion and consumption of reactants. They found that the size of the void depended on the length of the hole but not its radius, i.e., not on AR. Thus the attempt to decrease capacitance by increasing the thickness of dielectric films through which vias must etched and then filled, will exacerbate void formation. They also found that when the AR of the hole was less than one, including the surface curvature of the film at the top of the hole improves the calculated step coverage but had a negligible effect when AR is larger.

Hsieh (1993) modeled conformality using a two-parameter analytical mass balance method to couple molecular scattering and a two-step surface-activated reaction. The reaction consisted of the adsorption of the precursor molecule on the surface followed by the reaction of the sorbed species on the surface. The two parameters used in the simulations are the intrinsic sticking coefficient of the precursor, determined by the nature of the precursor species, and the degree of surface saturation, which is process-dependent. The sticking coefficient, if small, reduces the influence of geometric shadowing. However, the value of the sticking coefficient was determined experimentally as 0.48 (Yu et al., 1989) so that the only adjustable parameter is the saturation factor. Step coverage increased as this factor was increased. In practice, this means increasing the surface flux or decreasing the reaction rate, in agreement with experimental results.

However, it has been possible to simulate the step coverage/gap fill of CVD W and obtain results agreeing with the observed profiles by assuming a sticking coefficient close to zero (Rey et al., 1991; Hsieh, 1993). Thus, the agreement between a simulation and experimental results does not guarantee the correctness of the model.

The poor step coverage observed when W was deposited by the WF_6 + SiH_4 reaction was explained by the fact that although the reaction rate was increased, the diffusion rate was unchanged, resulting in depletion of WF_6. The reaction rate is first order in SiH_4. This implies worse step coverage than a reaction of lower order: the reduction of WF_6 by H_2 is half-order in H_2. Finally, since SiH_4 is a bigger molecule that H_2, its transport into and within the hole is inhibited (Chatterjee and McConica, 1990; Hasper et al., 1991).

Etchback

When blanket W is used to fill contact/via holes, the excess metal must be removed. In this section, etchback processes will be discussed; CMP will be discussed later.

Many processes, using F-based etchants (NF_3, but more usually SF_6) have been proposed. One of the earliest was simply to RIE the blanket film using an isotropic etchant at high pressure to increase the etch rate of W with respect to the underlying oxide surface. The etchant attacked the seam in the via leaving a wide gap but the sides of the vias are now sloped so that they are easily covered by the next conductor layer (Smith, 1985). By reducing the loading effect, the amount of W remaining after etch was maximized. A number of processes have been proposed; they involve, for example, the use of different etchants or multiple steps (Lee, 1987), precision-determination of the end-point (Clark et al., 1990), different etchers, e.g., magnetron (Clark et al., 1990; Berthold and Wiecorek, 1989), triode (van Laarhoven et al., 1989), and transformer coupled reactor (Allen, 1993), use of a clamp ring (Wilson et al., 1992) or an electrostatic wafer clamp (Marx et al., 1994), and coating the oxide with nitride (van Laarhoven et al., 1989). Nowicki (1994) patented the use of a sacrificial ring whose etch rate equalled that of metal in the vias and thus avoided microloading.

A sacrificial layer was used in an etchback process to improve uniformity, to insure that no plug was submerged, but severe microloading was observed. By

Chip Integration 423

using a consumable cathode (i.e., carbon) and a CCl_2F_2 mixture, the uniformity was improved further and there was minimal microloading. The equality of etch rates (resist/W) was maintained by adjusting the pressure (Saia et al., 1987a,b).

Mihara and Nakamura (1994) studied the effect of wafer temperature on loss of W in the plug during RIE etchback in SF_6/N_2 of W on a TiN adhesion layer. Reduced wafer temperature during overetch suppressed the loss of W because of the formation and subsequent deposition on the W surface of TiFx, sputtered from the exposed TiN.

Etchback may also be required when the W is deposited selectively. Small protrusions above the vias may be formed because of different via depths or inadequate process control during deposition. But this presents fewer difficulties than etchback of blanket W. The loading effects are minimized because the area of W is very much smaller than that of the dielectric layer. Saia et al. (1987) used a sacrificial layer which compensated for the nonuniform W thickness.

15.2 CVD ALUMINUM, ALUMINUM-COPPER, AND COPPER

15.2.1 Introduction

Aluminum and its alloys, and more recently copper, films are used for interconnection metallization. Therefore, the use of such metals, prepared by CVD, would not be expected to be restricted to hole filling, as is the case for tungsten with its much higher resistivity. But their use as plugs has the advantage over tungsten of reduced interlevel resistance. Both blanket and selective deposition have been reported.

15.2.2 Selective CVD

Selective deposition is, as already noted, intrinsically a hole filling process. For example, Ohta et al. (1994) were able to demonstrate, by taking SEMs at various times during the deposition process, that CVD Al grew from the lower Al surface, filling the vias, when the lower surface was cleaned of its native oxide (by etching in a BCl_3 plasma).

After plug formation, the next interconnection level has been fabricated by etching the same metal formed by blanket PVD or CVD processes so that the surface is no longer planar. For example, Amazawa et al. (1988), Ohta et al. (1994), and Takeyasu et al. (1994), sputtered the interconnection Al; Tsubouchi et al. (1990, 1992) used the same process for both Al levels but insured blanket deposition of the second one by exposing the surface to an rf plasma after hole-fill.

After selective hole fill by Cu, Awaya et al. (1990) used sputtered Cu which was patterned by RIE in $SiCl_4/N_2$ at 250°C; Kim et al. (1994) used blanket CVD Cu on a Ta seed layer.

To recover the planarity, one could envision a process in which the second metal pattern would be formed using a damascene CMP process after CVD deposition of the same metal into the stencil etched into the interlevel dielectric layer.

15.2.3 Coverage by Blanket CVD Films

The mechanism of step coverage by blanket films of these metals has not been studied extensively.

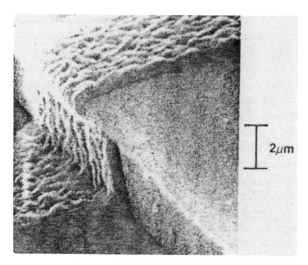

FIGURE VI-50 SEM of step coverage by CVD Al of a vertical step with an overhang. (From Cooke et al., 1982)

The coating of an overhang structure and vertical wall by CVD Al, shown in Fig. VI-50, (Cooke et al., 1982) is an indication of superior step coverage.

Sugai et al. (1993) used blanket CVD Al to fill via holes; sputtered AlCu was used for the interconnection metallization.

The coverage of an overhang structure, seen in Fig. V-8, demonstrated the capability of blanket CVD Cu. The results were consistent with a very low sticking coefficient (~0.015).

Figure VI-51 shows the simulation, by Cho et al. (1992), of void-free filling by CVD Cu of high AR trenches; their experimental evidence of the fill is shown

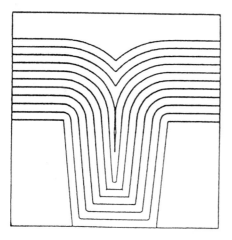

FIGURE VI-51 Simulation of step coverage and hole-fill by CVD Cu. (From Cho et al., 1992)

Chip Integration

in Fig. VI-52; similar results have been reported by Gelatos et al. (1993) and by Awaya and Arita (1993a,b).

After the via holes have been filled by blanket CVD Cu, the excess metal on the surrounding insulator can be removed by RIE, ion milling, or CMP (Cho et al., 1992; Gelatos et al., 1993). Or additional sputtered Cu can be deposited and the interconnect completed by RIE etching resulting in a nonplanar surface (Awaya and Arita, 1992). Blanket CVD Cu has also been used both to fill the vias and form the planar interconnection using a dual damascene CMP process (Krishnan et al., 1992).

16.0 ELECTROCHEMICAL DEPOSITION OF COPPER

Because it is a selective process, electroless Cu film grows from the bottom of a properly activated hole. It is a good choice for filling high AR vias without voids or seams. This capability has been demonstrated by Dubin et al. (1995) for 0.35 μm vias with an AR ~4. The Cu film penetrated beneath the oxide ILD layer increasing the mechanical stability of the metallization system, but might cause shorting or reliability problems when vias are closely spaced. The deposition rate was a function of the via size; a smaller hole (higher AR) was filled more slowly than the larger one (lower AR). This same size dependence was observed when plating Ni(P) on Al (Schwartz and Platter, 1974). However, there was no barrier layer on the sides of the Cu plugs. If migration of Cu through the dielectric films must be avoided, this process is of questionable value.

A process in which selective electroless Cu is fully encapsulated and the surface planarized was described by Cho et al., 1993; a diagram of the process is shown in Fig. VI-53. There are several drawbacks to this process. It is complex. The use of spacers in addition to the barrier metal liners restricts the conductor width so that it is not applicable to submicron structures. Also, the assumption of planarity when different size features are present on a chip is not valid, so that additional planarization by some other method is required.

Another approach is the use of blanket electroless plating, after collimated sputtering of a liner, Ti/TiN and a thin Cu seed layer. This process was used to

FIGURE VI-52 SEM after fill and removal of excess CVD Cu. (From Cho et al., 1992)

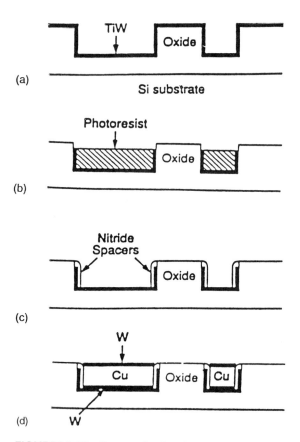

FIGURE VI-53 Process for forming planar encapsulated Cu lines using electroless Cu deposition. (From Cho et al., 1993)

fill 0.35-μm holes, AR ~ 3, without voids. Hole filling capability depended on the thickness of the Cu seed layer, if it was too thin, the results were unacceptable (Shacham-Diamand et al., 1995). Removal of the excess Cu, planarizing the surface at the same time, would have to follow.

Copper electroplating has also been used for hole filling. Pulsed-voltage plating in a flow-plating apparatus, was used to achieve good "throwing power," i.e., enhanced coverage in deep recesses. To provide the electrical continuity as well as a diffusion barrier, a layer of Ta, Cr, or TiW and a thin layer of Cu was sputtered into the features etched into the dielectric, preferably using collimation. Before plating, it was necessary to wet the sample to assure filling of the cavities. Upon completion, excess metal was removed by CMP, leaving Cu only in the vias (Contolini et al., 1995).

17.0 EMBEDMENT

17.1 INTRODUCTION

In this section we discuss the embedment or inlaid processes which are forerunners of the so-called "damascene" process of chemical mechanical polishing, CMP (discussed below). The surface of a dielectric film is coated with a resist layer, a pattern is created by standard lithography, the features are etched into the film, usually by RIE, metal is then deposited into the recesses, and, finally, any excess metal is removed by some suitable process.

17.2 METAL LIFT-OFF

Using a lift-off stencil, metal was evaporated to fill holes etched in an oxide matrix. Since the metal does not fill the holes completely, SOG was used to fill the gaps before deposition of the ILD (Sumitomo and Ohashi, 1978). In the same way, metal was deposited into recesses etched in a polyimide layer, but in this case, the gaps were filled when the next polyimide layer (the ILD) was spun on (Rothman, 1983). The process was repeated for each vertical (plug) and horizontal (conductor) metal layer to build a planar four-level structure.

17.3 SPUTTERED METAL

Wu (1986, 1987) described a process for embedding sputtered metal in oxide, building layer by layer a coplanar conductor/insulator composite for dense multi-layer metallization. The process was called "PRAIRIE," patterns recessed by anisotropic reactive ion etching.

17.4 CVD METAL

Blanket CVD W was used to fill deep grooves in a dielectric layer to form thick films for the primary levels of interconnection in a multilevel structure. A thin metallic layer was first sputtered into the grooves to promote the adhesion of W. After deposition, the W was etched back (in NF_3/O_2, using laser interferometry for end point detection) to expose the dielectric layer, thus producing a planar structure of W embedded in the insulator (Broadbent et al., 1988).

Processes called "reverse pillar" (Yeh et al., 1988) and self-aligned contact, SAC (Ueno et al., 1992), are etch-back versions of the "dual damascene" CMP process. The interconnect patterns and then the contacts were etched into the dielectric layer and filled simultaneously with CVD (but not sputtered) metal; the excess metal is etched back, leaving the recessed metal. If blanket instead of selective CVD metal is used, the metal lines should be limited to a single minimum width; if wider lines are used, resist etchback is required (Yeh et al., 1988).

17.4 ELECTROCHEMICAL DEPOSITION

This is one of the earlier techniques used for depositing the metal into a cavity. The use of electroless Cu for forming interconnection patterns avoids the need for RIE of Cu and can thus simplify processing (Ting, 1988; Kiang et al., 1991). By controlling the thickness of the deposit to equal the depth of the groove, planar surfaces can be produced if only one size of groove is used. Other applications of electroless and electrolytic plating were discussed in Section 16.0.

17.5 OXIDE LIFT OFF

A different approach to embedment is oxide lift-off. Ehara et al. (1984) formed the metal pattern by RIE, and leaving the resist mask in place, deposited the oxide in an ECR reactor. The oxide was etched briefly to expose the resist which could then be dissolved, lifting the oxide over the metal, leaving it embedded in the surrounding oxide. The subsequent ILD filled the gaps in the lower oxide encapsulating the metal. Another version (Rothman et al., 1985) used sputtered oxide over an AlCu/Hf conductor, capped by MgO. The oxide was sputter etched to expose the edges of the MgO layer which was then dissolved, lifting the overlying oxide. The Hf layer protected the underlying conductor from attack during dissolution of the MgO.

18.0 CHEMICAL MECHANICAL POLISHING (CMP)
18.1 INTRODUCTION

This technique, used in semiconductor fabrication for many years for wafer polishing and surface treatment, has been used only relatively recently for multilevel interconnection (e.g., Chow et al., 1988; Beyer et al. 1990). The driving force behind the use of CMP is the ability of the technique to achieve better planarization over much longer distances and over a much wider range of pattern factors and feature sizes than other techniques. That is, CMP can produce global planarization, but there are other issues to reckon with, as will be seen.

18.2 PRINCIPLES

CMP combines two processes: (1) a chemical reaction to form a surface layer which is easily removed (basis for selectivity) and (2) mechanical removal of the converted surface layer by polishing grains, suspended in a slurry, under the influence of applied pressure of a polishing pad. The latter is responsible for the planarizing action, since the removal rate, being proportional to pressure, polishes high points more rapidly than low areas. Runnels and Eymann (1994) stated that "CMP enhances the natural etching caused by the slurry through abrasion of the wafer surface." The mechanical removal rate during polishing is derived from the Preston equation for *mechanical* polishing of glass (Cook, 1990);

$$R = \frac{\Delta H}{\Delta t} = K \left(\frac{L}{A}\right)\left(\frac{\Delta s}{\Delta t}\right)$$

where $\Delta H/\Delta t$ = change of height H over time t, K = Preston coefficient (area/force) and is process-dependent; it is related to Young's modulus, hardness of glass; L = total load, A = area, Δs = relative travel between glass and pad.

The removal rate R has also been expressed as (Kolenkow and Nagahara, 1992)

$$R \sim NVT$$

where N = force normal to surface, V = relative speed, T = time.

Typical parameters are V = 50–150 ft/min; N = 1–10 psi; slurry flow rates 50–250 ml/min. Typical removal rates (of oxides, e.g.) are ~60–200 nm/min. There is no intentional heating of the wafer or slurry.

Chip Integration

18.3 EQUIPMENT

A schematic view of the essential part of the polisher (Patrick et al., 1991) is shown in Fig. VI-54. The heads are flat plates attached to templates which have recesses for the wafer to constrain the wafers mechanically. They contain wet inserts which set the wafers slightly above the surface of the templates; the wafers are held in place by surface tension. A downward force is applied to the head. The polishing pad is mounted on the table. The slurry is pumped to the center of the table. Both table and heads are rotated at a constant angular velocity.

There are many suppliers of polishers, slurries, and pads. Some of these are listed in Table VI-1 to VI-4. There are both single and multiple head polishers, usually computer controlled with load/unload cassettes. An example of a commercially available polisher, the Speedfam CMP-V is shown in Fig. VI-55.

The interaction between the polishing pad and the wafer to be polished has been illustrated, diagrammatically in Fig. VI-56 (Ali et al., 1994).

There is significant activity in pad and slurry development both by the suppliers and by the users; the recent patent literature reflects this, e.g., Cadien and Feller (1994), Tuttle (1993), Budinger and Jensen (1990). Cast polyurethane or polyurethane-impregnated felt are the most common pad materials. Particles are added to the pad material to modify its mechanical properties. The porosity of the pad (both macro and micro) is controlled by the cure dynamics during pad manufacture and by the use of mechanically punched perforations. Although some of the pad spec-

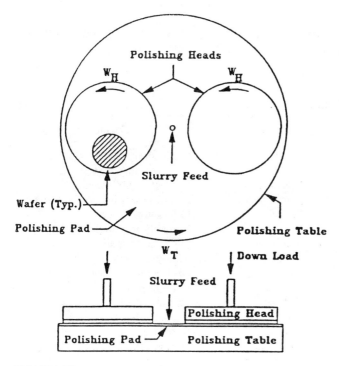

FIGURE VI-54 Diagram of the essentials of polishing equipment. (From Patrick et al., 1991)

TABLE VI-1 Suppliers of Polish for CMP

Supplier	Model	Comments
Westech	372M, 472	Single head, dual platen, cass-cass
R. H. Strassbaugh	6DS-SP	Dual spindles, cass-cass
Speedfam	CMP-V	5 wafer batch, cass-cass
Cybeq	Isoplanar 7000	6 wafer batch, cass-cass
Presi	Mecapol E2000-II B	Dual station, single head, cass-cass
Applied Materials, Inc.	Mirra CMP	4 head carousel, 3 polishing platens, cass-cass

ifications are covered by patents, many of them are proprietary. Slurries are usually premixed in large containers and pumped through a dispenser; an in situ slurry delivery system which mixes the components just prior to delivery to the table has been proposed by Murphy et al. (1995).

18.4 CMP OF SiO$_2$

18.4.1 Principles

The chemical component of oxide polishing is the slurry which contains colloidal particles in an alkaline medium (aqueous KOH or NH$_3$ to reduce ionic contami-

TABLE VI-2 Suppliers of Materials for CMP

Supplier	Item	Comments
Cabot	Slurry	Oxide slurry Semi-sperse 25, Cab-O-Sperse SC-1, 12
Rodel	Pads, slurry	Pads—IC 1000/Suba IV
Fujimi	Slurry, pads	Industrial lapping, wafer polishing
DuPont	Slurry	Monsanto's Syton
Hoechst	Slurry	Primarily in Europe
Nalco	Slurry	For Si wafer polish
Moyco	Slurry	High-purity silica source
Rippey	Slurry, pads	Distributor for Cabot slurries, Teijin pads
Teijin	Pads	Japanese market leader
Dai-ichi	Pads	Primarily Japan
Freudenberg Non-Woven	Pads	Interest in pads
Universal Photonics	Pads, slurry	Optical grinding and finishing

TABLE VI-3 Suppliers of Slurry Delivery System

Supplier	Model	Comments
Applied Chemical Solutions	2000 S 2200 S	Uses pressurized cylinders instead of pumps
Megasystems	OXYI and II	5-, 55-, and 220-gallon slurry
Semco	SMS 5530	System for mix and pump; 15-gallon reservoir

nation). Polishing is critically dependent on the presence of water, since interaction with water is the primary reaction. The attack of siloxane bonds (Si-O-Si) by water to form a hydrated surface controls the rate of surface removal. It is, therefore, not unexpected that silanol groups are found on the surface after polishing (Kaufman et al., 1995). Cleavage of Si-O-Si bonds below the surface is controlled by the diffusion of water in silica. Polishing occurs when the reversible hydration (polymerization) reaction

$$(SiO_2)_x + 2 H_2O \longleftrightarrow (SiO_2)_{x-1} + Si(OH)_4$$

proceeds in the forward direction, i.e., in the direction of hydration (depolymerization).

Material is removed as single silica tetrahedra or small clusters (Izumitani, 1979; Cook, 1990; Sivaram et al., 1992; Ali et al., 1994). The role of polishing compounds had been conceptualized as a "chemical tooth" expediting bond shearing at the surface. After reaction, the products must be transported from the surface, avoiding redeposition.

Wallace et al. (1996), using x-ray reflectivity, found that CMP decreased the surface roughness of PECVD SiO_2 and increased the density of the near-surface region. They concluded that the SiO_2 network was compacted under the applied pressure of the polishing pad which led to enhanced dissolution in the slurry as well as a very smooth surface.

TABLE VI-4 Suppliers of Post-CMP Cleaning Equipment

Supplier	Model	Comments
OnTrak Systems	DSS 200	Double-sided brush with megasonic option, chemical option
Dainippon Screen	SP-W813-AS	Sequential back, frontside clean, spin dry, DI only ultrasonic option
Solid State Equipment	Model 50	Modular, frontside clean, megasonic system
Semitool	NA	Spray processor for chemical cleaning

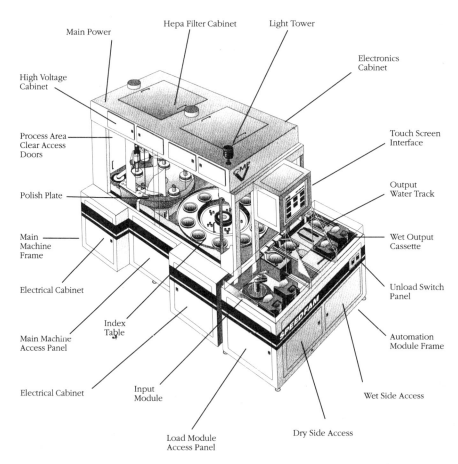

FIGURE VI-55 Diagram of a commercially available polisher. (Courtesy of Speedfam Corporation.)

18.4.2 Polishing Rate

The choice of the polishing agent is not determined solely by polishing rate. For example, ceria particles in the slurry polish SiO_2 at a higher rate than do silica, but silica is preferred because smoother surfaces are produced. Fumed silica has es-

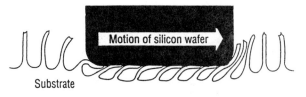

FIGURE VI-56 Diagrammatic representation of the interaction between a wafer and a polishing pad (slurry not shown). (From Ali et al., 1994)

sentially no polishing ability except at high pH. At pH ~ 7.5, the slurry is highly viscous; at a pH > 7.5, surface charge repulsion results in the dispersion of the particles. The polishing rate of fumed silica slurry increases with increasing pH up to the pH value at which the silica particles, themselves, dissolve. Although the Preston equation does not include particle size of the abrasives, this is important. The penetration depth of water (essential for the reaction) or the hydrated surface layer increases with increasing particle size; the effect is more pronounced at higher pad pressures. Therefore the polish rate increases with increasing particle size (and particle concentration). The particle shape also plays a role (Cook, 1990) with rounder particles increasing surface solubility. Another factor influencing the polish rate is the hardness of the surface oxide; soft films polish at a higher rate than do hard films (Izumitai, 1979; Dai et al., 1995). Addition of surfactants to a slurry may also affect the polishing results. Addition of a surfactant, a molecule with a large polar component (Achuthan et al., 1995) produced better uniformity but lowered the polish rate. Increased particle concentration improves the rubbing action and usually increases the polishing rate.

Polishing Rate/Uniformity Dependencies

For oxide polishing, hard, noncompressible (stiffer) pads are best for planarity, whereas improved uniformity and smooth surface are obtained by the use of softer pads (Morimoto et al., 1993). This has led to the use of stacked pads (i.e., combinations of soft and hard pads). A key problem encountered in using CMP for oxide planarization was the rapid decrease in polish rate of the oxide with successive wafers. This has been shown to be due to pad "glazing," i.e., the pads become smooth so that they no longer hold slurry as effectively. This is a surface phenomenon, and is not due to any chemical interactions between pad and slurry (Ali et al., 1995). Pad conditioning, i.e., "scrubbing," regenerates and restores the pad surface thereby restoring the polish rate. Pad conditioning has been done using carbide, diamond emery paper, blade, or knife to rub or scrape the polish pad surface. This procedure removes polish debris from the pad surface, reopens the pores, and forms microscratches in the surface of the pad; this results in longer useful pad life. The pad surface, before and after conditioning, are shown in the SEMs of Fig. VI-57.

Conditioning during polish (concurrent conditioning) improves pad consistency as it prolongs pad life (Holland et al., 1995). Adjustment of the pad/wafer rotation

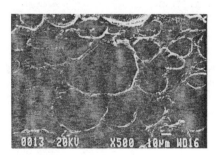 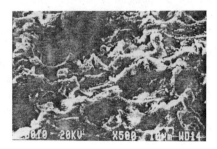

FIGURE VI-57 SEMs of polishing pads: (left) new, (right) used. (From Ali et al., 1995)

velocity has also been found to be useful in counteracting rate nonuniformity due to pad use.

The polish rate can be adjusted by changing the operating parameters. The rate increases, but the uniformity is degraded when the pad pressure is increased (Renteln et al., 1990). Increasing the table speed increases both rate and planarity (Morimoto et al. 1993), but vibration and other machine considerations limits the maximum usable table speed. Decreasing the temperature of the slurry and the polishing table improves the uniformity without changing the blanket polish rate (Morimoto, 1992).

The systematic nonuniformity, from the center to the edge of the wafer was shown to be minimized by shaping or contouring the template or carrier (Currie and Schulz, 1993). Jansen and Hanestad (1996) used variable back-side pressuring of the wafer and reported that it improved polish uniformity and pad lifetime.

Although CMP has been described as a process for achieving global planarity, a pattern sensitivity does exist. The polish rate of a feature can be influenced by the topography of surrounding features (Warnock, 1991). The initial rate of material removal depends on the width of the feature. At the start of the process, raised features are polished at a greater rate than recessed areas (the basis for planarization). As polishing continues, decreasing the step height differential, the rates tend to become equal (Renteln et al., 1990). Narrowly spaced features are polished at a greater rate than widely spaced ones as seen in Fig. VI-58. Small isolated raised features polish at a higher rate than groups of features, and wide gaps are polished at a higher rate than small ones as illustrated in Fig. VI-59. This figure also shows the rounding of the edges of lines (due to higher local pressure), although harder pads tend to produce less and more gradual corner rounding than soft pads (Burke, 1991). The influence of structure size on the final polish result is shown in Fig. VI-60.

The polish rate of blanket films, often used in preliminary experiments to determine the processing parameters and as a rate monitor; it is equivalent to that of a large raised area.

Other Polishing Issues with CMP

Large raised features and *large recessed* areas polish at about the same rate as a blanket film (or background). The intrinsic high polish rate of raised features and nonpolishing of depressions is not realized in these cases and therefore, planarization by just polishing is extremely difficult. Further, clusters of closely spaced narrow features act as a large feature, with a corresponding slower polish rate, which also works against achieving global planarization.

Landis et al. (1992) described other limitations as well: nonuniformity, rounding, "dishing," erosion, which are illustrated in Fig. VI-61 (the original surfaces are indicated by the dotted lines).

The severity of these effects can be reduced by process modifications, e.g., adjustment of the pad/slurry parameters. For example, harder pads polish recesses areas more slowly than soft pads. Reducing the variation in pattern density is another approach but more difficult to implement since it involves changing the chip layout.

Chip Integration

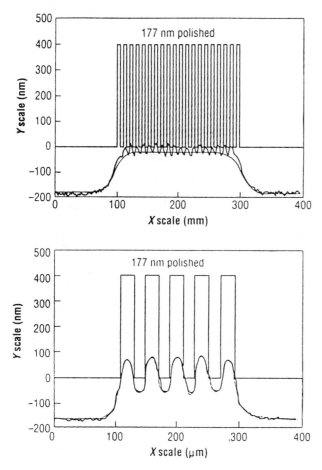

FIGURE VI-58 Influence of pitch on polishing results: (top) 5 μm × 5 μm, (bottom) 20 μm × 20 μm. (From Ali et al., 1994)

18.4.3 Applications of Oxide CMP

An example of the application of dielectric planarization to VLSI circuit fabrication is shown in Fig. VI-62. The horizontal and vertical AlCu interconnects were formed by a lift-off process, coated with bias-sputtered SiO_2 and then polished (Patrick et al., 1991). Another example is planarizing a dielectric layer before etching grooves to be filled with metal.

18.5 CMP of Metals

18.5.1 Introduction

In the recessed metal process, shown in Fig. VI-63, the metal layer outside of the groove is removed by polishing (i.e. planarized). This process has become known as the "damascene" process; the name comes from the ancient (inlay) process used to make enamelware and was introduced by the early users of CMP. The concept

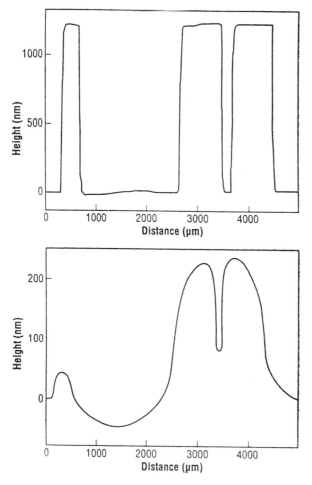

FIGURE VI-59 Stylus traces showing the effect of feature spacing on polishing rate and rounding of line edges: (top) before polishing, (bottom) after polishing. (From Sivaran et al., 1992)

is similar to the embedment or inlaid process described earlier in Section 17 of this chapter, but CMP is an essential component in the damascene process.

18.5.2 Principles

The principles are analogous to those for CMP of SiO_2, i.e., cycles of surface film formation and removal by mechanical means. In this case, the surface film is a passivating layer, inhibiting purely chemical etching so that high spots can be abraded whereas the low spots are protected. The total process must be noncorrosive and the final surface defect-free. The mechanism proposed by Kaufman et al. (1991) is illustrated in Fig. VI-64.

Chip Integration

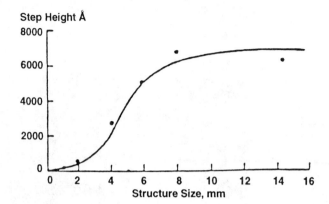

FIGURE VI-60 Step height vs. structure size in CMP. (From Morimoto et al., 1993)

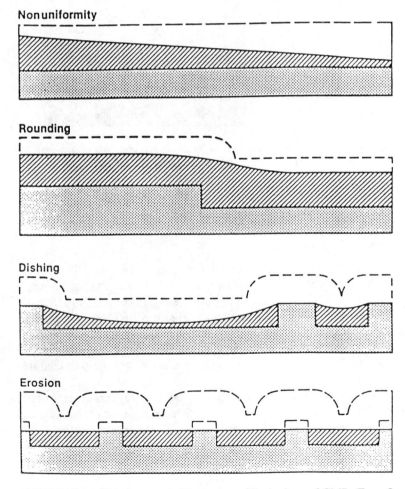

FIGURE VI-61 Diagrammatic illustration of limitations of CMP. (From Landis et al., 1992)

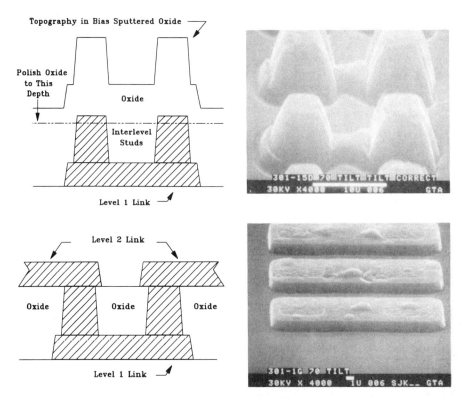

FIGURE VI-62 Application of CMP to fabrication of VLSI circuit interconnections before and after polishing: (left) diagram, (right) SEM. (From Patrick et al., 1991)

18.5.3 Applications

Polishing of CVD Tungsten

This was one of the early applications of CMP of blanket metal films. The slurry used was alumina or silica with a weak oxidant, $K_3Fe(CN)_6/KH_2PO_4$, plus ethylenediamine at a pH of ~6.5; at higher pH, wet etching occurred. Addition of ethylenediamine improved the surface passivation. Figure VI-65a shows a cross section of via holes in SiO_2 after filling them with blanket CVD W. Figure VI-65b shows top-view SEMs taken at several times during CMP of the W; in the final SEMs it can be seen that the W surfaces are coplanar with that of the surrounding insulator.

There was no effect on the resistance between the W and next level metal; either there was no CMP contamination of the interface or the contamination was removed by the sputter cleaning preceding deposition (Kaufman et al., 1991). An example (Uttrecht and Geffken, 1991) of a planar structure combining planarization of oxide over Ti/AlCu interconnects with formation of vertical interconnects using CMP of W is shown in Fig. VI-66.

Rutten et al. (1995) examined pattern density effects in CMP of W. They concluded that erosion of dense W wiring (dishing) resulted from the loss of se-

Chip Integration

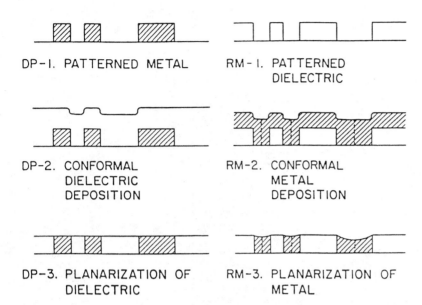

FIGURE VI-63 Comparison of CMP processes: (left) dielectric planarization, (right) planarization using the recessed metal process. (From Kaufman et al.)

lectivity between the metal and the underlying oxide layer. By changing the design rules so that the maximum line width is limited and using harder pads, thicker W, lower downforce, and higher speeds of the table and platen, the dishing was reduced.

Another application is the so-called "dual damascene" process in which the recesses for the vertical (stud) and the follow-on horizontal wiring are formed sequentially by RIE of the dielectric layer(s) and then filled with metal and polished (Kaanta et al., 1991; Dalal et al., 1993). A diagrammatic example is given in Fig. VI-67.

Zettler and Scheler (1993) described a similar process, which they called "planar self-reconstructing metallization"; in this process CMP of W was an alternative to an anisotropic metal etch.

Polishing of Al

CMP of Al has been more difficult to accomplish. It was first used as a final touch-up after laser planarization (Yu et al., 1991) and after high-temperature sputtering/laser exposure (Yu et al., 1992).

Yu et al. (1993) patented a process for polishing Al using $H_3PO_4 + H_2O_2$ (or other, unspecified oxidants) in an aqueous slurry. W capped Al has been used to avoid scratching and corrosion encountered when Al itself is polished (Roehl et al., 1992; Cote et al., 1993; Joshi et al., 1994).

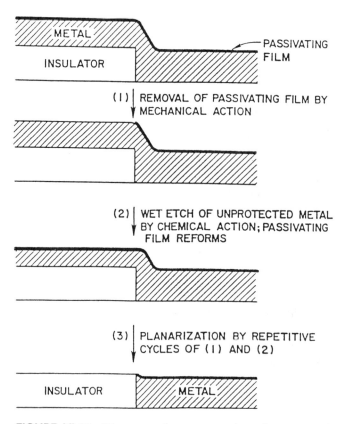

FIGURE VI-64 Diagrammatic representation of a proposed mechanism for metal CMP. (From Kaufman et al., 1991)

Polishing of Cu

There has been greater success in polishing Cu than Al. Both TiW and Ta have been used to line the dielectric before Cu deposition; liners serve as diffusion barriers, adhesion layers, and as polish stops. The first reported success (Krishnan et al., 1992) gave no details of the CMP process itself, although they reported that both sputtered and CVD Cu (with sputtered TiW liner) were polished. Luther et al. (1993) described a fully integrated planar four-level structure of Cu (with Ta liner) in a PI matrix, but again no process details about the slurry, abrasive, or pad were given. Each interconnect, horizontal (wiring) or vertical (stud) was formed by etching the appropriate recess into the PI (using RIE), filling it with Cu (deposition technique not given), and removing the excess with CMP. The process was repeated, for each wiring and each stud level, until all the levels were completed. Lakshminarayanan et al. (1994a,b) applied the dual damascene process to sputtered Ti/Cu to form a two-level structure. The slurry was alumina in a NH_4OH solution; any Ti remaining was etched in an HF solution.

Steigerwald et al. (1994a) studied the CMP process using blanket and patterned films. They observed dishing of sputtered Cu and erosion of the SiO_2 underlying the Ti liner during CMP using a Rodel 500 Pad in a Strasbaugh 6Cu polisher, and

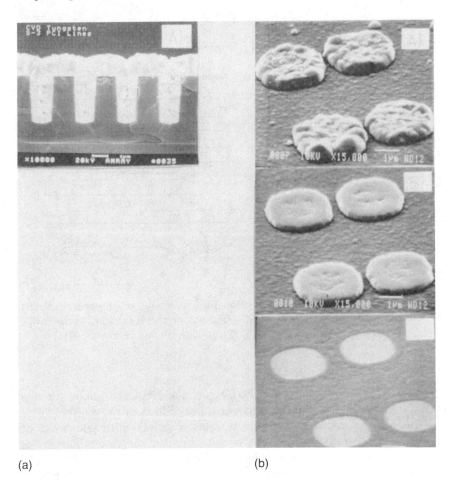

(a) (b)

FIGURE VI-65 (a) SEM of via holes filled by blanket CVD W. (b) SEMs taken at several times during CMP of the W.

an NH_4OH-based slurry. Dishing was dependent on the width of the Cu structure and only minimally on the structure density, but the SiO_2 polish rate increased with increasing density. The CMP process in an NH_4OH slurry is very mechanical; the static etch rate of Cu is very low and the surface film (Cu_2O) is very protective. Therefore, they concluded that dishing was not due to chemical attack but is more mechanical in origin; the pad was not completely rigid and thus conformed to the wafer surface, it exerted pressure on the recessed areas. Therefore a hard pad was recommended although marked scratching occurred. Also, since the selectivity between Cu and SiO_2 depends on the difference in hardness of the two films, a harder pad decreases the selectivity as well. The solution to minimizing both Cu dishing and SiO_2 erosion appeared to consist of using a hard pad but changing the slurry so that there is a higher chemical component in polishing Cu in high spots (without increasing the chemical removal of oxide) but good passivation in the low areas (protecting the oxide). Although uniformity is degraded by the use of a hard pad, if the process is highly selective, overpolishing is acceptable. However, the problem of metal scratching by a hard pad still remained to be solved.

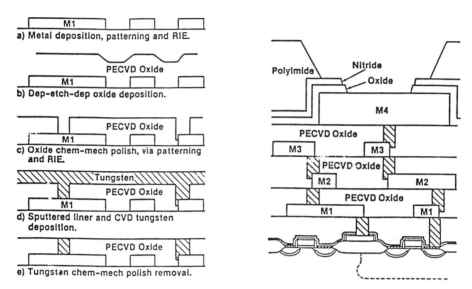

FIGURE VI-66 Diagram of (left) the process flow for planarizing an interlevel oxide and forming W studs on AlCu interconnects by CMP; (right) a multilevel structure formed using CMP and AlCu interconnects patterned by RIE. (From Uttrecht and Geffken, 1991)

Steigerwald et al. (1994b) added $K_3Fe(CN)_6$ to the NH_4OH solution for rate enhancement, but this emphasized the electrochemical effects of the polishing process. They found that presence of Cu ions in solution increased the polish rate of Ti, thereby reducing its effectiveness as a polish stop. The increase is due to the electrochemical reduction of Cu ions (plating Cu on the Ti surface) and the subsequent oxidation of Ti, thereby decreasing the effectiveness of Ti as a polish stop. An alternative to the NH_4OH-based slurry was patented by Yu and Doan (1993). It consisted of H_2O, abrasive, and a third component from the group of HNO_3, H_2SO_4, $AgNO_3$ or mixtures of them, with the possible addition of an oxidant such as H_2O_2, HOCl, $KMgO_4$, CH_3COOH. Hirabayashi et al. (1996) used a mixture of 0.1 wt% glycine, 5–11 wt% hydrogen peroxide and 5 wt% colloidal silica to pattern copper inlays in SiO_2/SiN. Using IC1000 and Suba800 polishing pads, with 29 kpa and at table speed of 160 rpm, they reported polish rates of about 40 nm/min and dishing less than 60 nm for feature widths ranging up to 100 μm. Luo et al. (1996) used a low pH (acidic) slurry with $Fe(NO_3)_3$, benzotraizle inhibitor and a surfactant and achieved polish rates of copper in the range of 80–160 nm/min with a selectivity to silicon dioxide of 15:1 to 45:1, depending on specific slurry composition.

18.6 CMP OF SOGs

Forester et al. (1995) reported that at low pH the polish rate of a siloxane SOG, Accuglass, and of FLARE is slower than that of thermal oxide and insensitive to the down pressure. They suggested that SOGs could be used as a polish stop in an

Chip Integration

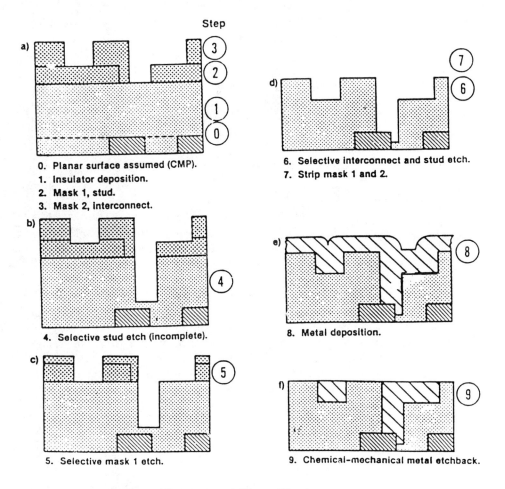

FIGURE VI-67 "Dual damascene": schematic of the essential steps of the process. (From Kaanta et al., 1991)

oxide CMP process. In the case of P-doped SOGs, the polish rate increased with increasing P-content of the SOG.

On the other hand, Homma et al. (1995) reported a very high polish rate of a reflowable low ε organic SOG in an alkaline slurry containing ceria abrasive with very small crystallites. The polish rate of SiO_2 in this slurry was almost negligible, i.e., the selectivity of SOG to SiO_2 was ~50. The polishing pad selected was a foamed fluoro-carbon pad which caused almost no scratches on the SOG.

18.7 REPRODUCIBILITY

Control of polishing to remove a desired thickness and achieve desired planarity in the case of insulators such as oxide is different from metal polishing where the objective is remove the metal outside of the grooves completely, but not subject the metal within groove to excessive chemical exposure. Each application has slightly different issues: in the case of oxide, the polishing is to be stopped within the oxide and too little polishing leads to excess insulator thickness or incomplete planarity, whereas overpolishing leads to thinning of insulator below specification. In metal polishing, the polishing is stopped over a composite surface of metal and insulators; too little polishing can lead to unwanted metallic residues whereas overpolishing can lead to "dishing" or depressions in the metal features. In oxide polishing, "pilot" runs are usually made, using the same pad/slurry/polisher to determine the time to reach the desired planarity and thickness. Subsequent runs are "blind," i.e., the end-point is determined by time, relying on the assumption of reproducibility and the effectiveness of regular monitoring of polish rates. However, the polish rate can change due to changes in the wafer contact surface as the topography is reduced and by changes in the surface of the polish pad. The latter is compensated by conditioning, but this may not be sufficient.

In the case of metal polishing, slower polishing etch stop layers can be easily integrated, which along with choice of pads and slurries can be used to minimize the residues while avoiding "dishing." A slower polishing etch stop layer, widely used in silicon polishing, is Si_3N_4 (Beyer et al., 1987) which has the disadvantage of high dielectric constant. Others are BN Poon and Gelatos (1991); Poon et al. (1993) whose dielectric constant is lower than that of Si_3N_4 and SiBN (Neureither et al., 1993) with an even lower dielectric constant. However, addition of Si to BN reduces the selectivity (by altering the hardness of the film), so that for small pattern factors, CMP erodes the stop layer.

Therefore, an end-point detection system has been desirable. One system senses the change of friction between the rotating wafer and the polishing surface by measuring the change in the motor current and the signal can be used to adjust or stop the process (Cote, 1990; Sandu et al., 1991; Cote et al., 1994; Litvak, 1995). Better results have been obtained for metal polishing than for oxide.

Another technique is the capacitance method (Miller and Wagner, 1992; Lai et al., 1993) used for oxide polishing. The method is based on the fact that the oxide on the wafer forms a capacitor with the slurry acting as the counter-electrode. However, if the process is not uniform, the output will not represent an average thickness. Thus it works well with blanket oxide films, but topography and pattern density affect the output so that calibration for each type of device is needed. Lustig et al. (1994) claimed to improve on Miller's endpoint technique by use of a center and guard electrodes embedded in the polishing table and connected to the polish wafer through the slurry and by applying a high-frequency sinusoidal ac voltage (0.5 voltage). The response current flowing through the center electrode is monitored by a differential amplifier. The output voltage is inversely proportional to the remaining insulating film thickness on the wafer.

Another technique for controlling oxide uses optical thickness measurement. In one implementation (Schultz, 1993), the wafer polish head overtravels the polishing table and projects the polishing wafer outside of the polishing table over an optical thickness measurement tool. In a more ambitious implementation, the optical system for interferometric measurement is integrated into polishing table and pad (Tolles et al., 1996). Lustig et al. (1995) describe the use of light reflected from the front side of the wafer to determine polish end point in situ. Sum et al.

Chip Integration

(1996) describe the use of infrared radiation through the back-side of the wafer (IR emitter and detector located in the polishing head assembly) to measure the insulator thickness change. Fukuroda et al. (1995) showed that polish head vibration, measured using an accelerometer can be used to determine the smoothing of the insulator peaks; this technique, however, can not determine the remaining insulator thickness. Other techniques such as acoustic waves (Meikle and Doan, 1995) or pad temperature (Chen and Diao, 1996) have also been suggested as useful for monitoring the progress of the polishing process.

18.8 DEFECTS

18.8.1 Types of Defects

Initially, CMP was regarded as a "dirty" process; however, low defect levels have been achieved, despite the many types of defects that have been observed. Fury (1995) and Kaufman et al. (1995) have assembled a "catalog" of CMP defects:

Metal roughness
Residual passivating films on metals (high contact/via resistance unless removed by sputtering or etching)
Residual slurry particles
Alkali ion contamination
Microcracking of insulator due to stress-induced diffusion of water into oxide; can be minimized by using non-K^+ slurry and softer pads
Trapped metal particulates
Scratching of oxide surface (possibility of metal deposition into scratches causing shorts when metal is etched by RIE but not when metal removed by CMP)
Corrosive attack
Gouging
Dried slurry
Uniformity degradation (with increasing pad use)

18.8.2 Post-CMP Cleaning

Removal of surface debris remaining after CMP presents a challenge. Both the slurry and the pad are primary sources of particles, adhered to the surface or embedded in it (Landis et al., 1992). One way of minimizing or eliminating polishing debris is by keeping the wafer wet until it is finally cleaned. Eliminating the flow of air onto the wafers prevents the introduction of particles from the environment. If opposite charges are developed on the wafer and the contaminant, impurity deposition is likely. There may be van der Waals forces acting as well. A postpolish rinse in DI water lowers the pH, leading to gellation (i.e., agglomeration of the colloidal particles in the slurry) and finally deposition. Therefore, the use of a basic solution, rather than DI water, has been recommended (Roy et al., 1995). Several other procedures for postpolish cleaning have been proposed: use (1) a different pad/platen, (2) double-sided scrubbers, (3) a high-speed water spindle (shear particles from the surface), (4) ultra (mega-)sonic cleaning, (5) surfactants to reduce surface tension, and (6) HF to remove ionic and metallic contaminants.

However, CMP often eliminates existing defects, such as etch residues and inclusions, as illustrated in Fig. VI-68.

18.9 PROCESS IMPLEMENTATION ISSUES

CMP may not be always the process of choice; that will depend on the requirements and may be somewhat subjective. For successful CMP, good step coverage of steep steps and void-free hole-fill of high AR holes must precede it. When the highest degree of planarization is important, regardless of cost, CMP individually or in combination with other process steps, seems to be the answer. However, an etchback process using a sacrificial layer with excellent planarizing properties may be adequate in some cases. Neureither et al. (1994) demonstrated that such a process (using Accuflo) was almost equal to CMP, up to 100-μm gaps, as shown in Fig. VI-69.

Pai and Konitzer (1992) suggested that if simplicity is more important, the use of flowable oxide in a nonetchback process should be considered since simplicity results in lower cost, as seen in Fig. VI-70.

The cost-of-ownership of CMP has been analyzed by Jairath et al. (1994). If the reliability of the equipment is poor, that dominates the cost, although consumables (pads, slurries) add to the cost. The latter cost becomes more significant as equipment reliability improves. As the throughput increases, the total cost of CMP decreases. The stability of the slurry may be an important factor; if the shelf life is short, there is either wastage or the need for frequent restocking. Although

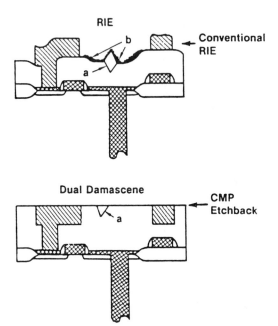

FIGURE VI-68 Defect reduction using CMP. (top) After RIE: tip of embedded defect (a) protrudes from surface; residual metal (b) on surface. (bottom) After CMP: both protruding tip and metallic residue removed by polishing. (From Kaanta et al.)

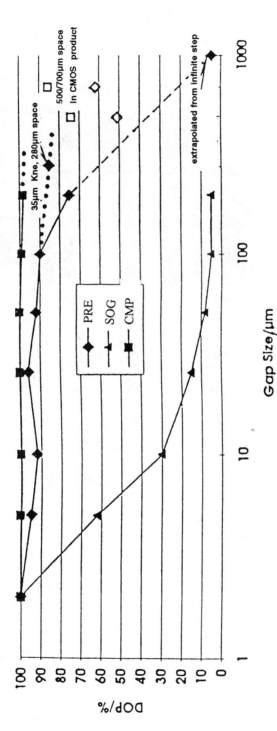

FIGURE VI-69 Comparison between CMP and other planarizing processes; PRE stands for an etchback process using a long range spin-on polymer (Accuflow) and SOG stands for SOG planarization. (From Neureither et al., 1994)

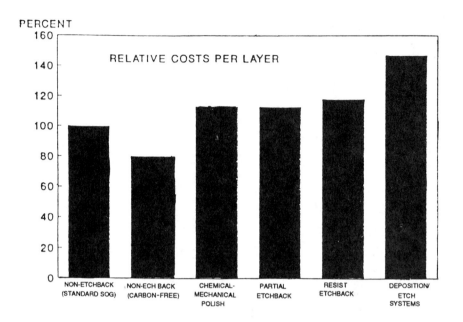

FIGURE VI-70 Comparison of processing costs of planarizing processes for (left to right) nonetchback SOG, flowable oxide, CMP, partial etchback, resist etchback, deposition/etch of oxide with a table comparing the technical merits of each process. (From Pai and Konitzer, 1992)

limited pad life was not considered to be an important cost, per se, improved life improves the ease of manufacturing.

18.10 IMPACT OF CMP

The use of CMP can reduce or even eliminate some defects, but does create new ones (see above). CMP forces reduction in the range of feature sizes since the best

Chip Integration 449

overall planarity is achieved when the features are approximately the same size. However, the use of larger numbers of vertical interconnects and narrower pitches are possible. The approach to global planarity minimizes film thickness variations (capacitance variations), minimizes via overetch and reduces or even eliminates depth-of-focus variability.

In addition, CMP is a low temperature process (unlike, e.g., high-temperature sputtering or flowage of metals). A single kind of equipment is needed for both dielectric and metal polishing. It is relatively inexpensive compared to RIE and ion milling. CMP of metal or the "damascene process" has opened up capability to pattern different materials into rectangular conductors that were not previously considered feasible because of difficulty in patterning. The need for complete selectivity in CVD and plating is eliminated. But CMP is a new technology with much to be learned. The compatibility with cleanrooms and cleanroom procedures has been questioned. To date, the successful implementation of CMP processes for oxide polishing and tungsten damascene in manufacturing is continuing to grow and use of copper wiring using CMP damascene process is widely discussed. Many suppliers of polishing equipment, slurries, pads, and other control devices have emerged.

18.11 MODELS OF CMP

There have been a number of models of CMP relating the feature size and pattern factor to polish rate (e.g., Burke, 1991; Warnock, 1991; Runnels and Olavson, 1995; Hayashide et al., 1995; Renteln, 1996) as well as a model of the deformation of the polishing pad during polishing (Ohtani et al., 1995). The models are primarily empirical and theoretical aspects are few. A critical review of some of the models can be found in a paper by Nanz and Camilletti (1995).

By the sheer enormity of the effort of many different parties, strong understanding of the use, control, and limitations of CMP have emerged. The starting choice of the CMP process, i.e., polisher, pad, slurry, or operating parameters is often based on the advice of the manufacturers of the polisher, slurry, and pads. Final learning and tuning of the processes are usually done in-house with test structures that represent products.

19.0 CONCLUSIONS, PART I

The need for planarization to cope with lithography and etching requirements and to ensure the reliability of the product is clear.

The processes involving metal and oxide reflow, SOG, spin-on organic/inorganic films, various metal and oxide deposition methods, sequential deposition/etch oxide sequences, etchback, etc., in many cases fill gaps well, without voids and are a prerequisite for CMP. Some of these techniques are planarizing as well, but usually achieve only local planarization. However, CMP shows the best planarization, the widest global range.

CMP, despite the limitations discussed above, it is being used more widely in chip fabrication; it is becoming accepted as a manufacturable process. This increasing use will be accompanied by more activity in the improvement/development of CMP equipment and processes. The successful implementation of chemical mechanical polish in the manufacture of microprocessors and DRAMS is indicative of the rapid progress and growing interest in the technology.

PART II: REMAINING ISSUES FOR CHIP INTEGRATION
1.0 OVERVIEW

Integration issues were discussed in Part I by detailed analyses of the individual process elements, their advantages and concerns. The elements were discussed from the point of view of the specific requirements they satisfy in building multilevel interconnection structures. A single process or material, however, can not be selected for inclusion in a process sequence without consideration of the mutual interactions with other processes or materials. It is important to remember that the overall manufacturability and quality of the completed device depend on the successful execution of *all* elements. These mutual interaction issues can be grouped broadly as: (1) conflict between the choice of a process and the required structure, (2) process (in)compatibility, (3) reliability (defects), and (4) manufacturability (yield, productivity). These will be discussed in the context of the need for lower RC delay in the interconnection wiring to reap the benefits of the smaller, faster-switching devices (in the silicon) and the increased number of circuits (and their density) on the chip, made possible by the advances in lithography and etching.

2.0 PROCESS/STRUCTURE CHOICE CONFLICTS

Many of these issues have been discussed in the individual chapters, but are repeated here to put them into an overall perspective.

The parts of the interconnection processes, the materials and structure are closely interrelated. The choice of one invariably will affect the others. Some of these scenarios will be discussed here, but the reader will do well to remember that there are likely to be many others.

The faster devices/circuits require minimum propagation (RC) delay in the interconnection wiring and the denser circuits require high wireability, i.e., the ability to wire a large number of circuits in a given area. These requirements drive the interconnection technology toward use of several wiring planes (known as multilevel metallization, MLM, or multilevel interconnects, MICs), (Berndlmaier, 1980; Fried et al., 1982; Sasaki et al., 1983), low-ε dielectrics for intra- and interlevel insulation, and high conductivity metals for wiring. The advantages gained thereby are illustrated in Fig. VI-71 which shows the decrease in cycle time as the number of wiring levels is increased; as well as the reduction due to both lower ε [e.g., using a fluorinated polymer (FP) in place of BPDA-PDA PI] and ρ (although some of the delay is due to the higher-ε SiO_2 as well as the higher-ρ Al). The relative chip size also decreases as the number of levels is increased; thus it is seen that the use of MLM increases the wiring efficiency and decreases chip size.

One way of increasing density is to decrease the wiring pitch. However, as seen in Fig. VI-72, although decreasing the pitch reduces the capacitance to silicon it increases the line-to-line capacitance, so that eventually, the overall capacitance and hence the delay, is increased. Increasing the line-to-line spacing lowers the intralevel capacitance but increases chip size. Increased chip size requires increased

Chip Integration

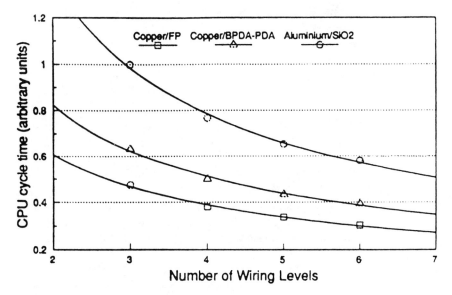

FIGURE VI-71 CPU cycle time vs. number of wiring levels for (top) Al/SiO$_2$, (middle) Cu/BPDA:PDA, (bottom) Cu/FP (fluorinated polymer). (From Paraszczak, et al., 1993)

wiring length, resulting in longer delay because both resistance and capacitance to the silicon are increased, and as noted above, the use of several wiring planes has long been an accepted solution to these problems. Once the choices of insulator and metal are made, the horizontal geometry of conductor features and separations

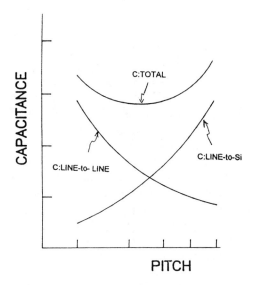

FIGURE VI-72 Capacitance vs. pitch (line width + space between adjacent lines): line-to-line, line-to-Si, total capacitance.

are optimized for wireability and minimum propagation delays by modelling and simulation. The use of several wiring planes introduces the need to keep the capacitance between successive metal levels as low as possible. Thus, thicker insulators might be preferred for reducing both interlevel capacitance and interlevel shorts. However, thicker insulators are difficult to etch, anisotropically, and the etched holes have high aspect ratio (AR); etching is complicated, still further, by RIE lag which requires significant overetching.

To maintain acceptable wiring pitches at higher wiring levels, thicker insulators necessitate the use of vertical interconnects (studs/plugs) instead of tapered vias to be filled when depositing the next level of metal. But it is difficult to fill a high AR hole by most metallization processes. Vertical studs require more process steps (e.g., an extra deposition step) and introduce other process complexities, and therefore, add to the cost of processing. A combination of processes that utilizes single metal deposition cycles with integral wire-stud interfaces has been proposed (Kaanta et al., 1988). Sloped vias are used, increasingly, only at the uppermost of the multilevel wiring levels.

To increase conductivity without sacrificing wiring pitch, tall narrow lines made of low resistivity metals, can be used. RIE of such metal features presents difficulties similar to those encountered in etching high AR holes in insulators. This also means that the insulator deposition process needs to fill high aspect ratio gaps if the conductors are first defined using RIE processes. Biased-high-density-plasma PECVD, SACVD and a dep/etch PECVD process have all been used for improved gap-fill. Alternatively insulating materials with good gap-fill properties such SOGs or spin-on organic films can be used. However, if a damascene metal process is used, it eliminates the need for metal RIE. Thus, the process involves filling high AR grooves in the dielectric layer by metal, rather than filling a high AR metal space by an insulator.

If organic insulators or SOGs are used for their improved gap-filling properties, one has to be concerned with their thermal stability and with moisture absorption which increases the dielectric constant of the film (Harada et al., 1990). The available low-dielectric-constant insulators are organic or highly fluorinated inorganic oxides. These materials also suffer from degradation and moisture absorption problems that may lead to delamination of subsequent metal or insulator layers.

A high conductivity metal such as Cu can be used; but issues such as adhesion, oxidation, and diffusion through insulators, discussed in Chapter V, will necessitate the use of glue layers and barrier layers which reduce the effective conductivity, as shown in Fig. VI-73 (Edelstein et al., 1995). Since Cu, unlike Al and its alloys, is not easily patterned by RIE, the choice of copper restricts the patterning options.

An underlying requirement for advances in semiconductor integration is the continued evolution of lithography process for printing smaller and smaller features. Optical exposure systems now use mid-UV to deep UV radiation. The shorter wave length, together with associated improvements in lenses, makes possible printing of images <0.25 μm. Phase-shift masks are being developed to push the capability to <0.2 μm. Printing these fine features exacts a price; at every exposure level, the exposure field has a tight planarity requirement, i.e., a small depth of focus (DOF) over the field of exposure; the field can be a 20 mm × 20 mm area. The DOF requirement can translate into a topography of less than about 0.1 μm for

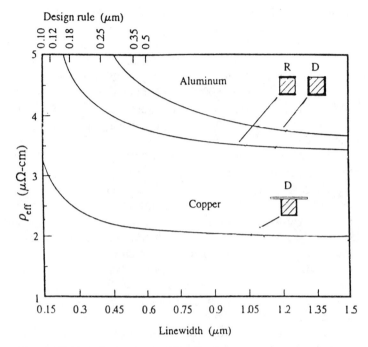

FIGURE VI-73 Effective resistance of encapsulated metal conductors vs. linewidth. (From Edelstein et al., 1995)

good focus. This requires a good planarization process at every level. Many of these have been described in the earlier part of this chapter.

There is both an upper and lower limit to the thickness of the resist used to transfer the mask image to the material to be etched. An upper limit exists since the smallest thickness of resist provides the best resolution. But a resist cannot be too thin; that limit is determined by the ability to spin-on continuous films without pinholes. RIE, used to etch anisotropically, in order to produce vertical features and control process bias, consumes resist masks at a finite rate. If thick films must be etched, then processes that have high selectivity to resist must be developed or else complicated multilayer resist schemes must be used to decouple etch masking and lithographic print requirements (Bushnell et al., 1986).

When the semiconductor devices and circuits require mixed feature sizes, the differential between the RIE rates of small and large features, RIE lag (Lee and Zhou 1991), will increase the need to overetch.

3.0 PROCESS COMPATIBILITY

Compatibility of later process steps and materials with earlier ones is essential. Plasma processes, especially deposition at high rates using high bias can increase the wafer temperature well above acceptable temperatures for Al wiring, unless adequate heat transfer equipment is used which complicates the design and operation of the plasma systems. As will be discussed in the reliability chapter (Chapter VII), thermal stresses are a major source of failure for thin conducting film. Laser and thermal reflow and CVD W deposition for stud levels can increase temperatures substantially to cause problems in underlying metals. If SOGs or organic insulators are used as gap fillers or as primary insulators, they can undergo irreversible changes or degradation at temperatures in the 500°C range. Outgassing of adsorbed moisture from SOGs and other dielectrics during subsequent process steps can result in blistered metal (Hirashita et al., 1990) and high via resistance (so-called "poisoned vias") (e.g., Romero et al., 1991; Hamanaka et al., 1994). Desorption can have other effects as well. For example, it takes very little outgassing of absorbed moisture to scatter metal atoms in vacuum and cause tapered studs, which appear black in an optical microscope (Schwartz, 1991). Too much outgassing can prevent metal deposition within the via, since the outgassing molecules can create high local pressure and scatter the arriving molecules.

If the deposition of insulators require oxidizing ambients, the metal surfaces can be oxidized, increasing the resistance of the contact. Reflow or densification of BPSG or PSG for planarization or improving oxide contours requires high temperatures, in excess of 700°C. Depending on the exact temperature, devices and gate electrodes or local silicide/polysilicon wiring can degrade. Films stresses can affect adhesion and cracking of underlying layers.

The patterning of any layer must be compatible with underlying layers. Exposure to an O_2 plasma during plasma stripping of resist can result in cracking of some SOGs. Etch selectivity is essential since overetching is required to accommodate process nonuniformity. If corrosive gases are used in etching a conductor, then the underlying insulator should be impervious. However, organic insulators can hold chlorine species (used in an Al alloy RIE process) which can lead to subsequent corrosion. Even small amounts of residual chlorine are known to cause corrosion in different metal systems (Parekh and Price, 1990).

The insulators must not absorb species that may degrade the insulator quality from the processing or fab ambient. Examples of the deleterious effect of absorption of moisture from the environment on some deposited oxides are increases in the dielectric constant and stress hysteresis. Dilute hydrofluoric-based solutions, isopropyl alcohol, and phosphoric-chromic cleaning mixtures are commonly used many times during the fabrication of the multilevel interconnection and both conductors and insulators used should be inert to these chemicals.

4.0 RELIABILITY

4.1 DEFECTS

Defects invariably cause premature or early reliability failures in use, which can be distinguished from the statistically distributed wear-out failures during extended use. Choice of interconnect materials, their thicknesses and specific processes used can also lead to reduced wear-out life times, which will be discussed later. A "defect" can be defined as an unwanted feature created by missing material such as pin-holes in insulators, voids or notches in the metal conductor, or by the presence of extraneous particulates, metallic or insulating. The effect of the defect is to alter the line/space dimension to a value below the minimum design value,

Chip Integration

outside of the usual distribution. In the extreme case, the defect can lead to an electrically short or an open circuit. The defects can be created from a variety processes, masks, and equipment. Here, we are mostly concerned with how the choice of a deposition or patterning process can lead to a defect at a later or earlier level. Deposition processes, equipment, and clean room ambient can all add particulates to the surface, as discussed in the last chapter. A metallic particle on an insulator surface can become a potential shorting path between conductors at a subsequent level as shown in Fig. VI-74. Upon planarization, the metallic particle simply becomes an extraneous stud which shorts conducting lines above it; however, this effect of the extraneous particle can be reduced by use of redundant insulating layer after the planarization step. Similarly a defect in the insulator such as a hole or a depression can lead to traps for unwanted metal films in a subsequent damascene process, which can lead to metal shorts between two levels as shown in Fig. VI-75. Sometimes the choice of the process can make the structure more immune to the defect, as was illustrated in Fig. VI-68. Damascene CMP is sensitive to pinholes and voids in the insulator. In etchback planarization, defects (a pinhole or hard particle) in the planarizing layer can be translated into defects in the insulator (a pinhole or a bump). These defects are not created if insulator planari-

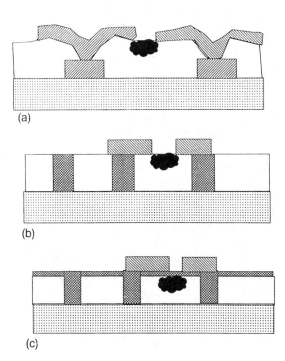

FIGURE VI-74 Diagrammatic representation of the effect of metallic defects causing shorts: (a) a via interconnection, (b) a stud interconnection, (c) the benefit of a dual insulator on a stud interconnection.

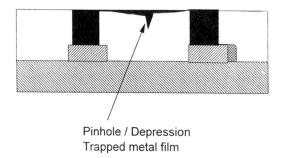

Pinhole / Depression
Trapped metal film

FIGURE VI-75 Diagrammatic representation of the effect of a defect in an insulator, in a damascene process; metal trapped in pinholes or depressions (e.g., scratches) causing shorts.

zation is not pursued or if an additive process is used for forming the stud or wiring. Similarly, if the chosen insulator tends to absorb moisture and release it during the hot metal deposition step, this can prevent the nucleation and growth of the metal leading to voids or porous metal films. Sputtering processes and equipment can add particles. Some sources are: impurities in the processing gases and delivery systems, transport and clamping/unclamping wafers within the reactor, spalling of poor quality material deposited on the chamber walls, as well as metals sputtered from the walls themselves. Load locks, soft pumping, and frequent in situ or ex situ chamber wall cleaning as well as better design of mechanical parts minimize some of the problems. Another source of particulate contamination arises in the plasma by gas phase nucleation and growth by condensation of species generated in the plasma (Selwyn et al., 1989; Weiss et al., 1995). Highly stressed metallic films, such as CVD W can delaminate locally and be transported to regions where they are unwanted. The delamination problem is aggravated if the adhesion and nucleation layer such as TiN is inadequate. Residues from RIE (e.g., deposited polymer, unetched film due to either inadequate overetch or unetchable impurities in the film) are a problem; clean-up processes/chemicals used after the etch process can lead to attack of the exposed metal or insulator. Post-RIE corrosion of Al (alloys) by residual Cl can be controlled but require the use of extra plasma or wet chemical treatments (Mayumi et al., 1990).

Defects in resist layers and masks can result in defective patterns; if the defect occurs in a sensitive area it may result in yield loss or compromised reliability. Point of use filtering, with extremely fine filters, is increasingly preferred with spin-on layers, to remove *gels* and other extraneous particles. In the same vein, some materials are more prone to create or be affected by defects. Incorporating refractory metals into AlCu conductors by layering or cladding, has been preferred since defects and voids in the Al-Cu film have less effect on the total structure. Aluminum forms a good passivating oxide but high copper additions make the film more susceptible to corrosion. Copper does not have a passivating oxide and is more susceptible to oxidative damage. Copper and Al are scratched easily during CMP. CVD metal growth is a *heterogeneous process* and any impurity in the surface of the seed film can lead to voids or other defects in the growth film.

4.2 INSULATOR RELIABILITY

A direct short or a leaky path in the insulator, even a momentary breakdown can affect satisfactory circuit operation. The intrinsic quality of the insulator material can change the breakdown statistics. If organic based insulators or F-doped SiO_2 films are used because of their low dielectric values, they may also have lower breakdown strength as illustrated in Figs. IV-7 and IV-23.

The molecular structure of some organic insulators such as PIs are different in the plane of the film and in the vertical direction; this has been shown to result in orientation dependent mechanical and electrical properties, such as the TCE and leakage. Thus, minimum thickness specifications need to take into account the difference in properties due to orientation. In the case of dep/etch SiO_2, using PECVD TEOS-based oxide (Schwartz and Johns, 1992), the material in the gap etched more rapidly in BHF than the oxide on the top surface and an unidentified filamentlike structure was also seen near the sidewalls. In structures made using the same oxide (whether or not a dep/etch sequence was used) the line-to-line capacitance was higher than expected from the value of ε determined on a blanket film. The capacitance of similar structures, using SiH_4-based or sputtered oxide, was what was expected from the ε determined on blanket films (Schwartz et al., 1992).

The choice of material and process can affect the adhesion of the insulators to each other and to the metallic layer. Local delaminations can lead to hot spots and early failures. The poor adhesion of PI to metal coupled with its tendency to absorb moisture makes possible the formation of a monolayer film of water, said to be necessary for initiation of corrosion (Wilson, 1980). The insulator film stress magnitude and sign has been known to affect the electromigration lifetimes of the conductors. Lloyd (1982) concluded that the open failure rate during temperature/current stressing (electromigration) in AlCu conductors became lower when they were coated with a PI film instead of SiO_2; the reason suggested was the lower level of microdefects in PI film. Later Lloyd and Steagall (1986) attributed the improvement to an increase in activation energy, due, possibly, to residual hydrogen in the PI providing hydrogen to the grain boundaries of the AlCu film. Compressive stress of the insulator has been cited as a reason for prolonging electromigration life in Al systems (Schafft et al., 1984, Yau et al., 1985).

Yamaji et al. (1991) found that stress-induced voiding and electromigration lifetime of Al conductors were *improved* when the deposition temperature of PECVD SiN and thus the compressive stress in the film was *reduced*. The samples were annealed in H_2 for 30 min at 400°C, i.e., at a temperature 200°C higher than the deposition temperature of the low-temperature SiN but only 20°C higher than that of the high temperature SiN. On the other hand, Tokunaga and Sugaware (1991) found *more* voids in PECVD SiN-coated Al conductors when the SiN was deposited at *lower*, rather than at a higher temperature. The samples were annealed in N_2 for 60 min; there was no voiding on any sample annealed at 350°C. But for higher temperature heat treatment, the larger the difference between the deposition and anneal temperatures, the higher the density of voids. They postulated that stress relaxation, due to evolution of H_2 from the PECVD SiN, induced different stress states in the Al lines when the low and high temperature SiN-encapsulated samples were heated. Since films deposited at the lower temperatures had a higher H-content, more voids were to be expected.

The insulator deposition process temperature should be compatible with the metal beneath it, since excess heat can lead to thermally driven voids and extrusions in the conductor that can cause failures. Grivna et al. (1993) found a correlation between absorption/desorption of moisture by the TEOS-based PECVD SiO_2, which resulted in a drift in stress and in-process stress voids in Al-1.5% Cu. By optimization of the deposition parameters (use of high power and temperature and a high O_2/TEOS ratio) metal line voiding was eliminated. Finally, the planarization process may determine the choice of insulators. For example, in CMP, a material such as silicon nitride is hard to polish and can act as a polish stop, and may be used despite its high ε. Organic insulators tend to scratch easily when polished and may, therefore, be unsuitable to be used with CMP.

4.3 METALLIZATION RELIABILITY

The major concerns about thin film metallization are corrosion, electromigration, and stress-induced failures such as voids, extrusion, cracking; each of the problems can aggravate the other. The choice of insulators, the geometrical structure of the interconnect, and finally the choice of metal and deposition process which determine its microstructure, all influence and affect the extent of the reliability failures. Organic insulators usually absorb moisture which can permeate readily to reach the thin metal film, leading to corrosion especially since the thin film (wiring) is likely to be electrically biased. Nishida et al. (1985) reported that moisture within the bulk of PIQ resulted in corrosion of Al-based metallurgy. Lin (1990) concluded that a higher temperature cure for PI-2555 can increase life time of AlCu metallization in a temperature-humidity-voltage stress. It may be necessary to use AlCu layered with or capped by a refractory metal, to improve electromigration performance although this increases the line resistance. When W studs are used to connect two levels of AlCu wiring, location of the failure shifts away from the stud or stud-wire interface (Estabil et al., 1991), unlike the case of aluminum vias or studs. As the thickness of the metal line is increased the failure mode has been known to shift from extrusions to opens. Another issue needing consideration is whether the choice of metal and its method of deposition and/or patterning has any adverse effect on the insulator. Metal solubility, migration in the insulators, reaction with and adhesion to the insulators, are some areas of concern. For example, polyamic acid is known to dissolve copper from the interface and clusters of copper redeposited within the polyimide body (LeGoues et al., 1988; Kim et al., 1990). Copper diffusivity in silicon dioxide has been found to be sufficiently high to cause concerns of insulator or device degradation (McBrayer et al., 1986; Raghavan et al., 1995).

4.4 DEVICE RELIABILITY

By and large, the devices formed in the silicon remain unaffected by MIC materials and processes except for the introduction of ionic impurities such as Na^+, plasma damage, moisture evolution, and impurities that emit alpha particles.

Chip Integration

Ionic impurities are usually introduced as trace impurities from wet processing using chemicals such as alcohols and acids, resist processing including developers, and polishing slurries used in oxide planarization by CMP. In all cases, it is essential to devise methods for eliminating these impurities, by purity specification or other actions. The choice of the insulator, especially adjacent to device layers, should be one that is an effective diffusion barrier to ionic impurities and/or moisture (such as silicon nitride) or one that actively getters the ionic impurities such as PSG.

Plasma damage to gate oxides occurs in both etching and deposition. The damage has been said to be due to plasma nonuniformity leading to a local imbalance between electron and ion currents from the plasma, resulting in gate charging and subsequent gate oxide degradation (Fang and McVittie, 1992). The damage becomes more serious as the thickness of the gate dielectric is decreased. In an ECR plasma, it was shown that the charging occurred at the time the discharge was turned off (Samukawa, 1990); to reduce damage, the power can be ramped down at the end of processing. Modification of the magnets in an ECR reactor was shown to improve the magnetic flux and current densities at the wafer surface, thus reducing charge damage (Nakahira et al., 1996). Other kinds of RIE damage occurs during etching contact holes which exposes the Si surface. Some of these affects are surface residues (e.g., polymer films), loss of dopant or dopant activity, impurity penetration (particularly hydrogen), lattice damage, and heavy metal contamination. This subject has been discussed more completely by Oehrlein (1989, 1990). Deposition of SiO_2 by sputtering has been shown to cause threshold shifts in gate oxides (Larsen, 1980). The damage is probably caused by production of x-rays in the gate, generated by secondary electrons produced at the target (Groswald et al., 1971); reducing their energy (Logan et al., 1977) or placing a dc-biased mesh between target and substrate (Hazuki and Moriya, 1987) reduces the damage.

Flat-band shifts have been attributed to high-energy (VUV) photons in plasmas.

Outgassing of moisture from insulators cause n-channel field inversion (Yamaha et al., 1993) and hot-carrier degradation (Doki et al., 1994).

Alpha-particle radiation can affect storage states in memory devices; many of the metals used in the interconnection and cleaning solutions used in processing can introduce trace amounts of thorium and other radioactive impurities that can cause unacceptable failure rates, often requiring the use of error-correction circuits or radiation hardened device designs.

The use of compressive insulating films has been found to delay the onset of emitter current gain (beta) degradation in bipolar devices (Zalar 1981; Hemmer et al., 1982).

5.0 MANUFACTURABILITY

When selecting the materials and processes, issues of manufacturing and volume production should be kept in mind. Two key issues are yield and productivity and both affect the cost of products. Yield is improved if the materials and processes are compatible and the individual processes have large process windows. Clustering of reactor chambers and processes, in situ cleaning, real time process monitoring, all help in achieving better process control and high throughput. The process/

application requirement is often tempered by the availability of process controls and equipment capable of high throughput and low defect levels. For example, CMP of oxide for planarization has clearly the conceptual advantage of providing excellent local and global planarity, but its increase in use is fueled equally by the wide choice of commercially available polishers, pads, and slurry materials, alternate cleaning methods, and finally modifications that lead to better uniformity and process monitoring. Similarly, new SOG materials that seem to be more crack-resistant and less prone to absorb/desorb moisture increase their use as gap-fill dielectrics. With the increased use of studs in place of vias, CVD W, and more recently CVD Al and Cu, have become more commonly used since these processes provide good fill behavior. Deposition processes for Al-Cu or Al-Si are being modified to improve hole-filling capabilities; among them are elevated temperature and directional sputtering. Another path for enhanced hole-filling is the use of substrate-biased high-density plasma systems for metal and PECVD oxide deposition. Improved RIE reactors, using high-density plasmas, have been introduced to meet the need for etching vertical, high AR features; process modifications have been designed to minimize pattern-factor effects.

But the improvements obtained by using the advanced equipment and processes must be cost-effective.

6.0 CONCLUSIONS, PART II

The starting point for choosing conducting and insulating films as well as the processes for use in multilevel interconnection, in order to achieve the best performance from the device or chip, is often determined by modelling and simulation. But quickly, other issues of integration become an important considerations for deciding which processes and materials are feasible. The issues range from compatibility of materials and processes to manufacturability and affordability. Often, the changes are made in small increments to improve predictability of the results and begin the learning process; to recover the maximum benefit from the existing infrastructure and minimize new investment. It is often wise to wait for the new processes and equipment to mature.

REFERENCES

Abe, K., Y. Harada, H. Onada, *1995 VMIC*, 308 (1995).
Abraham, T., *1987 VMIC*, 115 (1987).
Achuthan, K., D. R. Campbell, S. V. Babu, *1995 DUMIC*, 177 (1995).
Adams, A. C., C. D. Capio, *J. Electrochem. Soc.*, 128, 423 (1981).
Ali, I., S. R. Roy, G. Shinn, C. Tipton, *1995 DUMIC*, 311 (1995).
Ali, I., S. R. Roy, G. Shinn, *Solid State Technol.*, 10/94, 63 (1994).
Allen, L. R., in "Proc. Symp. Highly Selective Dry Etching and Damage Control" (S. Mathad and H. Horiike, eds.), *Electrochem. Soc. Proc. Vol.*, PV 93-21, 255, 1993.
Amazawa, T., Y. Arita, *IEDM 91*, 265 (1991).
Amazawa, T., H. Nakamura, Y. Arita, *IEDM 88*, 442 (1988).
Ames, I., F. M. D'Heurle, R. E. Horstmann, US Pat. 3,725,309, 1973.
Asamaki, T., R. Mori, A. Takagi, *Jpn. J. Appl. Phys.*, 33, 2500 (1994).
Awaya, N., K. Ohono, M. Sato, *1990 VMIC*, 254 (1990).
Awaya, N., Y. Arita, *J. Electronic Mat.*, 21, 959 (1992).

Awaya, N., Y. Arita, *1993 Symp. on VLSI Technol.*, 125 (1993a).
Awaya, N., Y. Arita, *Jpn. J. Appl. Phys.*, *32*, 3915 (1993b).
Bacchetta, M., L. Bacci, N. Iazzi, N. Limburn, L. Zanotti, *1993 VMIC*, 153 (1993).
Bader, H. P., M. A. Lardon, *J. Vac. Sci. Technol.*, *A3*, 2167 (1985).
Bader, H.P., M. A. Lardon, *J. Vac. Sci. Technol.*, *B4*, 833 (1986a).
Bader, H. P., M. A. Lardon, *J. Vac. Sci. Technol.*, *B4*, 1192 (1986b).
Bader, H. P., H. A. Lardon, K. J. Hoefler, in "Multilevel Metallization, Interconnection, and Contact Technologies" (L. B. Rothman and T. Herndon, eds.), *Electrochem. Soc. Proc. Vol.*, *PV 87-4*, 185, 1987.
Bader, M. E., R. P. Hall, G. Strasser, *Solid State Technol.*, *5/90*, 149 (1990).
Bai, P., G.-R. Yang, T.-M. Lu, L. M. W. Lau, *J. Vac. Sci. Technol.*, *A8*, 1465 (1990).
Bang, D. S., J. P. McVittie, M. M. Islamraja, K. C. Saraswat, Z. Krivokapic, S. Ramaswami, R. Cheung, in "Proc. 10th Symp. Plasma Processing" (G. S. Mathad and D. W. Hess, eds.), *Electrochem. Soc. Proc. Vol.*, *PV 94-20*, 557, 1994.
Barklund, A. M., S. Berg, I. V. Katardjiev, C. Nender, P. Carlsson, *Vacuum*, *44*, 197 (1993).
Barth, H. J., M. Frank, S. Rohl, M. Schneegans, H. Wendt, C. D. Dobson, P. Rich, K. E. Buchanan, M. G. M. Harris, *1995 VMIC*, 52 (1995).
Bartush, T. A., US Pat. 4,541,169, 1985.
Bartush, T. A., *1987 VMIC*, 41 (1987).
Baseman, R. J., *J. Vac. Sci. Technol.*, *B8*, 84 (1990).
Baseman, R. J., F. E. Turene, *J. Vac. Sci. Technol.*, *B8*, 1097 (1990).
Baseman, R. J., J. C. Andreshak, R. H. Schnitzel, J. E. Cronin, *J. Vac. Sci. Technol.*, *B8*, 1158 (1990).
Berndlmaier, E., *IEEE Conference on Circuits & Computers*, *ICCC 80*, 1112 (1980).
Bernhardt, A. F., R. J. Contolini, D. B. Tuckerman, A. H. Weisberg, in "Laser and Particle-Beam-Processing on Surfaces" (G. L. Loper, A. W. Johnson, T. W. Siemon, eds.), *Mat. Res. Soc. Symp. Proc. Vol. 129*, 559, 1989.
Berthold, J., C. Wiecorek, *Appl. Surf. Sci.*, *38*, 506 (1989).
Beyer, K. D., W. L. Guthrie, S. Makarewicz, E. Mendel, W. J. Patrick, K. Perry, W. Pliskin, J. Riseman, P. M. Schaible, C. L. Standley, US Pat. 4,944,836, 1990.
Beyer, K. D., J. S. Makris, E. Mendel, K. A. Nummy, S. Ogura, J. Riseman, US Pat. 4,671,851, 1987.
Blech, I. A., *Thin Solid Films*, *6*, 113 (1970).
Blewer, R. S., V. A. Wells, *1984 VMIC*, 153 (1984).
Blumenthal, R., G. C. Smith, p. 47 in *Tungsten and Other Refractory Metals for VLSI Applications II* (E. K. Broadbent, ed.), Mat. Res. Soc., Pittsburgh, Penn., 1988.
Bogle-Rohwer, E., J. Nulty, *Advanced Techniques for IC Proc.*, SPIE Vol. 1392, 280 (1990).
Bornside, D. E., *J. Electrochem. Soc.*, *137*, 2589 (1990).
Bornside, D. E., R. A. Brown, S. Mittal, F. T. Geyling, *Appl. Phys. Lett.*, *58*, 1181 (1991).
Bothra, L., D. Baker, M. Weling, *1995 DUMIC*, 66 (1995).
Bothra, S., M. Weling, *1996 DUMIC*, 201 (1996).
Brader, S. J. H., J. Rogers, S. C. Quinlan, *1986 VMIC*, 93 (1986).
Broadbent, E. K., J. M. Flanner, W. G. M. van den Hoek, I.-W. H. Connick, *IEEE Trans. on Electron Devices*, *35*, 952 (1988).
Broughton, J. N., C. J. Backhouse, M. J. Brett, S. K. Dew, G. Este, *1995 VMIC*, 201 (1995).
Budinger, W. D., E. W. Jensen, US Pat. 4,927,432, 1990.
Bui, M. D., T. A. Streit, K. E. Schoenberg, R. H. Dorrance, P. P. Proctor, *1987 VMIC* 385 (1987).
Burke, P. A., *1991 VMIC*, 379 (1991).
Bushnell, L. P. Mc., L. V. Gregor, C. F. Lyons, *Solid State Technology*, *6/86*, 133 (1986).
Cadien, K. C., D. A. Feller, US Pat. 5,340,370, 1994.
Cale, T. S., G. B. Raupp, *J. Vac. Sci. Technol.*, *B8*, 649 (1990).

Calc, T. S., G. B. Raupp, *J. Vac. Sci. Technol.*, *B8*, 1242 (1990).
Cale, T. S., G. B. Raupp, T. H. Gandy, *J. Vac. Sci. Technol.*, *A10*, 1128 (1992).
Cale, T. S., G. B. Raupp, M. B. Chaara, F. A. Shemansky, *Thin Solid Films*, *220*, 66 (1992).
Carey, P. G., B. J. Woratschek, F. Bachman, *Appl. Phys. Lett.*, *57*, 1499 (1990).
Chang, C., J. P. McVittie, K. C. Saraswat, *Electrochem. Soc. Ext. Abstr.*, 413 *PV 91-1*, 634 (1991).
Chang, C., J. P. McVittie, K. C. Saraswat, *Electrochem. Soc. Ext. Abstr.*, 127 *PV 91-2*, 211 (1992).
Chao, C. C., W. V. Wang, p. 783 in *Polyimides* (K. I. Mittal, ed.), Plenum Press, New York, 1984.
Chapman, B., *Glow Discharge Processes*, John Wiley & Sons, New York, 1980.
Chatterjee, S., C. M. McConica, *J. Electrochem. Soc.*, *137*, 328 (1990).
Chen, L.-J., *1995 VMIC*, 283 (1995).
Chen, S., E. Ong, *SPIE V. 1190*, 207 (1989).
Chen, S. N., Y. C. Chao, J. J. Lon, Y. H. Tsai, F. C. Tseng, *1988 VMIC*, 306 (1988).
Chen, L. J., C. C. Diao, *1996 CMP-MIC*, 241 (1996).
Cheng, L-Y., J. P. McVittie, K. C. Saraswat, in "ULSI Science and Technology, 1989" (C. M. Osburn, J. M. Andrews, eds.), *Electrochem. Soc. Proc. Vol. PV 89-9*, 586, 1989.
Cheng, L.-Y., J. P. McVittie, K. C. Saraswat, *1990 VMIC*, 404 (1990).
Cheng, P. F., S. M. Rossnagel, D. N. Ruzic, *J. Vac. Sci. Technol.*, *B13*, 203 (1995).
Chiang, E. J. H., K. C. Wang, D. Lee, J. Carmody, V. Hoffman, A. Helms, Jr., *1994 VMIC*, 201 (1994).
Cho, J. S. H., H.-K. Kang, S. S. Wong, Y. Shacham-Diamand, *MRS Bulletin*, *6/93*, 31 (1993).
Cho, J. S. H., H.-K. Kang, I. Asano, S. S. Wong, *IEDM 92*, 297 (1992).
Cho, J. S. H., H.-K. Kang, M. A. Beiley, S. S. Wong, C. H. Ting, 1991 Symp. VLSI Technol., *Digest of Technical Papers*, 39, 1991.
Chow, M., J. Cronin, W. Guthrie, C. Kaanta, B. Luther, W. Patrick, K. Perry, C. Standley, US Pat. 4,789,648, 1988.
Chu, J. K., J. S. Multani, S. K. Mittal, *1986 VMIC*, 474 (1986).
Clark, T. E., M. Chang, C. Leung, *J. Vac. Sci. Technol.*, *B9*, 1478 (1991).
Clark, T. E., P. E. Riley, M. Chang, S. G. Ghanayem, C. Leung, A. Mak, *1990 VMIC*, 478 (1990).
Contolini, R. J., L. Tarte, R. T. Graff, L. B. Evans, J. N. Cox, J. D. Gee, X.-C. Mu, C. Chang, *1995 VMIC*, 322 (1995).
Cook, L. M., *J. Non-Crystalline Solids*, *120*, 152 (1990).
Cooke, M. J., R. A. Heinecke, R. C. Stern, *Solid State Technology*, *12/82*, 62 (1982).
Coronell, D. G., K. F. Jensen, *J. Electrochem. Soc.*, *141*, 2545 (1994).
Cote, W. J., US Pat. 4,910,155, 1990.
Cote, W. J., J. E. Cronin, W. R. Hill, C. A. Hoffman, US Pat. 5,308,438, 1994.
Cote, W. J., K. L. Holland, T. M. Wright, US Pat. 4,796,360, 1988.
Cote, W. J., P. Lee, T. Sandwick, B. Vollmer, V. Vynorius, S. Wolff, US Pat. 5,262,354, 1993.
Crapella, S., F. Gualandris, *J. Electrochem. Soc.*, *135*, 683 (1988).
Cuomo, J. J., S. M. Rossnagel, *J. Vac. Sci. Technol.*, *A4*, 393 (1986).
Currie, J. E., R. N. Schulz, US Pat. 5,267,418, 1993.
Dai, B.-T., C.-W. Liu, C.-F. Yeh, *1995 DUMIC*, 149 (1995).
Dalal H., R. V. Joshi, H. S. Rathore, R. Fillipi, *IEDM 93*, 273 (1993).
Daubenspeck, T. H., J. K. DeBrosse, C. W. Koburger, M. Armacost, J. B. Abernathy, *J. Electrochem. Soc.*, *138*, 506 (1991).
deBruin, L., J. M. F. G. Laarhoven, *1988 VMIC*, 404 (1988).
Demaray, E., J. van Gogh, R. Kolenkow, *1987 VMIC*, 371 (1987).

Denison, D. R., C. Chiang, D. B. Fraser, in "ULSI Science and Technol., 1989" (C. M. Osburn and J. M. Andrews, eds.), *Electrochem. Soc. Proc. Vol., PV 89-9*, 563, 1989.
Dew, S. K., *J. Appl. Phys.*, 76, 4857 (1994).
Dew, S. K., D. Liu, M. J. Brett, T. Smy, *J. Vac. Sci. Technol., B11*, 1281 (1993).
Dew, S. K., T. Smy, R. N. Tait, M. J. Brett, *J. Vac. Sci. Technol, A9*, 519 (1991).
Dixit, G. A., M. F. Chisholm, M. K. Jain, T. Weaver, L. M. Ting, S. Poarch, K. Mizobuchi, R. H. Havemann, C. D. Dobson, A. I. Jeffryes, P. J. Holverson, P. Rich, D. C. Butler, J. Helms, *IEDM 94*, 105 (1994).
Doki, M., H. Watatani, S. Okuda, Y. Furumura, *1994 VMIC*, 235 (1994).
Dubin, V. M., Y. Shacham-Diamand, B. Zhao, P. K. Vasudev, C. H. Ting, *1995 VMIC*, 315 (1995).
Edelstein, D. C., G. A. Sai-Halasz, Y. J. Mii, *IBM J. Res. Develp.*, 39, 384 (1995).
Ehara, K., T. Morimoto, S., Muramoto, S. Matsuo, *J. Electrochem. Soc.*, 131, 419 (1984).
Elkins, P., K. Reinhardt, R. Tang, *1986 VMIC*, 100 (1986).
Estabil, J. J., H. S. Rathore, F. Dorleans, *29th IEEE/IRPS*, 57 (1991).
Faith, T. J., *J. Appl. Phys.*, 52, 4630 (1981).
Fang, S., J. P. McVittie, *J. Appl. Phys.*, 72, 4865 (1992).
Farkas, J., M. J. Hampden-Smith, T. T. Kodas, *J. Electrochem. Soc.*, 141, 3539 (1994).
Feng, US Pat. 3,976,524, 1976.
Forester, L., B. Coenegrachts, M. Stone, H. Meynen, J. Grillaert, L. Van den hove, *1994 VMIC*, 172 (1994).
Forester, L., D. K. Choi, R. Hosseini, J. Lee, B. Tredinnick, K. Holland, T. Cale, *1995 VMIC*, 482 (1995).
Fried, L. J., J. Havas, J. S. Lechaton, J. S. Logan, G. Paal, P. A. Totta, *IBM J. Res. Dev.*, 26, 62 (1982).
Friedrich, L. J., D. S. Gardner, S. K. Dew, M. J. Brett, T. Smy, *1996 VMIC*, 213 (1996).
Fritzsche, J., V. Grewal, W. Henkel, *1986 VMIC*, 45 (1986).
Fujii, S., M. Fukumoto, G. Fuse, T. Ohzone, *IEEE Trans. on Electron Devices*, 35, 1829 (1988).
Fujino, K., Y. Egashira, Y. Shimogaki, H. Komiyama, *J. Electrochem. Soc.*, 140, 2309 (1993).
Fujiwara, H., K. Hara, M. Kamiya, T. Hashimoto, K. Okamoto, *Thin Solid Films*, 163, 387 (1988).
Fukuroda, A., K. Nakamura, Y. Arimoto, *IEDM 95*, 469 (1995).
Fury, M., *Solid State Technol.*, 4/95, 47 (1995).
Gardner, D. S., D. B. Fraser, *1995 VMIC*, 287 (1995).
Gardner, D. S., R. B. Beyers, T. L. Michalka, K. C. Saraswsat, T. W. Barbee, Jr., J. D. Meindl, *IEDM 84*, 114 (1984).
Geiger, W., A. Sharma, *1986 VMIC*, 128 (1986).
Gelatos, A. V., S. Poon, R. Marsh, C. J. Mogab, M. Thompson, *1993 Symp. VLSI Technol.*, 123, 1993.
Gelatos, A. V., R. Marsh, M. Kottke, C. J. Mogab, *Appl. Phys. Lett.*, 63, 2842 (1993).
Georgiou, G. E., K. P. Cheung, R. Liu, *1989 VMIC*, 315 (1989).
Gimpelson, G. E., C. L. Russo, *1984 VMIC*, 371 (1984).
Gittleman, B., P. Bai, G.-R. Yang, T.-M. Lu, C.-K. Hu, *J. Vac. Sci. Technol., A8*, 1514 (1990).
Gokan, H., M. Mukainaru, N. Endo, *J. Electrochem. Soc.*, 135, 1019 (1988).
Grivna, G., R. Goodner, *J. Electrochem. Soc.*, 141, 251 (1994).
Grivna, G., C. Leathersich, H. Shin, W. G. Cowden, *J. Vac. Sci. Technol., B11*, 55 (1993).
Grosewald, P., L. V. Gregor, R. Powlus, Paper 3.7 in *Proc. of the Intl. Electron Devices Mtg.*, Washington, D.C., 1971.
Gross, M., C. M. Horowitz, *J. Vac. Sci. Technol., B11*, 242 (1993).
Gu, J., M. D. Bullwinkel, G. A. Campbell, *J. Electrochem. Soc.*, 142, 907 (1995).
Gupta, S., I. Wagner, S. Hurwitt, L. Wharton, *Semiconductor Intl.*, 9/87, 126 (1987).

Haberland, H., M. Mall, M. Moseler, Y. Qiang, T. Reiners, Y. Thurner, *J. Vac. Sci. Technol.*, *A12*, 2925 (1994).
Hamaguchi, S., S. M. Rossnagel, *J. Vac. Sci. Technol.*, *B13*, 183 (1995).
Hamanaka, M., S. Dohmae, K. Fujiwara, M. Shishino, S. Mayuml, *32nd IEEE/IRPS*, 405 (1994).
Harada, H., I. Kato, T. Takada, K. Inayoshi, *Electrochem. Soc. Ext. Abstr. 185*, PV 90-1, 285 (1990).
Hariu, T., K. Watanabe, M. Inoue, T. Takada, H. Tsuchikawa, *27th IEEE/IRPS*, 210 (1989).
Harper, J. M. E., *Solid State Technol.*, *4/87*, 129 (1987).
Hasper, A., J. Holleman, J. Middelhoek, C. R. Kleijn, C. J. Hoogendoorn, *J. Electrochem. Soc.*, *138*, 1728 (1991).
Hausamann, C., P. Mokrisch, *1988 VMIC*, 293 (1988).
Hayashide, Y., M. Matsuura, M. Hirayama, T. Sasaki, S. Hirada, H. Kotani, *1995 VMIC*, 464 (1995).
Hazuki, Y., T. Moriya, K. M. Kashiwagi, *1982 Symp. VLSI Technol.*, *Digest of Papers*, Jpn. Soc. Appl. Phys., IEEE, Paper 2-1, 18 (1982).
Hazuki, Y., T. Moriya, *IEEE Transactions of Electron Device*, *ED-34*, 628 (1987).
Hemmert, R. S., G. S. Prokop, J. R. Lloyd, P. M. Smith, G. M. Calabrese, *J. Appl. Phys*, *53*, 4456 (1982).
Hirabayashi, H., M. Kinoshita, H. Kaneko, N. Hayasaska, M. Higuchi, K. Mase, J. Oshima, *1996 CMP-MIC*, 119 (1996).
Hirashita, N., I. Aikawa, T. Ajioka, M. Kobayakawa, F. Yokoyama, Y. Sakaya, *28th IEEE/IRPS*, 216 (1990).
Hirose, K., K. Kikuta, T. Yoshida, *IEDM 94*, 557 (1994).
Hoffman, V., J. Griswold, D. Mintz, D. Harra, *Thin Solid Films*, *153*, 369 (1987).
Holber, W. M., J. S. Logan, H. J. Grabarz, J. T. C. Yeh, J. B. O. Caughman, A. Sugerman, F. E. Turene, *J. Vac. Sci. Technol.*, *A11*, 2903 (1993).
Holland, K., B. Tredinnick, M. Hoffman, M. Hitchel, H. Nguyen, B. Kraus, *1995 DUMIC*, 338 (1995).
Homma, Y., S. Harada, T. Kaji, *J. Electrochem. Soc.*, *126*, 1531 (1979).
Homma, Y., S. Tsunekawa, *J. Electrochem. Soc.*, *132*, 1466 (1985).
Homma, T., Y. Murao, R. Yamaguchi, *J. Electrochem. Soc.*, *140*, 3599 (1993).
Homma, Y., S. Tsunekawa, A. Satou, T. Terada, *J. Electrochem. Soc.*, *140*, 855 (1993).
Homma, Y., T. Furusawa, K. Kusukawa, M. Nagasawa, Y. Nakamura, M. Saitou, H. Morishima, H. Sata, *1995 VMIC*, 457 (1995).
Hopwood, J., F. Qian, *J. Appl. Phys.*, *78*, 758 (1995).
Horn, M. W., *Solid State Technol.*, *11/91*, 57 (1991).
Hou, T. W., C. J. Mogab, R. S. Wagner, *J. Vac. Sci. Technol.*, *A1*, 1801 (1983).
Hsieh, J. J., D. E. Ibbotson, J. A. Mucha, D. L. Flamm, *1989 VMIC*, 411 (1989).
Hsieh, J. J., *J. Vac. Sci. Technol.*, *A11*, 78 (1993).
Ikegawa, M., J. Kobayashi, *J. Electrochem. Soc.*, *136*, 2982 (1989).
IslamRaja, M. M., C. Chang, J. P. McVittie, M. A. Cappelli, K. C. Saraswat, *J. Vac. Sci. Technol.*, *B11*, 720 (1993).
Ito, H., N. Kajita, S. Yamah, Y. Minowa, *Jpn. J. Appl. Phys.*, *30*, 3228 (1991).
Iyer, R., *J. Electrochem. Soc.*, *141*, 3151 (1994).
Izumitani, T., p. 115 in vol. 17 of "Treatise on Mat. Sci. Technol." (M. T. Tomazawa and R. H. Doremus, eds.), Academic Press, New York, 1979.
Jairath, R., J. Farkos, C. K. Huang, M. Stell, S.-M. Tzeng, *Solid State Technol.*, *7/94*, 71 (1994).
Jang, C., S. R. Chen, T. F. Klemme, J. Lerma, H. M. Naguib, *1987 VMIC*, 357 (1987).
Janzen, J. W., R. J. Hanestad, *Semiconductor International*, *6/96*, 147 (1996).
Johnson, L. F., K. A. Ingersoll, J. V. Dalton, *J. Vac. Sci. Technol.*, *B1*, 487 (1983).

Joshi, R. V., *IEEE Electron Device Lett.*, *16*, 233 (1995).
Joshi, R. V., J. J. Cuomo, H. Dalal, L. Hsu, US Pat. 5,300,813, 1994.
Joshi, R. V., M. J. Tejwani, K. V. Srikrishnan, US Pat. 5,420,069, 1995.
Kaanta, C., S. G. Bombardier, W. J. Cote, W. R. Hill, G. Kerszykowski, H. S. Landis, D. J. Poindexter, C. W. Pollard, G. H. Ross, J. G. Ryan, S. Wolff, J. E. Cronin, *1991 VMIC*, 144 (1991).
Kaanta, C., W. Cote, J. Cronin, K. Holland, P. Lee, T. Wright, *1988 VMIC*, 21 (1988).
Kamoshida, K., H. Nakamura, *Ext. Abstr. 19th Conf. on Solid State Devices and Materials*, 439 (1987).
Kaufman, F. D., D. B. Thompson, R. E. Broadie, M. A. Jaso, W. L. Guthrie, D. J. Pearson, M. B. Small, *J. Electrochem. Soc.*, *138*, 3460 (1991).
Kaufman, F. B., S. A. Cohen, M. A. Jaso, in "Ultraclean Semiconductor Processing Technology and Surface Chemical Cleaning and Passivation" (M. Liehr, M. Hirose, M. Heyns, H. Parks, eds.), *Mat. Res. Soc. Symp. Proc. Vol. 386*, 85 (1995).
Kawai, M. K. Matsuda, K. Miki, K. Sakiyama, *1988 VMIC*, 419 (1988).
Kern, W., J. Hartman, *Thin Solid Films*, *206*, 64 (1991).
Kiang, M.-H., C. A. Pico, M. A. Lieberman, N. W. Cheung, X. Y. Qian, K. M. Yu, in "Low Energy Ion Beam and Plasma Modification of Materials" (J. M. E. Harper, K. Miyake, J. R. McNeil, S. M. Gorbatkin, eds.), *Mat. Res. Soc. Symp. Proc. Vol. 223*, 37, 1991.
Kikuta, K., T. Nakajima, K. Ueno, T. Kikkawa, *IEDM 93*, 285 (1993).
Kim, D.-H., R. H. Wenyorf, W. N. Gill, *J. Vac. Sci. Technol.*, *A12*, 153 (1994).
Kim, J., S. P. Kowalczyk, Y. H. Kim, N. J. Chou, and T. S. Oh, in "Advanced Electronics Packaging Materials" (A. Barfknecht, J. Partridge, C. J. Chen, C.-Y. Li, eds.), *Mat. Res. Soc. Symp. Proc. Vol. 167*, 137, 1990.
Kinsbron, E., W. E. Willowbrook, H. J. Levenstein, in "VLSI Science and Technology/1982" (C. J. Dell'Oca, W. M. Bullis, eds.), *Electrochem. Soc. Proc. Vol.*, *PV 82-7*, 116, 1982.
Kishimoto, K., M. Suzuki, T. Hirayama, Y. Ikeda, Y. Numasawa, *1992 VMIC*, 149 (1992).
Knauer, W., R. L. Poeschel, *J. Vac. Sci. Technol.*, *B6*, 456 (1988).
Kojima, H., T. Iwamori, Y. Sakata, t. Yamashita, Y. Yatsuda, *1988 VMIC*, 390 (1988).
Kolenkow, R., R. Nagahara, *Solid State Technol.*, *6/92*, 112 (1992).
Kotani, H., H. Yakushiji, H. Harada, K. Tsukamoto, T. Nishioka, *J. Electrochem. Soc.*, *130*, 645 (1983).
Koyama, L., M. Thomas, *1985 VMIC*, 45 (1985).
Krishnan, A., C. Xie, N. Kumar, J. Curry, D. Duane, S. P. Murarka, *1992 VMIC*, 226 (1992).
Kubo, T., US Pat. 4,614,563, 9/30/86 1986.
Kulper, L. L., US Pat. 3,609,470, 1971.
Kwakman, L. F., Tz, D. Huibretgtse, H. W. Piekaar, E. H. A. Granneman, K. P. Cheung, A. H. Labun, *J. Vac. Sci. Technol.*, *B12*, 3138 (1994).
Lai, W. Y.-C., G. L. Miller, R. J. Schutz, G. Smolinsky, E. R. Wagner, *1993 VMIC*, 147 (1993).
Lakshminarayanan, A., J. Steigerwald, D. Price, M. Bourgeois, T. P. Chow, R. J. Gutmann, S. P. Murarka, *1994 VMIC*, 49 (1994a).
Lakshminarayanan, S., J. Steigerwald, D. T. Price, M. Bourgeois, T. P. Chow, *IEEE Electron Device Lett.*, *15*, 307 (1994b).
Landis, H., P. Burke, W. Cote, W. Hill, C. Hoffman, C. Kaanta, C. Koburger, W. Lange, M. Leach, S. Luce, *Thin Solid Films*, *220*, 1 (1992).
Lardon, M. A., H. P. Bader, K. J. Hoelfer, *1986 VMIC*, 212 (1986).
Larsen, R. *IBM J. Res. Dev.*, *24*, 268 (1980).
Lassig, S., J. Li, J. McVittie, *1995 DUMIC*, 190 (1995).
Lee, P.-I., J. Cronin, C. Kaanta, *J. Electrochem. Soc.*, *136*, 2108 (1989).
Lee J.-J., D. Hartman, *1987 VMIC*, 193 (1987).

Lee, S., H. Yoo, Y.-W. Kim, B. Kim, *1995 DUMIC*, 106 (1995).
Lee, J.-H., K. Kim, W. Kim, S. Shin, H.-J. Kim, K. Suh, Y.-K. Jun, *1993 VMIC*, 182 (1993).
Lee, Y. H., N. J. Zhou, *J. Electrochem. Soc.*, *138*, 2429 (1991).
LeGoues, F. K., B. D. Silverman, P. S. Ho, *J. Vac. Sci. Technol.*, A6, 2200 (1988).
Leon, E. A., *IEEE Trans. on Computer-Aided Design*, 7, 168 (1988).
Levin, R. M. & K. Evans-Lutterodt, *J. Vac. Sci. Technol.*, B1, 54 (1983).
Liao, H., G. Thallikar, F. R. Meyers, T. S. Cale, *1995 DUMIC*, 204 (1995).
Lii, Y. T., H. Y. Ng, *ECS Ext. Abstr.* 222, PV 90-2 (1990).
Lim, S. W., Y. Shimogaki, Y. Nakano, K. Tada, H. Komiayama, *Ext. Abstr. 1995 Intl. Conf. Solid State Devices and Materials*, Osaka, 163 (1995).
Lin, T., K. Y. Ahn, J. M. E. Harper, P. N. Chaloux, *1988 VMIC*, 76 (1988).
Lin, B. J., p. 219 in "Introduction to Microlithography" (L. F. Thompson, C. G. Wilson, M. J. Bowder, eds.), *Amer. Chem. Soc. Symp. Series*, 1983.
Lin, Y-C., A. J. Purdes, S. A. Saller, W. R. Hunter, *J. Appl. Phys.*, 55, 1110 (1984).
Lin, A. W., *IEEE Trans. Components, Hybrids and Manuf. Tech.*, 13, 207 (1990).
Litvak, H., *1995 DUMIC*, 171 (1995).
Liu, D., S. K. Dew, M. J. Brett, T. Janacek, T. Smy, W. Tsai, *J. Appl. Phys.*, 74, 1339 (1993).
Liu, R., K. P. Cheung, W. Y.-C. Lai, R. Heim, *1989 VMIC*, 329 (1989).
Lloyd, J. R., *Thin Solid Films*, 91, 175 (1982).
Lloyd, J. R., R. N. Steagall, *J. Appl. Phys.*, 60, 1235 (1986).
Logan, J. S., M. H. Hait, H. C. Jones, G. R. Firth, D. B. Thompson, *J. Vac. Sci. Technol*, A7, 1392 (1989).
Logan, J. S., J. M. Keller, R. G. Simmons, *J. Vac. Sci. Technol.*, 14, 92 (1977).
Luo, Q., D. R. Campbell, S. V. Babu, *1996 CMP-MIC*, 145 (1996).
Lustig, N. E., R. M. Feenstra, W. L. Guthrie, US Pat. 5,337,015, 1994.
Lustig, N. E., K. Saenger, H. M. Tong, US Pat. 5,433,651, 1995.
Luther, B., J. F. White, C. Uzoh, T. Cacouris, J. Hummel, W. Guthrie, N. Lustig, S. Greco, S. Zuhoski, P. Agnello, E. Colgan, S. Mathad, L. Saraf, E. J. Weitzman, C. K. Hu, F. Kaufman, M. Jaso, L. P. Buchwalter, S. Reynolds, C. Smart, D. Edelstein, E. Baran, S. Cohen, C. M. Knoedler, J. Malinowski, J. Horkans, H. Deligianni, J. Harper, P. C. Andriacaos, J. Paraszczak, D. J. Pearson, M. Small, *1993 VMIC*, 15 (1993).
Machida, K., H. Oikawa, *J. Vac. Sci. Technol.*, B4, 818 (1986).
Malik, F., R. Solanski, *Thin Solid Films*, 193/194, 1030 (1990).
Marella, P. F., D. B. Tuckerman, R. F. Pease, *Appl. Phys. Lett.*, 54, 1109 (1989).
Marella, P. F., D. B. Tuckerman, R. F. Pease, *Appl. Phys. Lett.*, 56, 2625 (1990a).
Marella, P. F., D. B. Tuckerman, R. F. Pease, *J. Vac. Sci. Technol.*, B8, 1780 (1990b).
Mariu, T., K. Watanabe, M. Inoue, T. Takada, H. Tsuchikawa, *27th IEEE/IRPS*, 210 (1989).
Marks, J., K. Law, D. Wang, *1989 VMIC*, 89 (1989).
Marx, W. F., Y. Ra, R. Yang, C.-H. Chen, *J. Vac. Sci. Technol.*, A12, 3087 (1994).
Matsuda, T., M. J. Shapiro, S. V. Nguyen, *1995 DUMIC*, 22 (1995).
Mayumi, S., K. Fujiwara, S. Nishida, S. Ueda, M. Inque, *Jpn. J. Appl. Phys.*, 27, 280 (1988).
Mayumi, S., M. Shishino, Y. Hata, S. Ueda, W. Inoue, *1987 VMIC*, 78 (1987).
Mayumi, S., Y. Hata, K. Hujiwara, Seiji Ueda, *J. Electrochem. Soc.*, 137, 2534 (1990).
McBrayer, J. D., R. M. Swanson, T. W. Sigman, *J. Electrochem. Soc.*, 133, 1242 (1986).
McConica, C. M., S. Churchill, p. 257 in Tungsten and Other Refractory Metals for VLSI Applications III (V. A. Wells, ed.), *Mat. Res. Soc.*, Pittsburgh, Penn., 1988.
McEachern, R. L., W. L. Brown, M. F. Jarrold, M. Sosnowski, G. Takaoka, H. Usui, I. Yamada, *J. Vac. Sci. Technol.*, A9, 3105 (1991).
McInerney, E. J., *1986 VMIC*, 467 (1986).
McInerney, E. J., S. C. Avanzino, *IEEE Trans. on Electron Devices*, ED-34, 615 (1987).
Mehta, S., G. Sharma, *1989 VMIC*, 80 (1989).
Mei, S.-N., T.-M. Lu, S. Robert, *IEEE Electron Device Lett.*, EDL-8, 503 (1987).

Mei, S.-N., S.-N. Yang, T.-M. Lu, S. Roberts, *AIP Conf. Proc.*, *167*, 299 (1988).
Meikle, S., T. T. Doan, US Pat. 5,439,551, 1995.
Mercier, J. S., H. M. Naguib, V. Q. Ho, H. Nentwich, *J. Electrochem. Soc.*, *132*, 1219 (1985).
Mihara, S., M. Nakamura, in "10th Symp. Plasma Processing" (G. S. Mathad and D. W. Hess, eds.), *Electrochem. Soc. Proc. Vol.*, *PV 94-20*, 449, 1994.
Miller, G. L., E. R. Wagner, US Pat 5,081,421, 1992.
Misawa, Y., N. Kinjo, M. Hirao, N. Homma, *IEEE Tran. on Electron Devices*, *ED 34*, 621 (1987).
Mitsuhashi, M., *SEMI*, Japan, *Paper E1-1* (1986).
Mogami, T., M. Morimoto, H. Okabayashi, E. Nagasawa, *J. Vac. Sci. Technol.*, *B3*, 857 (1986).
Mogami, T., H. Okabayashi, E. Nagasawam, M. Morimoto, *1985 VMIC*, 17 (1985).
Molnar, L. D., *Semiconductor Intl.*, *8/89*, 92 (1989).
Monkowski, J. R., M. A. Logan, L. F. Wright, *Electronic Packaging Materials Sci. V. Symp. 1991*, 221 (1991).
Monkowski, J. R., M. A. Logan, US Pat. 5,104,482, 1992.
Moreau, W. M., p. 364 in *Semiconductor Lithography*, Plenum Press, New York, 1988.
Morgan, W. L., *Appl. Phys. Lett.*, *55*, 107 (1989).
Morimoto, S., US Pat. 5,104,828, 1992.
Morimoto, S., R. Breivogel, R. Gasser, S. Louke, P. Moon, R. Patterson, M. Prince, *Electrochem. Soc. Ext. Abstr. 297*, *PV 93-1*, 449 (1993).
Morimoto, S., S. Q. Grant, *1988 VMIC*, 411 (1988).
Mukai, R., S. Ozawa, *J. Vac. Sci. Technol.*, *B12*, 2826 (1994).
Mukai, R., N. Sasaki, M. Nakano, *IEEE Electron Device Lett.*, *EDL-3*, 76 (1987).
Mukai, R., K. Kobayashi, M. Nakano, *1988 VMIC*, 101 (1988).
Mukai, R., M. Iizuka, H. Kudo, M. Nakano, *1991 VMIC*, 192 (1991).
Murphy, J. J., J. Farkes, C. L. Markert, R. Jairath, US Pat. 5,478,435, 1995.
Naguib, N., C. Jang, T. F. Klemme, K. Wong, A. Rangappan, W. W. Yao, R. T. Fulks, *1987 VMIC*, 93 (1987).
Nakahira, J., Nakagi, M. Yamada, Y. Furumura, *1996 DUMIC*, 160 (1996).
Nanz, G., L. E. Camilletti, *IEEE Trans. on Semiconductor Manuf.*, *8*, 382 (1995).
Neureither, B., F. Binder, E. Fisher, Z. Gabric, K. Koller, S. Rohl, *1994 VMIC*, 151 (1994).
Neureither, B., C. Basa, T. Sandwick, K. Blumenstock, *J. Electrochem. Soc.*, *140*, 3607.
Nishida, T., K. Mukai, T. Inata, I. Tezuka, N. Horie, *23rd IEEE/IRPS*, 148 (1985).
Nishida, T., M. Saito, S. Iijima, *1989 VMIC*, 19 (1989).
Nishimoto, Y., T. Tokumasa, K. Maeda, *1995 DUMIC*, 15 (1995).
Nishimura, H., T. Yamada, R. Sinclair, S.-i. Ogawa, 1992 Symp. VLSI Technol., *Digest of Tech. Papers*, 74 (1992).
Nishimura, H., S. Ogawa, T. Yamada, *J. Vac. Sci. Technol.*, *B13*, 198 (1995).
Nishio, H., A. Shimizu, K. Watanabe, K. Kobayashi, *1994 VMIC*, 165 (1994).
Nowicki, R. S., US Pat. 5,330,607, 1994.
Oehrlein, G. S., *Mat. Sci. and Engr.*, *B4*, 441 (1989).
Oehrlein, G. S., in Chap. 8, *Handbook of Plasma Processing Technology*, (S. M. Rossnagel, J. J. Cuomo, W. D. Westwood, eds.), Noyes Publications, New Jersey, 1990.
Ohashi, N., H. Nezu, T. Fujiwara, N. Owada, *1994 VMIC*, 137 (1994).
Ohta, T., N. Takeysau, E. Kondoh, Y. Kawano, H. Yamamoto, *1994 VMIC*, 329 (1994).
Ohtani, H., M. Murota, M. Norishima, K. Shibata, M. Kakumu, *1995 VMIC*, 447 (1995).
Olmer, L. J., C. A. Daverse, *Electrochem. Soc. Ext. Abstr. 132*, *PV 90-1*, 193 (1990).
Ong, E., H. Chu, S. Chen, *Solid State Technol.*, *8/91*, 63 (1991).
Ono, T., H. Nishimura, M. Shimada, S. Matsuo, *J. Vac. Sci. Technol.*, *A12*, 1281 (1994).
Ono, H., Y. Ushiku, T. Yoda, *1990 VMIC*, 76 (1990).
Onuki, J., M. Nehei, *Appl. Phys. Lett.*, *63*, 1798 (1993).

Pai, P.-L., C. G. Konitzer, *1992 VMIC*, 213 (1992).
Pampalone, T. R., J. J. DiPiazza, D. P.Kanen, *J. Electrochem. Soc.*, *133*, 2394 (1986).
Paraszczak, J., D. Edelstein, S. Cohen, E. Babich, J. Hummel, *IEDM 93*, 261 (1993).
Parekh, N., R. Allen, W. Yao, R. Fulks, in "Multilevel Metallization, Interconnection and Contact Technologies" (L. P. Rothman, T. Herndon, eds.), *Electrochem. Soc. Proc. Vol. PV 87-4*, 221, 1987.
Parekh, N., A. Butler, W. Doedal, W. Heesters, L. Forester, *1990 VMIC*, 453 (1990).
Parekh, N., J. Price, *J. Electrochemical Soc.*, *137*, 2199 (1990).
Park, N.-H., J.-W. Park, S.-C. Chung, W.-S. Kim, K.-G. Rha, W. S. Kim, *1995 DUMIC*, 73 (1995).
Park, I. S., S. I. Lee, W. S. Jung, G. H. Choi, C. S. Park, S. T. Ahn, M. Y. Lee, Y. K. Kim, R. Reynolds, *IEDM 94*, 109 (1994).
Park, Y. H., F. T. A. Zold, J. F. Smith, *Thin Solid Films*, *129*, 309 (1985).
Park, C. S., S. I. Lee, J. H. Park, J. H. Sohn, D. Chin, J. G. Lee, *1991 VMIC*, 326 (1991).
Patrick, W. J., private communication (1991).
Patrick, W. J., W. L. Guthrie, C. L. Standley, P. M. Schaible, *J. Electrochem. Soc.*, *138*, 1778 (1991).
Pease, R. F. W., P. F. Marella, D. B. Tuckerman, A. H. Weisberg, *Electrochem. Soc. Ext. Abstr. 239, PV 91-2*, 348 (1991).
Pennington, S., D. Hallock, *1990 VMIC*, 71 (1990).
Pennington, S. L., S. E. Luce, D. P. Hallock, *1989 VMIC*, 355 (1989).
Poon, S. S., A. V. Gelatos, US Pat. 5,064,683 (1991).
Poon, S., A. Gelatos, A. H. Perera, M. Hoffman, *1993 Symp. VLSI Technol.*, Kyoto, 115, 1993.
Posadowski, W. M., Z. J. Radzimski, *J. Vac. Sci. Technol.*, *A11*, 2980 (1993).
Qian, L., M. C. Schmidt, G. Nobinger, J. Cassillas, J. T. Pye, H. W. Fry, *1995 DUMIC*, 50 (1995).
Rabinzohn, P., C. Villalon, F. Pasualini, S. Gourrier, *Microelectronic Engr.*, *11*, 599 (1990).
Raghavan, G., C. Chiang, P. B. Anders, S. M. Tzeng, R. Villasol, G. Bai, M. Bohr, D. B. Fraser, *Thin Solid Films*, *262*, 168 (1995).
Ramanarayanan, R., J. Wong, T.-M. Lu, D. Skelly, *J. Vac. Sci. Technol.*, *B4*, 1180 (1986).
Ramanarayanan, R., K. Polasko, D. Skelly, J. Wong, S.-N. Mei and T.-M. Lu, *J. Vac. Sci. Technol.*, *B5*, 359 (1987).
Ray, G. W., P. J. Marcoux, *1985 VMIC*, 52 (1985).
Renteln, P., *1996 CMP-MIC*, 217 (1996).
Renteln, P., M. E., Thomas, J. M. Pierce, *1990 VMIC*, 57 (1990).
Rey, J. C., L.-Y. Cheng, J. P. McVittie, K. C. Saraswat, *1990 VMIC*, 425 (1990).
Rey, J. C., L.-Y. Cheng, J. P. McVittie, K. G. Saraswat, *J. Vac. Sci. Technol.*, *A9*, 1083 (1991).
Riley, P. E., E. D. Castel, in "Multilevel Metallization, Interconnection and Contact Technologies, (L. B. Rothman and T. Herndon, eds.), *Electrochem. Soc. Proc. Vol., PV 87-4*, 259, 1987.
Riley, P. E., E. D. Castel, *IEEE Trans. Semiconductor Manuf.*, *1*, 154 (1988).
Riley, P. E., P. Bayer, *J. Electrochem. Soc.*, *137*, 2227 (1990).
Roehl, S., L. Camilletti, W. Cote, E. Eckstein, K. H. Froehner, P. T. Lee, D. Restaino, G. Roeska, V. Vynorius, S. Wolff, B. Vollmer, *1992 VMIC*, 22 (1992).
Romero, J. D., M. Khan, H. Fatemi, J. Turlo, *J. Mater. Res.*, *6*, 1996 (1991).
Ross, R. C., J. L. Vossen, *J. Appl. Phys.*, *45*, 239 (1984).
Rossnagel, S. M., D. Mikalsen, H. Kinoshita, J. J. Cuomo, *J. Vac. Sci. Technol.*, *A9*, 261 (1991).
Rossnagel, S. M., *J. Vac. Sci. Technol.*, *B13*, 183 (1995).
Rossnagel, S. M., J. Hopwood, *J. Appl. Phys.*, *63*, 3285 (1993).

Rothman, L., *J. Electrochem. Soc.*, *130*, 1131 (1983).
Rothman, L., P. M. Schaible, G. C. Schwartz, *1985 VMIC*, 131 (1985).
Rothman, L. B., *J. Electrochem. Soc.*, *127*, 2216 (1980).
Roy, S. R., A. Ali, G. Shinn, N. Furusawa, R. Shah, S. Peterman, K. Witt, S. Eastman, *J. Electrochem. Soc.*, *142*, 216 (1995).
Runnels, S. R., L. M. Eyman, *J. Electrochem. Soc.*, *141*, 1698 (1994).
Runnels, S. R., T. Olavson, *J. Electrochem. Soc.*, *142*, 2032 (1995).
Rutten, M., P. Feeney, R. Cheek, W. Landres, *1995 VMIC*, 491 (1995).
Saia, R. J., B. Gorowitz, D. Woodruff, D. M. Brown, *J. Electrochem. Soc.*, *135*, 936 (1987).
Saia, R. J., B. Gorowitz, D. Woodruff, D. M. Brown, in "Proc. 6th Symp. Plasma Processing" (G. S. Mathad, G. C. Schwartz, R. A. Gotscho, eds.), *Electrochem. Soc. Proc. Vol.*, PV 87-6, 173, 1987.
Samukawa, S., *Jpn. J. Appl. Phys.*, *29*, 980 (1990).
Sandu, G. S., L. D. Schultz, T. T. Doan, US Pat. 5,069,002, 1991.
Sasaki, N., A. Anzai, K. Uehara, *IEDM 83*, 546 (1983).
Sato, M., K. S. Harada, A. Saiki, T. Kimura, T. Okubo, K. Mukai, *IEEE Trans. Parts, Hybrids, Packaging*, PHP-9, 176 (1973).
Sato, M., Y. Arita, *Jpn. J. Appl. Phys.*, *25*, L764 (1986).
Sato, K., T. Oi, H. Matsumaru, T. Okubo, T. Nishimura, *Met. Trans.*, *2*, 691 (1971).
Schafft, H. A., C. D. Younkins, T. C. Grant, C.-Y. Kao, A. N. Saxena, *22nd IEEE/IRPS*, 250 (1984).
Schlitz, A., *Microelectronic Engr.*, *5*, 413 (1986).
Schlitz, A., M. Pons, *J. Electrochem. Soc.*, *133*, 178 (1986).
Schlitz, A., P. Abraham, E. Dechenaux, *J. Electrochem. Soc.*, *134*, 190 (1987).
Schmitz, J. E. J., R. C. Ellwanger, A. J. M. vanDijk, p. 55 in *Tungsten and Other Refractory Metals for VLSI Applications III*, (V. A. Wells, ed.), Mat. Res. Soc., Pittsburgh, Penn., 1988.
Schmitz, J. E. J., A. J. M. vanDijk, M. W. M. Graef, in "Proc. 10th Intl. Conf. CVD" (G. W. Cullen, ed.), *Electrochem. Soc. Proc. Vol.*, PV 87-8, 625, 1987.
Schmitz, J. E. J., A. Hasper, *J. Electrochem. Soc.*, *140*, 2112 (1993).
Schroder, K.-W., J. Schlote, S. Hinrich, *J. Electrochem. Soc.*, *138*, 2466 (1991).
Schultz, L. D., US Pat. RE34425, 1993.
Schwartz, G. C., V. Platter, unpublished (1974).
Schwartz, G. C., *J. Electrochem. Soc.*, *138*, 621 (1991).
Schwartz, G. C., P. Johns, *J. Electrochem. Soc.*, *139*, 927 (1992).
Schwartz, G. C., Y.-S. Huang, W. J. Patrick, *J. Electrochem. Soc.*, *139*, L118 (1992).
Selwyn, G. S., J. Singh, R. S. Bennett, *J. Vac. Sci. Technol.*, A7, 2758 (1989).
Shacham-Diamand, Y., V. M. Dubin, C. H. Ting, P. K. Vasudev, B. Zhao, *1995 VMIC*, 334 (1995).
Sheldon, D. J., C. W. Gruenschlaeger, L. Kammerdiner, N. B. Henis, P. Kelleher, J. D. Hayden, *IEEE Trans. Semiconductor Manuf.*, *1*, 140 (1988).
Shih, Y. C., C. S. Pai, K. G. Steiner, W. G. Wilkins, *1992 VMIC*, 109 (1992).
Shimogaki, Y., S. W. Lim, M. Miyata, Y. Nakano, K. Tada, H. Komiyana, *1996 DUMIC*, 36 (1996).
Shterenfeld-Lavie, Z., I. Rabinovich, J. Levy, A. Haim, C. Dobson, K. Buchanan, P. Rich, D. J. Thomas, *1995 VMIC*, 31 (1995).
Singh, B., O. Mesker, D. Devlin, *J. Vac. Sci. Technol.*, B5, 567 (1987).
Sivaram, S., H. Bath, R. Leggett, A. Maury, K. Monnig, R. Tolles, *Solid State Technol.*, 5/92, 87 (1992).
Skelly, D. W., L. A. Gruenke, *J. Vac. Sci. Technol.*, A3, 457 (1986).
Smith, G. C., *1985 VMIC*, 350 (1985).
Smith, G. C., A. J. Purdes, *J. Electrochem. Soc.*, *132*, 2721 (1985).

Smith, J. F., F. T. Zold, W. Class, *Thin Solid Films*, 96, 291 (1982).
Smith, J. F., *Solid State Technol.*, 1/84, 135 (1984).
Smy, T., S. Dew, M. J. Brett, *1992 VMIC*, 465 (1992).
Smy, T., S. K. Dew, M. J. Brett, W. Tsai, M. Biberger, K. C. Chen, S. T. Hsai, *1994 VMIC*, 371 (1994).
Smy, T., K. L. Westra, M. J. Brett, *IEEE Trans. Electron Devices*, 37, 591 (1990).
Sorita, T., S. Shiga, K. Ikuta, Y. Egashira, H. Komiayama, *J. Electrochem. Soc.*, 140, 2952 (1993).
Sorlie, C., M. J. Brett, S. K. Dew, T. Smy, *Solid State Technol.*, 6/95, 101 (1995).
Speedfam Corporation, Chandler, Arizona.
Spindler, O., B. Neureither, *Thin Solid Films*, 74, 67 (1989).
Steigerwald, J. M., S. P. Murarka, R. J. Gutmann, D. J. Duquette, *J. Electrochem. Soc.*, 141, 3512 (1994a).
Steigerwald, J. M., R. Zirpoli, S. P. Murarka, D. Price, R. J. Gutmann, *J. Electrochem. Soc.*, 141, 2842 (1994b).
Stillwagon, L. E., *Solid State Technol.*, 6/87, 67 (1987a).
Stillwagon, L. E., R. G. Larson, G. N. Taylor, *J. Electrochem. Soc.*, 134, 2030 (1987b).
Stillwagon, L. E., R. G. Larson, *J. Appl. Phys.*, 63, 5251 (1988).
Stillwagon, L. E., R. G. Larson, *Phys. Fluids, A4*, 895 (1992).
Sugai, K., T. Shinzawa, S. Kishida, H. Okabayashi, Y. Murao, T. Kobayashi, N. Hosokawa, T. Yako, H. Kadokura, M. Isemura, K. Kamio, *1993 VMIC*, 463 (1993).
Sukanek, P. C., *J. Electrochem. Soc.*, 136, 3019 (1989).
Sum, J. C., G. W. Ray, S.-H. Hsu, P. J. Matcoux, J. Kruger, C.-H. Liu, S. Peng., in "Multilevel Metallization, Interconnection, and Contact Technologies" (L. B. Rothman and T. Herndon, eds.), *Electrochem. Soc. Proc. Vol. PV 87-4*, 259, 1987.
Sumitomo, Y., Y. Ohashi, US Pat. 4,123,565, 1978.
Sun, M., H. M. Tzeng, H. Litvak, D. Glenn, *1996 CMP-MIC*, 169 (1996).
Suzuki, M., T. Homma, H. Koga, T. Tanigawa, Y. Murao, *IEDM 92*, 293 (1992).
Taguchi, M., K. Koyama, Y. Sugano, *1992 VMIC*, 219 (1992).
Tait, R. N., T. Smy, M. J. Brett, *Thin Solid Films*, 187, 375 (1990).
Tait, R. N., T. Smy, M. J. Brett, *Thin Solid Films*, 226, 196 (1993).
Takagi, *Vacuum*, 36, 27 (1986).
Takeyasu, N., Y. Kawano, E. Kondoh, T. Katagiri, H. Yamamoto, H. Shinriki, T. Ohta, *Jpn. J. Appl. Phys.*, 33, 424 (1994).
Talieh, H., E. Ong, H. Kieu, A. Tepman, *1993 VMIC*, 211 (1993).
Taniguchi, H., O. Sugiura, *Jpn. J. Appl. Phys.*, 33, L1845 (1994).
Thallikar, G., H. Liao, T. S. Cale, F. R. Myers, *J. Vac. Sci. Technol.*, B13, 1875 (1995).
Thomas, M. E., US Pat. 5,091,048, 1992.
Thompson, L. F., *Introduction to Microlithography*, 2nd ed., (L. F. Thompson, C. G. Wilson, M. J. Bowden, eds.), American Chemical Society, Washington, DC, 1994.
Ting, C. Y., V. J. Vivalda, H. G. Schaefer, *J. Vac. Sci. Technol.*, 15, 1105 (1978).
Ting, C. H., in "Proc. Symp. Electroless Deposition Metals Alloys" (M. Paunovic, I. Ohna, eds.), *Electrochem. Soc. Proc. Vol.*, PV 88-12, 223 (1988).
Tisier, A., J. Khallaayoune, A. Gerodolle, B. Huizing, *J. dePhysique IV*, C2, 437 (1991).
Toi, S. S., F. Choi, *1985 VMIC*, 138 (1985).
Tokunaga, K., K. Sugawara, *J. Electrochem. Soc.*, 138, 176 (1991).
Tolles, R., B. Guthrie, S. H. Ko, P. Cheng, T. Osterheld, S. Askari, M. Birang, *1996 CMP-MIC*, 201 (1996).
Tsai, Y. H., S. L. Hsu, F. C. Tseng, C. S. Yoo, *Thin Solid Films*, 185, 363 (1990).
Tsubouchi, K., K. Masu, *J. Vac. Sci. Technol.*, A10, 856 (1992).
Tsubouchi, K., K. Masu, N. Shigeeda, T. Matano, Y. Hiura, N. Mikoshiba, S. Matsumoto, T. Asaba, T. Mauri, T. Kajikawa, *1990 Symp. on VLSI Technol.*, 5 (1990a).

Tsubouchi, K., K. Masu, N. Shigeeda, T. Matano, Y. Hiura, N. Mikoshiba, *Appl. Phys. Lett.*, 57, 1221 (1990b).
Tuckerman, D. B., R. L. Schmidt, *1985 VMIC*, 24 (1985).
Tuckerman, D. B., A. H. Weisberg, *IEEE Electron Device Lett.*, *EDL-7*, 1 (1986a).
Tuckerman, D. B., A. H. Weisberg, *Solid State Technol.*, 4/86, 129 (1986b).
Turner, G. M., S. M. Rossnagel, J. J. Cuomo, *J. Vac. Sci. Technol.*, *A11*, 2796 (1993).
Tuttle, M. E., US Pat. 5,177,908, 1993.
Ueno, K., K. Ohto, K. Tsunenari, K. Kajiyana, K. Kikuta, T. Kikkawa, *IEDM 92*, 305 (1992).
Umimoto, H., S. Odanaka, S. Imai, *Symp. VLSI Technol.*, Digest of Papers, 99 (1991).
Usami, T., K. Shimokawa, M. Yoshimura, *Jpn. J. Appl. Phys.*, 33, 408 (1994).
Uttrecht, R. R., R. M. Geffken, *1991 VMIC*, 20 (1991).
van den Hoek, W. G. M., T. W. Mountsier, *Semicon Japan* (1994).
van Laarhoven, J. M. G. F., H. J. W. van Houtum, L. deBruin, *1989 VMIC*, 129 (1989).
Vossen, J. L. *J. Vac. Sci. Technol.*, 8, S12 (1971).
Vossen, J. L., G. L. Schnable, W. Kern, *J. Vac. Sci. Technol.*, 11, 60 (1974).
Wagner, I., *1995 VMIC*, 226 (1995).
Wallace, W. E., W. L. Wu, R. A. Carpio, *Thin Solid Films*, 280, 37 (1996).
Wang, S.-Q., E. Ong, *J. Vac. Sci. Technol.*, *B10*, 160 (1992).
Warnock, J., *J. Electrochem. Soc.*, 138, 2398 (1991).
Watanabe, K., H. Komiyama, *J. Electrochem. Soc.*, 137, 1222 (1990).
Weiss, C., A. Ghanbari, G. Selwyn, *1995 VMIC*, 412 (1995).
Weling, M., V. Jain, *1992 VMIC*, 204 (1992).
Weling, M., V. Jain, C. Gabriel, *1993 VMIC*, 188 (1993).
White, L. K., N. A. Miszkowski, W. A. Kurylo, J. M. Shaw, *J. Electrochem. Soc.*, 139, 822 (1992).
White, L. K., *J. Electrochem. Soc.*, 130, 1543 (1983).
White, L. K., *J. Electrochem. Soc.*, 132, 168 (1985a).
White, L. K., *J. Electrochem. Soc.*, 132, 3037 (1985b).
White, L. K., N. Miszkowski, *J. Vac. Sci. Technol.*, *B3*, 862 (1985).
Whitwer, F., D. Milligan, J. Garner, S.-P. Sun, M. Shenasa, T. Davies, C. Lage, *1990 VMIC*, 49 (1990).
Widmann, D. W., H. Binder, *IEEE Trans. on Electron Devices*, *ED-22*, 467 (1975).
Wilson, L., L. Shen, Y. Cu, in "Proc. 9th Symp. Plasma Processing", (G. S. Mathad and D. W. Hess, eds.), *Electrochem. Soc. Proc. Vol. PV 92-18*, 398, 1992.
Wilson, R. H., P. A. Piacente, *J. Electrochem. Soc.*, 133, 981 (1986).
Wong, J., T.-M. Lu, S. Mehta, *J. Vac. Sci. Technol.*, *B3*, 453 (1985).
Woratschek, B., P. Carey, M. Stolz, F. Bachman, *1989 VMIC*, 309 (1989a).
Woratshek, B., P. Carey, M. Stolz, F. Bachman, *Appl. Surf. Sci.*, 43, 264 (1989b).
Woratschek, B., P. Carey, F. Bachman, *1990 VMIC*, 83 (1990).
Wu, A. L., US Pat. 4,617,193 (1986).
Wu, A. L., in "Multilevel Metallization, Interconnection and Contact Technologies" (L. B. Rothman and T. Herndon, eds.), *Electrochem. Soc. Proc. Vol. PV 87-4*, 239, 1987.
Wulu, H. C., K. C. Saraswat, J. P. McVittie, *J. Electrochem. Soc.*, 138, 1831 (1991).
XMR, Inc., Santa Clara, Calif.
Xu, Z., H. Kieu, T.-y. Yao, I. J. Raaijmakers, *1994 VMIC*, 158 (1994).
Yamada, I., T. Takagi, *IEEE Trans. on Electron Devices*, *ED-34*, 1018 (1987).
Yamaha, T., Y. Inoue, O. Hanagasaki, T. Hotta, *1993 VMIC*, 302 (1993).
Yamaji, T., Y. Igarashi, S. Nishikawa, *29th IEEE/IRPS*, 84 (1991).
Yang, C., J. Multani, D. Paine, J. Bravman, *1987 VMIC*, 200 (1991).
Yao, W. W., I. W. Wu, R. T. Fulks, H. A. VanderPlas, *1985 VMIC*, 38 (1985).
Yapsir, A. S., L. You, T.-M. Lu, M. Madden, *J. Mat. Res.*, 4, 343 (1989).
Yau, L., C. Hong, D. Crook, *23rd IEEE/IRPS*, 115 (1985).

Yeh, J. L., G. W. Hills, W. T. Cochran, *1988 VMIC*, 95 (1988).
Yen, D., M.-H. Wang, J. Y. Chee, B.-Y. Jin, L. Forester, *1995 VMIC*, 240 (1995).
Younger, P. R., *Solid State Technol.*, *11/84*, 143 (1984).
Younger, P. R., *J. Vac. Sci. Technol.*, A3, 588 (1985).
Yu, M. L., B. N. Eldridge, R. V. Joshi, p. 221 in *Tungsten and Other Refractory Metals for VLSI Applications IV*, (R. S. Blewer and C. M. McConica, eds.), Mat. Res. Soc., Pittsburgh, Penn., 1989.
Yu, C., T. T. Doan, S. Kim, *1990 VMIC*, 444 (1990).
Yu, C., T. T. Doan, S. Kim, G. S. Sandhu, in "Electronic Packaging Materials Science V" (E. D. Lilliem, P. S. Ho, R. Jaccodine, K. Jackson, eds.), *Mat. Res. Soc. Symp. Proc. Vol.*, *203*, 357, 1993.
Yu, C., G. S. Sandhu, V. K. Mattheews, T. T. Doan, *SPIE V. 1958*, 186 (1991b).
Yu, C., T. T. Doan, M. Grief, *1991 VMIC*, 199 (1991c).
Yu, C., A. Laulusa, M. Grief, T. T. Doan, *1992 VMIC*, 156 (1992).
Yu, C., G. S. Sandhu, T. T. Doan, *J. Appl. Phys.*, 72, 1599 (1992).
Yu, C. C., T. T. Doan, US Pat. 5,225,034, 1993.
Yu, C. C., T. T. Doan, A. E. Laulusa, US Pat. 5,209,816, 1993.
Zalar, S., *19th IEEE/IRPS*, 257 (1981).
Zettler, T., U. Scheler, *1993 VMIC*, 359 (1993).

VII
Reliability

K. P. Rodbell
IBM Research Division
Yorktown Heights, New York

1.0 INTRODUCTION

The evaluation of reliability of an interconnect/insulator system in a multilevel structure is complex because of the small dimensions of the structures, the relatively high-use temperatures in the case of interconnects, and the existence of several different failure mechanisms. Even with the decrease in both the number and size of defects, the sensitivity to defects is heightened by the decrease in feature sizes and increase in number of features.

The total wiring length in state of the art VLSI and ULSI chips is in the neighborhood of hundreds of meters. Such chips have two to six levels of wiring and millions of interlevel wiring connections. Competitive reliability targets of chip failure rates are about one per thousand during years of field use. The reliability demands on each of the components, i.e., on the fine wiring lines, on each contact, etc., are enormous. The capability of producing even more complex wiring with even better reliability can be achieved by understanding potential failure mechanisms and using this understanding to select appropriate materials, optimize process procedures and controls, and implement design limitations. Burn-in is routinely and increasingly used to weed out early fail defects, so that the field failures are determined by wear-out processes that are easier to model and predict. This chapter will extensively review thin interconnect reliability, dominated by two critical wear-out mechanisms, electromigration (EM) and mechanical-stress-induced voiding (SV). The subjects of corrosion and insulator reliability are also discussed, but to a lesser extent.

2.0 THIN FILM INTERCONNECT RELIABILITY
2.1 OVERVIEW

The reliability of Al and Al-alloys, used as device interconnections in the semiconductor industry, has been of concern for many years. As device dimensions were scaled down and the current density increased, Al was found to be susceptible to electromigration, excessive interdiffusion (e.g. Si alloying) and hillock formation. An effective reduction in hillock formation and enhanced electromigration performance was achieved by (1) using Al-alloys (Ames et al., 1970; d'Heurle et al., 1972; Kwok and Ho, 1988), and (2) using multilayered structures of Al (or Al-alloys) with a layer (or layers) of a refractory metal (Finetti et al., 1986; Gardner et al., 1985, 1987; Howard and Ho, 1977; Howard et al., 1978, Rodbell et al., 1991, Wada, 1986). High creep strength aluminum alloys, e.g., Al-Pd or Al-Sc, have been found to have excellent electromigration and stress-induced-migration resistance as well as being less susceptible to corrosion than Al-Cu films (Onuki et al., 1990; Rodbell et al., 1993). Cu and Cu-alloys are being seriously considered for use as device interconnects (Luther et al., 1993), and the limited reliability studies on Cu thin films are reviewed. The chapter will discuss reliability techniques applicable to thin film interconnects and dielectric materials as well as review the relevant literature on these subjects.

2.2 THIN FILM WEAR-OUT
2.2.1 Electromigration

Two critical VLSI wiring wear-out failure mechanisms are electromigration (EM) (Scorzoni et al., 1991) and mechanical-stress-induced voiding (SV) (commonly referred to as stress migration) (Okabayshi, 1994). Both mechanisms involve the gradual formation of voids in metal wires eventually causing circuit failures due to interconnect resistance increases and electrical opens, and extrusions of metal that can cause electrical shorts. The primary driving force for EM is the electrical current that is carried in the wires; for SV it is the thermal expansion mismatch between the wires and both the insulating materials which enclose them and the Si substrate.

EM and SV are influenced both by macrostructural factors such as test structure geometry (Rathore et al., 1994), materials selection (d'Heurle and Ho, 1978), layering sequence, patterning dimensions, topography, and inter-layer connection methodology (Scorzoni et al., 1991; Hu et al., 1995b) and by microstructural factors such as the metal grain size and grain size distribution (Oates, 1990; Vaidya et al., 1980; Vaidya and Sinha, 1981, crystallographic grain orientation (texture) (Knorr, 1993), alloy solute distribution and precipitation (Colgan et al., 1994), dislocation densities (Livesay et al., 1992), and the number and quality of interfaces.

Electromigration is a wear-out mechanism of metal lines in which conducting electrons impart a drag force, due to ion/electron collisions ("electron wind"), on metal atoms causing them to move. This electron wind dominates over the electric field force for Al and the noble metals. There are numerous methods for measuring electromigration in thin films, a few of which will be addressed in this section.

2.2.2 Electromigration Testing Techniques

There are many experimental techniques available for measuring atomic transport in thin films; these can generally be divided into three categories: (1) measurements of compositional or structural variations, (2) direct determination of the ion flux, and (3) wear-out projections. Compositional and structural variations can be obtained, for example, by in situ scanning electron microscopy (SEM) or transmission electron microscopy (TEM) techniques and mass accumulation/depletion experiments. Measurements of either hillock or void growth rates lead to values for the rate and direction of atomic transport.

Experiments in the second category utilize marker diffusion techniques to determine the direction of transport, grain boundary diffusivity (D_{gb}) and the effective charge (Z^*). A similar method is used to calculate the drift velocity; in which a material with a low susceptibility to atomic transport and a high resistance (compared to the material of interest) is formed under a metal line (stripe); when current is passed through the sample one can measure the displacement of the stripe and calculate both the drift velocity (v) and the atomic flux (J):

$$v = J/N = (j\rho e Z^*/kT) \, D_0 \exp(-\Delta H/kT)$$

where v is the drift velocity of a metal ion, j is the current density, ρ is the resistivity, N is the metal density, and Z^* is the effective valence of the metal ion. This is a simple and fast technique; although the role of interface and surface diffusion (in unpassivated samples) needs to be accounted for, especially in narrow lines.

The techniques used to project wear-out lifetimes use test structures of lines and stud chains that are comparable to the minimum or worst case conditions in product and try to simulate conditions of topography and insulator thicknesses similar to product (Fig. VII-1). Figure VII-1a shows a top view of a line test site, where extrusion monitoring lines are defined adjacent to the test structure to indicate extrusion fails. A cross section of a via chain test structure is shown in Fig. VII-1b, which shows two levels of layered Al-Cu connected by a tungsten stud. Extrusion monitors are usually designed in one or both wiring levels (not shown in the cross-sectional figure). Techniques that compare the integrity of the film, such as lifetests and resistivity studies, are routinely used to compare the rates of atomic transport between different sets of films. The median-time-to-failure (MTF) of an interconnect subject to electromigration is often described empirically as (Black, 1969):

$$\text{MTF} = t_{50} = A j^{-n} \exp\left(-\frac{\Delta H}{kT}\right)$$

where t_{50} is the time to 50% failure of a group of identical samples, A is a sample-dependent constant, j is the current density, n is a constant (n is typically between 1 and 2), ΔH is the electromigration activation energy, k is Boltzmann's constant, and T is the absolute temperature.

In practice a large number of identically prepared thin-film interconnects are tested at elevated temperature and current density until failure (in most cases both open- and short-circuit failures are recorded). Due to the statistical nature (log-

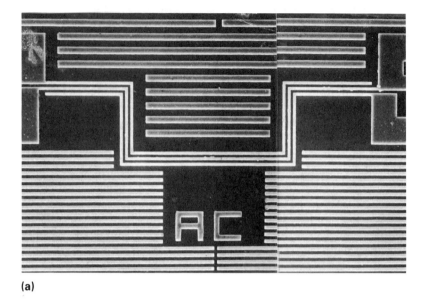

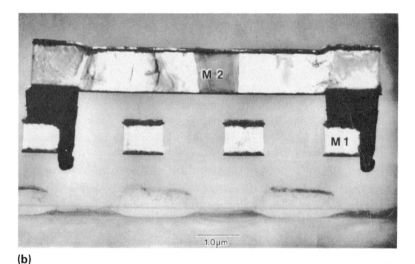

FIGURE VII-1 Electromigration testing structures: (a) Single level—SEM top view. (b) Multiple level—cross-sectional SEM. (Courtesy L. Gignac IBM)

normal or Weibull distribution) of electromigration, a range of failure times occurs at a given temperature and current density, see Fig. VII-2. Typically, a number of test structures are stressed under accelerated temperature and current density conditions. A test specimen is considered to have failed if a preset criterion is met; e.g., the criterion can be a change in resistance or an electrical short to the adjacent monitor. The fail time (log life) is plotted against cumulative fraction failed to determine a t_{50} and the variance or shape parameter for the test population. Usually t_{50} is measured at constant current density as a function of temperature from which

Reliability

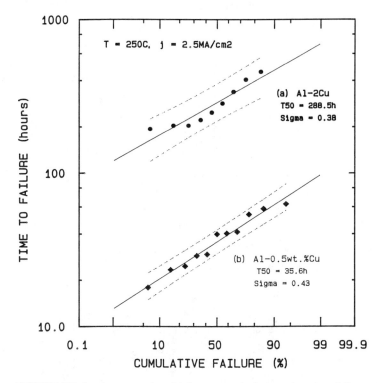

FIGURE VII-2 An example of a log normal electromigration failure plot.

ΔH is determined; t_{50} determined from tests at different current densities at the same temperature is used to determine the current exponent (n). Furthermore, the activation energy determined is an "aggregate" activation energy since it effectively averages over a group of samples, which can have a variety of failure modes. Even though it offers little insight into either the atomic diffusion mechanism(s) involved or the details of possible failure mechanisms, it is a useful tool to get estimates of design guidelines and failure rates in use conditions.

The lifetesting variables considered important are (1) geometric (such as shape; line length and width, multilevel vias and studs, and the type of passivation), (2) solute concentration(s), (3) temperature, and (4) current density (DC, AC, pulsed DC). The lognormal, Weibull, and Gamma life distribution are usually sufficient to plot the data, from which the shape parameter (*sigma* = ln (t_{50}/t_{16}), for a log normal distribution) is calculated. A *sigma* > 2 implies a high early failure rate, while *sigma* < 0.5 implies a wear-out type failure rate. Typically the use lifetime t_{50} projections are determined from the accelerated data (Fig. VII-3). Once the acceleration factor (AF) is known, then the projected use $t_{50\text{spine}}$ ($t_{50\text{use}}$) is calculated as

$$t_{50\text{use}} = \text{AF} \times t_{50 \text{ accelerated}}$$

In this analysis it is assumed that *sigma* at both the use condition and use population are equal to that in the field, i.e., the failure mechanisms are identical in both the test (lab) and use (field) conditions. As illustrated in Fig. VII-3, small variations

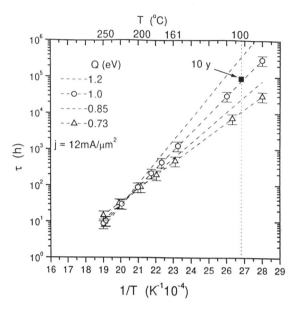

FIGURE VII-3 Illustration of effect of extrapolation to use conditions from experimental conditions. Small differences in experimental results get magnified, suggesting that a wide temperature range be used.

in failure rates can lead to big difference in activation energy, which in turn can lead to large difference in life times under use conditions. Many times, the designer is interested in failures of small sample sizes, such as 0.1% of the population or less rather than 50% of the population. The value of *sigma* of the test sample population and any inaccuracy in the determination of the same from a small number of test samples can exaggerate field failure rates (projected) by projecting to low or too high a value. Sometimes the weakness of this assumption is compensated for by using a higher *sigma* for calculating early or instant failure rates, in order to project conservative failure rates.

The present methods for determining interconnect reliability are both costly and time consuming, especially for metallization that has very low electromigration rates. An example is the use of refractory metals as discrete layers in Al metallization. This often results in >5000 h of accelerated lifetime testing at high oven temperatures (typically greater than 200°C) and high current densities (e.g., >10^6 A/cm^2). This lengthy process wears out both the annealing ovens and power supplies and delays the incorporation of "new and improved" metallurgical structures. Therefore, faster techniques have been proposed; including SWEAT (Standard Wafer-Level-Electromigration Accelerated Test-Root and Turners 1985), BEM (Breakdown Energy of Metal—Hong and Crook, 1985), WIJET (Wafer-Level Isothermal Joule Heated Electromigration Test—Jones and Smith, 1987), TRACE (Temperature Ramp Resistance Analysis—Pasco and Schwarz, 1983), assorted pulsed-current techniques (see, for example, Scorzoni et al., 1991), and noise measurements (Chen et al., 1985; Koch et al., 1985). In the $1/f$ noise technique, the noise spectrum, $S(f)$, of a thin metal film may be described as a sum of its thermal and current (excess) noise components (Vossen, 1973);

$$S(f) = 4kTR + \frac{KV^2}{f^a}$$

where R is the resistance of the film, K is a constant dependent on film microstructure, V is the voltage across the film, f is the frequency of measurement, and a is a constant which describes the frequency dependence of the excess noise term. When the dc biasing current is small and/or the film temperature is low, the measured noise spectra exhibit a $1/f$ frequency dependence; however, as one increases the film temperature and/or the current density, the $1/f^2$ noise component increases very rapidly and finally dominates the entire low-frequency spectrum. Measuring the temperature dependence of the $1/f$ noise allows one to determine the peak of the noise spectrum, from which an activation energy can be calculated. There is some evidence that this activation energy is identical to that measured for grain boundary diffusion and electromigration. The fact that it takes only a few hours to measure the $1/f$ noise spectrum as a function of temperature makes this technique particularly attractive, although not enough data has been taken to confirm that the $1/f$ noise activation energy is indeed identical to ΔH.

2.3.1 Thermal Voiding

Stress experienced by the thin film plays a critical role in the stability of fine line interconnects. Thermal voiding is a major problem of passivated Al metallization, especially with stiff dielectrics such as sputtered SiO_2 or SiN_x. First, Al films used as interconnections are subjected to many thermal cycles between room temperature and 400°C, during deposition, lithography, and etching. Some but not all of the film stress is relieved. Insulator deposition processes can exceed 300°C. For example, commonly used inorganic insulators, such as PECVD oxide, APCVD oxides and nitrides, are deposited at temperatures in the range of 300–500°C and on cool down, the rigid insulators restrict the ability of the metal film to relax thermal stresses.

2.3.2 Thermal Voiding Testing Techniques

The driving force for Thermal (Stress) Voiding is the thermal expansion mismatch between the wires and the insulating materials which enclose them, and the Si substrate (Flinn et al., 1987; Gardner and Flinn, 1990; Sullivan et al., 1991). It is thought that thermal-stress-induced failure is due to confinement of aluminum lines by insulators with thermal expansion coefficients smaller than that of aluminum. Traditionally, thermal voiding is studied by annealing samples at a temperature where grain boundary diffusion is rapid and the stress in the film is close to zero. The sample is typically cycled between room temperature and a higher temperature (300–500°C) and examined periodically for signs of void nucleation and/or void growth. The extent of stress voiding has been measured using electron microscopy and by monitoring the change in line resistance during constant temperature anneals. The main drawback with resistance techniques, however, is the lack of sensitivity to the nucleation and growth of small voids within the metal line. Recent work (Simpson et al., 1994) using positron annihilation has shown that it is possible to examine in situ void formation in SiO_2 passivated, 1 μm $\times$ 1 μm Al-0.5wt%Cu lines.

3.0 STRUCTURE/BEHAVIOR OF THIN FILMS
3.1 Al-ALLOYS/LAYERED STRUCTURES

Many factors are known to influence the electromigration lifetimes of aluminum films such as alloying, film geometry, and microstructure. An effective reduction in hillock formation and enhanced reliability is obtained with Al-alloys, e.g., Al-Cu (d'Heurle et al., 1972), and multilayered structures of Al (or Al-alloys) with a layer (or layers) of a refractory metal (Howard et al., 1978; Gardner et al., 1985). In Al-Cu alloys the electromigration lifetime is a function of Cu concentration, increasing as the Cu content increases from 0 to 4 wt%Cu(Ames, et al., 1970). The role of Cu in increasing the electromigration lifetimes in Al-Cu alloys is not well understood. It has been proposed that Cu atoms in solid solution preferentially migrate before Al in an electric field, thereby improving the electromigration behavior; e.g., Al atoms will not electromigrate until all of the Cu atoms are depleted (Rosenberg, 1972). The formation of θ-phase (Al_2Cu) precipitates at Al grain boundaries inhibits void growth and coalescence, limits vacancy migration, and stabilizes the Al grain boundaries (Frear et al., 1990). Furthermore, annealing conditions affect the shape and distribution of precipitates in AlCu alloys (Frear et al., 1991; Colgan and Rodbell, 1994) which control the resultant film stress, grain size, and has a strong influence on electromigration behavior. In pure Al (Knorr et al., 1991; Knorr and Rodbell, 1996) and AlCu (Rodbell et al., 1992) film texture was shown to impact the EM reliability of single-level structures (e.g., no interlevel vias or studs). The role of local grain misorientation in Al electromigration (Hurd et al., 1994; Longworth and Thompson, 1991) and stress voiding (Kordic et al., 1993) show that local flux divergences are often seen at, or near, damage sites. This data implies that one may want to control the local film texture by suitable control of the film deposition process. High creep strength aluminum alloys, e.g., AlPd, have recently been found to have excellent electromigration and stress-induced-migration resistance as well as being less susceptible to corrosion compared to AlCu films (Onuki et al., 1990; Rodbell et al., 1993). However the benefit of highly textured lines is overshadowed by the significant degradation of lifetimes (Estabil et al., 1991) measured for multilevel structures (e.g., separated with CVD vias). Joule heating, solute diffusion, electromigration, and flux divergences at via/conductor interfaces must be considered (Filippi et al., 1996; Korhonen et al., 1993).

Sputtered Al-low Cu (i.e., a Cu concentration <2wt%Cu), multilayered, submicron, device interconnect metallurgy is in common use today. It generally consists of $TiAl_3$ layers (approx. 50 nm thick) under and over an Al-Cu alloy conductor (approx. 0.5 μm thick) with a TiN cap layer (approx. 30 nm thick). Alternatively a layer of TiN is placed between the top and bottom Ti layers and the AlCu conductor to limit $TiAl_3$ formation. These films are patterned by reactive ion etching (RIE), and show both a low susceptibility to corrosion and a low resistivity. The electromigration behavior in single level lines in SiO_2 is good (e.g., Rodbell et al., 1991). However, the multilevel stack (consisting of either Ti/AlCu/Ti-TiN or Ti/TiN/AlCu/TiN lines and CVD W studs) has a significantly reduced electromigration behavior due to Cu depletion adjacent to the W stud resulting in a region of pure Al which fails quickly. This failure mechanism is not seen in single-level samples because (1) there are, generally, few abrupt flux divergence sites in single-level samples (which cause current crowding and, in the case of W studs, also prevent Cu from diffusing from one Al line to the next) and (2) there is an unlimited supply of Cu solute from adjacent large lines and pads in single level metal lines. Additional process problems have been found with this metallurgy during fabrication, including (1) thermal voiding, seen after SiO_2 deposition and repeated thermal cycling, and (2) undercutting of the bottom Ti layer during RIE of lines above W studs. Table VII-1 contains electromigration T_{50} data for several Al conductor systems reported in the literature.

TABLE VII-1 A Survey of Electromigration Lifetimes for Al Metallization

Alloy	Resistivity ($\mu\Omega$-cm) 1 h, 400°C forming gas	$t_{(50\%)}$ (hours) @ 250°C, 2.5 MA/cm^2
Al-0.5% Cu[a]	3.5	9000
Al-0.5% Cu[b]	3.4	12000
Evap[c]	3.7	400–500
Evap[d]	3.8	400–500
Cr/Al-4% Cu	3.0	400
Al	2.8	15
Al-0.5% Cu	2.9	50
Al-1.2% Si-0.15% Ti	3.1	23
Al-1.2% Si[e]	2.9	156*
Al-1% Ti	6.6	2
Al–Si/Ti[f]	3.1	300*

*150°C, 1 MA/cm^2, unpassivated.
[a] Sputtered 4250 Å Al0.5%Cu/1500 Å TiAl$_3$/4250 Å Al0.5% Cu; annealed in forming gas at 400°C.
[b] Sputtered 700 Å TiAl$_3$/8500 Å Al0.5%Cu/700Å TiAl$_3$/250 Å Al0.5%Cu and annealed in forming gas at 400°C.
[c] Evaporated 4250 Å Al0.5%Cu/1500 Å TiAl$_3$/4250 Å Al0.5%Cu and annealed in forming gas at 400°C.
[d] Evaporated 700 Å TiAl$_3$/8500 Å Al0.5%Cu/700 Å TiAl$_3$/250 Al0.5% Cu and annealed in forming gas at 400°C.
[e] F. Fisher, Siemens Forsch-U. Entwickl-Dec. 13, 21 (1984).
[f] D. S. Gardner, T. L. Michalka, P. A. Flinn, T. W. Barbee Jr., K. C. Saraswat, and J. D. Meindl, Proc. 2nd IEEE VMIC, pp. 102–113 (1985).

The reaction between thin films of Al, AlCu, and Ti has been extensively investigated; see Colgan (1990) for a comprehensive review. In most reports the initial phase formed was found to be $TiAl_3$ (see Table VII-2); however, Howard et al. (1976) observed Ti_9Al_{23} (in addition to $TiAl_3$) in evaporated films, annealed between 325 and 440°C. The Ti_9Al_{23} subsequently transformed to $TiAl_3$ at 440°C. A cubic metastable phase of $TiAl_3$ has also been reported by Hong et al. (1988). This phase was found in bi- and multi-layered, evaporated, and coevaporated samples which had been annealed at 350°C. At 400°C the metastable phase was found to transform to the equilibrium tetragonal phase. Van Loo and Rieck (1973) observed two crystalline modifications in annealed 20–100 μm thick evaporated Al/Ti films: (1) $TiAl_3$ alone at 638°C, and (2) extra peaks in the XRD patterns at 585°C which they interpreted as a mixture of $TiAl_3$ and Ti_8Al_{24} although it resembles the compound Ti_9Al_{23} indexed by Raman and Schubert (1965). Once this superstructure compound was formed at 585°C it could not be altered by subsequent long time (138 h) anneals at 638°C. Noting that AB_3 compounds are sensitive to impurities, which can affect the electronic configuration and, consequently the phase formed, van Loo and Rieck (1973) suggest that a certain transition range exists between the two crystalline modifications of $TiAl_3$ that is likely influenced by strain and/or impurities.

TABLE VII-2 Thin Film Ti-Al(Cu) Intermetallic Formation

Film(s)	Phases formed/$\Delta H(eV)$[a]	Range	Reference
Evap. Al/Ti/Si	$TiAl_3$/1.9(Al), Ti_7-$Si_{12}Al_5$	350–475°C	Bower (1973)
Evap. Al/Ti	$TiAl_3$, Ti_9Al_{23}(<440°C)	325–475°C	Howard (1976)
Evap. Al/Ti	$TiAl_3$, Ti_8Al_{24}	580–640°C	van-Loo (1973)
Evap. AlCu/Ti	$TiAl_3$/1.6(Al), 2.1(1%Cu)	375–450°C	Wittmer (1985)
Evap. AlCu/Ti	$TiAl_3$/1.7(Al), 2.2(0.25at%Cu)	350–500°C	Tardy (1985)
Evap. AlCu/Ti	$TiAl_3$/1.8(Al), 2.4(3at%Cu)	300–500°C	Krafcsik (1983)
Evap. Al/Ti	$TiAl_3$/1.9–2.0 (Al)	460–515°C	Zhao (1988)
Evap. Al/Ti	$TiAl_3$	400–550°C	Thuillard (1988)
Sput. Ti/Al/Si	$TiAl_3$, Ti_8Al_{24}, unknown[b]	380°C	Han (1985)
Sput. AlCu/Ti	$TiAl_3$/1.7(Al), 2.1(4%Cu), Ti_9Al_{23}	320–550°C	Ball (1987)
Evap. Al/Ti	$TiAl_2$(>350°C), Ti_9Al_{23}(>500°C)	300–500°C	Fujimura (1989)
Evap. Al/Ti	Cubic metastable $TiAl_3$, $TiAl_3$	350–400°C	Hong (1988)
Evap. AlCu/Ti	$TiAl_3$	400–425°C	Slusser (1989)
Sput. Al/Ti	$TiAl_3$	300–520°C	Ben-Tzur (1990)
Sput. Al/Ti	$TiAl_3$ 1.8 (Al), 2.2(2%Si)	400–500°C	Nahar (1988)

[a] Activation energy for reaction (from Arrhenius plots).
[b] Tetragonal (a = b = 5.782 Å, c = 6.713 Å).

The effect of Cu solute additions to Al has been shown to increase the activation energy for the growth of $TiAl_3$ from 1.6 eV (no Cu) to 2.1 eV (1 wt%Cu). Cu was found to influence the microstructure obtained, by (1) promoting Al grain growth, and (2) segregating to the grain boundaries of the growing $TiAl_3$ resulting in a smooth reaction interface (Wittmer et al., 1985). Tardy and Tu (1985) showed that the effect of Cu is much greater on Ti diffusion than on Al diffusion in $TiAl_3$, for evaporated Al-0.25at%Cu/Ti bimetallic thin films. The effect of oxygen contamination on $TiAl_3$ formation has also been investigated by Zhao et al. (1988) and Thuillard et al. (1988). $TiAl_3$ growth was found to be diffusion limited with an activation energy of 1.9–2.0 eV independent of the oxygen concentration in evaporated Ti/Al bilayers. However, the $TiAl_3$ growth rate was found to decrease due to the presence of oxygen in the Ti/Al samples. It was also found that $TiAl_3$ growth occurred at the free Ti surface when a thin oxide layer was present at the Al/Ti interface, suggesting that Al is the diffusing species (Han and Bene, 1985).

Ball and Todd (1987) measured the reaction kinetics of dc Magnetron sputtered Al/Ti and Al-4wt%Cu/Ti thin film reaction couples. Upon annealing Al and Al-4wt%Cu-Ti samples, Ti_9Al_{23} was indexed and was found to be stable to at least 550°C. When oxidized Al substrates were used, Ti_9Al_{23} also formed with annealing; showing that Al-oxide can prevent the Al-rich phase (i.e. $TiAl_3$) from forming. Fujimura et al. (1989) examined the effects of Ti film texture on solid state reactions between Ti and Al films on Si (100) substrates. Al (200 nm) was evaporated onto r.f. ion plated Ti (100 nm), with the Ti grain texture controlled by varying the bias voltage. Without bias a strong (0002) texture resulted. With a 1500 V dc bias a randomly oriented Ti film formed. $TiAl_3$ resulted in either case with >350°C anneals, although the rate of intermetallic formation was slower for highly textured Ti. It was also found that Ti_9Al_{23} formed with >500°C anneals.

Examination of the literature reveals a number of major trends: (1) A stable Ti_9Al_{23} phase (in addition to $TiAl_3$) can be formed in evaporated films; (2) Ti_9Al_{23} can be formed in sputtered films via high-temperature anneals; and (3) the phase(s) formed and intermetallic growth rate will depend on factors such as impurities, film texture, and/or defects present in the Al and/or Ti films.

It has been shown that films containing only $TiAl_3$ show improved electromigration behavior compared with films containing Ti-Al superlattice structures (Rodbell et al., 1991). The formation of superlattice structures in the Ti-Al system results in Ti-rich unit cells, i.e., Al-deficient structures. Such intermetallics tend to form large unit cells due to slight shifts in the atomic positions of the Ti and Al atoms from the symmetric positions found in the closely related line compounds, e.g., $TiAl_3$. A larger point defect density, which more readily allows for atomic diffusion, would also be expected in Ti-Al superlattices than in $TiAl_3$ stoichiometric compounds. This would result in faster diffusion of Al through the superlattice redundant layer(s), than through similar $TiAl_3$ layers, resulting in inferior electromigration performance. Furthermore, the intrinsic electromigration behavior of Ti_9Al_{23} versus $TiAl_3$ intermetallic(s) would also be expected to be different.

Another benefit of Ti underlayers is the improved crystallographic texture of the subsequently deposited Al-alloy films. Bragg-Brentano scans are often used to infer the texture strength of sputtered Al films (Vaidya and Sinha, 1981). Knorr (1993) subsequently showed that this ratio can be misleading and that a complete pole figure is indeed required to determine a film's crystallographic texture strength. It has been demonstrated (Rodbell et al., 1996) that even the more commonly used rocking curve technique is inadequate since one is unable to distinguish random textures from near-(111) fiber textures. In general, one cannot accurately determine the crystallographic texture of a thin film unless one measures the complete pole figure, or a fiber plot.

Correctly determining the film texture is important since the crystallographic texture determines the nature and extent of mass transport in a thin film. By relating grain boundary angle to diffusivity both along grain boundaries and on grain surfaces one may be able to determine the optimum microstructure required to minimize mass transport in fine lines. A tight (111) texture in Al means that most grains have (111) planes parallel to the substrate. Viewed in the direction normal to the film plane, most grains differ from each other only by a rotation, which would produce (111) tilt grain boundaries perpendicular to the film plane. For low-angle misorientations (less than 15°), these boundaries are composed of a vertical array of parallel edge dislocations. Since in this situation the direction of fastest diffusion (along dislocation cores) is orthogonal to the film plane, grain boundary diffusion is expected to be slower than in a film with poorer texture (e.g., a wider range of misorientations). High-angle misorientations generally produce faster diffusivity paths [except for a limited number of special boundaries, known as coincident site lattices (CSL) which behave like low-angle boundaries].

A second way texture can influence mass transport is through plastic deformation. Grain orientation (in Al films) with respect to the stress field determines the resolved shear stress on the (111) ⟨110⟩ slip systems. Susceptible grains which are deformed will have a higher dislocation density than the remaining population. These grains can be sources of flux divergence for electromigration and atomic sinks for stress-induced void growth; they may also act as void nucleation sites.

The formation of thermally induced hillocks (compressive stress) has been found to occur more readily in off-(111) oriented grains (Schwarzer and Gerth; 1993) suggesting that these grains are weaker than their neighbors, and thus more easily deform. Stronger film texture is usually accompanied by a slightly smaller average grain size and a tighter grain size distribution. Hillocks and grains near both electromigration voids and stress-voids tend to be more randomly oriented than the undamaged grains along a line. Recently, the role of the oxide surface roughness (Onoda et al., 1995) and deposition conditions on film texture has been explored. Careful control of both these variables will allow one to obtain uniform grain size and grain orientations in thin films.

For electromigration and stress-voiding, damage tends to occur heterogeneously at grain boundaries, suggesting that the important aspect of texture is its influence on the distribution of grain boundary misorientation angles. In electromigration, damage occurs at flux divergence sites where material flow into and away from a damage region is not balanced. This can be due to aspects of the film microstructure, including grain size, precipitates, and crystallographic texture, or to physical constraints such as the presence of blocking boundaries (e.g., W vias). Table VII-3 summarizes the activation energy determined from electromigration studies on aluminum with different grain sizes and structures. From a crystallographic texture viewpoint one would like to minimize the number of off-(111) grains since flux divergences will form at all off-(111) grains in an otherwise (111)

TABLE VII-3 Electromigration Activation Energy for Aluminum Thin Films

Activation energy (Q) (eV)	Grain size (μm)	Method
0.74 ± 0.08	4	Transported volume
0.51 ± 0.10	2	Life test
0.73 ± 0.05	8	Life test
0.48	1.2	Life test
0.84	8	LIfe test
0.3–1.2*	—	Life test
0.55	—	Life test
0.51 ± 0.10	—	Life test
0.34	~1	Life test
0.46	~5	Life test
0.41	~1	Life test
0.70 ± 0.20	0.5–Several	Transported volume
0.63	—	Transported volume
0.70	—	Transported volume
1.22	Single crystal	Transported volume
0.5–0.6	—	Resistance
0.5–0.6	—	Resistance

*Thickness dependent
Source: From A. J. Learn, *J. Electrochem. Society*, 123, 6 (1976).

textured film. In stress-voiding, however, one may want a uniform grain structure with regions of off-(111) oriented grains to serve as locations for strain relief. In a perfect fiber texture one can expect a small uniform strain in all grains; therefore a local nonuniformity in strain may result in rapid void formation in nearby grains or grain boundaries. In a weakly textured film, however, many sites exist to absorb strain making an individual grain boundary or grain less susceptible to void nucleation and growth.

3.2 COPPER ALLOY/LAYERED FILMS

Copper has been considered an attractive replacement for Al-Cu, not only for its lower resistivity, but also since its higher melting point implies better electromigration resistance. The data on copper is limited to date (Table VII-4), but a review is in order because of its potential for use as an interconnect in semiconductors.

Zielinski et al. (1995) studied sputtered copper films on Si, Al, and Cu substrates. The film was thermally cycled to 300°C. The authors found abnormally

TABLE VII-4 Electromigtration Activation Energies for Cu Thin Films

Film	Technique	Activation energy	Reference
Cu-W	Drift velocity	0.66 eV	Hu (1992)
Cu-TiW	Electromigration	0.54 eV	Kang (1993)
Cu (large grain)	Electromigration	1.25 eV	Nitta (1993)
Cu (small grain)	Electromigration	0.86 eV	Nitta (1993)
Cu	$1/f$ noise	1.1 eV	Rodbell (1991)
Cu-Sn, Zr	Drift velocity	1.1–1.3 eV	Hu (1995)
Cu	Drift velocity	0.75 eV	Hu (1995)

large (100) oriented grains were favored on both compressive (Si substrate) and tensile (Al substrate); however, Cu films on a copper substrate had a strong (111) orientation. The authors suggest that the (100) orientation was preferred when high biaxial strain is present. The higher biaxial modulus of Cu for the (111) orientation makes it the preferred orientation in the stress free situation such as a Cu film on a copper substrate. The same authors (Zielinski et al., 1995b) studied Cu films deposited on a Ta layer. In this study, Ta films were deposited on a substrate at 30°C as well as over 100°C. Cu films were deposited at 30, 150, and 250°C. The authors found that Ta deposited at 30°C had a slight (001) fiber texture, which became more prominent at higher temperature. For the case of a Ta underlay deposited at 30°C and a Cu film deposited at 30°C, a (111) fiber texture developed, which decreased at higher deposition temperatures. However, Ta deposited at 100°C had a strong (001) fiber texture and Cu films deposited on top showed a strong (111) fiber texture. Higher deposition temperature led to a larger average Cu grain size as one would expect. Keller et al. (1995) studied grain growth as a function of thermal cycling from RT to 600°C in Cu films capped with a SiN layer. The cap layer inhibited grain growth and suppressed diffusion processes. It was also found that dislocation movement controlled the mechanical behavior in uncapped films. Murarka et al. (1993) built inlaid Cu with a Ti liner using a damascene CMP process. Li and Mayer (1993) used Cu/Cr and Cu/Mo bilayers on SiO_2 and nitrided the Cr and Mo layers by heat-treating at 700°C in NH_3.

Frankovic and Bernstein (1995) measured the electromigration lifetime of evaporated Ti/Cu films using a 50 μm long, 8 nm Ti and 82 nm Cu thick film at a current density of 5×10^7 A/cm^2 and found no fails at 100 h. They extrapolated a minimum lifetime based on the zero fails and compared it with other published data on Al and Al-Cu. They concluded that the MTF of evaporated Ti/Cu is at least 2–4 orders of magnitude better than Al-Cu. The EM data is limited and the comparison is not rigorous, but the main observation that there were no fails in Cu at the stress current density and temperature is valid. Rodbell et al. (1991) used a two-frequency ac bridge to measure the $1/f$ noise in Cu films as a function of temperature. The activation energy determined was 1.1 eV.

Hu et al. (1992) used drift velocity measurement on Cu/W structures, using sputtered 0.4 μm thick copper. They determined that the activation energy for drift velocity was 0.66 eV comparable with vacancy migration in grain boundaries. Ding et al. (1992) showed that boron (B) implanted Cu films oxidized less and proposed this as a possible technique to improve the corrosion resistance of Cu. Russell et al. (1992) showed that Cu(Ti) films in contact wth SiO_2, when heated to a temperature range of 400–600°C, showed that Ti preferentially segregates to the Si/SiO_2 interface, leaving a pure Cu matrix of low resistivity. This is an interesting way to encapsulate Cu films; however, the effectiveness of the approach in actual use is not known.

Nitta et al. (1993) studied the electromigration properties and grain growth (using ion bombardment and thermal anneals) of 1 μm thick Cu (111) and Cu (100) films. The film resistivity was 1.76 $\mu\Omega$-cm, close to bulk Cu. The test structures were 4–5 μm wide, 80 μm long and were cycled between 0 and 10^7 A/cm^2 with active cooling of the test specimen. The resistivity was measured at room temperature using lower currents. The maximum temperature from Joule heating was estimated to be between 100–200°C. For large grain copper they found the activation energy for electromigration to be 1.25 eV; for films as deposited the activation energy was 0.86 eV. This led the authors to conclude that both large and small grain Cu films have much higher electromigration resistance compared to Al-Si-Cu (an activation energy of 0.6 eV). Hu et al. (1995) used drift velocity measurements on evaporated and CVD deposited copper films and found their activation energy to be 0.75 eV. Addition of magnesium to copper degraded the electromigration lifetime, which was attributed to its inhibiting effect on grain growth. Small additions of Sn and Zr to copper films increased the activation energy of copper films (from drift velocity) from 0.75 eV to 1.1–1.3 eV. Gardner et al. (1995) found that encapsulating copper with Mo improved the resistance to hillock formation as compared to layered structures. In layered structures, W or TiN underlayers degraded the electromigration lifetime of Cu, presumably due to a smaller Cu grain size. Molybdenum layering provided a modest improvement in lifetime. Hoshino et al. (1990) reported an increase in mean time to failure for Cu films with small additions of Ti. However, increased additions of Ti to copper led to a sharp decrease in lifetimes. They used 2 μm wide test specimens of Ti/TiN/Cu-Ti films and stress tested at 5×10^6 A/cm^2, at 250°C. They compared the lifetime of Cu-Ti films with TiN layering with Cu-Ti films that were encapsulated with TiN. The TiN encapsulation was achieved by nitriding the Cu-Ti film at about 800°C. The Ti in Cu-Ti diffused to the surface and formed TiN. The TiN encapsulated film showed much higher electromigration resistance compared to layered Cu-TiN films Kang et al. (1993) studied CVD copper in a variety of structural forms: Cu/TiW; Cu/SiO_2; TiW/Cu/TiW, W/Cu/TiW, SiO_2/Cu/TiW, and a buried Cu/TiW. As deposited, Cu grains were in the 0.1–0.2 μm range and grew to the 0.2–0.4 μm range on annealing at 350°C. The EM activation energy for Cu/TiW was 0.54 eV and the sandwich structures showed an activation energy ranging from 0.32 to 0.41 eV. These structures in spite of lower activation energies, showed an increase in EM lifetimes of 2 orders of magnitude or higher than comparable Al-Si/TiW structures. Kang et al. (1992) also reported an activation energy of 0.81 eV for electroless plated copper on a thin Pd_2Si underlayer. Clearly, the published results from different references show strong variability in activation energies and fail times, consistent with diverse film structures and test conditions. However, an underlying general trend is that all copper results show improved EM lifetimes compared to layered Al-Si and Al-Si-Cu films. The Cu EM data on the effect of layering seem mixed depending on the exact material used for the layering, although encapsulation seems to provide better resistance to both hillock formation and electromigration. Microstructure and in particular grain size seems to have a strong effect on EM lifetime, larger grain size films providing longer EM lifetimes.

3.3 ROLE OF PASSIVATING FILMS

There is *almost* general agreement that a *rigid, defect-free dielectric overcoat increases* the electromigration lifetime. Since the conductors at every level of a MLM structure must be isolated by such dielectrics, this is a fortunate circumstance. Even the topmost level is passivated, if only for scratch protection.

It may be noted that, in the work described below, many of the passivating films have not been described adequately in terms of the deposition method/parameters or physical properties.

Increased lifetime due to overcoating with a passivating film was correlated to an increase in activation energy, which was attributed to a reduction in surface diffusion (Black, 1969). Anodization of Al stripes, instead of depositing a film on them, was another technique for increasing lifetime (Learn and Sheperd, 1971) and the improvement was explained in terms of reduced vacancy supply and/or reduced void-nucleation sites. Grain growth during glassing can also increase electromigration lifetimes (Blair et al., 1970). Reduction of surface diffusion was questioned by Ainslie et al. (1972) who pointed out the practically nonexistent electromigration rate in single-crystal Al thin films. Ainslie et al. (1982) stated that it was the *mechanical constraint* exerted by a coating that was important in increasing lifetimes and the benefit was due to the effect of pressure (backflow) on reducing diffusion. To reduce electromigration, the coating must have good mechanical properties and should be defect-free (Lloyd et al., 1982). Stated somewhat differently, a restraining (strong) passivation layer would permit accommodation of higher stresses, which are built up in the conductor during the passage of current (Blech and Herring, 1976), and in this way, reduce diffusivity of metal atoms (Lloyd and Smith, 1983). It has been shown that a more rigid passivation, which should resist the formation of the local hillocks and voids formed during electromigration, did improve electromigration lifetime (Schafft et al. 1984). Thicker rigid films were more effective (e.g., Lloyd and Smith, 1983); with thick passivation, the failure mode is more likely to be extrusions, causing short circuits to adjacent or overlying conductors. If there are defects in the thick passivating film, the metal film stress will be relieved at those sites and extrusions will form.

The nature of the insulating film was held to be unimportant (Felton, 1985; Yau et al., 1985). However, Schlacter et al. (1970) found that the lifetime of a conductor was diminished when passivated by a tensile CVD SiO_2; when the oxides were P-doped, the tensile stress was reduced and lifetimes of the PSG-coated conductors were now longer than those of the unpassivated ones. On the other hand, tensile PECVD SiN was found to increase the lifetime by about an order of magnitude over that of uncapped films (Levy et al., 1985). Improved lifetimes and reduced drift velocity were found when the films were encased in sputtered SiO_2, but the extent of improvement depended on the deposition conditions. But neither the specifics of the deposition conditions nor the differences in the film properties were described (Grabe and Schreiber, 1983). Planarization of the dielectric overcoat was said to reinforce the sidewalls and suppress crack formation and in this way increased the electromigration resistance of the underlying conductor (Isobe et al., 1989). The lack of damage in an M1 layer in a two-level structure, in which the identical upper layer was unpassivated and failed catastrophically, was attributed to the encapsulation of the first-level metal in a thick layer of (uncharacterized) deposited oxide (Martin and McPherson, 1989; May, 1991). Pramanik et al. (1994) reported that the failure *mode* depended on the mechanical restraint imposed by the overcoat but the kinetics of mass accumulation and depletion were unchanged. A thick rigid overcoat (in this case PECVD oxide) prevents relief of the stresses generated during EM, so they build up to the point of dielectric cracking and the metal extrudes. With a thinner, more flexible overcoat (in this case SOG plus a thinner oxide), stress relief can occur and the line resistance increases due to void growth.

Any benefit of a yielding film, such as Polyimide, was attributed to its ability to supply H_2 to the grain boundaries of the metal film (Lloyd and Stegall, 1986). A photoresist coating was found to be ineffective (Yau et al., 1985.)

It is difficult to reconcile, with these earlier results, those of Hishimura et al. (1995) who reported that the thicker passivation (PECVD SiN/PSG) on a tapered via was responsible for the decreased lifetime. They correlated the degradation resulting from increased thickness with increased thermal tensile stress which increases the diffusivity of the Al atoms. The role of tensile stress in the passivated lines was also invoked by Witvrouw et al. (1995) to explain the shorter lifetimes of AlCuSi lines passivated with PETEOS SiO_2 when compared to unpassivated lines. Another possible explanation is the structure differences due to heating during oxide deposition. They also pointed out, as did Lloyd (1995) and Clement et al. (1995), the role of thermal stresses, i.e., stress voiding (in the tensile, encapsulated metal lines), in lowering the threshold for damage nucleation and the need to take into account the increased stress at lower temperatures when extrapolating from accelerated test conditions to use conditions.

Kahn and Thompson (1991) stated that "it has been proposed that the *beneficial* effect of passivation is due to the *high compressive stress induced* by the dielectric coating" after Ainslie et al. (1972), who favored the term "mechanical constraints." To test this hypothesis, there have been several experiments in which bare Al interconnects on oxidized Si were subjected to mechanical bending stresses. Kahn and Thompson (1991) and Rosenmayer et al. (1991) found that *applied tensile stress* had a small, but measurable, *degrading* effect on electromigration resistance, although Lowry et al. (1993) did not see this effect.

4.0 ELECTROMIGRATION BEHAVIOR OF VIA CHAINS

4.1 OVERVIEW OF INTERLEVEL CONNECTORS

There are two basic types of interlevel connectors linking sequential metal levels through the interlevel dielectric layer. Figure VII-4 shows the tapered via structure, into which a metal is evaporated or sputtered to directly contact the underlying

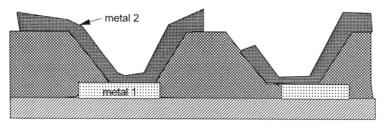

FIGURE VII-4 A schematic of a tapered via structure.

metal at the via opening, but the hole is *not filled*. The stud (plug) structure is shown in Fig. VII-5; the walls are vertical and the holes *filled* by a process designed for good gap-fill, e.g., CVD, thermal flow, etc. The CVD W-filled stud is used increasingly because a vertical-walled connector is required to satisfy the need for denser chips (see Figure VII-6). CVD W is an established technology from available tools and processes and for most applications CVD W has acceptable electrical resistivity for via applications. The connector may be Al, an Al alloy, or a multi-layer film which includes a barrier layer (e.g., TiW) at the interface between the two metal levels. Or it may, itself, be a barrier to mass/vacancy transport (e.g., W). Whether it be a via or a stud, electromigration resistance usually decreases as the diameter of the hole is decreased.

Two kinds of test structures have been used: the Kelvin or four-point test structure which probes a single via, and via (stud) chains containing different numbers of vias (studs) and intervia (stud) spacing. The upper level may or may not be passivated.

4.2 VIAS WITH NO BARRIER LAYER IN M2 METAL

Even in a "clean" via structure, i.e., one in which there is no barrier film, i.e., a contaminant such as a native oxide or F-containing layer on the Al surface at the interface between the first (M1) and second (M2) metals, can cause degradation of electromigration resistance somewhat similar to the effect of bends in a line. Kwok et al. (1987a,b), using finite element methods, studied the current crowding and

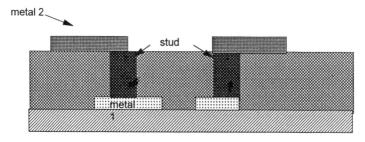

FIGURE VII-5 A schematic of a vertical stud interconnect between two wiring levels.

Reliability 491

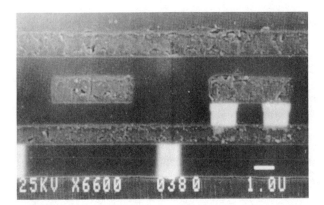

FIGURE VII-6 An SEM cross section of Figure VII-5, where a tungsten stud connects two levels of Al-Cu wiring.

local heating in the via region; these effects contribute to increased mass transport rates, i.e., to reduced electromigration resistance.

The current density peak increases with via step angle, since the thickness of metal at the via side wall is decreased, when deposition is done by sputtering or evaporation. As the aspect ratio of the via is increased, the cross-sectional metal area at the base of the via can eventually become smaller than in the step region and thus shift the current density peak to that region. By increasing the thickness (and thus, reducing the current density) of the upper metal, the current density peaks and thus the hot spots in the via can be reduced. The vulnerable regions are, therefore, the sidewall of the via and the interface between the metal levels and the via.

The effect of step coverage by the second metal (M2) has been investigated experimentally. In the absence of any barrier layer, poor step coverage has sometimes been found to be responsible for the degradation of electromigration resistance (Rathore, 1982; Matsuoka et al., 1990). Rathore (1982) used in situ sputter cleaning of the vias to ensure the absence of a barrier (resistive) layer (Bauer, 1980, 1990) and therefore attributed high via resistance to poor step coverage. They correlated high initial via chain resistance to low electromigration lifetimes. They showed SEMs of the microcracks in M2 at the top rim of the via which confirmed the failure mechanism.

However, according to Yamaha et al. (1992), in the range of 8 to 22% step coverage of the M2 metal, the lifetime of via chains was independent of M2 coverage. They also showed that there was an optimum sputter-clean time for lifetime improvement and attributed the degradation after that time to deposition on the interface of a thin insulating film consisting of material sputtered from the dielectric sidewall as had been reported by Tomioka et al. (1989).

Bui et al. (1990) as well as Yamaha et al. (1992) found the failure to be an electrical open at the Al-to-Al via interface. The presence of an interfacial film, as well as the structural factors described by Kwok (1987a,b) could be responsible for the divergence in the current density at that site. Smaller vias failed first but the mechanism was the same for all sizes of via. The location of the open circuit was not affected by the stress in the passivation layer.

4.3 VIAS WITH A BARRIER LAYER IN M2

The barrier layers used in conjunction with Al alloys in vias are Ti, TiN, TiW, and WSi. There is a flux divergence at the interface between M1 and M2 due to the presence of the barrier. When electrons flow from M2 to M1, void formation at M1 would be expected since the electron flow results in metal transport away from the interface with no means of replenishing it; metal accumulates at M2. When the electrons flow from M1 to M2, metal accumulates at the interface at M1; if the barrier prevents penetration into the via there should be no via failures in this case (Martin and McPherson, 1989). But of course failures can occur on the stripes in the Kelvin structure as well as in the via which can complicate analysis of the results. Oates et al. (1992) found that open-circuit failure was not, at least in this case, associated with poor step coverage and that the interface at the barrier was most vulnerable. Voiding occurred by drift which determines the rate of mass transport away from the interface; the flux divergence dominates over any contributions of nonhomogeneous current and temperature distributions. They attribute the directional dependence to the current density of the stripes leading to the vias, since this determines the drift velocity. If there are differences in grain size between M1 and M2, this would influence the drift velocity as well. The current density threshold for electromigration can influence the location of voids and the threshold will be influenced by test site design. May (1991) found that the lifetime of a very long TiW/AlCu stripe (representative of an actual device interconnection) was significantly less than that of the vias, although the sigma of the distribution was lower.

Okuyama et al. (1990), who fabricated vertical walled vias in which the step coverage of the TiW/AlCuSi/TiW M2 layer was proper, reported the lifetimes of the chains superior to that of the lines. TiW suppressed the electromigration in via chains. For the via chains, they sound a very high value for the apparent activation energy for electromigration (3.0 eV), an indication of a reaction between Al and TiW. Since all lines were encased in TiW, it is difficult to understand the superior electromigration resistance of the via chains as compared to the lines.

4.4 STUDS

4.4.1 Introduction

Studs are increasingly replacing vias in order to achieve small wiring pitches required in shrinking ground rule devices. Effects related to the taper of the via are, of course, no longer important for vertical, completely filled studs.

4.4.2 Tungsten Stud

Although the discussion refers to the line as an AlCu lines, the line may also contain Si and/or may be a layered film. W is, essentially, electomigration-resistant. The flux divergence at the AlCu line/W stud interface is the main factor responsible for open circuit failure; voids are likely to form at or close to the stud/line interface. (e.g., Kwok, 1991; Hu et al., 1993; Korhonen et al., 1995). XRD has shown that, after accelerated current flow, Cu is found to be almost completely depleted near the stud (Kwok, 1991). Depletion of Al at the interface is preceded by an incubation period during which Cu is swept past a threshold distance. The kinetic process is controlled by electromigration of Cu along the grain boundaries and not by dissolution of precipitates (Hu et al., 1993). Since the W barrier prevents replenishment of the Cu supply, it is not surprising that higher concentrations of Cu in the lines extend the lifetime (Filippi et al., 1992).

The flux divergence from the different materials has been identified as the principle reason for the poor performance of studded interconnects. Current crowding at the stud-line interface has also been cited as a contributor, even though local heating calculations show that the temperature rise is insignificant (Kwok et al., 1987b).

Estabil et al. (1991) found that the mode of failure, void-open or extrusion-short, depended on the current density in M1; as the current density decreased at constant M1 thickness, there was an increased tendency for failure by extrusion shorts.

Although the W-stud structure (example, Fig. VII-6) satisfies many of the needs of VLSI, the electromigration performance has been disappointing. Several suggestions have been made for improving it. Grass et al. (1994) reported that using a layered structure conductor, in which the Al alloy was capped with Ti/TiN, improved the lifetime of the stud structure (as it does with stripe lifetime) and eliminated the differences due to the direction of current flow. The benefit, with respect to the plug, was postulated to be protection of the Al alloy surface by a continuous $TiAl_3$ layer. Ting et al. (1995) reported that a conformal coating of the via hole by CVD TiN ensured a good barrier between the W plug and M1 and thus improved the lifetime.

Because of the existence of a critical current-length threshold (current density × length) below which no metal transport occurs during current stressing due to the backflow resulting from stress build-up (Blech and Herning, 1976), the interstud (via) spacing would be expected to influence the electromigration performance of a given stud (or via) chain. This effect has been demonstrated by Filippi et al. (1993) and by Aoki et al. (1994) where it was found that for interstud distances below a critical value there were no failures.

4.4.3 Other Studs

Lower resistivity metals, such as Al(Cu) and Cu can be deposited by techniques which fill holes as well as CVD W; their use reduces the interlevel resistance in study structures. But there is little information about the electromigration performance of these other stud structures.

AlCu studs have been formed by the "Flowfill" process in conjunction with a Ti/TiN liner deposited using a long target-to-substrate sputtering geometry (Dixit et al., 1994); the stud was capped with TiN. The AlCu plug exhibited an order of magnitude improvement in reliability over a W stud. This was postulated to be a result of reduced current crowding due to fewer dissimilar interfaces and higher stud conductivity. The improved performance of the Al(Cu) stud interconnect formed during the process was attributed to its larger grain size resulting from the "Flowfill" process.

5.0 CORROSION

5.1 Introduction

Basic discussions on corrosion, the electrochemical reaction, its characterization, and common measurement techniques can be found in many textbooks (Uhlig, 1966; Evans, 1960). Therefore, the discussions here will be limited to the corrosion of thin films used in microelectronics and their characterization.

Corrosion in thin film wiring can be described as the (unintended) loss of conducting metal by dissolution in an electrolyte or conversion to a nonconducting form in response to an external stress, which is primarily chemical or electrochemical in nature. The external stress is usually made of a combination of the following:

1. Impurities external to the film, such as moisture or corrosive gases
2. Temperature
3. Applied voltages
4. Mechanical stresses including that caused by coatings

Corrosion reaction can be described by an electrolytic cell having two electrodes and an electrolyte. A metal in equilibrium with its ions in an electrolyte (half-cell) can be written as a reaction, with a characteristic free energy of reaction and a corresponding electrical potential. Corrosion cell potentials (electromotive forces) of many metals have been measured with respect to a hydrogen half-cell (0.0 V reference potential), and the list of values are known as the *galvanic series*. Corrosion cell reaction is broadly classified as:

1. *Galvanic*, where elements with dissimilar emfs form a cell
2. *Concentration*, where the differential local ionic concentration creates sites of different potentials or emfs
3. *Externally impressed* potential (emfs) on electrodes

Electrodes or metals with large positive values of emf are usually active in a corrosive reaction (anodic), whereas the ones that have negative emfs tend to be passive (cathodic). However, the corrosion kinetics even in a simple corrosion cell can be changed by passivity which is when the element forms a passive surface layer, or by polarization when the local ionic concentration adjacent to the electrode is changed. Thin film metallization used in interconnection suffers from the presence of all of the above mechanisms contributing to corrosion cell formation and to corrosion. For example, the corrosion susceptibility of thin film conductors is influenced by the choice of metal (passivating, galvanic potential); the presence of dissimilar metallurgical junctions caused by layers or phases, and compositional inhomogeneities; the presence of electrolytes in contact with the film surface are required for the onset and continuation of corrosion. Externally impressed voltages on the conductors (biases) can accelerate the corrosion reaction.

Corrosion in thin films can be caused during processing and between process steps when residual impurities can react with films surfaces—this is known as *in-process* corrosion. *In-process* corrosion can lead to yield losses or cause reliability failures prematurely. When a functional device at the end of all processing cycles, fails in use or storage due to corrosion of an interconnect, then it becomes a reliability failure.

Reliability   **495**

5.2 RELIABILITY MEASUREMENTS / T&H STRESSING

Unlike corrosion studies of bulk materials using metal blanks or coupons, thin film used in interconnections are usually studied using a comb-serpentine structure (Fig. VII-7a), or a double maze structure of parallel meandering lines (Fig. VII-7b), or maze structures separated by an insulating layer (Fig. VII-7c). The maze lines can be built to the same design rules of the circuit using similar processes and structures. In addition, the adjacent lines can be electrically biased with respect to each other. The structure is usually terminated in a bond pad and mounted on a substrate. Wire bonding or other techniques are used to complete electrical connections between the substrate and the bond pad. The test chip is usually encapsulated in an organic or inorganic material and test structures on the chip are electrically biased, while the chip is under a high temperature and humidity (THV) environment. The leakage current between adjacent lines (corrosion current) is sometimes monitored

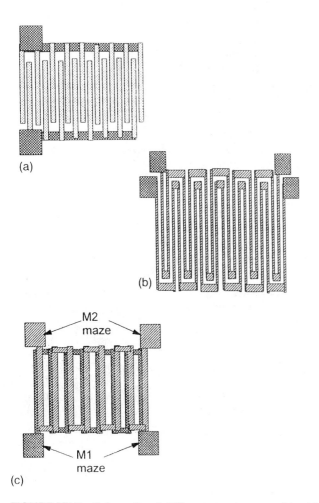

FIGURE VII-7 Schematic of different structures used in thin film THV corrosion studies.

and, alternatively, the test is interrupted periodically and the resistivity of the corroding lines are measured. The leakage current is a measure of the corrosion reaction rate whereas the change in resistivity is an indirect measure of the extent of corrosion. The test sample is examined using analytical tools to determine corrosive products and the extent and location of corrosion. Temperature and humidity testing is usually performed on packaged chips and is designed to test the integrity of the hermetic seal for preventing moisture diffusion and thus corrosion of the interconnects. The test conditions that are most widely used are 85°C, 85% relative humidity with a voltage bias. Although this is a rather simple "engineering" test, it is very useful in assessing the corrosive potential of the test structure in the context of the integrity of the encapsulant. Sometimes high-pressure and high-temperature tests (pressure cooker) are used to accelerate corrosion failures (Gunn et al. 1981). Typical corrosion models express fail time or lifetime, as an Arrhenius equation having an activation energy and field factors:

$$t = A \exp\left(-\frac{B}{T}\right) f(\text{RH}, V)$$

wherein t is the time for failure of a percent of a test population where failure is defined as exceeding a critical resistance change or leakage current, B is the activation energy of the corrosion rate limiting process, $f(\text{RH}, V)$ is some function of the RH (relative humidity), and V is the applied bias voltage; A is the preexponential value in the rate equation. Since the primary purpose of these equations is to determine the acceleration factors between stress and use conditions, the primary variables used are temperature, RH, and bias voltage. The sensitivity of each variable can be determined individually and a total acceleration factor calculated.

In an encapsulated test structure, the corrosive failure has many steps in sequence: absorption of moisture; migration of water to the film location; corrosive reaction at the metal/electrolyte interface; transport of ions under the electric field; etc. In most studies for structures that have insulator coatings, the diffusion of moisture determines the corrosion rate. Under this assumption, the activation energy is the energy for diffusion or permeation of water through the insulator. In most electronic applications, the semiconductor wires are usually well insulated, and most often the packaging elements and exposed pads experience corrosion failures, especially if they are not adequately encapsulated or the structures are exposed to aggressive environments.

5.3 CORROSION BEHAVIOR OF AL-CU CONDUCTORS

The most commonly used conductor for semiconductor interconnection is aluminum-based metallurgy. Often the metallurgy is alloyed with copper or silicon and occurs with layers of Ti, TiN, or TiW. In the context of the multilevel interconnection, even in the absence of externally impressed voltages, the lines are connected to silicon regions of different impurities and conductivities which make them susceptible for corrosion. Paulson and Kirk (1982) studied the corrosion of aluminum test structures coated with P-doped glass. In their study the corrosion rate of aluminum under electrical bias and humidity increased with the amount of P in the glass, presumably due to an increase in absorbed moisture from higher P_2O_5.

Humidity had the strongest effect on corrosion rate. The corrosion product was Al(OH)$_3$. Wada et al. (1986) studied aluminum corrosion using a double-layer metallization, where two mazes separated by an interlayer dielectric were stressed under a voltage bias and humidity. Vertical electrolytic current between the two mazes was monitored. The corrosion was always on the cathode, with an activation energy of 0.9 eV, irrespective of which maze, top or bottom, was biased as the cathode. When the top insulator and/or the interlayer are SiN, the corrosion diminished considerably indicating that SiN was an excellent barrier to moisture. P-doped oxide, undoped silicon dioxide, and PECVD SiN were compared for their passivation effectiveness over sputter deposited and patterned Al-2%Si in a THV stress test (Wada et al. 1989). It was found that electrolytic leakage and cathodic corrosion occurred for both doped and undoped oxide passivated samples. The corrosion was negligible for the case of SiN. The study also showed that applied bias had a strong effect on the corrosion current and the current leakage was primarily at the plastic resin and passivating layer interface. Will et al. (1987) studied in situ corrosion of aluminum metallization using a polymer passivation layer with pinholes. They found that electrolyte in the gap has a resistivity 100 times lower than distilled water, which they attributed to the leaching of ionizable impurities from the polymer coating.

Intergranular corrosion in Al-4.5%Cu, "missing Al" was seen when "theta" particles are overaged and made larger (Totta 1976). Kawal et al (1996) studied in situ corrosion behavior of age hardened aluminum-copper alloys in aqueous (dilute) hydrochloric acid and found extensive pit formation in the vicinity of second-phase particles; the dissolved aluminum redeposits as hydrated oxide at the local cathode; however, the main cathodic reaction becomes hydrogen evolution rather than Al deposition. Mayumi et al. (1990) observed a strong occurrence of corrosion-induced contact failure in double-level Al-Si-Cu metallization, which they attributed to the formation of large precipitates in the via holes, which served as a cathode and promoted the dissolution of adjacent Al (anode). Corrosion susceptibility of Al-Cu and Al-Cu-Si was studied by Lawrence et al. (1990) using unpassivated metal structures. The test structures consisted of an Al alloy overlaying a Ti-W barrier layer. The copper and silicon contents were varied from 0 to 2% and 0 to 1%, respectively. A controlled concentration of NH$_4$Cl was used to affect corrosion. The study found that corrosion sensitivity is greatly increased when the copper content is increased to between 1 and 2%. Addition of 1% Si to Al-2%Cu resulted in delaying the onset of corrosion, consistent with improved native oxide formation protecting the film. Weston et al. (1990) similarly observed that adding Si to Al-2%Cu reduced corrosion. This study also attributed localized surface corrosion to the galvanic action due to the presence of theta phase particles. Koelmans (1974) studied the corrosion of Al metallization in plastic-encapsulated ICs and concluded that water which a film absorbed at high relative humidity lead to passivation or destructive corrosion depending on the ionic impurities present. The presence of Cl$^-$ ions lead to poor quality oxide at the anode, whereas Na$^+$ moves to the cathode and increases the pH thereby promoting cathodic corrosion. Parekh and Price (1990) studied the effect of residual Cl levels on the corrosion of Al-Si metallization with copper addition and TiW layering. They found that in the range of Cl contamination studied, the corrosion rate was strongly influenced by the amount of residual Cl, and the galvanic cell effects of Cu and TiW were less significant.

Rotel et al. (1991) using potentiodynamic polarization measurements compared Al-Cu lines and Al-Cu lines containing Ti and Hf layers patterned by wet and RIE processes. The observations were: a decrease in copper reduced the corrosion rate, reactive ion etched Al-Cu showed more corrosion compared with wet etched Al-Cu, Ti and Hf layered structures had lower corrosion resistance compared to Al-Cu alone.

5.4 CORROSION BEHAVIOR OF COPPER CONDUCTORS

Copper is increasingly being considered as an alternative interconnect material in place of Al-Cu due to its lower resistivity. Even though corrosion studies with thin copper films in integrated circuits is limited, extensive observations exist on corrosion of thick films and bulk copper. Copper has an emf of 0.521 V in the galvanic series which makes it passive. Copper forms cuprous oxide (Cu_2O) cathodically in slightly acidic solutions, cupric oxide (CuO) anodically in alkaline solutions. Cupric oxide dissolves in dilute acids and ammonia. Cupric hydroxide is sometimes formed. Copper forms numerous complexes such as copper (of valency 1) with Cl^-, CN^-, NH_3, $S_2O_3^{2-}$, which are colorless. In chlorine solutions, complexes between Cu^+ and Cl^- can lead to appreciable dissolution of copper. A summary of liquid solutions and gases that react with copper can be found in Uhlig and Revie (1985). Copper is also known to corrode in oxidizing acids and aerated solutions. In seawater and fresh water, the presence of copper oxide films determines the corrosion rate (i.e., diffusion of oxygen). The corrosion rate of copper in aerated chloride solutions is roughly an order of magnitude higher than nonaerated chloride solutions (Bjorndahl and Nobe 1984). This study, using cyclic voltammetry with disc electrodes, concluded that in the presence of oxygen much higher concentrations of $CuCl_2^-$ ions are in equilibrium with Cu, which increased the Cu corrosion rates in oxygenated solutions. Rice et al. (1981) studied the atmospheric corrosion of copper and silver using coupons in a controlled ambient chamber with predetermined trace impurities of different gases such as sulfur dioxide, nitrogen dioxide, hydrogen sulphide, chlorine, hydrogen chloride, ammonia, and ozone. Copper corroded in SO_2, H_2S, Cl_2, HCl, and O_3, but the corrosion rate was very sensitive to the relative humidity and substantially increased at high humidity. The effect of organic inhibiting molecules on copper such as benzotriazole (Poling, 1970) and N-heterocyclic (Thierry and Leygraf, 1985) have been studied for passivating against corrosion. These and other similar studies will not be discussed here, but interested readers can consult relevant articles since they may suggest means for protecting thin copper lines in multilevel structures.

6.0 INSULATOR RELIABILITY

6.1 GENERAL REMARKS

Insulators for interconnect application are primarily used in a passive mode, to insulate and isolate conductors from adjacent ones and underlying silicon where so desired, and further act as a passivant to hermetically seal the conductor and devices from impurities such as water or ions that can cause corrosion of the conductors, or device instabilities. Current leakage and breakdown with use

Reliability

(time dependent dielectric breakdown—TDDB) are two important reliability or quality concerns for an insulator. For a good passivating film, the insulator should be a barrier to moisture with few pinholes or defects. If the insulator is used adjacent to silicon, it is desirable that it be a barrier or a scavenger of alkaline ions. Silicon dioxide is the most widely used insulator for interconnect applications and silicon nitride is used to a much lesser extent, usually in conjunction with an oxide. Organic insulators such as polyamide have been used as scratch coating and as final passivants. Increasingly the very small features and spaces in the silicon devices are necessitating the use of organics, spin-on-glass, and other materials as gap-fillers. Most breakdown fields are orders of magnitude higher than the average field experienced by insulating films used in interconnect applications. However, such studies are likely to explain how defect locations start conducting and breakdown.

6.2 ELECTRICAL CONDUCTION/LEAKAGE

A good quality insulator without impurities or defects easily satisfies the leakage requirement of less than 1 μA typically specified for interconnect applications, as it has a large band gap and it takes very large fields to excite electrons across the gap. However, insulator properties are usually degraded by contact with electrodes, which modify the energy barriers, and by the presence of traps, ionic impurities, and charges from compositional deficiencies. Manufacturing defects lead to thinning of insulators. Poor control of deposition processes can produce off-stoichiometry films which can result in conducting islands interspersed with insulating regions. Electrical conduction mechanisms in insulators (Sze 1981) are grouped as follows: (1) Schottky emission, the thermionic emission of electrons across the electrode-insulator interface; (2) Poole-Frenkel, the field-enhanced thermally excited trapped electrons or holes into the conduction band; (3) Fowler-Nordheim, the tunneling of electrons from the metal fermi-level into the insulator conduction band; (4) injection of charge without a compensating charge across a space-charge region; (5) ionic, the migration of ions under temperature and electrical field; and (6) ohmic. All of these mechanisms are present in some combination at the temperature and electrical field used. Studies on thin oxide film suggests that at device temperatures, electron injection (Fowler-Nordheim) dominates and there is negligible hole or electron conduction. The electrode material has some effect, presumably due to the influence on electron injection. Postmetallization anneals reduce leakage. For the thickness of insulators employed for interconnection, it would be accurate to say that any measurable current is caused by defects.

Conduction in thin silicon nitride has been proposed to be caused by the Poole-Frenkel mechanism (Sze, 1967) and by trapped holes. The current is bulk controlled and shows a weak electrode and a strong temperature effect. Both silicon nitride and aluminum oxide have higher conduction at low and moderate fields compared to silicon dioxide. Most circuit applications can withstand leakage that does not contribute to unintended device switching or unacceptable voltage drops. The primary concern is when there is leakage in the insulator, since it usually leads to breakdown within a very short period.

6.3 TIME-DEPENDENT DIELECTRIC BREAKDOWN (TDDB)

6.3.1 General Remarks

Dielectric breakdown is a major concern for applications involving thin dielectrics such as gate or storage capacitors. TDDB is conveniently divided into intrinsic breakdown and defect-related breakdown. The second is usually more of a concern for intermetal dielectrics, as the nominal thickness used is appreciable (0.5 to 1 μm). However, it is likely that very thin insulators can be present between adjacent conductors due to process and design tolerances in ULSI devices using submicron ground rules. Solomon (1977) has reviewed different models for intrinsic breakdown for silicon dioxide and favored the impact ionization—recombination model. According to this model, electrons tunnel into a thin SiO_2 region by the Fowler-Nordheim mechanism. A fraction of the electrons with very high energies ionize the lattice atoms creating electron-hole pairs; holes are locally trapped increasing the field at the cathode, which increase electron injection thereby causing a localized breakdown.

6.3.2 Test Methodology/Results

TDDB is studied using a capacitor structure wherein the desired insulator is sandwiched between a bottom electrode (usually silicon) and a top electrode which is usually aluminum MOS or MIM test structures. Commonly used test sites are built on a degeneratively doped silicon substrate with the back side metallized. The top electrode usually has varying areas (dots) which is usually aluminum unless the effect of the electrode is also being studied. Plasma processes and e beam processes are usually avoided because of the concern of damage and the accompanying degradation in breakdown. The most commonly used test subjects the capacitors to a constant electrical voltage (field) at different temperatures; the failure behavior with time is observed and analyzed. The breakdown population is divided into two groups, intrinsic breakdown and extrinsic breakdown. Oxides of different thickness have been studied and in one of the models, the logarithmic failure rate has been found to be proportional to the applied field where the activation energy is also affected by the electrical field (Crook, 1979; Berman, 1981). The corresponding physical model describes "breakdown" as the activated process. A more commonly subscribed-to model predicts the failure rates to depend on the reciprocal of the field. In this $1/E$ model (Chen et al., 1985; Lee et al., 1988; Moazzami et al., 1989), holes are created by electron injection and impact ionization, and the trapped holes increase the electric field at the cathode interface leading to an increase in electron injection, which lead to local breakdown. In another method, the breakdowns were treated as a stochastic process (Shatzkes and AvRon, 1981); the test structures were ramped from zero to breakdown voltage at different ramp rates and at different temperatures. The fails in SiO_2 stressed under different temperatures, electrical fields, and voltage ramp rates were analyzed using a single plot of $R \ln(1 - F_{fail})$ versus F_c(cathode field). F_c is the applied field E, corrected for electron capture and release, as a function of ramp rate. Other methods of stressing without an electrode are corona discharge, laser breakdown, and e-beam-EBIC mode. Internal photoemission is used to determine barrier inhomogeneities. Prendergast et al. (1995) studied the breakdown of thermally grown oxide in the temperature range of 175 to 400°C at constant fields ranging from 7 to 11 MV/cm. The breakdown data showed a bimodal distribution at moderate fields and temperatures and concluded that intrinsic breakdown data exhibited a direct E dependence and the extrinsic fail population exhibited a reciprocal E dependence. Breakdown results with defects vary widely. Shatzkes et al. (1980) has studied the variability of the Si-SiO$_2$ interface and identified regions of low-barrier heights as defects. As much as agreement on models are important for failure rate projections, there is wide variation in the breakdown results obtained from insulators with different thicknesses. Ting et al. (1991) reported that oxidizing Si in N$_2$O resulted in lower oxide growth and the resulting oxide showed some pileup of nitrogen at the interface similar to nitrided silicon dioxide films. This film showed superior TDDB under hot electron stressing. Kim et al. (1994) studied nitriding native oxide in ammonia followed by depositing nitride from silane-ammonia mixtures and reported that these films had higher breakdown and lower leakage compared to ONO films. The degradation of TDDB of stacked ONO films and the role of nitride deposition temperature was studied by Tanaka et al., (1993) who concluded that lower temperature deposited Si$_3$N$_4$ films had improved TDDB due to reduced film roughness. Clearly, the TDDB behavior of thin dielectrics of composite films, is strongly affected by the structure and atomic distribution in the thin insulating layers. In thicker composite layers, bulk defects and inhomogeneities add to the complexities of the conduction and breakdown behavior.

6.3.3 TDDB: A Reliability Issue for Thick Insulators?

The average electrical field experienced by the insulator in the interconnection is quite small; however, defects are the primary concerns. In fact, it will be justified that ramp breakdown stress tests for thick insulators in low fields is usually indicative of defects. Such use is common in semiconductor manufacturing to quantify the defect levels by use of ramping a capacitor test structure with electrodes formed of two levels of interconnection, with the dielectric sandwiched in between. Two types of defects tend to affect the field reliability: the first, environmentally caused defects such as from particulates; the second, caused by processes or designs. The first type invariably causes pinholes or local thinning. The latter causes local damage off-stoichiometry, islands of conducting materials, narrow spaces, ionic contaminations, etc. The general trends of how these affect the breakdown are known. For example, Fe impurities reduced the breakdown voltage with an increase in metallic concentration (Henley et al., 1993). The understanding of how these defects affect acceleration factors and projection of failure rates from accelerated to use conditions is entirely unknown. The most powerful approach is to use redundant insulating layers (Gati et al., 1986; Joseph and Wong, 1986) and create a composite layer structure to reduce the overall defects. Equally important is to prescribe design rules for given process tolerances to guarantee some minimum insulator thickness between wiring.

7.0 CONCLUDING REMARKS

This chapter has reviewed interconnect reliability—electromigration, stress-voiding, corrosion, and insulator reliability. The understanding of the interaction between microstructure, physical structure (layering), and behavior of conducting films under applied currents and elevated temperature has been reviewed. Al-based systems have served the needs of semiconductor applications for over two decades and only recently has a serious alternative, namely, copper, being considered. The factors used to improve the reliability of aluminum are being applied to copper conductors today, namely, alloying additions, microstructure control, and the use of cladding layers. In insulators, a switch from deposited silicon dioxide to lower dielectric materials is being seriously pursued. In many designs, defects can overwhelm the field failure rate. Test methods for rapid determination of wear out as well as defect assessment is important. Burn-in techniques, even though not reviewed here, continue to play a critical role in guaranteeing a minimum product performance; but burn-in selection still requires an understanding of the defects and failure mechanisms involved. In determining the reliability of a packaged structure, many assumptions need to be made, which can determine the accuracy of the projected product reliability. These assumptions, if they are to be trusted, must be based on an understanding of the various failure mechanisms involved.

REFERENCES

Ainslie, N. G., F. M.d'Heurle, O. C. Wells, *Appl. Phys. Lett.*, 20, 173 (1972).
Ames, I., F. M. d'Heurle, R. E. Horstmann, *IBM J. Res. Develop.*, 14, 461 (1970).
Aoki, T., U. Kawano, T. Nogami, *IEEE/VMIC*, 266 (1994).
Ball, R. K., A. G. Todd, *Thin Solid Films*, 149, 269 (1987).
Bauer, H. J., *8th Intl. Vac. Conf.*, Vol. I, 649 (1980).
Bauer, H. J., *Journal of Vacuum Science and Tech.*, B12, 2405 (1990).
Ben-Tzur, M., R. Fastow, M. Eizenberg, J. Rosenberg, M. Frenkel, *J. Vac. Sci. Technol.*, A8, 4069 (1990).
Berman, A., in *IEEE/IRPS*, p. 204, 1981.
Bjorndahl, W. D., Ken Nobe, *Corrosion*, 40(2), 82 (1984).
Black, J. R., *IEEE Trans. Electron Devices*, ED-16, 338 (1969).
Black, J. R., *Proceedings of the IEEE*, 57, 1587 (1969).
Blair, J. C., P. B. Ghate, C. T. Haywood, *Appl. Phys. Lett.*, 17, 281 (1970).
Blech, I. A., C. Herring, *Appl. Phys. Lett.*, 29, 131 (1976).
Bower, R. W., *Appl. Phys. Lett.*, 23, 99 (1973).
Bui, N. D., V. H. Phan, J. T. Yue, D. L. Wollensen, *IEEE/VMIC*, 142 (1990).
Chen I.-C, S. E. Holland, C. Hu, *Trans of Electron Devices*, ED-32(2), (1985).
Chen, T. M., T. P. Djeu, R. D. Moore, *IEEE/IRPS* (1985).
Clement, J. J., J. R. Lloyd, C. V. Thompson, *MRS Symp. Proc.*, 391, 423 (1995).
Colgan, E. G., *Materials Science Reports*, 5, 1 (1990).
Colgan, E. G., K. P. Rodbell, *J. Appl. Phys.* 75, 3423 (1994).
Crook, D. L., in *IEEE/IRPS*, 1 1979.
d'Heurle, F. M., N. G. Ainslie, A. Gangulee, M. C. Shine, *J. Vac. Sci. Technol.*, 9, 289 (1972).
d'Heurle, F. M., P. S. Ho, *Thin Films: Interdiffusion and Reactions* (J. Poate, K. N. Tu, J. Mayer eds.), John Wiley, New York, 1978.
Ding, P. J., W. A. Lanford, S. Hymes, S. P. Murarka, *MRS Proceedings*, 265, 199 (1992).
Dixit, G. A., M. F. Chisolm, M. K. Jain. T. Weaver, L. M. Ting, S. Poarch, K. Mizobuchi, R. H. Havemann, C. D. Dobson, A. I. Jeffryes, P. J. Holerson, P. Rich, D. C. Butler, J. Hems, *IEEE/IEDM*, 105 (1996).
Estabil, J. J., H. S. Rathore, F. Dorleans, *IEEE/IRPS*, 57 (1991).

Evans, U. R., *The Corrosion and Oxidation of Metals: Scientific Principles and Practical Applications*, Edward Arnold Pub., London, 1960.
Felton, L. E., J. A. Schwarz, R. W. Pasco, D. H. Norbury, *J. Appl. Phys.*, 58, 723 (1985).
Filippi, R. G., G. A. Biery, M. H. Wood, *MRS Proceedings*, 309, 141 (1993).
Filippi, R. G., R. A. Wachnik, H. Aochi, J. R. Lloyd, M. A. Korhonen, *Appl. Phys. Lett.*, 69, 2350 (1996).
Filippi, R. G., H. S. Rathore, R. A. Wachnik, D. Kruger, *IEEE/VMIC*, 359 (1992).
Finetti, M., H. Ronkainen, M. Blomberg, I, Suni, *Mat. Res. Symp. Proc.*, 54, 811 (1986).
Flinn, P. A., D. S. Gardner, W. D. Nix, *IEEE Trans. on Electron. Devices*, ED-34, 689 (1987).
Frankovic, R., G. H. Bernstein, *MRS Proceedings*, 391, p 403 (1995).
Frear, D. R., J. E. Sanchez, A. D. Romig, J. W. Morris, Jr., *Metall. Trans.*, 21A, 2449 (1990).
Frear, D. R., J. R. Michael, C. Kim, A. D. Romig, Jr., J. W. Morris, Jr., *IEEE/SPIE Proc.*, San Jose, CA 72-82. (1991).
Fujimura, N., N. Nishida, T. Ito, Y. Nakayama, *Materials Science and Engi.*, A108, 153 (1989).
Gardner, D. S., J. Onuki, K. Kudoo, Y. Misawa, Q. T. Vu, *Thin Solid Films*, 104 (1995).
Gardner, D. S., K. Saraswat, T. W. Barbee, US Pat. 4 673 623, June 16, 1987.
Gardner, D. S., P. A. Flinn, *J. Appl. Phys.* 67, 1831 (1990).
Gardner, D. S., T. L. Michalka, K. C. Saraswat, T. W. Barbee, J. P. McVittie, J. D. Meindl, *IEEE Electron Devices*, ED-32 (1985).
Gati, G. S., A. P. Lee, G. C. Schwartz, C. L. Standley, 1986, US Pat. 4,601,939, July 22, 1996.
Grabe, B., H.-U. Schreiber, *Solid State Electronics*, 26, 1023 (1983).
Grass, C. D., H. A. Le, J. W. McPherson, R. H. Havemann, *IEEE/IRPS*, 173 (1994).
Gunn., J. E., S. K. Malik, P. M. Mazumdar, *Reliability Physics*, IEEE 81CH1619-6, (1981).
Han, C. C., R. W. Bene, *Appl. Phys. Lett.*, 47, 1077 (1985).
Henley, W. B., L. Jastrzebski, N. F. Haddad, *IEEE/IRPS*, p. 22, 1993.
Hong, C. C., D. L. Crook, *IEEE/IRPS* (1985).
Hong, Q. Z., D. A. Lilienfeld, J. W. Mayer, *J. Appl. Phys.*, 64, 4478 (1988).
Hoshino, K., H. Yagi, H. Tsuchikawa, *IEEE/VMIC* (1990).
Howard, J. K., J. F. White, P. S. Ho, *J. Appl. Phys.*, 49, 4083 (1978).
Howard, J. K., P. S. Ho, US Pat. 4 017 890, April 12, 1977.
Howard, J. K., R. F. Lever, P. J. Smith, P. S. Ho, *J. Vac. Sci. Technol.*, 13, 68 (1976).
Hu C-K, M. B. Small, P. S. Ho, *MRS Proceedings*. 265, p. 171 (1992).
Hu C-K., B. Luther, F. B. Kaufman, J. Hummel, C. Uzoh, D. J. Pearson *Thin Solid Films*, 262, 84 (1995).
Hu, C.-K., K. P. Rodbell, T. Sullivan, K. Y. Lee, D. P. Bouldin, *J. IBM Research & Development*, 39, 465 (1995).
Hu, C. K., N. J. Mazzeo and C. Stanis, *Materials Chemistry and Physics*, 35(1), 95 (1993).
Hurd, J. L., K. P. Rodbell, D. B. Knorr, N. L. Koligman, *MRS Proceedings*, 343, 653 (1994).
Isobe, A, Y. Numazawa, M. Sakamoto, *1989 VMIC*, 161 (1989).
Jones, R. E., L. D. Smith, *J. Appl. Phys.*, 61, 4670 (1987).
Joseph, R. R., M-C. Wong, US Pat. 4600624, 1986.
Kahn, H., C. V. Thompson, *Appl. Phys. Lett.*, 59, 1308 (1991).
Kang, H.-K, I. Asano, S. S. Wong, J.A.T. Norman, VMIC-1992, 280 (1992).
Kang, H. K., I. Asano, C. Ryu, S. S. Wong, J.A.T. Norman, *IEEE-VMIC*, 223 (1993).
Kawal, K. J. DeLuccia, J. Y. Josefowicz, C. Laird, G. C. Farrington, *JECS*, 143(8), 2471 (1996).
Keller, R. M., S. Bader, R. P. Vinci and E. Arzt, *Mat. Res. Soc. Proc.*, 356, 435 (1995).
Kim, K. H., D. H. Ko, S. H. Kang, S. T. Kim, S. J. Shim, E. S. Kim, S. T. Ahn, *J. Electronic Mater.* 23(12), 1273 (1994).
Knorr, D. B., D. P. Tracy, K. P. Rodbell, *Appl. Phys. Lett.*, 59, 3241 (1991).

Knorr, D. B., *MRS Symp. Proc. 309*, 127 (1993).
Knorr, D. B., K. P. Rodbell, *J. Applied Physics*, 79, 2409 (1996).
Koch, R. H., J. R. Lloyd, J. Cronin, *Phys. Rev. Lett.*, 55, 2487 (1985).
Koelmans, H. *IEEE Reliability Physics*, (1974).
Kordic, S., R.A.M. Wolters, K. Z. Troost, *J. Appl. Phys.*, 74, 5391 (1993).
Korhonen, M. A., P. Borgensen, K. N. Tu, C.-Y. Li, *J. Appl. Phys.* 73, 3790 (1993).
Korhonen, M. A., T. Liu, D. D. Brown, C.-Y. Li, *MRS Symp. Proc. Vol.*, 391, 411 (1995).
Krafesik, I., J. Gyulai, C. J. Palmstrom, J. W. Mayer, *Appl. Phys. Lett.*, 43, 1015 (1983).
Kwok, T. *IEEE/SPIE Proc.*, San Jose, CA V1596, 60 (1991).
Kwok, T., C. Tan, D. Moy, J. J. Estabil, H. S. Rathore, S. Basavaiah, *IEEE/VMIC*, 106 (1990).
Kwok, T., P. S. Ho, *Diffusion Phenomena in Thin Films and Microelectronic Materials*, (D. Gupta, P. S. Ho, eds.) Noyes Publ., Park Ridge, NJ, p. 369, 1988.
Kwok, T., T. Nguyen, S. Yip, P. Ho, *IEEE/IRPS* (1987a).
Kwok, T., T. Nguyen, S. Yip, P. Ho, *IEEE/VMIC*, 252 (1987b).
Lawrence, J. D., J. W. McPherson, V. T. Cordasco, *JECS*, 137(12), 3879 (1990).
Learn, A. J., *J. Electron. Mater.*, 3, 531 (1974).
Learn, A. J., W. H. Sheperd, *IEEE/IRPS*, 129 (1971).
Lee, J. C., I.-C. Chen, C. Hu, *IEE/IRPS*, 26, 131 (1988).
Levy, R. A., L. C. Parrillo, L. J. Lecheler, R. V. Knoell, *J. Electrochem. Soc.*, 132, 159 (1985).
Li, J., and J. W. Mayer, *MRS Bulletin, June*, 52 (1993).
Livesay, B. R., N. E. Donlin, A. K. Garrison, H. M. Harris, J. L. Hubbard, *Proc. of 30th IRPS*, 217 (1992).
Lloyd, J. R., *MRS Symp.*, 391, 231 (1995).
Lloyd, J. R., F. M. Smith, G. S. Prokop, *Thin Solid Films*, 93, 395 (1982).
Lloyd, J. R., P. M. Smith, *J. Vac. Sci. Technol.*, A1, 455 (1983).
Lloyd, J. R., R. N. Steagall, *J. Appl. Phys.*, 60, 1235 (1986).
Longworth, H. P., C. V. Thompson, *J. Appl. Phys.*, 69, 3929 (1991).
Lowry, L. E., B. H. Tai, J. Mattila, L. H. Walsh, *MRS Symp. Proc. 309*, 205 (1993).
Luther, B., et al., *IEEE/VMIC*, 1993; p. 15.
Martin, C. A., J. W. McPherson, *IEEE/VMIC*, 168 (1989).
Matsuoka, F., H. Iwai, K. Hama, H. Itoh, R. Nakata, T. Nakakubo, K. Maegichi, K. Kanzaki, *IEEE Trans. on Electron Devices*, 37, 562 (1990).
May, J. S., *IEEE/IRPS*, 91 (1991).
Mayumi, S., I. Murozono, H. Nanatsue, S. Ueda, *JECS*, 137(6), 1861 (1990).
Moazzami, R., J. C. Lee, C. Hu, *Trans. Electron Devices*, ED-36(11), 2462 (1989).
Murarka, S. P., J. Steigerwald, R. J. Gutmann, *MRS Bulletin*, June 46 (1993).
Nahar, R. K., N. M. Devashrayee, W. S. Khokle, *J. Vac. Sci. Technol.*, B6, 880 (1988).
Nishimura, H., Y. Okuda, K. Yano, *J. Electrochem. Soc.*, 142, 3565 (1995).
Nitta, T., T. Ohmi, T. Hoshi, S. Sakai, K. Sakaibara, S. Imai, T. Shibata, *J. Electrochemical Society*, 140(4), 1131 (1993).
Oates, A. S., *IEEE/IRPS*, 28, 20 (1990).
Oates, A. S., F. Nkansah, S. Chittipeddi, *J. Appl. Phys.*, 72, 2227 (1992).
Okabayshi, H., *Second International Stress Workshop on Stress Induced Phenomena in Metallization*, (K. Aizawa, P. Ho, P. Totta, C.-Y. Li, eds.), America Institute Physics, New York, 1994, p. 33.
Okuyama, K., T. Fujii, Y. Tanigki, R. Nagai, *1990 Symp. on VLSI Technol.*, 35 (1990).
Onoda, H., K. Touchi, K. Hashimoto, Jpn. J. Appl. Phys., L1037 34, (1995).
Onuki, J, Y. Koubuchi, S. Fukada, M. Suwa, M. Koizumi, D. S. Garder, H. Suzuki, E. Minowa, *IEDM*, 43 (1990).

Paraszczak, J., D. Edelstein, S. Cohen, B. Babich, J. Hummel, *Proc. of IEDM*, IEEE, New York, 1993.
Parekh, N., J. Price, *JECS*, *137*(7) 2199 (1990).
Pasco, R. W., J. A. Schwarz, *Solid State Electronics*, *26*, 445 (1983).
Paulson, W. M., R. W. Kirk, *IEEE/IRPS*, 172 (1982).
Poling, G. W., *Corrosion Science*, *10*, 359 (1970).
Pramanik, D., V. Chowdhury, V. Jain, *IEEE/IRPS*, 261 (1994).
Prendergast, J., J. Suehle, P. Chaparala, E. Murphy, M. Stephenson, *IEEE* (1995).
Raman, A., K. Schubert, Z. *Metallkde.*, *56*, 44 (1965).
Rathore, H. S., *IEEE/IRPS*, 77 (1982).
Rathore, H. S., R. G. Filippi, R. A. Wachnik, J. J. Estabil, T. Kowk, *Second International Stress Workshop on Stress Induced Phenomena in Metallization*, America Institute Physics, 1994, p. 165.
Rice, D. W., P. Peterson, E. B. Rigby, P.B.P. Phipps, R. J. Cappell, R. Tremourex, *J. Electrochemical Society*, 275 (1981).
Rodbell, K. P., J. D. Mis, D. B. Knorr, *J. Electronic Materials*, *22*, 597 (1993).
Rodbell, K. P., J. L. Hurd, P. W. DeHaven, *MRS Proceedings*, *403*, 617 (1996).
Rodbell, K. P., D. B. Knorr, D. P. Tracy, *MRS Proceedings*, *265*, 107 (1992).
Rodbell, K. P., P. A. Totta, J. F. White, U.S. Pat. 5 071 714, December 10, 1991.
Rodbell, K. P., P. W. Dehaven, J. D. Mis, *MRS Proceedings*, *225*, 91 (1991).
Rodbell, K. P., R. H. Koch, *Physical Review B*, *44*, 1767 (1991).
Root, B. J., T. Turner, *IEEE/IRPS* (1985).
Rosenberg, R., *J. Vac. Sci. Technol.*, *9*, 263 (1972).
Rosenmayer, C. T., F. R. Brotzen, J. W. McPherson, C. F. Dunn, *IEEE/IRPS*, 52 (1991).
Rotel, M., J. Zahavi, H. C. W. Huang, P. A. Totta, G. C. Schwartz, in "Corrosion of Electronic Materials and Devices", (J. D. Sinclair, ed.), The Electrochemical Society, New Jersey, 1991, vol 91-29 p. 387.
Russell, S. W., Jian, Li, J. W. Strane, J. W. Mayer, *MRS Proceedings*, 265, p. 205 (1992).
Schafft, H. A., C. D. Younkins, T. C. Grant, C.-Y. Kao, a. N. Saxena, *IEEE/IRPS*, 250 (1984).
Schlacter, M. M., E. S. Schlegel, R. S. Keen, R. A. Lathlaen, G. N. Schnable, *IEEE Trans. on Electron Devices*, *ED-17*, 1077 (1970).
Schwarzer, R. A., D. Gerth, *J. Elect. Mat.*, *22*, 607 (1993).
Scorzoni, A., B. Neri, C. Caprile, F. Fantini, *Mat. Sci. Reports*, *7*, 143 (1991).
Shatzkes, M., M. AvRon, R. Gdula, *IBM J of R&D.*, *24*, p. 469 (1980).
Shatzkes, M., M. AvRon, *IEEE/IRPS*, 210 (1981).
Simpson, P. J., M. T. Umlor, K. G. Lynn, K. P. Rodbell, *Applied Physics Letters*, *65*, 52 (1994).
Slusser, G. J., J. G. Ryan, S. E. Shore, M. A. Lavoie, T. D. Sullivan, *J. Vac. Sci. Technol. A*, *7*, 1568 (1989).
Solomon, P, *Journal of Vacuum Science and Tech*, *14*(5), 1122 (1977).
Spitzer, S. M., S. Schwartz, *IEEE Trans. on Electron Devices*, *ED-16*, 348 (1969).
Spitzer, S. M., S. Schwartz, *J. Electrochem. Soc.*, *116*, 1368 (1969).
Sullivan, T. D., J. G. Ryan, J. R. Riendeau, D. P. Bouldin, *IEEE SPIE Proc*, San Jose, California, 1991.
Sze, S. M., *Physics of Semiconductor Devices*, John Wiley & Sons, New York, 1981.
Sze, S. M., *J. Appl. Phys. 38*, 2951 (1967).
Tanaka, H., H. Uchida, T. Ajioka, N. Hirashita, *IEEE Trans. of Electron Devices*, *40*(12), 2231 (1993).
Tao, J., K. Konard, N. Cheung, C. Hu, *IEEE Trans. Electron Dev.* *40*, 1398 (1993).
Tardy, J., K. N. Tu, *Phys. Rev. B*, *32*, 2070 (1985).
Thierry D., C. Leygraf, *J. Electrochemical Society*, *132*(5), 1009 (1985).

Thuillard, M., L. T. Tran, M.-A. Nicolet, *Thin Solid Films*, *166*, 21 (1988).
Ting, W., G. Q. Lo, J. Ahn, D. L. Kwong, *Int. Symp. of VLSI Technology, System and Applications*, IEEE, 1991, p. 47.
Ting, L. M., G. Dixit, M. Jain, K. A. Littau, H. Tran, M. Chang and A. Sinha, *Mat. Res. Soc. Proc.*, *391*, 453 (1995).
Tobias, D. B., D. Trindale, *Applied Reliability*, Van Nostrand Reinhold, New York, 1986, p. 140.
Tomioka, H., S. Tanabe and K. Mizukami, *IEEE/IRPS*, 53 (1989).
Totta, P. A., *J Vacuum Science & Tech.*, *13(1)*, 26 (1976).
Uhlig, H. H., R. Winston Revie, "Corrosion and Corrosion Control", John Wiley & Sons, New York, 1985, p. 330.
Uhlig, H. H., "Corrosion Handbook", John Wiley & Sons, New York, Ninth Printing, 1966.
Vaidya, S., A. K. Sinha, *Thin Solid Films*, 75, 253 (1981).
Vaidya, S., T. T. Sheng, A. K. Sinha, *Appl. Phys. Lett.* 36, 464 (1980).
van Loo, F.J.J., G. D. Rieck, *Acta Metallurgical*, 21, 61, 73 (1973).
Vossen, J. L., *Appl. Phy. Lett.*, 23, 287 (1973).
Wada, T., H. Higuchi, T. Ajiki, *JECS*, *133(2)*, 362 (1986).
Wada, T., M. Sugimoto, T. Ajiki, *JECS*, *136(3)*, 732 (1989).
Wada, Y., *J. Electrochem. Soc.*, *133*, 1432 (1986).
Weston, D., S. R. Wilson, M. Kottke, *JVST A8(3)*, 2025 (1990).
Will, Fritz G., K. H. Janora, J. G. McMullen, A. J. Yerman, *IEEE IRPS*, 34 (1987).
Wittmer, M., F. LeGoues, H.-C. W. Huang, *J. Electrochem. Soc.*, *132*, 1450 (1985).
Witvrouw, A., Ph. Roussel, B. Deweerdt, K. Maex, *MRS Symp. Proc. 391*, 447 (1995).
Yamaha, T., M. Naitou, T. Hotta, *IEEE/IRPS*, 349 (1992).
Yau, L., C. HOng, D. Crook, *IEEE/IRPS*, 115 (1985).
Zhao, X.-A., F.C.T. So, M.-A. Nicolet, *J. Appl. Phys.*, *63*, 2800 (1988).
Zielinski E. M., R. P. Vinci, J. C. Bravman, *MRS Proceedings*, *391*, 103 (1995).
Zielinski, E. M., R. P. Vinci, J. C. Bravman, *MRS Proceedings*, *391*, 303 (1995).

VIII
Contamination Control in Multilevel Interconnection Manufacturing

A. Rapa
Jacobs Engineering
Phoenix, Arizona

Arthur Bross
IBM Microelectronics*
Hopewell Junction, New York

1.0 INTRODUCTION
1.1 OVERVIEW

Random defects and their control during manufacturing are major factors affecting the cost and reliability of semiconductor microelectronics products. Most random defects are caused by contamination introduced during wafer processing. Contamination-related defects result in scrapped wafers, lower yields, and poor reliability which, at worst, may severely constrain the ability of a manufacturing line to deliver the required volumes of high quality products at the right price and on time. The importance of controlling contamination is well recognized in the semiconductor industry and is represented by large capital investments in cleanrooms and process equipment specifically designed to reduce defects. The need for controlling contamination can also be seen in the increasingly stringent specifications for high-purity raw materials (e.g., evaporator sources, chemicals, gases, etc.) and in the cleanroom protocol demands placed on manufacturing personnel. These efforts exact a high price, and a significant portion of microelectronic chip costs can be attributed, directly or indirectly, to contamination control. As devices are decreasing in size and increasing in density they are becoming increasingly more sensitive to defects caused by contaminants. As a result, the competitiveness of a semiconductor manufacturing line (often referred to as a wafer fabrication facility or fab) will be

---

*Retired.

based, in large measure, on its ability to control product defects and the various costs associated with controlling contaminants.

This chapter examines the various interactive factors relative to the sources, control and prevention of contamination-related defects during thin film (interconnection) processing. Practical examples and applications are given to further aid the reader in addressing day-to-day on-line contamination problems.

1.2 CONVENTIONS AND DEFINITIONS

Contamination control in semiconductor manufacturing is an emerging technology where new terms are evolving and the meanings of words are changing to fit specific needs. As a result, contamination control terminology may have slightly different meanings at different locales. Every attempt has been made to use the most commonly used definitions. In some cases terms and definitions are modified to help clarify the discussion of a particular topic. These modifications are defined within the text wherever they occur. The following are definitions of commonly used words as they apply to this chapter.

Cleanroom. Cleanroom (single word) is becoming accepted spelling for specially designed controlled areas where air is filtered to control particle contamination. This term differentiates a cleanroom from a room which is described as "clean" but is not specifically designed for the purpose of controlling contamination. In a cleanroom, the word "clean" could also mean control of various other contaminants such as ionics, viable particles (bacteria, viruses), volatile organic compounds, heat, static electricity, and others.

Process equipment and process tools. These terms are often used interchangeably for describing apparatus used to fabricate device structures (e.g., wafer stepper, ion implanter, metal sputtering system, etc.). The term *process equipment* is used in this text predominantly. It should be noted, however, that there are many other tools used in manufacturing which are not directly involved in affecting change to the wafer. These include wafer handling aids such as vacuum tweezers, carriers, boxes, sorters, and storage enclosures. From a contamination control perspective, all types of apparatus used during wafer fabrication, whether used directly or indirectly, are capable of causing defects. Therefore process equipment and tools are defined in this chapter as *any* apparatus used in wafer fabrication.

Support equipment. Ancillary items used to support process equipment (e.g., vacuum pumps, chemical and gas dispensing systems, etc.) are referred to as *support equipment*. Recently, these items have increased significantly in technology, quantity, and size. Support equipment plays an important role in the control of contamination-related defects and, therefore, it is given its own identity in this chapter.

Defects and electrical faults. A *defect* is defined here as a pattern deformity or abnormality that *may or may not* cause the device to fail electrically. For example, a particle obstruction may result in a circuit deformity such as a void in metal line. Although this void is a *defect*, it may not be large enough to cause a failure in the device's functional performance. If the particle obstruction is large enough, however, the metal void may span the width of the line causing an *electrical fault* rendering the device inoperative. Sometimes a large defect may not be of sufficient size to cause the chip to be rejected during final electrical testing. The chip may function normally for a period of time but fail later while installed in operational electronic equipment. Chip abnormalities which cause reliability problems are referred to as *latent defects* or *latent faults*. It is important to note that a contaminant such as a particle on a wafer is neither a defect nor an electrical fault. For example, many particles may deposit on the wafer during processing but few will actually cause defects and electrical faults.

Protocols and procedures. Although these terms are often used interchangeably, they have a slightly different meaning in contamination control. *Procedures* are the steps required to perform a function (e.g., repairing inoperative equipment) while *protocols* are the special work practices and rules which must be strictly adhered to in order to achieve effective contamination control. Protocols include the wearing of cleanroom garments and other cleanroom entry requirements such as the washing of hands and the use of tacky mats. Other protocols relate to how work is performed inside the cleanroom. For example, servicing procedures for a specific piece of equipment may be identical regardless of whether or not the equipment is installed in a cleanroom. However, if the equipment is installed in a cleanroom the execution of the procedures must conform to cleanroom protocols which relate to how the work is to be performed in order to assure contamination control.

2.0 AN OVERVIEW OF CONTAMINATION CONTROL DURING INTERCONNECTION PROCESSING

2.1 INTRODUCTION

Contamination in semiconductor manufacturing is defined as *any unwanted matter or energy* which results in *product defects* or *process instabilities*. The effects of contamination are most obvious during processing prior to final test, sort, and chip packaging. Contaminants, or the flaws they create, may also cause electrical failures later, after the packaged chip has been shipped and installed as a component in a finished product. Many manufacturers of electronic products perform "burn-in" tests before shipping them to their customers. During these tests, electronic circuits are exercised under conditions which stress devices. Although most latent faults will occur during the burn-in period, a few may remain undetected and cause product failures during customer use. These failures may not occur until many weeks or months after the chip has been tested and mounted in its component package. The source, however, is almost always traceable to a defect or contaminant introduced during wafer processing. Latent defect problems are of special concern in high circuit density products. These complex products cannot be tested for all of the functions and operational conditions they are likely to encounter during their expected life. The trend toward products with decreased feature sizes, and increased circuit densities and number of interconnection levels, will demand a continuing emphasis on the control of contamination-related defects during chip manufacturing.

2.2 FORMS OF SEMICONDUCTOR CONTAMINATION

The following is a brief description of the common forms of semiconductor contamination. Additional details are provided in the later sections in this chapter.

Particles are minute bits of matter capable of obstructing or deforming semiconductor device features. Particles come in an infinite variety of shapes, sizes, and materials. Their shapes include chunks, flakes, agglomerates, rods, and fibers. Particle contaminants range in size from 0.001 μm to 1 mm. Most discrete particle contaminants causing semiconductor defects during manufacturing are between 0.1 and 100 μm.

Ionic refers to chemically active substances such as sodium or other metal ions which can diffuse into silicon and act as unwanted dopants causing device instabilities. Ionic contaminants can also corrode metal device lines and contacts.

Surface films are trace amounts of unwanted chemical substances on the wafer surface. They may be residues from previous processes or deposits from ambient *volatile organic compounds* (*VOC*). The resulting defects are usually layer delamination or high contact resistance. The films may only be a few atomic layers in thickness and are difficult to detect during wafer fabrication.

Static electricity can directly affect semiconductor devices by causing electrostatic discharge (ESD) damage. ESD occurs when electrical charges result in a high electric field between the very small distances separating device features. Electrical opens or shorts may occur in the device if the charge is allowed to bleed-off rapidly. Masks and reticles are also affected in a similar manner. Static electricity on various surfaces and production materials may also cause attraction of unwanted particles or disruptions in the operation of process equipment.

Heat and moisture in combination, refers to ambient temperature and humidity. Unwanted heat and moisture (water vapor) cause instability of process materials such as photoresist. Moisture can also influence the control of static electricity and contamination generation in vacuum process equipment.

Vibration is energy which affects the performance of sensitive process equipment such as wafer photo exposure equipment and measurement and inspection equipment. Vibration can also influence the generation and migration of particle contamination.

Some other examples of energy contamination include *alpha radiation* which causes unstable or unreliable performance of semiconductor devices, stray *UV radiation* during photolithography operations, and *electromagnetic interference* (*EMI*) which affects process equipment performance.

2.3 THE NATURE OF PARTICLE CONTAMINATION

Particles have always been the major source of contamination-related defects for semiconductor products during manufacturing. This is due primarily to the fact that particles can contaminate wafers at virtually any point during the manufacturing process. Particles can account for more than 75% of the yield loss (Fig. VIII-1) in volume VLSI production by causing random structural defects (Hattori, 1990). During thin film processing, particles on wafers can ultimately cause electrical shorts or opens within a single layer (intralevel) or between successive layers (interlevel). Given the key role they play in causing contamination-related defects, it is important to understand some basic particle characteristics.

Particle size is normally expressed as *particle diameter* (Dp). This arbitrary single dimension implies that all particles are spherical. Particles that cause defects, however, have an infinite variety of shapes and few are truly spherical. Chunks, flakes, fibers, rods, agglomerates, and clusters are just a few of many possible shape descriptions. Variety of shape notwithstanding, particle diameter has been accepted as a convenient way to describe the size of these minute irregular objects. It should be noted, however, that particle diameter is a relative term which is derived by various indirect methods such as light scattering intensity or aerodynamic characteristics. These sizing techniques are influenced by the various properties of the particles (shape, index of refraction, mass, etc.) and may not accurately reflect actual size. This is an important point to consider when analyzing particle size information obtained from various detection and counting instruments.

Particles also come in an infinite variety of compositions and physical characteristics. They are very pervasive and are found in virtually every facet of manufacturing from the cleanroom environment, to sophisticated process equipment, to

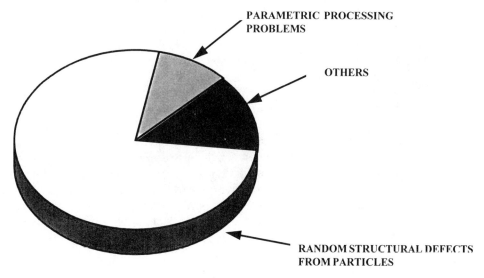

FIGURE VIII-1 Typical failure modes in VLSI devices. (*Contamination Control: Problems and Prospects*)

the carriers and containers used to store wafers. Particles may be relatively pure such as "freshly generated" aluminum flakes deposited during an aluminum metal deposition process. Single particles may also be composed of numerous elements. For example, individual environmental particles will generally contain many elements such as Al, Ca, K, Si, Cl, and Na (Rapa, 1980). Although environmental particles may begin their "life" as a pure material, they eventually pick up "impurities" during their migration lifetime before depositing on a wafer surface. This migration lifetime may be a few days or many months. In addition to composition, other physical characteristics such as size and shape provide valuable information for determining the source of the particles and how they were generated.

Thousands of particles may be deposited on a wafer's surfaces during processing, however, only a relative few develop into defects resulting in device failures. This phenomena may be better understood by examining the factors which are involved in the creation of an electrical defect from a particle.

Particle deposits in sensitive area. A sensitive area is one where a device feature is to be formed or where an anomaly caused by the particle could interfere with a subsequent level formed above it. It may seem that the requirement of a particle being in a sensitive area could easily be met given the circuit densities of current products. However, we generally perceive circuit densities in terms of completed devices with all levels in place. While there may be a few levels with a high percentage of sensitive area, many levels have large areas which are relatively insensitive.

Particle must be present during a critical process step. A process step is usually required to convert a particle into a defect. For example, if a particle is present on top or within the photoresist coating during the exposure cycle, it is likely to obstruct the light path resulting in the formation of a defect. However, if the particle falls on the wafer after the exposure takes place, it will have no affect. This "window of opportunity" is also present in thin film deposition and etch processes. Metal sputtering systems, for example, will generate most of their particulate contamination either during the pumpdown or vent portions of their process cycle. Particles deposited on the wafer during pumpdown are candidates for causing defects while particles added during the vent cycle will not produce defects. Whether particles deposited during venting eventually produce defects depends on subsequent processing.

Particle must survive the process. Many process environments are hostile to small particles making it difficult for them to survive. The high temperatures of hot processes (500 to 900°C) will burn off virtually all organic contaminants. Chamber process environments may vaporize organic particles or reduce their size before they have a chance to affect the wafer. Centrifugal forces in spin coaters and ion implanters can remove relatively large particles before any damage is done. The action of some chemical processes (e.g., wet etches) also have the effect of dissolving some particles before defects are formed.

Particle must have certain physical characteristics. Theoretically, a particle which is 1 μm long, as viewed normal to the wafer surface, could bridge a one micrometer gap between two metal interconnecting lines. The probability of a particle depositing precisely in this manner is low and it is more likely that it will only partially block the gap. This may result in a pattern irregularity but the flaw may not be large enough to cause an electrical fault. The "necking-down" of the device pattern, however, may be large enough to affect the reliability of the chip.

In the examples cited above, statistical probabilities favored a "nondefect" result. However, shape, orientation, and composition of particles are factors which can extend their area of influence thereby creating defects that are larger than the particles themselves. In some cases the height of the particle may be the primary influencing factor. For example, a relatively high particle on a wafer before photoresist application can result in an nonuniform film coating in the area of the particle's "shadow" (striation). Extremely thin films (<200 angstroms) used in CMOS processes suggest very small particles of <0.1 μm are capable of causing defects. As a result, the semiconductor industry has generally accepted the "1/10th rule" for controlling particle contamination. This rule implies that a particle 1/10th of the device minimum feature size has the potential for causing defects at critical process steps.

2.4 CHARACTERISTICS OF A CONTAMINATION PROBLEM

Semiconductor processing is very complex with many interactive factors which can lead to a contamination problem. Sometimes, what appears to be a process problem may actually be a contamination problem. For example, high contact or via resistance may be due to incomplete or improper processing during the forming of via holes. However, if the via holes were opened properly, the problem may be caused by an unwanted surface film (contamination) which has deposited or formed (e.g., oxides) in the via holes during the storage period between process steps. In this case, the problem is not directly related either to the via hole etch or metal deposition processes. The logical problem solving approach in this case is to identify and eliminate the source of the contamination.

Contamination problems can be very expensive in terms of the resulting product defects, production down time, and costs associated with investigating and accurately diagnosing the problem. Lost production time is very costly and there may be a tendency to continue normal production while applying trial-and-error methods to find a quick solution. Trial-and-error methods, however, do not always result in a significant improvement and any relief is often temporary.

Experience has shown that a contamination problem will generally exhibit some or all of the following characteristics.

Random events, with unpredictable "spikes" of high contamination levels.
Inconsistency and nonuniformity, with wide variations within a wafer, wafer to wafer, within a lot and/or lot to lot.
Clean and stable periods, lasting from several lots to days or weeks. It is not unusual to have contamination problems disappear after the application of a trial-and-error solution (or no action) only to return several weeks or months later.
Process "tweaking" is ineffective in obtaining an effective permanent solution.

Although these characteristics are not universal, they can serve as useful guidelines for determining whether the problem is related to contamination or some aspect of processing.

2.5 CONTAMINATION CONTROL CONSIDERATIONS

Many serious contamination problems are not single isolated events occurring in localized areas. When isolated events do occur, they are generally corrected quickly by manufacturing personnel. For example, a process engineer may see a significant increase in the number of large particles deposited on wafers processed through thin film deposition equipment. Since material build-up and flaking is a normal occurrence in certain types of equipment, line personnel follow established procedures to fix the problem and return the equipment to production quickly. The first step may be to perform several pump-down and vent cycles followed by a test to requalify the tool for cleanliness. Since particle contamination in vacuum equipment is sporadic, the pump-down and venting procedure may be all that is required to return the equipment to normal operation. If pump-down and vent cycles are not successful, the next step may require a partial dismantling to clean the internal chamber surfaces. This procedure is complex and often requires the equipment to remain out of service for many hours. If this procedure fails to reduce particle levels, the third step may be to perform a complete *parts change*, i.e., replacement of the chamber liner, shields, and various other removable components which may have become coated excessively. The coated parts are taken off-line where the coating is removed and components returned to their original clean state. Parts changes are expensive and time consuming. The equipment can be down for long periods while requalification tests are performed to ensure that all aspects of the equipment are performing satisfactorily. Unplanned parts changes have a significant effect on production and every attempt is made to predict when a parts change will be required. Knowing this, it is possible to plan product flow to compensate for the expected equipment downtime. While planned down-time may affect production, the disruption will be significantly less than that which results when the equipment is unexpectedly taken off-line. It should be noted, however, that material build-up to the point of generating contamination cannot be predicted accurately since there may be unknown factors which influence chamber shedding.

This example illustrates a typical contamination problem where cause and effect are localized and within the control of people who are responsible for the process. Many serious problems, however, are not localized and they persist in spite of manufacturing efforts to correct them. The following case histories are examples of contamination problems where control was beyond the responsibilities of manufacturing personnel.

> *Case History 1.* High levels of particles are deposited on wafers processed through a sputter deposition system. The normal recovery procedures, including a complete parts change and cleaning, fail to correct the problem. Other sputter deposition tools soon develop similar symptoms and have to be taken out of service. The problem is finally traced to a chemical used in the manufacture of the polyethylene packaging material used to wrap

cleaned chamber components. The outgassing chemical formed a thin chemical film on the chamber parts which resulted in premature flaking.

Case History 2. A large wafer fab experiences a *delamination* problem where separation occurs between interconnection layers causing the formation of microblisters which affect chip reliability. An organic contaminant is suspected but ongoing investigations fail to trace the source to either the process or process equipment. The contaminant is finally identified as a trace chemical impurity in a solvent used in the process. The solvent is stored in a remote bulk storage facility and is pumped as required to smaller storage tanks inside the fab building. A recurring leak in the pump diaphragm resulted in the leaching of the contaminant from the pump's organic polymer components. The diaphragm failure-rate is considered "normal" by support equipment maintenance personnel and the pump is routinely serviced without manufacturing notification.

In both of these cases, the source of the problem concerned *global* interactions which were outside the immediate control of line personnel. Although the characteristics of these problems were similar to those routinely experienced in the line, the application of standard measures failed to correct the problem. Problems with global interactions are more difficult to accurately diagnose and generally require an extensive investigative process.

2.6 APPLYING A SYSTEMATIC CONTAMINATION CONTROL METHOD

Serious problems often require an extensive investigation in order to accurately identify the prime contamination source or sources and to fully understand the defect-causing mechanism. Once these factors are known, however, many solution alternatives may become apparent of which only one will be the best option. Effective and practical contamination control requires the development of *balanced* solutions that are compatible with manufacturing and the fab's business objectives. A balanced solution is one that considers the many interactive factors which are present in a typical wafer fab during high-volume production. The process of developing these solutions can be greatly simplified by the application of a systematic approach which is helpful in evaluating their potential benefits and disadvantages in terms of a global perspective. The three-dimensional matrix shown in Fig. VIII-2 is a graphical representation of a systems approach showing the primary elements to be considered when troubleshooting a contamination problem. The matrix consists of three groups of elements: *contaminants*, *manufacturing elements* (the production essentials required to fabricate the product), and *control elements* (contamination control activities).

An ideal application of this approach would be to require coverage of each of the three groups and each of the individual elements in each group. However, complete coverage of each element in the matrix is neither practical nor necessary. Each manufacturing line will have different priorities and the most effective application of this approach is to develop a *balance* where focus is placed on the areas considered most important for the specific fab.

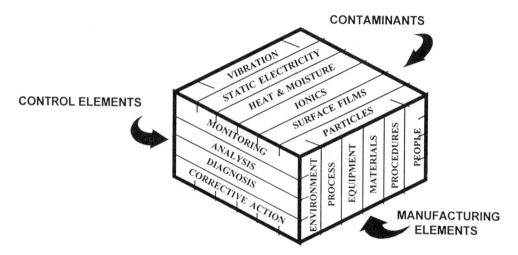

FIGURE VIII-2 The Contamination Control System Matrix. Interactions during manufactuirng occur between contaminants and manufacturing elements. Control elements are activities that are used to understand and correct contamination problems in manufacturing.

2.7 CONTAMINATION CONTROL PRIORITIES

Any contamination control solution will require the implementation of corrective action in one or more of the manufacturing elements (environment, processing, process equipment, raw materials, procedures and protocols, and people). These actions will have direct or indirect effects on production. Therefore, the effectiveness of a solution for controlling contamination must be weighted against its potential influence on production in order to determine its overall benefits to the wafer fab. Generally, the elimination of the source of contamination is the most effective solution which also achieves the greatest overall production benefits. However, it is not possible to eliminate all potentially contaminating materials and mechanisms from the semiconductor manufacturing line. For example, photoresist is an indispensable process material for pattern generation. However, any residual photoresist on wafers during sputter deposition of metal could result in a severe equipment contamination problem. Eliminating photoresist from the line would eliminate this material as a source of contamination in sputtering equipment but this is not a viable solution. When process materials develop into contaminants, they are considered *misplaced*; matter which is in the wrong place or in an unwanted form. When the source of contamination has been identified and its defect-causing mechanism understood, the application of the following prioritized approach can help to identify the most cost-effective corrective actions.

> ***Elimination of the source.*** Elimination of the source of contamination is the most effective method of control. It should be emphasized that what is meant by *effective* control is the elimination of the *known* or *proven* source. Inaccurate source identification generally leads to ineffective corrective actions and costly delays.
>
> ***Control of migration.*** When it is not possible or practical to eliminate the source of the contaminant, controlling its migration is usually the next most

cost-effective control method. Effective migration control requires that both the source and the migration path from the source to the wafer are known. Much of the information about migration paths and mechanisms is obtained through contamination investigations and analysis. The objective of migration control is to intercept, trap, and remove the contaminant somewhere between the source and the process location where it can ultimately form product defects. Filtration is an example of contamination migration control.

Isolation. Isolation is defined as the use of a physical *barrier* to block the movement of the contaminant from its source to the place where it may be detrimental to the product. Although isolation can be considered a form of migration control, it differs in that it is a barrier only acting as a *shield*. Contamination isolation, as defined here, does not contain equipment to remove the contaminant physically. The effectiveness of isolation depends largely on how contaminated the "dirty" side is and the interactions between the two sides of the barrier. Cleanroom walls, minienvironment equipment enclosures, and wafer boxes are examples of isolation.

Product Cleaning. Product or wafer cleaning is considered the least cost-effective approach to control contamination and it is generally used as a last resort. Removing particles from product wafer surfaces is complex and can result in undesirable side effects. Particles become increasingly difficult to remove as they become smaller. Submicron particles adhere very strongly to the wafer surface as a result of van der Waals and other molecular forces. In many cases the only way these particles can be removed is by chemical dissolution or by removing the underlying film to which they are adhered. The addition of a cleaning step can also add significantly to the final cost of the product.

In semiconductor manufacturing, the term *wafer cleaning* is also applied to a necessary step used to prepare the surface for the next layer. The purpose of this step is to remove contaminants *known* to be present such as native oxides and organic residues from a prior operation and procedures can be defined and implemented to remove them. To avoid confusion, wafer cleaning to remove *known* unwanted material is referred to as *wafer preparation* or *wafer prep.*

2.8 CONTAMINATION DETECTION, MONITORING, AND ANALYSIS

Monitoring, analysis, and diagnosis are the elements used in the contamination control process. Monitoring refers to routine on-line inspections and measurements used to detect and measure contamination levels and trends. The ability to quickly detect the presence of wafer contamination, and defects caused by contamination, is very important during wafer production. Without monitoring, a contamination problem may go undetected for several days or weeks resulting in a significant decrease in yields for the job lots processed during that period. Monitoring is costly and is, therefore, designed to acquire just enough information from which production "go/no go" decisions can be made.

Monitoring does not normally produce qualitative data but simply detects "events" that are indications of the existence of an abnormality. For example, it only takes a wafer scanner a few seconds to scan the surface of the wafer and determine a particle count and size distribution. However, further analysis using analytical techniques are usually required to determine the exact nature of what was detected. Analysis which can be performed quickly, such as optical microscopy, are typically done on the production line. If further investigation is necessary the wafers are taken to an off-line laboratory for more sophisticated analysis. On-line microscopic reviews can be accomplished in a few minutes while off-line analysis usually requires a few hours to several days to obtain. As wafer products have become increasingly complex, more powerful analytical equipment, such as the scanning electron microscope (SEM), are being employed on the production line. In addition, on-line detection equipment is being developed with more analytical capabilities to permit shorter turn-around times between detected events and problem diagnosis.

2.8.1 Detecting and Sizing Particles in Cleanroom Environments

The wide size range together with limitless combinations of shapes, mass, and material composition, makes accurate detection and sizing of environmental particles very complex and difficult. Various particle detection and sizing techniques have been developed, each having special characteristics and limitations. In most cases, particle size is derived indirectly from another physical property such as aerodynamic behavior, mass, diffusional characteristics, or electrical mobility. These methods are *relative* sizing techniques and do not necessarily represent *physical* size as determined by microscopic measurements. Optical microscopy also has limitations since it is two-dimensional and does not take into account the height of the particle. The most common cleanroom particle monitoring equipment, *airborne particle counters*, detect particles using *light scattering* principles providing the real-time information necessary for monitoring cleanroom performance. As a result, the airborne particle counter has become the accepted standard for the measurement of cleanroom environments. The operating principle of the airborne counter is shown in Fig. VIII-3. A small pump draws a quantity of air into the sampling probe which then travels through tubing to the instrument. A portion of the tube inside the instrument has a transparent window were a strong light beam is focused. A particle suspended in the air stream flowing inside the tube will scatter light as it passes through the window. The scattered light is then picked up by a detector which converts the signal into an electrical pulse which is counted as a particle. A laser is the typical light source used for most current airborne counters which can detect airborne particles as small as 0.1-μm light scattering size.

2.8.2 Cleanroom Classifications

The development of practical airborne particle counters led to the establishment of a federal government standard (Fed. Std. 209E) for classifying cleanrooms based on this method. Cleanroom classifications are based on the number of particles per cubic foot (or liter) of air equal to or greater than a given size as detected by the airborne counter. Since the minimum particle size detected by the early airborne particle counters was 0.5 μm, cleanroom classes were based on this detection limit. For example, a class 100 cleanroom was defined as one in which there were less

Contamination

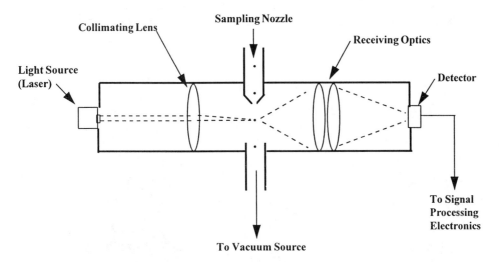

FIGURE VIII-3 Diagram of a light-scattering airborne particle counter.

than 100 airborne particles whose size was equal to or greater than 0.5 μm. The latest revision of the standard (FS-209E) includes the smaller particles detected by newer counters. Table VIII-1 lists a few cleanroom classifications as defined by FS-209E.

2.8.3 Detection and Monitoring of Particle Settling in Cleanrooms

Although the airborne counter and FS-209E are the standard method of certifying cleanrooms, they are insufficient for determining cleanroom performance in terms of particles which cause wafer defects. Environmental particles that cause defects are generally larger than 10 μm and are not detected by airborne counters. A method which measures large environmental particles in terms of particle settling is often used to complement airborne counters for monitoring cleanroom performance. Blank wafers are placed horizontally in the cleanroom for a specified time period. After exposure, a particle count is obtained using a *wafer scanner*. Wafer scanners employ light scattering principles to detect particles on blank and patterned wafer surfaces. The particles added are determined and the results are expressed in terms of a particle settling rate (particles per cm²/h) which represents the level of particles accumulating on surfaces in the cleanroom. The settling rate S_r is:

$$S_r = \frac{P_a - P_b}{t_e}$$

where P_a = wafer surface particle count after exposure to cleanroom environment in particles per cm²

P_b = wafer surface particle count before exposure to cleanroom environment in particles per cm²

t_e = time wafer is exposed to the cleanroom environment in hours

TABLE VIII-1 Airborne Particulate Cleanliness Classes (Federal Standard 209-E)

Class limits
Airborne particle concentrations per unit volume of equal to or larger than particle sizes shown.

Class name (English/SI)	0.1 μm ft³/m³	0.2 μm ft³/m³	0.3 μm ft³/m³	0.5 μm ft³/m³	5 μm ft³/m³
—/M 1	9.91/350	2.14/75.7	0.875/30.9	0.283/10	—
1/M 1.5	35/1240	7.50/265	3.00/106	1/35.3	—
—/M 2	99.1/3500	21.4/757	8.75/309	2.83/100	—
10/M 2.5	350/12400	75/2650	30/1060	10/353	—
—/M 3	991/35000	214/7570	87.5/3090	28.3/1000	—
100/N 3.5	—	750/26500	300/10600	100/3530	—
—/M 4	—	2140/75700	875/30900	283/10000	—
1000/M 4.5	—	—	—	1000/35300	7/247
—/M 5	—	—	—	2830/100000	17.5/618
10000/M 5.5	—	—	—	10000/353000	70/2470
—/M 6	—	—	—	28300/1000000	175/6180
100000/M 6.5	—	—	—	100000/3530000	700/24700
—/M 7	—	—	—	283000/10000000	1750/61800

S_r = particle settling rate in particles per cm²/h

Particle settling data provides a more accurate measurement of the type of environmental particles that are likely to contaminate wafers. In addition, particles on wafers exposed to the cleanroom and those placed in equipment (*equipment monitor wafers*) can be analyzed using SEM/EDX analysis (*particle fingerprinting*). Particle fingerprinting and equipment monitor wafers are discussed in Section 2.8.4. In general, both airborne counting and monitor wafers are used in a cleanroom monitoring program to provide a complete "picture" of cleanroom performance.

2.8.4 Equipment Contamination Detection and Analysis

There are two basic methods used for detecting contamination in process equipment; (1) *monitor wafers* and (2) *in situ particle monitoring*. Each technique has advantages and disadvantages but each method usually provides data which complements the other. In the monitor wafer method, clean blank silicon wafers are cycled through the equipment for the purpose of acquiring particle measurements. Where possible, the monitors are processed together with product wafers in order to obtain data which closely represents what the product is "seeing" during processing. The monitor wafers are then scanned on a wafer scanner which detects, counts and sizes the particles. When the monitor wafer particle level exceeds the allowable limit, the equipment is taken out of service for cleaning or repair.

Although the monitor wafer method is generally accurate in terms of detecting what product wafers pick up, it has several disadvantages. Monitor wafers cannot always be placed in with product wafers because of the specific nature of some equipment and their processes. In addition, monitor wafers are costly in terms of wafer usage, equipment nonutilization, and no-value-added work activity (work that is not directed toward actual wafer fabrication). This necessarily limits the amount of monitoring which can be accomplished as a practical matter. Another disadvantage is that blank silicon monitor wafers do not have the same surface characteristics as product (patterned) wafers. These differences may influence the behavior of particles such that the particle deposition on monitors differs from that on product wafers. Disadvantages notwithstanding, monitor wafers are indispensable for controlling equipment particle contamination and industry standards for the technique have been established (SEMI Std. E14-93).

The development of product wafer scanners capable of detecting particles and defects on patterned wafers, addresses some of the previously described drawbacks. Contamination can now be monitored directly by scanning product wafers. Product wafer scanners can also detect particles on plain silicon or blanket coated wafers where blank silicon wafer scanners have difficulties. Scanning product wafers, however, also has disadvantages. Since measurements are made directly on very expensive product wafers, there is always a concern that something detrimental will happen (e.g. additional contamination, wafer damage, broken wafers) as a result of a no-value-added monitoring procedure. Monitoring product wafers often results in processing delays since job lots need to be held-up for monitoring. In addition, product wafer scanning systems cannot discriminate between particles and other artifacts such as pattern defects. This may result in additional microscopic inspection in order to obtain meaningful data for practical follow-up action. Considering the advantages and disadvantages, both types of wafer scanners generally complement each other and a comprehensive picture of particle contamination is obtained when both scanner types are used.

Prior to the advent of wafer scanners, *oblique light* or *bright light* methods were the workhorses for inspecting production wafers. These systems consist primarily of a focused white light beam illuminating the wafer surface at some oblique angle (Fig. VIII-4). Most oblique light inspection devices are in-house designed using inexpensive readily available high-intensity light sources. The inspection routine consists of placing the wafer in the light beam and visually detecting "sparkles" which indicate the presence of particles or other abnormalities. As a discrete particle detector, oblique light is a poor performer detecting only very large events. Detection capability is also highly variable from operator to operator.

Significant advances have been made in recent years in techniques for monitoring contamination during the process cycle inside (in situ) equipment. In situ monitoring is particularly useful in vacuum and other chamber type systems were the monitoring of abnormal events is difficult to achieve by other methods. The insitu concept employs sensors designed to detect specific contaminants during processes such as plasma etching and sputter deposition. Although there are various types of sensors and in situ techniques under development, two particle detecting techniques have been used in manufacturing to some degree: *laser light scattering* (*LLS*) and *in situ particle monitoring* (*ISPM*). The LLS technique is used for observing the density and movement of particles suspended in a plasma inside the vacuum chamber during processing (Selwyn et al., 1989, 1991). The apparatus consists of a laser, a scanning mirror, and video camera with recorder and monitor (Fig. VIII-5). The laser light beam is raster scanned through a chamber view port. A video camera is positioned to view the process through a second chamber view

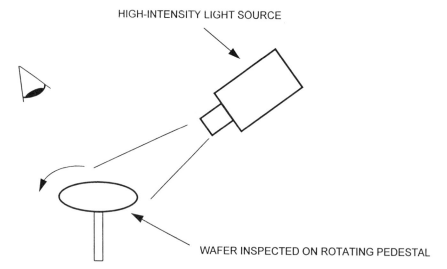

FIGURE VIII-4 Oblique light wafer inspection apparatus.

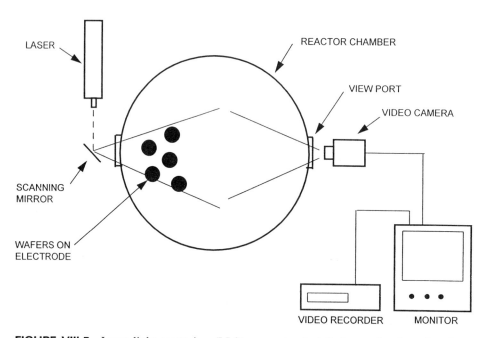

FIGURE VIII-5 Laser light scattering (LLS) apparatus installed on a batch wafer plasma processing system. A scanning mirror is used to sweep laser light across a portion of the reactor chamber. Particles made visible by light scattering are detected by the video camera and viewed through a video monitor and recorded.

Contamination

port. Light scattered from particles inside the chamber forms an image which can be viewed in real time on the monitor and recorded on a video cassette recorder (VCR) for later review and follow-up analysis. This technique provides an opportunity to observe the behavior of particle "clouds" which form during wafer processing. This data can be helpful in fine-tuning chamber design and processing parameters in order to reduce wafer contamination. Because of its size and configuration, the LLS technique has been used primarily as a troubleshooting technique rather than for routine monitoring.

ISPM incorporates a compact particle detector (sensor) which is installed in the exhaust line of vacuum process equipment (Fig. VIII-6). The sensor can provide real-time continuous monitoring of particles in the exhaust stream during dry etch, deposition, and ion implantation processes (Borden, 1991; Spanos et al., 1994). Although it has been difficult to obtain correlation with monitor wafers, recent studies indicate that ISPM is a valuable technique for process control and reducing defects caused by particles in reactors (Monkowski, Freeman, 1990; Hunter, 1993). In addition, ISPM can provide a more accurate determination of when vacuum equipment requires cleaning or other servicing (Peters, 1992). It can also be used to complement wafer monitoring, thereby providing more comprehensive particle contamination data (Stern et al., 1991).

The RGA (*residual gas analysis*) is beginning to see increased use as a real-time monitoring technique for chamber processes. Studies indicate that RGA monitoring can provide valuable process control information and detect the presence of trace contaminants long before wafer quality is affected or a malfunction occurs

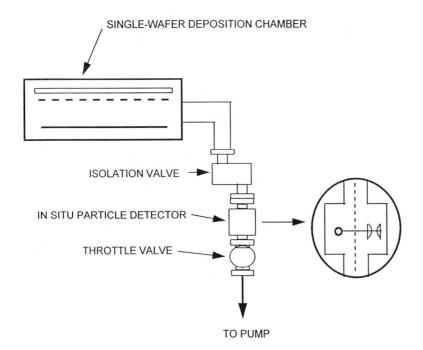

FIGURE VIII-6 Diagram showing the installation of a in situ particle detection system on the exhaust line of a single-wafer CVD reactor.

in the process equipment (Reath et al., 1994; Simpson, 1991). These benefits are particularly important during metal deposition where trace contaminants (e.g., moisture, hydrocarbons, gaseous impurities) can result in product defects and process instabilities.

When monitoring detects an increase in contamination levels or an abnormal event, it is often necessary to quickly obtain more information before a decision is made which affects production. In most cases, this information is from analytical equipment in the manufacturing area. Perhaps the most useful on-line (located in the production area) analytical instrument for this purpose is the *optical microscope*. The optical microscope is a low-cost instrument with many illumination options useful for characterizing particles and defects. Shrinking device features and increasing circuit densities, however, often require the use of higher resolution instruments such as the *scanning electron microscope* (*SEM*). The SEM employs a scanning electron beam which generates secondary and back-scattered electrons when bombarding the sample. The secondary and back-scattered electrons are amplified and used to create a high resolution image with magnifications up to $100,000\times$. In addition to secondary and back-scattered electrons, the sample also emits x-rays which can be used to determine the sample's elemental content. The x-rays are gathered by an *energy dispersive x-ray* (*EDX*) detector which provides qualitative and quantitative information for identifying the elements contained in the sample. The SEM/EDX combination has evolved into a powerful analytical instrument useful for analyzing particle contamination.

Occasionally, on-line analytical equipment is not sufficient to determine the nature and source of the contamination. There are numerous off-line analytical techniques available and a complete listing with descriptions is beyond the scope of this text. The following is a brief description of the more commonly used methods for addressing semiconductor contamination-related problems.

Secondary Ion Mass Spectroscopy (*SIMS*) is a technique that uses a strong ion beam to sputter material from the surface of the sample being analyzed. Secondary ions are also produced as the surface is bombarded with the ion beam. These secondary ions are analyzed by a mass spectrometer which provides elemental makeup information. This technique is especially useful for analyzing bulk materials, such as silicon, for impurities and contaminants near the surface. A depth profile can be obtained characterizing the concentration of contaminants through a depth of up to 10 μm.

Auger Electron Spectroscopy (*AES*) employs a focused electron beam which irradiates the sample surface creating a type of secondary electrons called *Auger electrons*. These electrons have energies characteristic of the material that emitted them and they are collected and analyzed by an electron spectrometer to obtain elemental information. As a consequence of the differences between Auger electrons and secondary x-rays, AES complements EDX elemental analysis. AES is used for surface analysis and is capable of analyzing the topmost 15 to 20 Å of material.

Electron Spectroscopy for Chemical Analysis (*ESCA*) is a technique where x-rays are used to bombard the sample surface causing an interaction which results in the release of photoelectrons or Auger electrons. The kinetic energy distributions of the ejected electrons are analyzed by a high-resolution electrostatic spectrometer to provide elemental analysis of the specimen. In addition to elemental data, molecular (organic and inorganic) and chemical bonding information can also be obtained. ESCA is primarily a surface analysis technique with a depth penetration of approximately 100 Å. ESCA is also referred to as X-ray Photoelectron Spectrometry (XPS).

Time of Flight Static Secondary Ion Mass Spectrometry (TOF-SSIMS) is the SIMS technique operated in the static mode (also referred to as Static SIMS). It is considered static because the total ion beam dose is below the threshold for damage due to sputtering. As a result, a time of flight mass spectrometer is able to obtain spectra and images while only sampling less than 1% of the top monolayer. TOF-SSIMS is an extremely sensitive technique for analyzing low levels of surface contaminants. It can identify virtually all elements in the periodic table and provide molecular information for both organic and inorganic materials. As such, this technique is very useful for identifying organic contaminants.

3.0 ENVIRONMENTAL CONTAMINATION CONTROL

3.1 THE SEMICONDUCTOR MANUFACTURING CLEANROOM

The manufacturing cleanroom is the "envelope" which houses all process equipment and it is the primary system for controlling particles and other environmental contaminants which can cause wafer defects. A cleanroom is defined as any controlled space that (1) provides isolation from external contaminants, and (2) controls contamination generated by sources within the cleanroom. Various filtered air flow and containment methods are used to achieve these objectives. The term "filtered" assumes either HEPA (high-efficiency particulate air) or ULPA (ultralow penetration air) filters are used for final filtration. HEPA and ULPA filters are "absolute" filters and are very effective in trapping virtually all airborne particles that can be deposited on wafer surfaces that are detectable by the SEM (>0.05 μm). In addition to removing airborne particles, incoming cleanroom air can be "processed" or conditioned to control other contaminants such as heat, moisture, and organic vapors. This process takes place in air handling systems called HVAC (heating, ventilating, and air-conditioning) units. Heat and moisture are controlled by first cooling and dehumidifying the air. The dry, chilled air is then reheated and humidified to maintain constant control of temperature and humidity. Airborne organic contaminants can be removed by the use of activated charcoal filters in the HVAC units. HVAC units may contain a variety of conditioning elements depending on their function in the overall air management scheme. Some air handling units may contain only fans for recirculating air through HEPA/ULPA filters while others include components for conditioning air to attain specific qualities. HVAC units are also used to supply *make-up air* to the cleanroom. Make-up air is drawn from outside the building and is used to replenish air exhausted from the cleanroom by process equipment. Make-up air is also used to pressurize the cleanroom. Figure VIII-7 shows the air-handling components that are used in basic cleanroom design.

The following are the basic air flow and filtered air confinement concepts used in cleanroom design.

> ***Turbulent.*** Air flows in random patterns. Purging of airborne particles is accomplished through a mixing and dilution process.

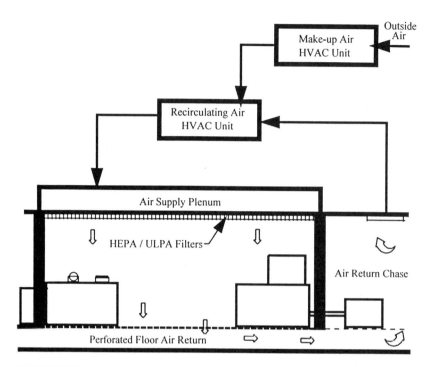

FIGURE VIII-7 Diagram of a basic cleanroom showing key air system components.

***Unidirectional* [*vertical laminar flow (VLF) or horizontal laminar flow (HLF)*]**. Air is made to flow in a uniform parallel pattern. Dilution is prevented and purging is accomplished by air streams carrying particles directly out of the controlled zone.

Mixed flow. A combination of turbulent and unidirectional air flow concepts are used in a given cleanroom environment.

Room level confinement. The entire room area is controlled and filtered air is confined by cleanroom walls.

Local isolation. Filtered air is confined to small areas or work zones.

Combination. Areas where both room level and local isolation methods are used.

These concepts can be "mixed and matched" to provide numerous cleanroom designs and configurations (Rapa, 1996). For example, the number of air changes (turbulent flow areas) and air flow rates (unindirectional flow areas) may be varied to suit specific needs. Local isolation can used in a few selected work zones or at all process stations.

3.1.1 Particle Behavior in Cleanrooms

Effective control of environmental particles requires an understanding their behavior in cleanrooms with respect to their potential for causing wafer defects. Figure

Contamination

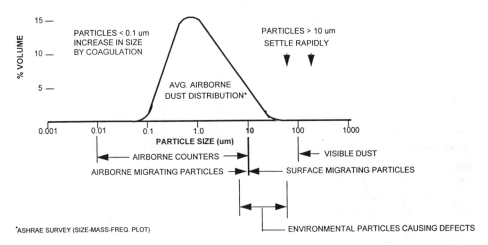

FIGURE VIII-8 Spectrum of environmental particles showing sizes and behavior.

VIII-8 shows the spectrum of particle sizes and their behavior. The size spectrum of particle contamination ranges between 0.001 and 1000 μm (1 mm). There are numerous forces that influence particle behavior and the effect that each force has is dependent on factors such as size, shape, mass, and material composition.

Figure VIII-8 also illustrates the mechanisms of particle growth and settling. Very small suspended particles, those less than 0.1 μm, are primarily influenced by diffusion and similar forces which promote particle coagulation (particle-to-particle attraction). As these small particles grow in size, they become part of the population which is larger than 0.1 μm. Particles greater than 10 μm in size tend to settle rapidly. Although these large particles may become temporarily airborne, they spend most of their "life" moving about from surface to surface (*surface migrating*). Particles between 0.1 and 10 μm tend to remain suspended in air for long periods of time and are called *airborne migrating* particles. Suspended liquid and solid airborne particles are also referred to as *aerosols*. It should be noted that the vast majority of environmental particles found to be causing wafer defects are in the surface migrating range (>10 μm).

An important element in the control of environmental particles is predicting their behavior in an active manufacturing cleanroom. The are numerous factors influencing the behavior of particles such as air flows, contamination load (concentration of environmental particles), types of process equipment, activity levels, and cleanroom protocols. A key factor is the design of the cleanroom and the air flow and containment concepts it uses. Figure VIII-9 illustrates particle behavior in a turbulent flow cleanroom. Filtered air enters the room through HEPA filters placed in various parts of the ceiling. Air flows throughout the cleanroom in random patterns and particles are removed (purged) via a dilution process. This process is described using a water analogy shown in the insert of Figure VIII-9. In this analogy, clean water is continuously fed into a tank to compensate for drain losses and to maintain a constant liquid level. When a "contaminant" is added in the form of a dye (represented by the flask), the dye quickly disperses (dilutes) throughout the tank. When addition of the dye is stopped, the water in the tank begins to clear

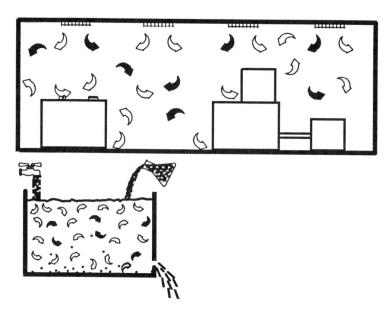

FIGURE VIII-9 Diagram showing particle behavior in a turbulent flow cleanroom.

(recover), gradually returning it to its original clear state. If we measure the amount of water required to clarify the tank water, we would see that several tank volumes of clean water are required for full recovery. This process is called *dilution recovery*. If the dye in the flask contains particles of various sizes, some of the larger particles will settle to the tank bottom during the purging process. Airborne particles in a turbulent cleanroom behave in a similar way. It should be noted that for an airborne particle to cause a semiconductor defect, it must first settle onto a surface. Although environmental particles may settle directly on a wafer surface, most of defect causing particles settle on process equipment, work surfaces, wafer boxes, and various other surfaces and objects in the manufacturing cleanroom. Work activity gradually causes these particles to migrate, some eventually reaching wafer surfaces at a critical time and place thereby causing defects. Classes of turbulent cleanrooms range between 100,000 and class 100, with some designs achieving better than class 100 under full-load operational conditions. However, particle settling can vary considerably, because of the way airborne particles behave in the turbulent flow conditions in these cleanrooms. Particle settling rates in turbulent cleanrooms, under full load operational conditions, can vary from 0.01 to > 10 particles per cm^2/h independent of cleanroom classification. Since the air is turbulent, particle settling is also generally independent of monitoring location and height.

The particle control limitations of turbulent cleanrooms led to the development of *laminar* or *unidirectional flow* cleanroom concepts. In these rooms, air is balanced to flow in parallel lines either vertically or horizontally, throughout the room. In a vertical laminar flow (VLF) cleanroom, the ceiling has maximum (>90%) HEPA or ULPA filter coverage. Filtered air enters the cleanroom at a uniform air flow rate and is balanced to flow in vertical parallel lines from the ceiling to the floor. Air in VLF cleanrooms is usually returned through a raised floor with perforated panels or open floor gratings. Particle behavior in VLF cleanrooms is shown in Fig. VIII-10. In this water analogy, there are multiple sources of clean water at the top and many openings at the bottom of the tank allowing the water to flow in parallel streamlines from top to bottom. If the dye/particle mixture is introduced, the contaminants remain localized near the origin, that is they are not dispersed throughout the tank. When introduction of these contaminants is stopped, recovery is rapid and the tank water returns to its original clear state in only one exchange of fluid. In the VLF cleanroom, both small and large airborne particles remain near their source locations and are quickly driven to the floor. Recovery is rapid, usually requiring less than 10 s.

VLF cleanrooms can be designed to operate under full load conditions at better than class 1 with particle-settling rates less than 0.001 particles per cm^2/h at the processing level (approximately 1 m above the floor). The traditionally accepted air flow velocity for VLF clean rooms has been 90 ft/min. Recent studies suggest that air flow velocities can be as low as 60 ft/min and still maintain acceptable performance during full load production (Carr et al., 1994). Many new semiconductor cleanrooms re being designed and built to operate at VLF air flow velocities between 60 and 75 ft/min.

Horizontal laminar flow (HLF) cleanrooms are constructed with full HEPA filter coverage on one wall with air returning through the opposite wall. Air is pulled horizontally through the cleanroom from the supply or filter wall to the return wall. The operation of HLF cleanrooms is cleaner than turbulent cleanrooms

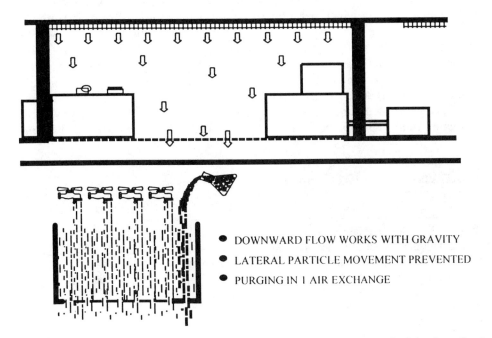

FIGURE VIII-10 Diagram showing particle behavior in a VLF (vertical laminar flow) cleanroom.

only in an area close to the filter wall. HLF cleanrooms have limited applications and are rarely used for semiconductor wafer fabrication. The principle, however, is used in HLF workstations and some equipment enclosures.

3.1.2 Contamination from People

In the 1960s and through the 1970s, people were considered the greatest source of environmental contamination on the semiconductor line (Morrison, 1973). There was ample evidence to support this belief in the form of defects caused by skin flakes, fingerprints and human hair. In addition to being a direct source, people transported environmental particles from uncontrolled spaces into the cleanroom. Thus, people activity was of great concern in the control of environmental contamination. An estimate of airborne particulates generated by people activity is shown in Table VIII-2. Advances in contamination control through the 1980s and early 1990s have virtually eliminated people as a direct contamination source in the modern semiconductor line fabricating leading-edge products. These advances have come through improvements in cleanroom design, cleanroom garments, personnel protocols, and procedures. Since early semiconductors were essentially fabricated in a chemical laboratory, the lab coat was the standard working garment. As wafer fabrication moved into turbulent flow cleanrooms, the lab coat evolved into the cleanroom *smock* (Fig. VIII-11) and additional accessories were added including light head coverings (caps), beard covers, shoe covers, and disposable gloves. Although each added accessory provided some additional protection, particles from people continued to be a significant problem. Particles released from the smock's bottom opening and leg areas, although below the work level, were adding significantly to the cleanroom's contamination load. In addition, smocks allowed contact with regular clothing which permitted people to gain access to handkerchiefs and other personal items in their pockets. Thus, smocks, including light head and shoe coverings, are referred to as *partial containment* cleanroom garments. That is, they are only partially effective in containing particles from the human body and regular clothing in a manner which prevent them from entering the cleanroom environment. *Full containment* garments (also known as *bunny suits*) are designed to cover virtually all areas of the human body except for the eyes (Fig. VIII-12). They typically consist of a jumpsuit type body suit, full hood head and face covering, and *booties*

TABLE VIII-2   Estimates of Personal Particle Generation

Activity	Particles emitted/min $\geqq 0.3$ μm
Standing or sitting, no movement	100,000
Sitting with light head, hand and forearm movement	500,000
Sitting with average body and arm movement, toe tapping	1,000,000
Changing position, sitting to standing	1,500,000
Slow walking, 2.0 mil/h	5,000,000
Average walking, 3.5 mil/h	7,500,000
Fast walking, 5.0 mil/h	10,000,000
Calisthentics	15,000,000 to 30,000,000

Source: *Environmental Control in Electronics Manufacturing*, (Morrison) 1973.

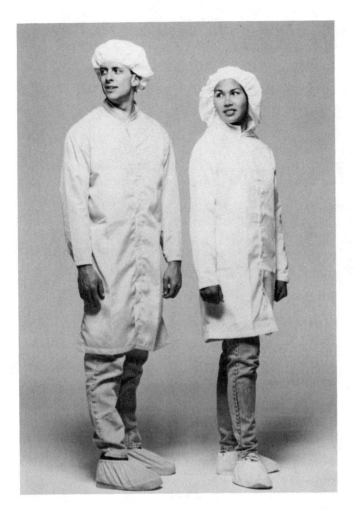

FIGURE VIII-11 Cleanroom smocks or partial containment garments. (Terra Universal, Inc.)

which cover the entire shoe and overlap the jumpsuit pant leg above the calf. The hood may be constructed of fabric or a rigid clear plastic bubble. Bubble hoods are equipped with an air pack which draws in air from the facial opening and passes it through a filter at the waist before expelling it back into the cleanroom environment. Full containment garments are constructed of various synthetic materials and are available in launderable and disposable designs.

Although full containment garments are a significant advancement in the control of direct human contamination, there is also an indirect aspect to contamination from people. For example, cleanroom gloves may be perceived to be very clean and acceptable for use in handling wafers directly. Actually, cleanroom gloves can contact transfer organic particles and plasticizers to sensitive surfaces. In addition, gloves do not offer any protection from contact transferring contamination from one surface to another. Therefore, wafers should never be handled directly with hands either with or without cleanroom gloves. In addition, gloves should be re-

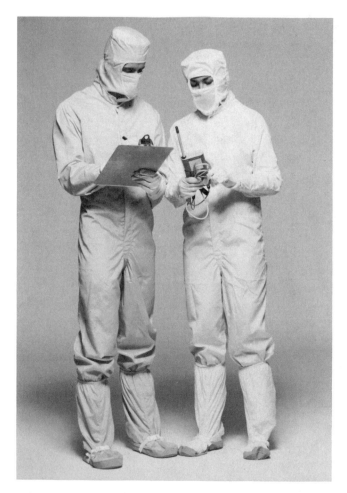

FIGURE VIII-12 Full containment cleanroom garments or bunny suits. (Terra Universal, Inc.)

placed often, especially after touching bare skin or other surfaces which are known to be heavy sources of contamination. It should be noted that care should also be exercised when choosing and wearing gloves to perform equipment maintenance since this activity may damage them and generate contamination from glove materials.

People represent an uncertain element in controlling contamination on the manufacturing line. Although most people want to do the right thing, the actions of individuals are driven by complex issues such as the needs of the moment, background training, understanding of what is required, and company and country culture. These reasons are partially responsible for driving many manufacturers towards greater process automation and automated materials handling.

3.1.3 Transport of Particles into Cleanrooms

If a cleanroom could operate without doors and remain tightly sealed with no people, activity, or other particle sources inside, it would be a simple matter to maintain a perfectly clean environment within. The recirculating filtered air systems would eventually remove all airborne particles and maintain the environment particle-free regardless of the type of cleanroom used. Manufacturing cleanrooms, however, are intended to provide a protective environment for contamination sensitive products where people and equipment are engaged in various production activities. All production activities generate and redistribute particles within the cleanroom. In addition, wafer fabrication results in continuous interaction between the cleanroom and uncontrolled areas external to the cleanroom. Environmental particles infiltrate the cleanroom as product, equipment, and people enter and leave the cleanroom. This influx, together with particles generated by activities inside the cleanroom, constitute the contamination load which must be controlled by the cleanroom's air systems. Particle infiltration in a poorly designed or malfunctioning cleanroom could be significant and result in it's inability to recover (*overload*). When overload occurs, particles cannot be removed faster than they are being introduced and continuous particle accumulation results.

It is apparent that effective control of particles within the cleanroom cannot be achieved without proper attention to the control of infiltration from external sources. Infiltration is facilitated by both airborne and surface particle migration mechanisms. Control of airborne migrating particles is accomplished by (1) maintaining particle-tight seals at all cleanroom wall and ceiling joints, (2) limiting the number of cleanroom entry and exit locations for both people and material, (3) the installation of air locks at entry and exit locations, and (4) maintaining a positive cleanroom air pressure with respect to all external areas.

Having an entry door open directly into the cleanroom from uncontrolled space would result in a significant infiltration of airborne contaminants. Small pressure-controlled clean anterooms with door interlocks are very effective in limiting this influx. Sometimes two anterooms (change room plus preentry room) are used as a double air lock (Fig. VIII-13). Air showers for personnel are also very effective air locks. These units use high-velocity HEPA filtered air to blow off large particulate matter from people's clothing before they enter the change room and/or cleanroom.

Controlling the infiltration of surface migrating particles is more difficult since much of this contamination is carried into the cleanroom on surfaces rather than directly by air flow. People entering the cleanroom wearing regular clothing would be the greatest source of these particles. Today's full-containment garments are very effective in keeping particulate matter on street clothing from entering the cleanroom. However, surface migration is not a direct mechanism but requires many individual events which, collectively, move particles into the cleanroom and, eventually, to sensitive areas. For example, a bunny suit may brush along the floor, pick up tracked-in particles and then transfer those particles to the cleanroom and, perhaps, ultimately to the interior of a reactor. This inadvertent transfer may seem to be an insignificant source of particles when viewed as an isolated incident. However, an active semiconductor line operates continuously and will experience many such events around the clock for nearly 365 days per year. The flow of people and

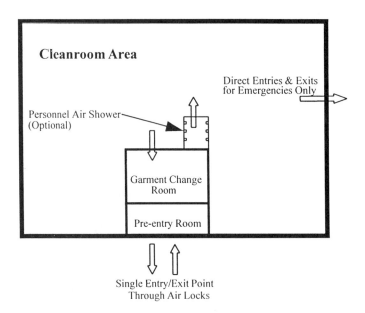

FIGURE VIII-13   A basic cleanroom entry arrangement.

materials between the cleanroom and uncontrolled space results in a continuous infiltration of particles. Although the techniques previously described for minimizing airborne infiltration are helpful, additional steps are needed to control infiltration by surface migration. The following is a listing of some of the techniques which are commonly used to control this type of infiltration.

Shoe cleaning machines are used to remove gross debris from street shoes prior to entering a cleanroom zone.

Tacky mats or tacky flooring are placed at cleanroom entrances to remove debris from the bottom of footwear.

Removal of street shoes at the building entrance and donning shoes dedicated to use inside the building housing cleanrooms.

Garment change rooms and protocols—designed to segregate areas where cleanroom garments are donned from areas where people are moving about in street clothing.

Double-bagging—of parts and materials brought into the cleanroom where the final wrapping is removed just inside the cleanroom.

Equipment entry cleanroom—for cleaning and preparing equipment prior to its entry into the processing cleanroom.

Dedicated equipment maintenance hand tools—that are kept clean and not removed from the cleanroom.

Clean maintenance shop—connected to the main cleanroom for component repairs which cannot be performed at process equipment location.

Nested cleanrooms and controlled perimeters or buffers—which surround the process cleanroom thereby reducing particle infiltration in stages.

Contamination 535

Although some of these techniques are also effective for airborne infiltration, their main purpose is to control surface migrating particle infiltration.

3.1.4 Minienvironments and Isolation Technology

In the early days of semiconductor manufacturing, wafers were processed inside laminar flow *clean workstations* or *clean hoods* which were located in uncontrolled environments. As the size of wafers and equipment increased and processes became more complex and sensitive, it became increasingly difficult to keep wafers isolated within the confines of the clean workstation. This led to the placement of clean workstations in a turbulent flow cleanroom environment and, eventually, wafer processing directly in VLF cleanrooms without clean workstations. The shortcomings of turbulent cleanrooms and laminar flow workstations, combined with the high cost of new VLF facilities, led to the development of *isolation technology* for semiconductor processing. In principle, this concept is similar to laminar flow workstations where a very clean environment is maintained only in the relatively small active processing area. Unlike workstations, however, the process zone in the isolation concept is physically isolated from the process operator. An example of isolation technology is the glove box shown in Fig. VIII-14. The glove box has a small HEPA filtered air environment which is totally isolated from the external ambient. Parts are introduced into the glove box enclosure in sealed containers and the glove box is then allowed to purge and recover. The parts are removed from

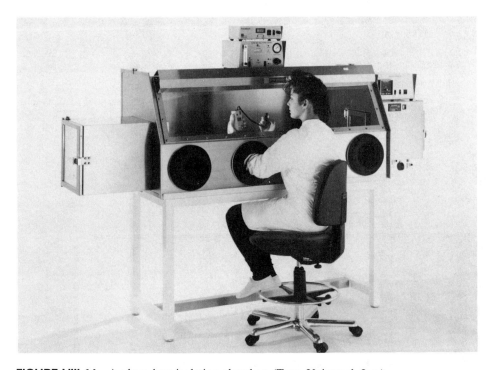

FIGURE VIII-14 A glove box isolation chamber. (Terra Universal, Inc.)

their container and manipulated using the large rubber gloves which are an integral part of the enclosure. After manipulation, the parts re returned to the container and removed from the glove box enclosure. Gloves boxes are limited to very simple operations due to their small work area and restricted movement permitted by the gloves. In principle, however, the concept permits sensitive processes to take place in an environment which is totally isolated from the external ambient. The development of the SMIF (Standard Mechanical Interface) concept combined with equipment enclosures (minienvironments) resulted in an isolation technology which is compatible with current day semiconductor processing. In this concept (Fig. VIII-15), wafers are stored and transported between process stations in containers called SMIF pods. Unlike standard wafer boxes, SMIF pods seal the cassette of wafers in an airtight container which is opened mechanically by means of a SMIF Arm of SMIF Indexer. The operator places a SMIF Pod containing wafers on the SMIF Arm or indexer which then automatically extracts the cassette and places it on the input station of the equipment. After processing, the wafer cassette is returned to the SMIF POD to be transported to storage or to the next process station. When process equipment is housed within a clean environmental enclosure (*minienvironment*), physical isolation is complete. The SMIF pod limits access to wafers thereby reducing the risk of wafer damage from manual handling. Since process equipment and wafers are contained in their own clean environments, the external ambient can be made into a less stringent cleanroom, typically a class 1000 environment. Studies have shown that minienvironments are effective in isolating environmental particulate (Baechle et al., 1992; Rothman et al., 1995). The effectiveness of minienvironment isolation, however, may be dependent on the frequency and extent of equipment servicing and the particle settling rate in the cleanroom (Hattori, 1994). In addition, hardware costs may negate any savings from building a class 1000 ambient cleanroom instead of a class 10 or class 1 VLF cleanroom (Lynn, 1994). The SMIF/minienvironment concept can provide a cost-effective alternative for upgrading an existing older facility (Grande, 1993). SMIF pods can also be designed to contain an inert gas environment thereby preventing the degradation of wafer surfaces between process steps (Elliott, 1995). This may become

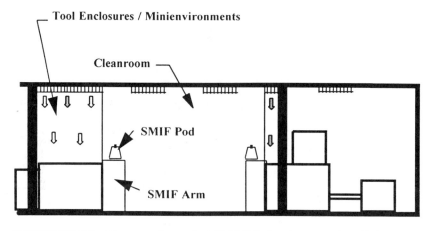

FIGURE VIII-15 Diagram of a typical SMIF/Minienvironment installation.

an increasingly desirable feature as wafers become more sensitive to volatile organic compounds. The semiconductor industry is currently in the process of developing standards for minienvironments and isolation technology (Purcel, 1994).

3.2 CLEANROOM SUPPORT SYSTEMS

Cleanroom support systems include Deionized (DI) water, chemical, and gas distribution systems. In addition, equipment used to control other contaminants (e.g., static electricity and vibration) in the general cleanroom environment are also considered cleanroom support systems. Due to their close relationship with contamination-related defects in thin film processes, gas systems are discussed in Section 5.3.1.

3.2.1 DI Water

DI water is considered a unique process chemical requiring special contamination control consideration. One aspect of DI water is that it is a common rinsing agent and the last chemical wafers are exposed to in any wet chemical surface preparation and cleaning process. Therefore, DI water is expected to have the highest cleanliness level independent of the cleanliness of preceding chemical operations. Poor or inconsistent quality DI water can significantly affect the cleanliness of wafer surfaces and the quality and stability of the subsequent process step. Another aspect of DI water is that it is a very good solvent which is very effective in scavenging ionic residues from wafer surfaces. When heated, DI water becomes very aggressive and its cleaning capabilities further enhanced. The cleaning attributes of high-purity DI water also makes it a difficult material to distribute to the point of use without degrading its purity. The design of the distribution system plays a key role in insuring DI water quality. Large distribution systems may degrade DI water purity, requiring additional polishing near the point of use. The ultra-high-purity DI water used in today's semiconductor lines is commonly called *ultra-pure water* (UPW). Figure VIII-16 is a diagram of a typical UPW system. It consists of various water treatment elements such as (1) activated carbon beds for removal of organics, (2) various types of filters and reverse osmosis (RO) membrane units for particulate removal, (3) heat exchangers to control water temperature, (4) vacuum degassifiers to remove oxygen and other residual gases, (5) ion exchange and mixed bed units to remove ionic contaminants, and (6) ultraviolet (UV) sterilization units to control bacteria. The characteristics of incoming water are different for each semiconductor facility and the UPW system is custom designed for each fab location. Regardless of configuration, the key objectives for semiconductor UPW systems are centered on controlling the following contaminants:

> ***Ionics.*** Anions and cations such as sodium, chloride, calcium, potassium, and various other metals. The basic measure of UPW ionic content is resistivity. Typical leading-edge fabs maintain UPW resistivity to above 18 mΩ/cm at 25°C with total ionic contaminants less than 10 parts per billion.
> ***Bacteria.*** Bacteria is viable (live) particulate that can multiply. Live bacteria is measured in terms of colony forming units (CFU) growing in a culture media.

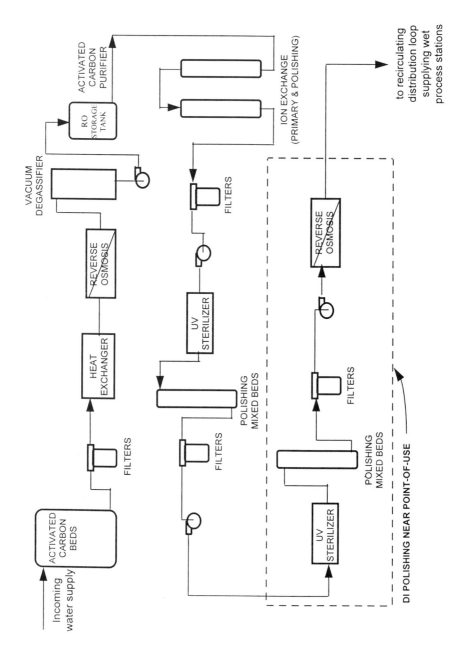

FIGURE VIII-16 Diagram of a typical semiconductor UPW system for a large fab. (Jacobs Engineering Group)

Organics. Trace organic contaminants may be present in UPW. These are not detectable by resistivity measurements. Total organic carbon (TOC) is measured either by an on-line continuous monitor or by off-line lab analysis.

Dissolved oxygen. Oxygen in UPW can contribute to accelerated native oxide growth causing process instabilities and spotting on wafers.

Particulate. In addition to inert particles, UPW may contain live and dead bacteria. Table VIII-3 lists current guidelines for UPW impurity levels for new semiconductor fabs.

The same design and maintenance principles also extend to process equipment. For example, effective control of bacterial growth requires that the water is kept continuously flowing through all parts of the distribution system. UPW systems are designed with closed recirculation loops without "dead zones" where water can become stagnant (Carmody, Martyak, 1989). If UPW stops flowing, significant bacterial growth can be expected. Once this occurs, bringing the UPW system back to its original bacteria-free state requires the introduction of sterilizing chemicals or ozonation. This may take several days to complete, during which time there is no wafer production. Wet process equipment which have dead zones or are designed not to allow continuous water flow will also experience bacteria build-up. The materials used in the construction of a UPW distribution system are also critical. Most metals are quickly attacked by UPW, causing ionic contamination and lower water resistivity. In addition, many plastics are unacceptable because of the ability of UPW to leach organics and ionics from these materials. Certain polymeric materials are also nutrients which can promote bacterial growth. Micro surface roughness on internal piping and component surfaces also provide microscopic dead zones where bacteria can proliferate and form a biofilm (Martyak, et al., 1993). These are some of the factors must be considered in the design, operation, and maintenance of any wet process equipment where UPW is used.

3.2.2 Liquid Chemicals

Liquid chemical distribution systems and application methods have evolved significantly since the early 1980s. Although large fabs had central storage and distribution facilities for a few solvents and other high-usage chemicals, most early wet processes simply used static processing tanks which were manually filled from bottles. Tank chemicals were replaced after processing a prescribed number of job lots or on the basis of work shifts. Large wafer sizes and the need for better process control has encouraged the development of central chemical dispensing systems and improved application methods (Deal, Granut, 1994). Centrifuge spray systems were developed using local chemical dispensing systems which where an integral part of the processing equipment. Today's chemical dispensing systems are capable of serving many wet processing stations and deliver high-purity chemicals with low levels of particulate (Grant et al., 1990; 1994).

Particles and trace impurities have always been major contaminants because of the intimate relationship wafers and chemicals have during processing. The drive toward smaller device features and increased circuit densities has created a need for ever increasing chemical quality and stability at the wafer surface during processing. Current requirements for leading-edge products are limiting total ionic im-

TABLE VIII-3 Allowable Levels of Impurities in UPW for Semiconductor Manufactuirng

Impurity	Maximum level
Total organic carbon (TOC)	<1 ppb
Particles 0.05 μm to 0.1 μm	<100/liter
Bacteria (cultured)	<0.1 CFU/liter
Silica	<0.5 ppb
Dissolved oxygen	1 ppb to 20 ppb
Na^+	<10 ppt
K^+	<25 ppt
Cl^-	<10 ppt
Br^-	<10 ppt

Source: Jacobs Engineering Group.

purities for some chemicals to less than 1 part per billion (ppb) and particles >0.1 μm in size to less than 20 per ml (Hashimoto et al., 1989). One approach under development, *point-of-use chemical generation* (*POUCG*), combines precise quantities of DI water with the appropriate UHP gases, to produce extremely pure chemical, such as NH_4OH, HF, and H_2O_2 at the point of use (Peters, 1994). In addition to generating wafer cleaning chemicals with metallic contamination levels in the low parts per trillion (ppt) range, this method eliminates the need to store and transport the chemicals it produces. Increasing chemical usage and environmental concerns are also making spent chemical disposal more difficult and expensive. This is placing increased pressures on production personnel to reduce chemical usage. Demands for higher purity chemicals and concerns for the environment has resulted in the development *chemical reprocessors* that are able to recycle spent chemicals and return them to their original high-purity state for reuse in the line.

Chemical reprocessors have introduced an evolutionary approach for achieving the necessary chemical contamination control while reducing usage. Initially, purity and particle control were maintained by simply replacing the used chemical with fresh chemical in a static tank. As chemical requirements increased, the chemical was replaced more often. In addition to increasing chemical usage, frequent chemical replacement also resulted in lost production time. It also became obvious that regardless of the purity of the incoming chemical, chemicals degraded with the first carrier load of wafers. At some point in time, it would become impractical to replace chemicals as often as processing requirements demanded. One solution is to provide a continual flow of purified and filtered chemicals to the wafer surface during processing. This is accomplished by utilizing an automated chemical distribution system combined with the appropriate wet process station design to provide *continuous flow chemical reprocessing* (Martyak, Rapa, 1995). This concept, illustrated in Fig. VIII-17, can be implemented in stages, each stage producing a degree of improvement. The first stage requires a processing tank designed to recirculate and flow chemical upwards through the wafers thereby aiding in the rapid removal of by-products produced by the chemical action at the wafer surface. A filtration package can then be added which continuously removes particulate matter as the

Contamination

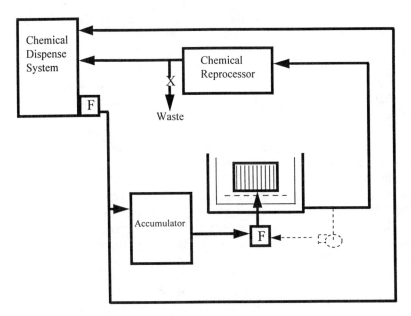

FIGURE VIII-17 Diagram of a continuous-flow chemical distribution and reprocessing system. This arrangement continuously flows purified and filtered chemicals over the wafers during processing.

chemical in the tank is recirculated (*recirculation filtration*). A central chemical distribution system can also be installed to eliminate manual filling of the processing tank. The system can be designed with a return loop and recirculation filtration to maintain particle-free bulk chemicals in the central system. The addition of a *chemical reprocessor* is the final element in the concept. The chemical can now be purified and recycled allowing consistently pure and particle-free chemicals to be in contact with the wafer surfaces during processing.

A continuous flow reprocessing system will likely provide a good return on investment from reduced chemical usage, improved yields, and reduced chemical waste disposal costs. However, these systems are complex and costly to purchase and maintain. In addition, a continuous flow reprocessing system may not be necessary for manufacturing many semiconductor products. Various factors such as chemical usage and device complexity will determine which evolutionary stage is best suited for the specific products being fabricated.

3.2.3 Static Electricity

The major generators of static electricity are (1) triboelectric charging caused by rubbing or contact and separation of dissimilar materials, and (2) induction or the transfer of electrical charges on surfaces in close proximity to each other (Steinman, 1992). In semiconductor manufacturing static electricity can cause (1) electrostatic discharge (ESD) damage to devices, masks, and reticles, (2) particle attraction to wafers and other sensitive surfaces, and (3) disruptions in electronic equipment controls. ESD occurs when electrical charges form a high voltage potential between small device features that are separated by dielectric material. If the electrical potential is large enough the charge may bleed off quickly across the dielectric gap forming a minute but intense arc of electricity that is hot enough to damage the device. ESD defects are most likely to occur between two interconnection levels as illustrated in Fig. VIII-18. In Fig. VIII-18a, a particle is encapsulated within the insulating layer between two metal device layers. Opposing charges build up on the metal layers as shown in part (b). In Fig. VIII-18c an event occurs which allows the charge to quickly bleed off across the dielectric layer. The discharge occurs above the particle since this area provides the path of least resistance. Fig. VIII-18d shows that hot electrical discharge arc has formed a short circuit tunnel containing metallic residues. This type of ESD failure is well known and special precautions are taken during electrical testing, chip packaging operations, and in the handling the finished package through to its final destination (Kendrick, 1994; Jacobs, 1988). The precautions may include the use of air ionization, grounding straps, static dissipative chip containers, and special handling procedures.

Static electricity can also influence the movement or migration of particles. This movement can be either detrimental or beneficial. For example, there are many plastic objects and surfaces in a semiconductor line which can acquire very high static charges. These surfaces will naturally attract airborne environmental particles having the opposite polarity. If the particles are attracted to active working surfaces, they increase the risk of particle-related defects. Particles could also be attracted directly to wafers or indirectly by transferring from highly charged plastic wafer

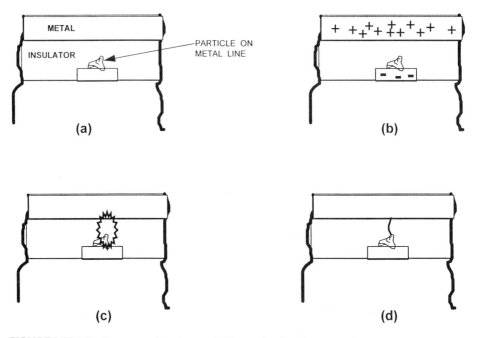

FIGURE VIII-18 Sequence showing an ESD mechanism between device layers. (a) A particle or flaw is present. (b) Electrical charges develop between layers. (c) ESD occurs. (d) A fault is created resulting in a short between levels.

Contamination

carrier, or box surfaces, or from the environment (Ohmi et al., 1989). Another problem caused by static electricity relates to the operation of process equipment. Automated input and output devices can promote static build-up which can cause wafers to deviate from their designated path resulting in wafer damage and breakage. Static charges on people can also cause equipment to malfunction.

Air ionization is a common method used to control static electricity in cleanrooms. In this method, air molecules are charged to form positive and negative air ions which then seek out and neutralize static charges of the opposite polarity (Steinem, 1994). Other static control methods include increasing relative humidity levels, use of static dissipative materials, and various grounding devices which allow static charges to slowly bleed to ground. Effective static control usually requires the use of several complementary methods that are compatible with specific conditions in the manufacturing area.

3.2.4 Vibration

Vibration primarily affects the performance and stability of photolithographic and metrology equipment. This type of contamination is very complex and the development of cost-effective control measures involves extensive engineering analysis of equipment, building structure, and specific site conditions where the fab building is located (Medearis, 1995). There are numerous techniques used in controlling vibration ranging from special equipment design and installation methods to building construction features. A comprehensive discussion of vibration control is beyond the scope of this text. The following discussion examines some practical vibration control considerations associated with process and support equipment in a typical semiconductor fab.

Figure VIII-19 illustrates various potential vibration sources that are present in process equipment installations. The figure shows a through-the-wall equipment installation where much of the process tool and its support equipment is installed in the service chase and only the portion where wafers are loaded and unloaded extends into the process cleanroom area. Process equipment (a) with support equipment (b) is a multichamber vacuum system with vibration sources such as motors and vacuum pumps. These sources are not properly isolated and, therefore, transmit vibration to raised floor (d) and cleanroom wall (e). Process equipment (c) is sensitive to vibration contamination and the recipient of vibration. The recipient can be located on an opposite wall in the immediate areas as shown, on the same wall as equipment (a), or some distance away in another process area. The raised floor, in both service chase or process areas, and cleanroom walls provide common paths for vibration transmission. In some cases, the vibration may be amplified as it resonates various floor and wall elements in its path. The vibration may also cause subtle effects such as *acoustic vibration* and particle generation from filters and other ceiling elements. Acoustic vibration occurs when the "skin" of the cleanroom wall vibrates, similar to the cone of a sound speaker. This may result in low-frequency sound waves that are sufficiently energetic to cause other cleanroom or process equipment elements to vibrate. The net effect is that the air environment acts as another vibration path. Another potential problem relates to the influence of vibration on HEPA and ULPA filters and other ceiling components. Vibration can cause relative motion between the various ceiling elements (filters, blank pan-

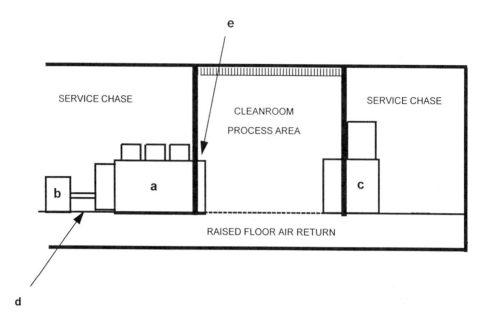

FIGURE VIII-19 Diagram showing potential sources of vibration transmission from equipment: (a) equipment processing section, (b) support equipment, (d) vibration transmitted to raised floor, (e) vibration transmitted to the cleanroom wall system, (c) potential process equipment vibration receiver.

els, support grids, etc.) resulting in a continuous "rain" of particles. Ceiling vibration can also cause particle sloughing from HEPA and ULPA filter media. The filter media is a fiberglass paper formed into multiple pleats. The pleats expose as much as 20 ft^2 of media surface, which can shed particles, for each square foot of filter face area. Particles generated at the ceiling by both mechanisms will generally be larger than 10 μm and be undetectable by airborne particle counters.

Like other forms of contamination, vibration is most effectively controlled by either eliminating the source(s) or controlling its migration or transmission. While much of equipment-generated vibration is unavoidable, there are various techniques for minimizing its transmission. Many of these techniques are relatively simple and inexpensive when implemented during equipment installation. Figure VIII-20 illustrates alternative methods to isolate vibration at or near the source as described below:

(a) *Vibration isolation mounts.* These can be relatively inexpensive rubber-in-shear devices and spring isolators used as the support points for equipment. Vibration isolation mounts are often used to isolate vibration transmission from rotating equipment such as motors and vacuum pumps. Although vibration isolation mounts are relatively simple, they must be carefully selected and "tuned" to vibration frequencies and other equipment characteristics to be effective. Vibration isolation mounts are also used as foundation elements for protecting vibration sensitive equipment (recipients). These isolation

Contamination

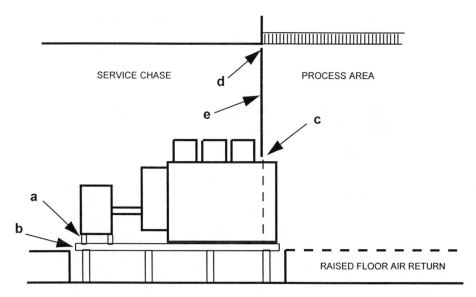

FIGURE VIII-20 Diagram illustrating methods of controlling vibration from process equipment: (a) vibration isolators, (b) support base independent of raised floor, (c) decoupling of wall and equipment, (d) decoupling of wall and filter ceiling, (e) non-load-bearing partitions.

mounts can range from the simple mounts described to sophisticated pneumatic and electrical feedback-loop systems.

(b) *Independent support bases.* Equipment can be isolated from floor vibration by simply isolating a portion of the raised floor or through the use of a specially fabricated base constructed from steel or concrete which rests on the subfloor.

(c) *Wall cut-out isolation.* The cleanroom wall is a key element in transmitting vibration from through-the-wall equipment installations. After equipment installation, the gap between the equipment and wall cut-out is typically closed off with rigid materials to prevent any significant air leakage or transfer from the service chase to the cleanroom. This provides a relatively tight coupling for vibration between the equipment and wall. A simple solution is to replace the rigid gap-filler with soft material such as heavy gauge vinyl. This technique provides the necessary air seal but prevents the transmission of vibration from equipment to wall.

(d) *Loosely coupled wall support at ceiling.* Wall to ceiling vibration transmission can be attenuated by providing a gap. Support for the wall is provided by a pin-in-hole connection which constrains horizontal movement but permits vertical movement allowing the load to be supported entirely by the floor. The gap between wall and ceiling can be closed off by using heavy gauge vinyl similar to the technique described for wall cut-outs.

(e) *Equipment installation partition.* In addition to loose coupling at the ceiling, walls can be designed as simple *partitions* designed solely for the purpose of providing a separation between the cleanroom and service chase environments. Items which could transmit vibration (e.g., piping, electrical distribution components, and various other process equipment accessories) which would normally be mounted on the cleanroom wall system would be mounted elsewhere.

4.0 INTERCONNECTION PROCESS AND EQUIPMENT CONTAMINATION CONTROL

4.1 INTRODUCTION

Process equipment is the final terminal point for contamination from all sources and knowing where and when the wafers become contaminated is essential information for controlling contamination. One early view suggested that most defect-causing contaminants are due to people and the environment (Morrison, 1973). However, most recent experience indicates that new cleanrooms are very effective in controlling environmental contamination so that most defect-causing particles are deposited on wafers during or just before the start of the process inside the equipment (Hattori, 1994). This fact has resulted in a trend toward monitoring process equipment routinely for particles. Inside the equipment, there may be a single direct source or a combination of interacting factors involved. A direct source is the shedding (sloughing) of particles from materials used to construct the chamber or other parts of the equipment. Another direct source is flaking of process material deposited on the chamber walls. An example of contamination from several interactive factors is unwanted chemical reactions caused by impurities in process chemicals and gases. While process reactions and other internal factors are complex, external factors which influence the movement of contamination from its source to the equipment are also complex. In some cases the migration process may take many weeks or months causing changes in the makeup of the contaminant (e.g., impurities acquired by environmental particles during their migration). The investigative process must consider these possibilities in order to correctly identify the contamination source.

4.2 GENERAL TROUBLESHOOTING

Troubleshooting is defined here as the systematic investigation of a problem until its cause is found and permanently corrected. Semiconductor contamination problems can be very difficult to troubleshoot since they are complex and may involve several simultaneous contamination sources and interactions. In addition, the chain of events which starts with the release of a contaminant from its source and culminates as a wafer defect, may take days or weeks. Since the process equipment is the final terminal point for these interactions, it serves as a good starting point for troubleshooting investigations.

4.2.1 Partitioning

Identifying the exact point in the process where most of the contamination is occurring is usually the first step in the troubleshooting process. As a rule, defect analysis will indicate the general area or process sector where the majority of

Contamination

defects are forming. The process sector may include several pieces of equipment, each having the potential of being the primary cause of the problem. Although there may be several significant contamination sources in the area, there is usually only one primary contributor to the current problem. Once this source is identified, other contributing sources may be discovered. The process of isolating the major contributing source, in terms of specific location and time, is called *partitioning*. Figure VIII-21 is a diagram showing potential contamination contributors in a typical process sector consisting of several types of process equipment and various wafer handling activities (logistics) which take place between process steps. Location partitioning is usually performed by temporarily instituting product inspections (microscope and/or wafer scanner) before and after each suspected process point. For example, temporary inspections may be placed immediately before and after equipment "A." If no problem is found, the inspection points may be moved to points before and after equipment "B." If the primary source is isolated to particular process step, the process may be partitioned to determine at what point in the equipment's cycle the problem is occurring. For example, assume that equipment "B," a metal sputtering system, is identified as the primary contributor. The first step may be to cycle blank wafer monitors through the input and output (wafer handling) portion of the operation. In addition to exposing the monitors to wafer handling, the next monitor group would be allowed to enter the process chamber. The following monitor group would then be exposed to chamber pump down and venting. Subsequent monitor wafer groups may be exposed to various processing times to see if particles are accumulating on wafers during the actual deposition process. The data obtained from the partitioning the process cycle can be charted (Fig. VIII-22) to indicate what portion of the process cycle contributes the highest level of contamination.

Partitioning experiments may indicate that the source of a contamination problem is in the wafer handling logistics area and not related to actual wafer processing. For example, the problem source may be isolated to the logistics area between process equipment "A" and "B" (Fig. VIII-21). Contamination problems associated

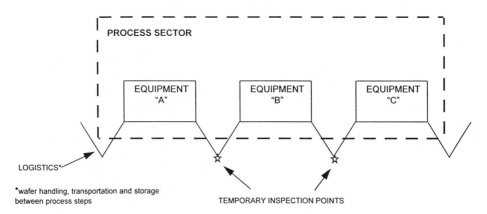

FIGURE VIII-21 Diagram illustrating the partitioning of process steps within a sector. Temporary inspection points are established to identify the specific process step causing the problem.

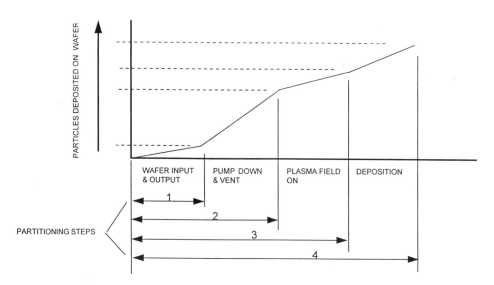

FIGURE VIII-22 Diagram showing partitioning of the process cycle of a typical vacuum deposition system. Each subsequent step includes the one that precedes it. In this example, the pumpdown and vent portion of the process cycle is the highest particle contributor.

with logistics may be related to activities which are not obvious contamination concerns (e.g., wafer boxes, storage temperature, etc.). Therefore, it is best to consider wafer logistics as another "process step" to be partitioned if necessary.

When using the partitioning process to troubleshoot a problem, it is important to remember that contamination occurs sporadically and unpredictably. As a result, it may take several partitioning experiments to finally capture a contaminating event. In addition, the experiments should be repeated several times to make sure that the captured event actually represents the problem under investigation. While this iterative process may be time consuming, it is usually far less costly than incomplete testing which results in an incorrect diagnosis.

4.2.2 Particle Analysis

Partitioning, carefully designed and implemented, is a powerful technique for solving contamination problems. In many cases, partitioning alone is all that is required to determine the cause and understand the contamination mechanisms. Other situations, however, may require extensive analytical work before an accurate diagnosis is obtained. SEM/EDX particle analysis (fingerprinting) has been found to be a very effective technique for this purpose. From the fingerprinting process, elemental, size, and shape information is obtained from a random sample of wafer particles. The detection frequency of each element is plotted in a consistent format to produce a histogram or signature representing the composition of the particles on the wafer. The histogram of elements, combined with particle size and shape information obtained from SEM photomicrographs, provides a fingerprint useful for identifying the contamination source. For example, particles originating from equipment, processes, or other nonenvironmental sources, are generally made up

Contamination

of only a few elements which tend to be similar from particle to particle. These similar combinations are referred to as *dominant groups*. Figure VIII-23 is a fingerprint example showing the elemental histogram with dominant groups and particle size and shape information. The various elements in the histogram are in a specific format in which the sequential order from aluminum (Al) to sodium (Na) remains constant. Particles containing four or more random elements from this group are typically environmental particles. The order for the remaining elements may change depending on the sample being analyzed and the variety of elements detected. The Y-axis indicates the frequency each element was found in terms of the total sample (e.g., Al was found in 25% of the particles sampled) This fingerprint is derived from particles picked up on a wafer cycled through scan-projection wafer exposure tool. In this fingerprint, most the elements are in distinct dominant groups which link them to the sources indicated in the pie chart in Fig. VIII-24. SEM/EDX fingerprinting has been found to be most effective in samples where there is a large particle population and several contamination sources.

4.2.3 Materials and Finishes

Construction materials and finishes are common sources of particulate contamination in process equipment. Particles originating from inappropriate exterior finishes can migrate via contact transfer to other areas and eventually to product surfaces. However, the appropriate finish is also dependent on the equipment's design. For example, standard enamel paint is considered a poor finish for equip-

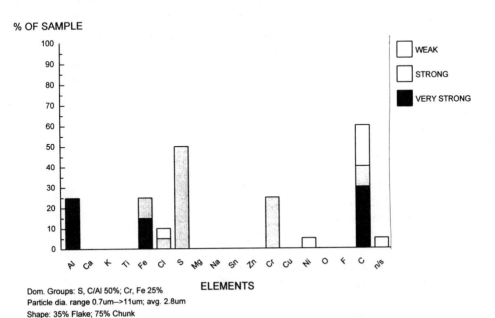

FIGURE VIII-23 SEM/EDX fingerprint graph of particles deposited on a monitor wafer cycled through scan-projection exposure equipment installed in a class 1 cleanroom. The graph indicates an absence of an environmental particle signature and distinct dominant element groups which identify equipment and process material sources.

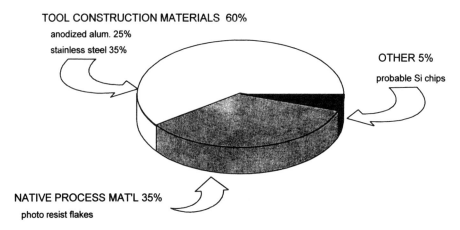

FIGURE VIII-24 Pie chart showing particle sources derived from the fingerprint example in Fig. VIII-23.

ment and other cleanroom surfaces. This is because standard paints oxidize over time causing loosely bonded particles to develop and migrate. Standard paints also have a tendency to chip and offer poor resistance to chemicals. These weaknesses make paint a poor choice for any surface exposed to even casual contact during manufacturing and maintenance operations. Standard paint, however, may not be a problem in areas where contact with people and chemicals is unlikely. This is an important factor to consider since it is impractical or prohibitively costly to use "risk-free" materials and finishes in all cases. The appropriate choice will be a balance between contamination control concerns and business factors such as costs and availability. Alternate finishes do not need to be more costly and there are usually options available which are less costly yet offer excellent contamination control characteristics. The following suggestions apply to cleanroom and equipment surfaces.

Avoid:

- Standard oil and latex based paints
- Crinkled finishes
- Brush finishes on stainless steel or other materials
- Unpassivated stainless steel
- Bare aluminum and steel
- Vapor (bead, sand, etc.) blasted surfaces
- Unclad wood
- Unclad pressed mineral board materials
- Unencapsulated foam plastics and rubber
- Cloth fabrics
- Unclad particle board

Favor:

- 2-part epoxy paints
- Polished finish stainless steel

- Smooth construction plastics (PVC, polycarbonate, polyethylene, etc.)
- Heavy anodized or hard-coated smooth aluminum
- Heavy chrome-plated steel
- Plastic laminate (totally clad particle board)

4.3 SPECIAL CONSIDERATIONS FOR THIN FILM EQUIPMENT AND PROCESSES

In general, vacuum deposition and etch systems generate the most particulate as compared with other types of processing equipment (Tab. VIII-4). A major problem relative to any vacuum deposition process equipment is particles flaking from chamber surfaces. Although these tools are designed to deposit materials on wafer surfaces, deposition also occurs on surrounding chamber surfaces. Etch processes can attack or remove material from o-rings and other system components as a result of high-energy plasma reactions. In deposition equipment, it is important that films deposited on chamber surfaces remain tightly bonded to the previously deposited layers for as long as possible to minimize production downtime. Eventually, the film on chamber surfaces begins to break down and flake, resulting in high-contamination levels. Methods are available to perform in situ cleaning thereby minimizing downtime resulting from equipment disassembly. Eventually, all dry deposition and etch systems require disassembly for cleaning in order to maintain low particle contamination levels. Most reactors are designed with removable shields and other internal parts which can be cleaned off-line. Additional manual cleaning may be required on nonremovable portions of the tool.

Vacuum deposition and etch reactors are either bulk tools processing many wafers in a single large chamber, or single-wafer, multichamber tools with load-locks and several process chambers. In general, wafer particle deposition levels are lower in single-wafer reactors than in bulk equipment. The improvement in single-wafer equipment may be attributed, in part, to maintaining the processing chamber under relatively constant temperature and pressure conditions thereby minimizing stress factors that often generate particles. In addition, bulk systems typically require the chamber to be vented and opened to the environment for wafer loading and unloading. This exposes chamber surfaces to ambient conditions and to thermal

TABLE VIII-4 Ranking of Process Equipment in Terms of Particle Generation

MOST ↑	High-current ion implanter
	Reactive-ion etcher
	Plasma etcher
	Plasma-enhanced CVD
	Sputterer
↓	Wet processing station
	Oxidation furnace
	Coater/developer
LEAST	Wafter stepper

Source: Deal, "Contamination Control: Problems and Prospects," (1994).

cycling and other stress factors causing flaking to occur more often. Wafers processed in bulk equipment are generally manually loaded using vacuum tweezers which add to contamination control concerns.

Studies show that there are numerous forces present during plasma processing which influence particle generation and movement. These forces include gravity, electrostatics, thermophoresis, photophoresis, ion drag, gas flows, and mechanical inertia. In situ particle monitoring and laser light-scattering visualization techniques (see Section 2.8.4) have helped significantly in our understanding of reactor particle contamination mechanisms. Trace amounts of water, however, are probably the greatest contamination concern in vacuum systems. Water is a potent contaminant in any vacuum process since it is reduced to ions as well as H_2 and O_2 gas molecules as a result of reactor chemistry or high energy plasmas (Harvell, Lessand, 1991). It also sticks tenaciously to all surfaces, outgasses slowly, and is a chemical poison in reactors. The tenacity of water sticking to chamber walls is so great that it is estimated there are 10,000 times more molecules of water on the chamber walls than in the vacuum space of a one cubic meter processing chamber. It is also well known that significant contamination-related activity takes place during the pump down and vent portions of a reactor's process cycle. It is possible that water may also play a role in the formation of particles during the pump down cycle. For example, Fig. VIII-25 shows the hypothetical steps which may be involved in particle formation during the pump down cycle (Liu, 1994). According to this hypothesis, water droplets formed during the pump down by adiabatic cooling and homogeneous nucleation first absorbs SO_2 from the surrounding air along with some trace chemical species that can oxidize the absorbed SO_2 in the liquid phase to form H_2SO_4 residue particles. Gas flows and mechanical agitation are additional factors which contribute to high particle contamination levels during pump down and venting.

Contamination in plasma reactors can be minimized and controlled by several techniques described in U.S. Patent 5,367,139 (Bennett et al., 1994). These techniques include the interruption of the plasma by pulsing the source of plasma energy periodically, or application of energy to provide mechanical agitation such as mechanical shock waves, acoustic stress, ultrasonic stress, vibrational stress, thermal stress, and pressure stress. Following a period of applied stress, the reactor is pumped down, vented, and opened for internal cleaning. The addition of small amounts of CF_4 or NF_3 to the feed gas was found to aid the cleaning process and significantly reduce the level of particulate in the plasma (Selwyn et al., 1990; 1991; 1992). A method to reduce contamination on wafers etched in a reactive ion etch chamber is described in U.S. Patent 5,221,425 (Blanchard et al., 1993). In this method, a radio frequency (RF) voltage is applied within the etch chamber while reducing pressure in the etch chamber to a base pressure. Particulate matter is reduced by varying the flow of high and low reactive gases and varying the RF voltage in a specific manner. There are other patented methods for reducing particle contamination in plasma reactors such as creating a particle-repelling electric field above the wafers during the time the wafers are actually undergoing processing (Savas, 1992, U.S. Patent 5,102,496) and preventing low temperature dry etch deposits by maintaining power to keep gases in the plasma state during evacuation (Cathey, Frankamp, U.S. Patent 4,992,137) The use of electric fields to control

Contamination 553

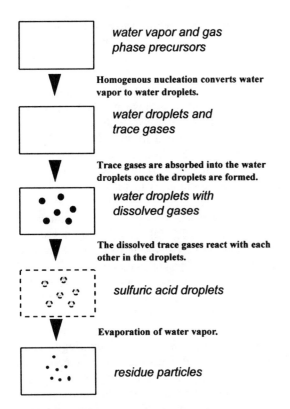

FIGURE VIII-25 The hypothetical steps involved in particle formation during the pump down cycle.

particle deposition on wafers during metal deposition has also been demonstrated (Periasamy et al., 1992).

In situ cleaning techniques have been developed which greatly reduce the need to disassemble process reactors for cleaning purposes, thereby avoiding the risk of water adsorption. However, process equipment still requires frequent and on-going servicing where the skills of maintenance personnel and the procedures they follow are key elements in the control of contamination. Most thin film process tools are highly complex and very sensitive to subtle events which can result in different contamination performance even between similar equipment performing the same process. The following suggestions may be helpful in maintaining stable processing systems and low contamination levels.

> ***Maintain good temperature and humidity control*** in the cleanroom environment where bulk deposition equipment is located. This is particularly important for bulk processing reactors where processing chambers are exposed to the ambient conditions when loading and unloading wafers. Although processes can be fine-tuned for a consistent ambient temperature and humidity level, process and contamination control are affected when there are excessive temperature and humidity excursions.

Control the equipment's pump and vent cycles to minimize turbulence and prevent particle generation when processing wafers. Hard pump down and vent cycles are also effective for removing particles during cleaning operations.

Choose abrasive cleaning materials carefully. Metal deposition equipment often requires the use of an abrasive material to remove residual metal deposits from portions of the chamber wall. An abrasive material sometimes used for this purpose consists of a grit imbedded in an polymer (plastic) fiber pad. This type of abrasive pad may leave behind small bits of polymer that cannot be removed by vacuuming. It is generally safer to use a hard metal mesh pad (e.g. stainless steel or aluminum) in place of a polymer abrasive pad for this purpose.

Use only exact replacement parts and materials. Manufacturing demands to quickly repair process equipment often places maintenance personnel under pressure to find a quick solution. When exact parts and materials are not available, there may be a natural tendency for the maintenance person to try come up with a creative solution. The creative solution may enable the equipment to perform functionally, satisfying immediate manufacturing demands, but the fix may also introduce a subtle latent contamination problem.

Use gas-handling components that are low-particle generators. Significant progress has been made in the design and construction of clean gas-handling components. Certain components, such as check valves, regulators, and automatic valves, can generate particulate matter during operation. Particle contamination test standards have been established to evaluate gas-handling components and should be used, if possible, for selecting the cleanest component (Periasamy et al., 1995).

4.3.1 Gases and Their Influence on Equipment and Processes

Gases are the raw materials used in dry deposition and etch processes and their purity plays a key role in contamination control and maintaining process stability. The gases used by semiconductor process equipment are classified in two groups: *bulk* gases and *specialty* gases (also referred to as *feed* or *cylinder* gases). Bulk gases are used in large quantities and are generally centrally distributed from large storage tanks. Bulk distributed gases generally consist of nitrogen, oxygen, hydrogen, and argon. Semiconductor grade bulk gases are referred to as *ultra-high purity (UHP) gases*. Specialty gases are reactive process gases (silane, HCL, etc.) which are used in smaller quantities and typically supplied from gas cylinders or bottles. The primary impurities or contaminants in UHP gases are trace water, gaseous impurities (e.g., oxygen in nitrogen), and hydrocarbons. Particle contamination is also important but direct particle sources are generally low in UHP gases and are considered a minor concern compared with the effects caused by the other contaminants. Trace amounts of water and other chemical impurities in a gas can cause an unwanted reaction which produces literally millions of particles, unstable processes, or unsatisfactory device structures. Table VIII-5 lists some gas impurity effects and causative mechanisms which can occur during plasma etching (Flamm, 1993). Water or hydrocarbons on UHP gases used in a metal deposition processes

TABLE VIII-5 Gas Impurity Effects, Mechanisms, and Examples

Impurity effect	Example/chemistry	Mechanism in example
Alter gas phase etchant concentration	O_2 or NF_1 in C_2F_6	Oxidant/unsaturate chemistry
Introduce competing etchant	Carbon species in polysilicon etch	Thermodynamic assist. SiO_2 forms CO via carbon surface species
Alter etchability of primary substrate	O_2 in Cl_2/Si etching	Forms surface SiO_2 which retards etching
Alter etchability of mask	O_2 in Cl_2/Si etching	Oxygen species attack photoresist carbon chains
Alter etchability of underlying substrate	O_2 in Cl_2/Si etching	Reform surface SiO_2 on "damaged" SiO_2
Cause residue	H_2O in Cl_2/Al etching	AlOx and Al containing sidewall films formed
Form, remove, alter films or pasivation on reactor surface	Carbon-containing silicide etchants	Unsaturated species for films on walls
Changes etch morphology	Some fluorine-containing species in GaAs etch feed gases	Passivates surface sites
Changes dominant ion	Molecular gas added to ultrapure rare gas plasmas	Lower energy ionization processes, charge exchange to form lower energy ions
Ion bombardment energy	Fluorocarbons added to halogen plasmas	Additive gases with low energy states "absorb" tail of electron energy and alter plasma potentials

Source: Solid State Technology, October 1993.

can result in poor film integrity or high electrical resistance. The higher priority given to nonparticulate contaminants is important since an emphasis on reducing particles in ultra-high purity (UHP) gases may result in the undesirable side effect of increasing water, gaseous, and hydrocarbon impurities. For example, it has been a common contamination control practice to install point of use filters in distribution lines supplying gases and liquids to process equipment. However, a properly designed UHP gas distribution system is intrinsically free of defect-causing particles for all practical purposes. Therefore, installing a point of use particulate filter is not likely to provide any meaningful improvement. In addition, point of use filters may actually cause a significant increase in contamination levels or other process problems. For example, the media in a particulate filter represents a significant increase in the surface area that the UHP gas is exposed to prior to entering the reactor. The media itself may be a source of impurities such as hydrocarbons. However, a greater risk is the filter media's ability to act as a "sponge" to acquire and hold trace amounts of water. Filters also require periodic maintenance and replacement thereby increasing the risk of exposing the UHP distribution system to contaminants from the ambient environment. While the risk may be relatively small when one filter is installed on a single reactor, the risk increases significantly when point of use filters are installed on all equipment as a general rule. These factors should be carefully considered prior to the installation of point of use filters on UHP gas distribution lines.

In general, line and support personnel responsible for critical vacuum depositing and etch processes are well aware of the importance of maintaining consistently high purity levels in gases. Other processes, however, are not influenced to a great degree by slight changes in gas purity. Wet processes, for example, may use gases for various purposes (agitation, drying, etc.) where there is little concern for trace gaseous, water, or hydrocarbon impurities in the gas. Therefore, gas line hook-up and repair protocols may not be as stringent as those used on vacuum systems. Nonetheless, these factors can influence the high-purity performance of the entire UHP gas system. Therefore, large facilities typically provide two grades of nitrogen, standard and UHP, in order to avoid these problems.

The purity requirements for both bulk and specialty gases have tightened significantly in recent years driven by device fabrication needs. Table VIII-6 lists the impurity levels for two grades of UHP bulk gases currently available for semiconductor processing. As this table indicates, UHP gases can be manufactured very pure. The difficulty is in delivering the gases to their point of use while consistently maintaining these ultra-high-purity levels. To accomplish this, very smooth electropolished stainless steel is used for all piping, valves, and other components with an emphasis on "building clean" (Sugiyama et al., 1989; Tomari et al., 1991). The piping is orbital butt welded in an inert gas environment forming a smooth joint. UHP gas piping preassembly and fit-up is performed under cleanroom conditions with cleanroom protocols. In addition, UHP gas distribution systems are designed leak-tight to standards which are similar to the leak-tight requirements of vacuum process equipment. After the distribution system is completed, it is heated (heat traced) while being purged with flowing nitrogen to remove any traces of water and hydrocarbons. Building a UHP gas delivery system is costly and can account for as much as 18 to 20% of building a large fab (Singer, 1994). Certification of UHP gas delivery systems usually requires sophisticated equipment such as con-

TABLE VIII-6 Impurity Levels in UHP Gases

	Oxygen (ppb)	Water (ppb)	CO_2 (ppb)	CO (ppb)	THC* (ppb)	Hydrogen (ppb)	Nitrogen (ppb)
Argon							
VLSI/ULSI	20	50	30	10	30	20	100
with purifier	1	1	1	1	1	1	1
Oxygen							
VLSI/ULSI	—	50	40	20	30	10	75
with purifier	—	5	5	5	5	5	500 ppm
Nitrogen							
VLSI/ULSI	N/A	N/A	N/A	N/A	N/A	N/A	—
with purifier	1	1	1	1	—	—	—
Hydrogen							
VLSI/ULSI	20	50	5	10	20	—	100
with purifier	1	1	1	1	10	—	1

*Total hydrocarbons. (Jacobs Engineering Group.)

densation nuclei counters (CNC) for measuring particles and atmospheric pressure ionization mass spectrometer (APIMS) for impurities (Siefering et al., 1992; Borkman et al., 1992). Specialty gas delivery systems are constructed and certified in a manner similar to bulk gas systems. However, the integrity of specialty gas systems is more difficult to maintain and measure due to the toxic and reactive nature of these gases and the need to manifold and change gas cylinders (Cestari, Yelverton, 1995; Iscoff, 1992; Kasper et al. 1989). Table VIII-7 lists the various specialty gases used in wafer fabrication and their applications (Hart et al., 1995).

Once certified, contamination problems associated with UHP gases are seldom the direct cause of the main distribution system itself. Although the distribution system may become contaminated, the cause of such problems are often traced to some aspect of equipment installation or system operation and maintenance factors. The following are some conditions which are likely to cause contaminated UHP gases.

1. UHP nitrogen is used as a backup for compressed air in a piping arrangement where the failure of a valve or other component results in the compressed air contaminating the nitrogen distribution system.
2. UHP nitrogen is used for bubbling agitation in a wet process (may cause back diffusion of water).
3. A UHP gas component is repaired or replaced in a manner with allows air to diffuse into the gas distribution system.
4. Process equipment hook-up is improperly performed allowing air or hydrocarbons into the system.
5. UHP nitrogen is used for operating air cylinders causing air and hydrocarbons to infiltrate the distribution system.
6. Use of plastic tubing and fittings to distribute UHP gases inside process equipment (diffusion through tubing).
7. The use of improper pipe fittings or poor connections on UHP lines inside process equipment.

TABLE VIII-7 Applications for Various Specialty Gases in Semiconductor Manufacturing

	Crystal growth	Thermal oxidation	Epitaxy	Thermal diffusion	Ion implant	Chemical vapor deposition	Metalization	Dry etching
Ammonia (NH₃)						X		X
Argon (Ar)*	X	X	X	X	X	X	X	X
Arsine (AsH₃)	X		X	X	X	X		
Boron trichloride (BCl₃)								X
Boron trifluoride (BF₃)				X	X			
Boron 11 trifluoride (B11F₃)					X			
Chlorine (Cl₂)		X						X
Diborane (B₂H₆)			X	X		X		
Dichlorosilane (SiCl₂H₂)			X			X		
Disilane (Si₂H₆)						X	X	
Halocarbon 23 (CHF₃)								X
Halocarbon 116 (C₂F₆)								X
Helium (He)*	X				X	X		X
Hydrogen (H2)*	X	X	X	X	X	X	X	X

Contamination

Gas					
Hydrogen bromide (HBr)					X
Hydrogen chloride (HCl)	X				X
Nitrogen (N$_2$)*	X	X	X		X
Nitrogen trifluoride (NF$_3$)	X	X			X
Nitrous Oxide (N$_2$O)	X		X		X
Oxygen (O$_2$)	X	X			X
Phosphine mixes (PH$_3$)	X	X	X		
Silane (SiH$_4$)	X	X	X	X	
Silicon tetrachloride (SiCl$_4$)	X	X			X
Sulfur hexafluoride (SF$_6$)					X
Tetrafluoromethane (CF$_4$)					X
Tungsten hexafluoride (WF$_6$)		X		X	

*Carrier/purge gas.
Source: J. J. Hart et. al., "Particle Measurement in Specialty Gases," *Solid State Technology*, September 1995.

As a rule, consistency and reliability of gas purity performance is dependent on the complexity of the system. Fewer components in the UHP gas distribution system will generally result in cleaner, more reliable performance. From a contamination control perspective, the UHP gas distribution system does not end at the supply connection to the equipment and the same stringent design and operational considerations should be applied for gas piping within vacuum process equipment.

The following are two additional factors to be considered relative to UHP gases and their use in process equipment.

- ***Procedures and protocols*** established for UHP gases should be applied equally and universally regardless of the differing contamination or purity requirements for specific process equipment.
- ***Gas purifiers*** are devices which are often used to improve the quality of gases. These devices are basically absorber-type filters which can be designed to remove a specific impurity such as water or hydrocarbons. Purifiers are available in various sizes and capacities and can be installed in the main distribution system or at the equipment (point of use). Although purifiers are very effective at removing a specific impurity, some may raise the level of another impurity as a side effect. In addition, purifiers need to be regenerated after acquiring a certain load and should be carefully monitored and maintained.

4.3.2 Molecular Contamination

Molecular contamination is defined as any trace chemical substance that causes defects and process instabilities. As a category, molecular contamination is relatively new and its affect on products and processes is only beginning to be discovered. It is not considered a "common" problem but one which is likely to become a serious concern as devices extend below the 0.5-μm feature size level. Molecular contaminants are grouped into four categories; acids, bases, condensables (organics), and dopants. The types of molecular contamination problems relating to interconnection technology primarily affect contact resistance and deep ultraviolet (DUV) photoresist performance. Molecular contaminants also promote metal corrosion and can also cause defects on wafers confined in wafer storage boxes. Environmental molecular contaminants have caused problems with DUV photoresists and current leading-edge photo processes integrate coating, curing, exposure, and development within an environmental enclosure which filters out trace chemical substances while maintaining tight temperature and humidity control. Wafers with recently deposited metal, especially aluminum, are also sensitive to trace levels of environmental acids and other chemicals. Wafers with aluminum metal patterns have been known to begin to corrode within a matter of minutes when exposed to an environment contaminated with trace levels of chlorine. It is anticipated that these types of contamination problems will increase as device features continue to shrink (Semiconductor Industry Assoc., 1994).

4.3.3 Alpha Particle Contamination

Alpha particle radiation contamination penetrating a chip can generate electron-hole pairs leading to erroneous data or soft-errors (momentary failures) in electronic equipment. DRAMS and certain bipolar memory devices are particularly sensitive to this form of contamination. The primary sources of alpha particle contamination are chip packaging materials and materials used in the fabrication of wafer interconnection layers (Ditali, Hasnain, 1993). These materials naturally contain radioactive impurities which emit alpha particle radiation. Table VIII-8 is a listing of these materials showing their alpha particle emission rates. Suppliers of encapsulating compounds have greatly reduced alpha radiation sources in there products. High-purity packaging materials combined with protective organic coatings have been successful in controlling alpha contamination from packaging material sources. Most materials, including processing chemicals, contain trace amounts of radioactive materials that are likely to have an increasing influence on semiconductor products as devices shrink and become more sensitive.

4.3.4 Polymer Insulators

The use of polymer dielectric materials (e.g., polyamide) in the fabrication of thin film insulators poses contamination problems which are similar in magnitude to

TABLE VIII-8 Alpha Emission Rate (AER) of Various Chip Packaging Materials

Processing films and materials	AER (alphas/cm^2/h)
Bare silicon (Si)	0.00020
Si + CVD oxide (TEOS)	0.00164
Si + plasma oxide	0.00188
Si + plasma nitride	0.00443
Si + tungsten	0.00308
Si + aluminum	0.00682
Si + polysilicon	0.00098
Si + field oxide	<0.00010
Si + BPSG	<0.00010
Si + CVD nitride	<0.00010
Fully processed without WSix	0.02400
Fully processed with WSix	0.04230
Diecoat (polyimide)	<0.00010
Leadframes	
1 Meg DIP	0.00677
1 Meg ZIP	0.00258
256k DIP	0.00124
64k DIP	0.00109
Packaging Material	
Plastic	0.00080
Ceramic DIP (vendor A)	0.02320
Ceramic LCC (vendor A)	0.02530
Ceramic DIP (vendor B)	0.03230
Ceramic DIP (vendor C)	0.02610

Source: Semiconductor Int. (June, 1993).

those found during metal conductor and inorganic thin film processes. Polymer dielectric materials are affected by contaminants such as particles, trace chemical impurities, and water (Czorny, 1996). In addition, polymer dielectric materials must be formulated from highly purified monomers and solvents in very precise proportions to prevent the retention of minuscule amounts of unwanted solvents and impurities after curing. During application, contamination can be a major factor influencing the adhesion and conformal performance of the polymer to a wide variety of surface materials and structures. After curing, residual solvents and other contaminants can significantly affect the thermal properties and dielectric characteristics of the finished dielectric layer. These contaminants could also cause unwanted reactions on metal conductors in addition to acting as unwanted polymer dopants. As a result, polymer dielectric processes have required the extensive use of chemical analysis and film characterization techniques, such as NMR (nuclear magnetic resonance), FTIR (Fourier transform infrared spectroscopy), and TGA (thermogravimetric analysis) to ensure a reliable finished product. In order to overcome these difficulties, there have been attempts in the past to vapor deposit dielectrics by reacting ultrapure monomers directly at the wafer surface to form the polymer dielectric layer. This process eliminates the need for a solvent carrier and the risk of introducing particles and trace chemical impurities commonly found in polymer dielectric solutions.

4.4 FUTURE OVERLOOK

Table VIII-9 shows key wafer fabrication projections for future semiconductor products. Major interconnection process changes influencing contamination control are expected to occur as we approach the manufacturing of devices with minimum feature sizes below 0.25 μm. Significant increases in chemical-mechanical polishing (CMP) operations are expected. These increases will have a significant impact on the facility and contamination control. In addition to floor space requirements, early experience confirmed the need for CMP cleanroom and environmental contamination control to be similar to that of other wafer processes ($<$ class 10). For example, environmental particles adhering to the wafer surface just before or during the polishing operation will result in scratches. The pristine metal lines and contacts were also found to be very susceptible to corrosion from environmental ionic contaminants (e.g., chlorine) during CMP. Since CMP polishing slurries are suspensions of very fine particulate, it is important that spillage is not allowed to dry on the floor where it can become a source of surface migrating particles. Preventing the migration of CMP process material becomes more important as additional types of slurry compounds and copper conductive layers are introduced. Physical segregation of the CMP from other processes will eventually become essential. This segregation will include the dedication of wafer carriers, pods, and boxes and special material handling procedures and personnel entry/exit protocols. DI water usage will increase significantly resulting in increased pressure to develop viable DI water reclaiming and recycling programs which will not compromise the purity and stability requirements of processing. These are but a few of the contamination control challenges which will be facing semiconductor manufacturing in the near future.

TABLE VIII-9 Projections for Semiconductor Products

Minimum feature size (μm)	0.35	0.25	0.18	0.13	0.10
Bits/chip (DRAM and flash memory)	64 M	256 M	1 G	4G	16 G
No. of logic transistors/cm^2 (high-volume microprocessor)	4 M	7 M	13 M	25 M	50 M
No. of logic transistors/cm^2 (low-volume ASIC)	2 M	4 M	7 M	12 M	25 M
Chip size (mm^2)					
DRAM	190	280	420	640	960
Microprocessor	250	300	360		
Minimum no. of interconnect wiring levels (logic)	4–5	5	5–6	6	6–7
Interconnect contact/via critical dimensions (μm)	0.40	0.28	0.20	0.14	0.11
Interconnect minimum killer particle size (μm)	0.12	0.08	0.06	0.04	0.04
Minimum no. of masking levels	18	20	20	22	22
Electrical defect density (defects/m^2)	240	160	140	120	100
Lithography defect density (per layer/in^2 @ defect size μm)	690 @ 0.12	320 @ 0.08	135 @ 0.06	TBD	TBD
Liquid defect particle size (nm) at density of 25 particles/ml of photo resist	300	200	150	110	75
Maximum wafer diameter (mm)	200	200	300	300	400

Source: The National Technology Roadmap for Semiconductors (SIA).

References

Baaechle, T., G. Marvell, M. Lynch, *Microcontamination*, 25, (May 1992).
Bennett, R. S., A. R. Ellington, G. G. Gifford, K. L. Haller, J. S. McKillop, G. S. Selwyn, J. Singh, U.S. Pat. 5,367,139, (1994).
Blanchard, G. W., C. R. Bossi, E. H. Payne, T. W. Weeks, U.S. Pat. 5,221,425, (1993).
Borden, P., *Microcontamination*, 43, (Feb. 1991).
Borkman, J. D., W. R. Couch, M. L. Malczewski, *Microcontamination*, 23, (Mar. 1992).
Bowling, R. A., "Contamination Control and Defect Reduction in Semiconductor Manufacturing III", *Electrochem. Soc.*, 15, (1994).
Carmody, J. C., and J. E. Martyak, *Microcontamination*, 28, (Jan. 1989).
Carr, P. E., A. C., Rapa, W. J. Fosnight, R. J. Baseman, D. W. Cooper, *J. Institute of Environ. Sci.*, 41, (May/June 1994).
Cathey, D. A., Jr., H. Frankamp, U.S. Pat. 4,992,137, (1991).
Cestari, J. and M. Yelverton, *Solid State Technology*, 109, (Oct. 1995).
Czorny, G., 1996, private communication.
Deal, D. B. and D. C. Grant, "Contamination control and defect reduction in semiconductor manufacturing III", *Electrochem. Soc.*, 167, (1994).

Ditali, A. and Z. Hasnain, *Semi. Int'l.*, *136*, (June (1993).
Elliott, D., *Semi. Int'.*, *109*, (Oct. 1995)
Fed. Std. 209-E, Institute of Environmental Sciences.
Flamm, D. L., *Solid State Tech.* 49, (Oct. 1993)
Grande, W. C., *Microcontamination*, 25, (Jan. 1993).
Grant, D. C., *J. Institute of Environ. Sci.*, *32*, (July/Aug. 1990).
Grant, D. C., D. Smith, P. Palm, et al., *J. Institute of Environ. Sci.*, *41*, (Nov./Dec. 1994).
Hart, J. J., W. T. McDermott, A. E. Holmer, and J. P. Hatwora, Jr., *Solid State Tech.*, 1111, (Sept. 1995).
Harvell, J., P. Lessard, *Semi. Int'l*, (June 1991).
Hashimoto, S., M. Kaya, and T. Ohmi, *Microcontamination*, 25, (June 1989).
Hattori, T., *Solid State Tech.*, 90, (July 1990).
Hattori, T., *Solid State Tech.*, 80, (Feb. 1994).
Hattori, T., "Contamination control and defect reduction in semiconductor manufacturing III", *Electrochem. Soc.*, *3*, (1994).
Hunter, J., *Semi. Int'l.*, 80, (Nov. 1993).
Iscoff, R., *Semi Int'l.*, 45, (Jan. 1992).
Kendrick, J. J., *Quality*, *51*, (Sept. 1994).
Kasper, G., H. Y. Wen, and H. C. Wang. *Microcontamination*, *18*, (Jan. 1989).
Liu, B. Y. H., *Semi. Int'l.*, 75, (Mar. 1994).
Lynn, C., *J. Institute of Environ. Sci.*, *41*, (Nov./Dec. 1994).
Martyak, J. E., G. R. Carmody, and G. R. Husted, *Microcontamination*, 39, (Jan. 1993).
Martyak, J. E. and A. C. Rapa, *Cleanrooms '95 West Proceedings*, (Nov. 1995).
Medearis, K., J. Institute of Environ. Sci., 35, (Sept./Oct. 1995).
Monkowski, J. R. and D. W. Freeman, Solid State Technology, 14, (July 1990).
Morrison, P. W., *Environmental Control in Electronics Manufacturing*, *337*, (1973).
Ohmi, T., H. Inaba, and T. Takenami, *Microcontamination*, 29, (Oct. 1989).
Periasamy, R., R. P. Donovan, A. C. Clayton, and D. S. ENsor, *Microcontamination*, 39, (Oct. 1992).
Periasamy, R., D. S. Ensor, A. C. Clayton, R. P. Donovan, and J. Riddle, *J. Institute of Environ. Sci.*, *29*, (Jan./Feb. 1995).
Peters, L., *Semi. Int'l.*, *52*, (Nov. 1992).
Peters, L., Semi. *Int'l.*, *62*, (Jan. 1994).
Purcell, X, Semi. *Int'l.*, 69, (Apr. 1994).
Rapa, A. C., *J. Institute of Environ. Sci.*, *17*, (Nov./Dec. 1980).
Rapa, A. C., *STP 850*, 163, American Society for Testing and Materials (1984).
Rapa, A. C., *CleanRooms '96 East Proceedings*, 299, (1996).
Reath, M., J. Brannen, P. Bakeman, and R. Lebel, *J. Institute of Environ. Sci.*, 57 (Sept./Oct. 1995).
Rothman, L. B., R. J. Miller, R. D. Wang, T. Baechle S. Silverman, and D. Cooper., *Solid State Tech.*, 80, (May/June 1995).
Savas, S. E., U.S. Pat. 5, 102,496, (1992).
Selwyn, G. S., J. Vac. Sci. Technol., *A. 7/4*, 2758, (1989).
Selwyn, G. S., J. E. Heidenreich, and K. L. Haller, *J. Vac. Sci. Technol.*, *A. 9/5*, 2817, (1991).
Selwyn, G. S., J. S. McKillop, K. L. Haller, and J. J. Wu, *J. Vac. Sci. Technol.*, *A, 8/3*, 1726, (1990).
Selwyn, G. S., J. E. Heidenreich, and K. L. Haller, *Appl. Phys. Lett.*, *57*, 18, (1990).
Selwyn, G. S., J. E. Heidenreich, and K. L. Haller, *J. Vac. Sci. Technol.*, *A9/5*, 2817, (1991).
Selwyn, G. S. and E. F. Patterson, *J. Vac. Sci. Technol.*, *A, 10/4*, 1053, (1992).
Semiconductor Industry Assoc., *The National Technology Roadmap for Semiconductors*, *18*, 149, (1994).

SEMI Std. E14-93, Semiconductor Equipment and Materials International (1993).
Seifering, K., H. Berger, and W. Whitlock, Microcontamination, 31, (Sept. 1992).
Simpson, D., *Semi.* Int'l., *96*, (Mar. 1991).
Singer, P., 1994 Sept., *Semi. Int'l.*, *64*,
Spanos, C. J., S. P. Cunningham and L. A. Smith, "Contamintion control and defect reduction in semiconductor manufacturing III", *Electrochem. Soc.*, 117 (1994).
Steinman, A., *Microcontamination*, *46*, (Oct. 1992).
Steinman, A., *Semi. Int'l.*, (Sept. 1994).
Stern, J. E., D. J. Dopp, and J. J. Wu, *Microcontamination*, (Nov. 1991).
Sugiyama, K., T. Ohmi, T. Okumura, and F. Nakahara, *Microcontamination*, 37, (Jan. 1989).
Sugiyama, K., F. Nakahara, and T. Ohmi, *Microcontamination*, *29*, (July 1989).
Tomari, H., H. Hamada, Y. Nakahara, K. Sugiyama, and T. Ohmi, *Solid State Tech.*, *S1*, (Feb. 1991).

Appendix

SELECTED CONTAMINATION CONTROL-RELATED STANDARDS AND PRACTICES

Available from:
Institute of Environmental Sciences
940 East Northwest Highway
Mount Prospect, IL 60056

- IES-RP-CC001   HEPA and ULPA Filters
- IES-RP-CC002   Laminar-Flow Clean-Air Devices
- IES-RP-CC003   Garment System Considerations for Cleanrooms and Other Controlled Environments
- IES-RP-CC005   Cleanroom Gloves and Finger Cots
- IES-RP-CC006   Testing Cleanrooms
- IES-RP-CC007   Testing ULPA Filters
- IES-RP-CC008   Gas-Phase Absorber Cells
- IES-RP-CC012   Considerations in Cleanroom Design
- IES-RP-CC013   Equipment Calibration and Validation Procedures
- IES-RP-CC015   Cleanroom Production and Support Equipment
- IES-RP-CC016   The Rate of Deposition of Nonvolatile Residue in Cleanrooms
- IES-RP-CC018   Cleanroom Housekeeping—Operating and Monitoring Procedures
- IES-RP-CC020   Substrates and Forms for Documentation in Cleanrooms
- IES-RP-CC021   Testing HEPA and ULPA Filter Media
- IES-RP-CC022   Electrostatic Charge in Cleanrooms and Other Controlled Environments
- IES-RP-CC023   Microorganisms in Cleanrooms
- IES-RP-CC024   Measuring and Reporting Vibration in Microelectronics Facilities

- IES-RP-CC026   Cleanroom Operations
- IES-RD-CC009   Compendium of Standards, Practices, Methods, and Similar Documents Relating to Contamination Control
- IES-RD-CC011   A Glossary of Terms and Definitions Related to Contamination Control
- FED-STD-209E   Airborne Particulate Cleanliness Classes in Cleanrooms and Clean Zones (Federal Standard)
- MIL-STD-1246C   Product Cleanliness Levels and Contamination Control Program (Military Standard)

Available from:
Semiconductor Equipment and Materials International
805 East Middlefield Road
Mountain View, CA 94043-4080

- SEMI C3-95   Specifications for Gases
- SEMI C3.42-90   Standards for Argon (Ar), VLSI Grade Bulk
- SEMI C3.30-96   Standards for Hydrogen (H_2), Bulk, 99.9997% Quality
- SEMI C3.29-96   Standards for Nitrogen (N_2), Bulk Gaseous, 99.9995% Quality
- SEMI C3.41-95   Standards for Oxygen (O_2), 99.9998% Quality
- SEMI C10-94   Guide for Determination of Method Detection Limits for Trace Metal Analysis by Plasma Spectroscopy
- SEMI C13-95   Test Method for Particle Shedding and Penetration Performance of Point-of-Use Gas Filters
- SEMI C15-95   Test Method for PPm and PPB Humidity Standards
- SEMI C1-95   Specifications for Reagents
- SEMI C1.8-95   Standard for Hydrofluoric Acid
- SEMI C1.25-95   Standard for *n*-Methyl-2-Pyrrolidone
- SEMI C1.12-96   Standard for Nitric Acid
- SEMI C1.7-95   Standard for Hydrochloric Acid
- SEMI C1.13-96   Standard for 80% Phosphoric Acid
- SEMI C1.16-06   Standard for Sulfuric Acid
- SEMI E14-93   Measurement of Particle Contamination Contributed to the Product from the Process or Support Tool
- SEMI E43-95   Recommended Practice for Measuring Static Charge on Objects and Surfaces
- SEMI E45-95   Test Method for the Determination of Inorganic Contamination from Minienvironments
- SEMI E46-95   Test Method for the Determination of Organic Contamination from Minienvironments
- SEMI E49-95   Guide for Standard Performance, Practices, and Sub-Assembly for High Purity Piping Systems and Final Assembly for Semiconductor Manufacturing Equipment
- SEMI E49.3-95   Guide for Ultrahigh Purity Deionized Water and Chemical Distribution Systems in Semiconductor Manufacturing Equipment

Appendix

- SEMI E49.5-95   Guide for Ultrahigh Purity Solvent Distribution Systems in Semiconductor Manufacturing Equipment
- SEMI E49.6-95   Guide for Subsystem Assembly and Testing Procedures—Stainless Steel Systems
- SEMI E49.8-96   Guide for High Purity Gas Distribution Systems
- SEMI E49.9-96   Guide for Ultrahigh Purity Gas Distribution Systems
- SEMI F1-96   Specification for Leak Integirty of High Purity Gas Piping Systems and Components
- SEMI F21-95   Classification of Airborne Molecular Contaminant Levels in Clean Environments

REFERENCE TRADE JOURNALS

- *Chemical Engineering*
  McGraw-Hill
  1221 Avenue of the Americas
  New York, NY 10020
- *CleanRooms*
  Pennwell Publishing Co.
  1421 South Sheridan Road
  Tulsa, OK 74112
- *Journal of Vacuum Science and Technology*
  American Institute of Physics
  One Physis Ellipse
  College Park, MD 20740-3843
- Electrochemical Society
  10 South Main Street
  Pennington, NJ 08534-2896
- *"EE"* (Evaluation Engineering)
  Nelson Publishing
  2504 N. Tamiami Trail
  Nokomis, FL 34275-3482
- *Journal of the IES*
  Institute of Environmental Sciences
  940 East Northwest Highway
  Mt. Prospect, IL 60056
- Materials Research Society
  9800 McKnight Road
  Pittsburgh, PA 15237
- *Medical Device and Diagnostic Industry*
  Canon Communications
  3340 Ocean Park Blvd., Suite 1000
  Santa Monica, CA 90405
- *Micro* (formerly *Microcontamination*)
  Canon Communications
  3340 Ocean Park Blvd., Suite 1000
  Santa Monica, CA 90405

- *NASA Tech Briefs*
  Associated Business Publications Co., Ltd.
  41 East 42nd Street
  New York, NY 10017-5391
- *Semiconductor International*
  The Cahners Publishing Co., Div. of Reed Publishing
  275 Washington Street
  Newton, MA 02158-1630
- *Solid State Technology*
  Pennwell Publishing Co.
  1421 South Sheridan Road
  Tulsa, OK 74112
- *Test and Measurement World—ESD Supplement*
  The Cahners Publishing Co., Div. of Reed Publishing
  275 Washington St.
  Newton, MA 02158-1630
- *Ultra Pure Water*
  Tall Oaks Publishing, Inc.
  P.O. Box 621669
  Littleton, CO 80162

Index

1/f noise, 478–479
Actinometry, 151
Activated reactive evaporation, SiO$_2$, 243
Activation energy, 8, 475–478
Adhesion, 102
Airborne migrating particles, 526
Al alloys/layered films, reliability, 480
Al-Cu corrosion, 496–497
Al-Cu Ti/TiN layers, reliability, 480
Al-NiSi contact, 205
Al-Si phase diagram, 171
Al/Si contact, 203
Alloy evaporation, 3
Alpha particles 510, 560
Aluminum, 302–306
  CVD, 303–305
Al-Cu, 306–308
Analysis of contaminants, 517, 520–524, 548
Angular dependence of sputter yield, 35, 37–41
Anisotropic etching, Al alloys, 335, 337
Anodic oxidation, aluminum, 328
APCVD, CVD SiO$_2$, 231
Argon in SiO$_2$, 258
Asymmetric plasma reactor, 16, 28
Auger electron spectroscopy (AES), 524
Auger electron spectroscopy (AES), 116–123

Backscattering, 34
Band gap, 165
Bardeen barrier, 164
Barrier height, 163, 173, 204

Berger test site, 191
Biased high-density PECVD SiO$_2$, 380–381
Bias sputtering of metals, 393
Bias-PECVD SiO$_2$, 373
Bias-sputter deposition SiO$_2$, 375–376
Bipolar chuck, 27, 29
Bipolar emitter contact, 186
Blanket deposition
  CVD Cu, 316
  W, 320
Blister technique, 104
Bordered contacts, 216
BPSG, 224, 225, 250–251
Breakdown strength, 101, 102
BSG, 224, 225, 247–249
Burn-in, 509

Cantilevered bending beam, 106
Capacitively-coupled glow discharge, 15
Capacitively-coupled reactors, 16, 18
CARIS, 82
Catcher plate, 34, 36
Cermet barriers, 189
Characterization techniques, 78, 79
Chemical distribution systems, 540
Chemical mechanical polish, 428–448
Chemical vapor deposition (CVD), 6
Chromatography, 147
Cleaning, post CMP, 445
Cleanroom, 508, 518, 526–534
  classification, 518
  garments, 529, 530
Cluster "tools," 10, 11

571

CMP
  defects, 445
  of SiO$_2$, 430
  of SOGs, 443
  reproducibility, 444–445
  suppliers, 430
  cost of ownership, 446
Cobalt silicide, 180, 207
Coefficient of thermal expansion, 112, 113
Collimated sputtering, 194, 403
Colloidal particles, 430
Conformal coverage, 371
Contact cleaning, 201
Contact hole, 161
Contact liners, 215
Contact resistance, 163, 166, 187
  specific, 192, 193
Contact studs, 214
Contacts, etching, 199
Contaminant migration, 516
Contamination, 507
  control, 509, 513–514
    deposition systems, 550, 551
    equipment, 553
    etchers, 550, 551
    methodology, 515
    plasma, processing, 551–553
    priorities, 516
    process equipment, 551–554
    standards, 565
    vacuum systems, 551
Contamination
  characteristics, 513
  detection, 517–520
  monitoring, 517, 518
Copper
  alloy/layered films, 485–487
  barriers, 311
  corrosion, 498
  properties, 310–313
Corrosion, 494
  control, Al etching, 341–343
Cr–Cr$_x$O$_y$ cermets, 205
Cu–CVD reactors, 315
Cu(hfac)$_2$, 314
Cu–Ti, reliability, 487
Curvature change, 106–108
CVD
  Al, Al-Cu, Cu/hole-fill, 423–425
  of AlCu, 308
  of copper, 314

[CVD]
  principles, 7
  reactors, 8–13
  SiO$_2$, 371
  SiO$_2$, 229–234
  of tungsten, 320
  W hole-fill, 420–422
  W step-coverage, 421–422
Cycle time, 451

Damage, contact etching, 200
Damascene, 426, 435
dc bias, 14
dc chuck, 27
DEERS, 229
Defects, 507, 508
  creation, 512
Deflection technique, for stress measurement, 108
Deposition methods, 2
Depth of focus, 366
Device contacts, 202
DI water, contamination control, 536–539
Dielectric constant, 99, 100
  organic insulators, 289
Differential scanning calorimetry (DSC), 113
Differential thermal analysis (DTA), 113
Digital CVD, SiO$_2$, 244
Dimethyl aluminum hydride DMAH, 303–305
Diode structure, 211
Dishing in CMP, 434
Dissipation factor, 100
Distributed ECR (DECR), 45, 47
Divergent field ECR, 43–46
Dopant detection by IR, 88
Doped/undoped oxide properties, 225
Doped oxides, 245
Driven substrate, 17, 22
Dual damascene, 439
Dual frequency, 15
Dual ion beam sputtering, 229

ECR, SiO$_2$, 239, 240
ECR sputtering of metals, 407
Eddy currents, 95
Elastic biaxial modulus
  oxides, 111
  substrates, 107
Electrical conduction in insulators, 499
Electrical faults, 508

Electrical measurement, contacts, 213
Electrochemical deposition of Cu, 318
  for hole-fill, 425–426
Electroless plating, 66, 67
  of Cu, 319
Electrolytic plating, 67, 68
Electromagnetic interference (EMI), 510
Electromigration, 474
  vias, 490–492
  W studs, 492–493
Electromigration of via chains, 489–493
Electron bias sputtering of Al, 402
Electron cyclotron resonance (ECR), 42–45
Electron-gun, 2
Electron microprobe, 128–130
Electron spectroscopy for chemical analysis (ESCA), 122, 524
Electroplating of Cu, 319
Electrostatic chuck, 27
Electrostatic discharge (ESD), 541
Ellipsometry, 78, 79, 81
Encapsulation, Cu, 313
Energetic cluster impact, 411
Energy dispersive x-ray (EDX), 523
Enhanced mobility of metals, 397–399
Environmental contamination, 529–531
Epitaxial Si precipitate, 171, 174
Equipment, particle contamination, 520–525, 548–554
Etch gases used, SiO_2, 260
Etch mechanism, SiO_2, 259
Etch selectivity, SiO_2, 260
Etchback
  organic sacrificial layers, 385
  planarization, 382–387
  sacrificial inorganic overcoat, 382–384
  selective masking, 384
  SOG sacrificial layers, 385–287
Etching Al/halogen based etchants, 332
Etching holes in SiO_2, 262
Etching methods, 3
Etching, polyimides, 286
Evaporation, 1–5
  SiO_2, 243
Evaporation/sputter etch hole-fill, 391–392
Evaporation of metals, step coverage, 389

F-doped SiN, 266
F-doped SiO_2, 252–256
Facet formation, 40, 41

Feature size
  Al etching, 338
  dependent etch rate, 64, 65
  effect in etching SiO_2, 263–264
Fermi-level, 167
Field emission, 167
Film growth, 4, 6
Film stress, 105
Flash evaporation, 3
Flexible diode, 17, 20
Flowable oxide, SOGs, 272–273
Flowage of blanket oxide, 387–388
Flow fill, CVD SiO_2, 230
Fluoroptic thermometry, 148–149
Focused ion beam (FIB), 142–143
FORCEFILL, 417
Forward recoil scattering, 136
Fourier-transform IR (FTIR), 86
Frequency, 15
Future defect requirements, 562

Gap-fill
  bias-PECVD SiO_2, 377–378
  F-doped SiO_2, 374
  PECVD SiO_2, 374
Gases in semiconductor manufacturing, 558–559
  contamination control, 554–560
  impurity levels, 556
Gas purifiers, 557
Glove box, 535
Guard rings, 208

Helicon plasmas, SiO_2, 241, 242
Helicon sources, 49, 50
Hexode, 16, 19
HI-FILL, 417
High density plasmas, 42, 50
High temperature
  bias sputter/metals, 400
  metal evaporation/step coverage, 390
Hillocks, 170, 349
Hole-fill metals, 389, 401
Hollow cathode enhanced sputtering, 405
Hollow cathode PECVD, 381
Hollow cathode reactor, 21, 26
Humidity, process instability, 510
Hybrid structures, contacts, 209–210
Hydrogen
  analysis, 131–133
  effect, 213
  reduction, CVD W, 320

[Hydrogen]
  in SiO_2, 257, 258

ICP, inductively coupled-SiO_2, 242, 243
Ideality, Schottky barrier diode, 173
Impurity detection by IR, 87, 91, 92
In situ particle monitoring (ISPM), 521, 522
In-situ planarization/gap fill, 375
In-situ sputter clean, 5
Inductive plasma sputtering of metals, 407–410
Inductively-coupled plasma (ICP), 48, 49
Inductively coupled plasma-atomic emission spectroscopy (ICP-AES), 147
Infra-red (IR) spectrometry, 84–93
Inorganic precursors, CVD SiO_2, 230
Interferometer temperature measurement, 148
Interferometry, 80, 82, 83
Intermetallic, contact, 206
Ion beam deposition, SiO_2, 243
Ion milling, 33
Ionic contaminants, 510
Ionized cluster beam, 410
IR frequency identification, 85
IR spectrometers, 85
Isolation of contaminants, 516
Isolation technology, 534–536

Johnson–Rahbek chuck, 27, 29
Junction penetration, 161, 170

Laminar flow, 525, 528–529
Langmuir probes, 152
Laser annealing, 191
Laser ionization mass spectroscopy (LIMS), 128
Laser light scattering (LLS), 521, 522
Laser microprobe mass analyzer (LMMA), 128
Laser probe for stress measurement, 107
Laser reflow of metal films, 414–417
Latent defects, 509
Lift-off patterning, 328–330
Liquid chemicals: contamination control, 539–541
Liquid phase oxidation, 244
Loading effect, 62, 63
Local interconnection, 201

Local isolation, 525
Log-normal distribution, 476
Long target to substrate sputtering, 404
LPCVD, SiO_2, 231

Magnetic confinement, 18
Magnetron, 20, 23, 24
Masking, 62
  Al etching, 343
Mass-transport limited reaction, 8
Mechanical polishing, 428
Median life, t_{50}, 476, 477
Metal–silicon reaction, 170
Microwave interferometer, 152
Minienvironments, 534–536
Mixed flow, 525
Modulation, 60, 61
Moisture absorption, SiO_2, 236
Molecular beam deposition plus annealing, 411–413
Molecular contamination, 560
Monitor wafers, 520
$MoSi_2$ contact, 206
Multiple-beam interferometry, 82, 83
Multipole, 20, 25, 26

n&k analyzer, 84
Neutron activation, 147
Ni–Cr, Ni–Cr–Ag, 189, 190
Nickel silicide, 180, 207
Nonequilibrium plasma, 15
Nuclear magnetic resonance (NMR), 133

Ohmic contact, 170, 186
Optical characterization of dielectric films, 77
Optical emission spectroscopy (OES), 149, 151
Organic dielectric films, 274–288
Organic precursors, CVD SiO_2, 231
Overhang formation, 372
Oxidation of Cu, 312
Oxide lift-off, 428

Pad conditioning, 433
Pad glazing, 433
Pad materials, 429
Palladium silicide, 175
Partial cluster beam, 411
Particle, 509, 510, 511
  analysis, 548

Index

[Particle]
  behavior in cleanrooms, 526–529
  characteristics, 510–512
  defect creation, 512
  infiltration, 532–534
  monitoring, 519–521
  size (diameter), 511, 518
  transport, 531–534
Partitioning, 546
Passivation coating, effect on reliability, 488–489
Pattern transfer by sputtering, 34
PECVD–SiO_2, 232–234
PECVD SiO_2–SiH_4 based, 235, 238
PECVD SiO_2–TEOS based, 235, 236
PECVD mechanisms, 52
PECVD reactors, 52–55
Peel test, 102, 103
Photo-CVD, 11
Photosensitive polyimides, 286
Physical vapor deposition, Cu, 313
Planarization, 363
  metals, 389
Plasma
  diagnostics, 149–153
  etchants, 58, 59
  potential, 15
  processing, 14
  reactors, 16, 18
Plasma-assisted etching of Al alloys, 331–336
Plasma-enhanced CVD (PECVD), 51
Plating, Cu, 426
Platinum silicide, 161, 175
Polishers, 429
Polishing
  of Al, 439
  of Cu, 440–442
  of CVD W, 436
  rate, 432
Polyimides, 275–288
  adhesion, 279
  dielectric constant, 285
  electrical properties, 282–284
  mechanical properties, 281
  thermal properties, 278
Polymers, 560–562
Polysilicates, SOGs, 268–270
Polysiloxanes, SOGs, 270, 271
Prism coupler, 83, 84
Procedures, 509

Process equipment, 508
  contamination control, 545
Process instabilities, 509
Process tools, 508
Product cleaning, 517
Product defects, 509
Profile control, 59, 60
Profile tailoring, SiO_2, 261
Propagation delay, 450
Protocols, 509
Proton-proton scattering, 132
PSG, 224, 225, 246
PtSi/Al contact, 203
Pulsed etching, 60
Pyrometry, 148, 150

Radial flow reactor, 17
Radio frequency induction (RFI), 47–49
Rapid thermal anneal, 189, 193, 212
Reactive ion etching (RIE), 57, 58
  of copper, 345–346
  of W, 347
Reactive plasma assisted etching, SiO_2, 259
Reactive sputter etching (RSE), 57, 58
Reactive sputtering, 32, 229
Reactive-plasma-enhanced etching, 56
  mechanisms, 57
  systems, 57
Redeposition, 34, 35
Reemission coefficient, 227
Refractive index, 77, 78
Remote PECVD (RPECVD), 55
Residual gas analysis (RGA), 523
Resistivity of metal films, 91
Resistors, 211
  contact, 188
Resonant nuclear reactions, 131
Resputtering, 395–396
Reverse pillar, 427
RFI–RF inductively coupled plasma, 408
Rutherford backscattering spectrometry (RBS), 133–138

SACVD–CVD SiO_2, 231
Salicide, 180
Scanning electron microscope (SEM), 137, 139–140, 523
Scattering cross section, 134
Schottky barrier, diode, 162, 164
Scotch tape test, 102
Scratch test, 103, 104

Secondary ion mass spectroscopy (SIMS), 126, 127, 524
Secondary neutral mass spectrometry (SNMS), 128
Selective CVD aluminum, 305
Selective deposition
   CVD Cu, 317
   CVD W, 321–325
Selective W, 206
Self-align structure, 195–199
Self-sputtering, 406
Sheet resistance, 92–94
Si nodules, 172
SiBN and SiBON, 266, 267
Sidewall tapering, bias-PECVD, 377
Sigma, 477–478
SiH_4 reduction, CVD W, 325–327
Silicon damage effects, 211, 212
Silicon dioxide, 224
Silicon nitride, 265
Silicon oxynitride, 265
Silsesquioxanes-SOGs, 271
Slurry, 429
SMIF pods, 535, 536
Snow-plow effect, 169, 181
Spacers, 388
Spike, 170
Spin-on films, planarization, 368
Spin-on-glass, 267, 274
Sputter deposition, 28, 30, 31
   of metals, 392
Sputtered SiO_2, 227, 228
Sputter etching, 33
   ion milling of Al alloys, 328
   SiO_2, 259
Sputtering target, 28
Sputtering yield, 30
Static electricity, 510, 541
Static SIMS, 128
Step coverage, 370
   metals, 389
   SiO_2, 237
Stopping cross section, 134
Stress deformation, 105
Stress voiding, 474
Structure of CVD Al, Cu films, 351
Structure of evaporated films, 350
Structure of sputtered Al films, 351
Substrate contacts, 201
Support equipment, 508
Surface films, 510
Surface migrating particles, 526

Surface-limited reaction, 8
Symmetruc reactor, 16

T and H stressing, 494–496
Tantalum silicide, 185, 206
TDDB, breakdown of insulators, 500–502
Temperature
   control in discharge, 25
   effect on Al etching, 344
   effects in discharge, 25, 31
   effects in RIE, 65, 66
   process instability, 510
TEOS-based oxide, 238
$TEOS/O_3$-CVD SiO_2, 231, 232
Testing, electromigration, 475
Thermal voiding, 479
Thermionic emission, 165, 167
Thermionic field emission, 168
Thickness monitors, 4
Thin-film intermetallic formation, 482
Threshold energy, 30
Time of flight SIMS (TOF-SIMS), 524
Ti-Al reaction/effect of Cu, 482–483
TiN/Ti contacts, 191
TiN_xO_y, 183, 184
$TiSi_2$, C49 phase, 177–180
$TiSi_2$, C54 phase, 177–180
Titanium nitride, 176, 183–184
Titanium silicide, 176–180, 197, 199
TiW contact, 207
TiW diffusion barrier, 182, 183
Topography, 364
   consequences, 364–366
Trace chemical, 560
Transformer-coupled plasma (TCP), 48
Trenching, 34, 36
Triisobutyl aluminum TIBA, 303–305
Triode, 17
Troubleshooting, 546–548
Tuned substrate, 17, 21
Tungsten, 319
Tunnelling, 165
Turbulent air flow, 525, 527

Ultra-high frequency source, 51
Ultra-pure water (UPW), 537
Unipolar chuck, 27, 28
UPW system, 538
UV radiation, 510

Vibration, 510, 542–545

Index

Voids, 371

Wafer clamp, 25, 27
Wafer scanners, 521
Weibull distribution, 476
Wet etching, SiO$_2$, 259

Wet etching of aluminum, Al alloys, 327
Wiring levels, 451
Work function, 166

Zr, Sn addition to Cu, reliability, 487